Handbook of Time Series Analysis

Edited by

Björn Schelter,
Mathias Winterhalder,
and Jens Timmer

Related Titles

Kedem, B., Fokianos, K.

Regression Models for Time Series Analysis

360 pages

2002

Hardcover

ISBN-13: 978-0-471-36355-2
ISBN-10: 0-471-36355-3

Pourahmadi, M

Foundations of Time Series Analysis and Prediction Theory

448 pages

2001

Hardcover

ISBN-13: 978-0-471-39434-1
ISBN-10: 0-471-39434-3

Pena, D

A Course in Time Series Analysis

496 pages

2000

Hardcover

ISBN-13: 978-0-471-36164-0
ISBN-10: 0-471-36164-X

Handbook of Time Series Analysis

Recent Theoretical Developments and Applications

Edited by
Björn Schelter, Matthias Winterhalder,
and Jens Timmer

WILEY-VCH Verlag GmbH & Co. KGaA

The Editors

Björn Schelter
University of Freiburg
Center for Data Analysis
Eckerstr. 1
79104 Freiburg

Matthias Winterhalder
Freiburger Zentrum für Datenanalyse und Modellbildung-FDM
Eckerstr. 1
79104 Freiburg

Jens Timmer
Albert-Ludwig-Univ. Freiburg
Zentrum f. Datenanalyse (FDM)
Eckerstr. 1
79104 Freiburg

Library of Congress Card No.:
applied for

British Library Cataloguing-in-Publication Data
A catalogue record for this book is available from the British Library.

Bibliographic information published by Die Deutsche Bibliothek
Die Deutsche Bibliothek lists this publication in the Deutsche Nationalbibliografie; detailed bibliographic data is available in the Internet at <http://dnb.ddb.de>.

Typesetting Da-TeX Gerd Blumenstein, Leipzig
Binding Litges & Dopf GmbH, Heppenheim
Cover Design aktivComm GmbH, Weinheim

Printed in the Federal Republic of Germany
Printed on acid-free paper

ISBN-13: 978-3-527-40623-4
ISBN-10: 3-527-40623-9

Contents

Handbook of Time Series Analysis. Björn Schelter, Matthias Winterhalder, Jens Timmer

ISBN: 3-527-40623-9

Preface

Reiterated measurements of an experimentally accessible quantity of a dynamical system result in a time series, and one may wonder, what this information can tell about the system on which the measurements are done. Time series analysis is, thus, a very obvious way of an attempt to understand nature—already Kepler did it when studying the observations of Tycho Brahe. He came up with a very simple synopsis formulated in his famous laws and Newton could ascribe these to a single law by postulating a fundamental gravitational force. This marks the beginning of modern science and then, in exploring the nature, fundamental laws or equations motivated by first principles played a dominant role.

Turning to more and more complex systems guidance by first principles became less fruitful for finding a mathematical model. Thus, observations cannot serve any more as indication or pointer to some fundamental underlay but have to be regarded only as a fingerprint of the system. First tasks in analyzing these fingerprints then are e.g. characterization or establishing a relation or correlation to other observations. Time series analysis in this sense, thus, has already a long history in fields where the systems to be studied are very complex such as meteorology or medical science. Sophisticated mathematical methods appeared first in late 19th century and during the last decades these methods have been utilized also by many scientists working in applied fields. This has led to many successes in understanding complex systems.

This handbook comprises a wide range of current topics in the field of time series analysis. The editors are well-known for both their theoretical work on time series analysis techniques and their applications. Therefore, the editors attached great importance to both theoretical work and applications. Especially, the interplay of theory and practice is included in this Handbook of Time Series Analysis. The editors brought together contributions of worldwide accepted experts of different branches, e.g. from Physics, Mathematics, Biology, Medicine, Neuroscience, and Engineering. With respect to the theory this Handbook covers a broad variety of presently used methodologies in different disciplines, ranging from linear stochastic systems to Nonlinear Dynamics, from univariate to multivariate time series analysis.

The Handbook of Time Series Analysis will provide guidance for all those working on time series analysis, from students to experienced investigators. I

Handbook of Time Series Analysis. Björn Schelter, Matthias Winterhalder, Jens Timmer

ISBN: 3-527-40623-9

hope that it develops into a standard textbook and that the editors find time to keep it up-to-date in future.

Josef Honerkamp July 26, 2006

Chair for "Stochastic Dynamical Systems" at the Physics Department of the University of Freiburg, and founder of the "Freiburg Center for Data Analysis and Modeling"

List of Contributors

- *Henry D. I. Abarbanel*
  Marine Physical Laboratory
  (Scripps Institution of Oceanography)
  Department of Physics
  Institute for Nonlinear Science
  University of California San Diego
  9500 Gilman Drive
  Mail Code-0402
  La Jolla, CA 92093-0402
  USA
  e-mail: hdia@jacobi.ucsd.edu
- *Claude Adam*
  Cognitive Neurosciences
  and Brain Imaging Laboratory (LENA)
  CNRS-UPR-640
  Hôpital de La Salpêtrière
  47 Bd. de l'Hôpital
  75651 Paris CEDEX 13
  France
- *Luiz Antonio Baccalá*
  Departamento de Engenharia
  de Telecomunicações e Controle
  Escola Politécnica
  Universidade de São Paulo,
  Av. Prof. Luciano Gualberto, Trav. 3, #158,
  São Paulo, SP, 05508-900
  Brazil
  e-mail: baccala@lcs.poli.usp.br
- *Boris P. Bezruchko*
  Department of Nano-
  and Biomedical Technologies
  Saratov State University
  83 Astrakhanskaya Street
  Saratov 410012
  Russia
  e-mail: bbp@sgu.ru
- *Katarzyna J. Blinowska*
  Department of Biomedical Physics
  Warsaw University
  ul. Hoza 69
  00-681 Warszawa
  Poland
  e-mail: Katarzyna.Blinowska@fuw.edu.pl
- *Stefano Boccaletti*
  CNR-Istituto dei Sistemi Complessi
  Florence
  Italy
- *Jose Miguel Bornot-Sánchez*
  Cuban Neuroscience Center
  Apartado 6880
  La Habana
  Cuba
- *Steven L. Bressler*
  Center for Complex Systems
  and Brain Sciences
  Florida Atlantic University
  USA
- *Erick Canales-Rodríguez*
  Cuban Neuroscience Center
  Apartado 6880
  La Habana
  Cuba
- *Thomas L. Carroll*
  Naval Research Laboratory
  Washington, DC 20375
  USA
- *Patrick Celka*
  Applied Cognitive Neuroscience
  Research Centre
  School of Engineering
  Griffith University
  Gold Coast Queensland 9726
  Australia
  e-mail: P.Celka@griffith.edu.au
- *Mario Chavez*
  Cognitive Neurosciences
  and Brain Imaging Laboratory (LENA)
  CNRS-UPR-640
  Hôpital de La Salpêtrière

Handbook of Time Series Analysis. Björn Schelter, Matthias Winterhalder, Jens Timmer

ISBN: 3-527-40623-9

47 Bd. de l'Hôpital
75651 Paris CEDEX 13
France
e-mail: mario.chavez@chups.jussieu.fr

- *Yonghong Chen*
  The J. Crayton Pruitt Family
  Department of Biomedical Engineering
  University of Florida
  USA
- *Laura Cimponeriu*
  Department of Physics
  University of Potsdam
  Am Neuen Palais 10
  14469 Potsdam
  Germany
  e-mail: laura@stat.physik.uni-potsdam.de
- *Manfred Deistler*
  Institut für Wirtschaftsmathematik
  Technische Universität Wien
  Argentinierstraße 8/105-2
  1040 Wien
  Austria
  e-mail: Manfred.Deistler@tuwien.ac.at
- *Mingzhou Ding*
  Department of Biomedical Engineering
  University of Florida
  149 BME Building/PO Box 116131
  Gainesville, FL 32611
  USA
  e-mail: mding@bme.ufl.edu
- *Michael Eichler*
  Department of Quantitative Economics
  University of Maastricht
  P.O. Box 616
  6200 MD Maastricht
  The Netherlands
  e-mail: m.eichler@ke.unimaas.nl
- *Ralf Engbert*
  Institute of Psychology
  University of Potsdam
  Postfach 601553, 14415 Potsdam
  Germany
- *David Engster*
  Applied Nonlinear Dynamics
  University of Göttingen
  Bürgerstraße 42-44
  37073 Göttingen
  Germany
- *Roland Fried*
  Department of Statistics
  University of Dortmund
  Vogelpothsweg 87
  44221 Dortmund
  Germany
  e-mail: fried@statistik.uni-dortmund.de
- *Ursula Gather*
  Department of Statistics
  University of Dortmund
  Vogelpothsweg 87
  44221 Dortmund
  Germany
  e-mail: gather@statistik.uni-dortmund.de
- *Elly Gysels*
  Biosense Webster
  Johnson & Johnson Medical Switzerland
  Johnson & Johnson AG
  Rotzenbuehlstraße 55
  8957 Spreitenbach
  Switzerland
- *Trevor J. Hine*
  Applied Cognitive Neuroscience
  Research Centre,
  School of Psychology
  Griffith University
  Mt Gravatt Queensland 4111
  Australia
  e-mail: t.hine@griffith.edu.au
- *Maciej Kamiński*
  Department of Biomedical Physics
  Warsaw University
  ul. Hoza 69
  00-681 Warszawa
  Poland
  e-mail: Maciek.Kaminski@fuw.edu.pl
- *Holger Kantz*
  Max Planck Institute
  for the Physics of Complex Systems,
  Nöthnitzer Str. 38
  01187 Dresden
  Germany
  e-mail: kantz@pks.mpg.de

- *Myron J. Katzoff*
  National Center for Health Statistics
  Centers for Disease Control
  Hyattsville, MD 20782
  USA
  e-mail: mjk5@cdc.gov
- *Reinhold Kliegl*
  Institute of Psychology
  University of Potsdam
  Postfach 601553
  14415 Potsdam
  Germany
- *Jürgen Kurths*
  Institute for Physics
  University of Potsdam
  Am Neuen Palais 10
  14469 Potsdam
  Germany
  e-mail: jkurths@agnld.uni-potsdam.de
- *Agustin Lage-Castellanos*
  Cuban Neuroscience Center
  Apartado 6880
  La Habana
  Cuba
- *Vivian Lanius*
  Department of Statistics
  University of Dortmund
  Vogelpothsweg 87
  44221 Dortmund
  Germany
  e-mail: vivian.lanius@uni-dortmund.de
- *Jacques Martinerie*
  Cognitive Neurosciences
  and Brain Imaging Laboratory (LENA)
  CNRS-UPR-640
  Hôpital de La Salpêtrière
  47 Bd. de l'Hôpital
  75651 Paris CEDEX 13
  France
- *Lester Melie-García*
  Cuban Neuroscience Center
  Apartado 6880
  La Habana
  Cuba
- *Theoden I. Netoff*
  Boston University
  Boston, MA 02215
  USA
- *Eckehard Olbrich*
  Max Planck Institute
  for the Mathematics in the Sciences
  Inselstr. 22
  04103 Leipzig
  Germany
- *Ulrich Parlitz*
  Applied Nonlinear Dynamics
  University of Göttingen
  Bürgerstraße 42–44
  37073 Göttingen
  Germany
  e-mail:
  U.Parlitz@dpi.physik.uni-goettingen.de
- *Louis M. Pecora*
  Naval Research Laboratory
  Washington, DC 20375
  USA
- *Arkady Pikovsky*
  Department of Physics
  University of Potsdam
  Am Neuen Palais 10
  14469 Potsdam
  Germany
  e-mail:
  Pikovsky@stat.physik.uni-potsdam.de
- *Martin Rolfs*
  Institute of Psychology
  University of Potsdam
  Postfach 601553
  14415 Potsdam
  Germany
- *Maria Carmen Romano*
  Institute of Physics
  University of Potsdam
  Am Neuen Palais 10
  14469 Potsdam
  Germany
- *Michael Rosenblum*
  Department of Physics
  University of Potsdam

Am Neuen Palais 10
14469 Potsdam
Germany
e-mail: mros@agnld.uni-potsdam.de

- *Koichi Sameshima*
Faculdade de Medicina,
Universidade de São Paulo
Av. Ovídio Pires de Campos s/n,
São Paulo, SP, 054030-010
Brazil
e-mail: ksameshi@usp.br

- *Björn Schelter*
Center for Data Analysis
and Modeling (FDM)
University of Freiburg
Eckerstrasse 1
79104 Freiburg
Germany
e-mail: schelter@fdm.uni-freiburg.de

- *Steven J. Schiff*
Dept. Neurosurgery &
Dept. Engineering Science and Mechanics
The Pennsylvania State University
212 Earth-Engineering Sciences Building
University Park, PA 16802
USA
e-mail: sschiff@psu.edu

- *Dmitry A. Smirnov*
Institute of Radioengineering
and Electronics
Russian Academy of Science
Saratov Department
38, Zelyonaya Street
Saratov 410019
Russia
e-mail: smirnovda@info.sgu..ru

- *David S. Stoffer*
Department of Statistics
University of Pittsburgh
Pittsburgh, PA 15260
USA
e-mail: stoffer@pitt.edu

- *Daniel Yasumasa Takahashi*
Faculdade de Medicina,
Universidade de São Paulo
Av. Ovídio Pires de Campos s/n,
São Paulo, SP, 054030-010
Brazil
e-mail: yasumasa@ime.usp.br

- *Marco Thiel*
Institute of Physics
University of Potsdam
Am Neuen Palais 10
14469 Potsdam
Germany

- *Jens Timmer*
Center for Data Analysis
and Modeling (FDM)
University of Freiburg
Eckerstr. 1
79104 Freiburg
Germany
e-mail: jeti@fdm.uni-freiburg.de

- *Pedro A. Valdés-Sosa*
Cuban Neuroscience Center
Apartado 6880
La Habana
Cuba
e-mail: peter@cneuro.edu.cu

- *Mayrim Vega-Hernández*
Cuban Neuroscience Center
Apartado 6880
La Habana
Cuba

- *Rolf Vetter*
Systems Engineering Division
Control and Signal Processing Section
Rue Jaquet-Droz 1, 2007 Neuchatel
Switzerland
e-mail: rolf.vetter@csem.ch

- *Matthias Winterhalder*
Center for Data Analysis
and Modeling (FDM)
University of Freiburg
Eckerstr. 1
79104 Freiburg
Germany
e-mail: matthias.winterhalder@fdm.uni-freiburg.de

1 Handbook of Time Series Analysis: Introduction and Overview

Björn Schelter, M. Winterhalder, and J. Timmer

Mathematics, Physics, and Engineering are very successful in understanding phenomena of the natural world and building technology upon this based on the first principle modeling. However, for complex systems like those appearing in the fields of biology and medicine, this approach is not feasible and an understanding of the behavior can only be based upon the analysis of the measured data of the dynamics, the so-called time series.

Time series analysis has different roots in Mathematics, Physics, and Engineering. The approaches differ by their basic assumptions. While in Mathematics linear stochastic systems were one of the centers of interest, in Physics nonlinear deterministic systems were investigated. While the different strains of the methodological developments and concepts evolved independently in different disciplines for many years, during the past decade, enhanced cross-fertilization between the different disciplines took place, for instance, by the development of methods for nonlinear stochastic systems.

This handbook written by leading experts in their fields provides an up-to-date survey of current research topics and applications of time series analysis. It covers univariate as well as bivariate and multivariate time series analysis techniques. The latter came into the focus of research when recording devices enabled more-dimensional simultaneous recordings. Even though bivariate analysis is basically multivariate analysis, there are some phenomena which can occur only in three or more dimensions, for instance, indirect interdependences between two processes.

The aim of this handbook is to present both theoretical concepts of various analysis techniques and the application of these techniques to real-world data. The applications cover a large variety of research areas ranging from electronic circuits to human electroencephalography. The interplay between challenges posed by empirical data and the possibilities offered by new analysis methods has been proven to be successful and stimulating.

In the first chapter, Henry D. I. Abarbanel and Ulrich Parlitz present different approaches to nonlinear systems. By means of a real-world example of a recording from a single neuron, they discuss how to analyze these data. Concepts such as the Lyapunov exponent, i.e., a measure for chaos, prediction, and modeling in

Handbook of Time Series Analysis. Björn Schelter, Matthias Winterhalder, Jens Timmer

ISBN: 3-527-40623-9

nonlinear systems, are introduced with a critical focus on their limitations. Ready to apply procedures are given allowing an immediate application to one's own data.

Local modeling is being dealt with by David Engster and Ulrich Parlitz. Local models are amongst the most precise methods for time series prediction. This chapter describes the basic parameters of local modeling. To show the efficiency of this procedure, several artificial and real-world data, for instance experimental friction data sets, are predicted using local models. As an alternative to strict local modeling, cluster weighted modeling is also discussed using an expectation-maximization (EM) algorithm as a parameter optimization procedure.

Holger Kantz and Eckehard Olbrich present concepts, methods, and algorithms for predicting time series from the knowledge of the past. Thereby, they especially concentrate on nonlinear stochastic processes which have to be dealt with by probabilistic predictions. They calculate a certain prediction range in which future values are going to fall. They complete their chapter by discussing verification techniques for their forecasted values, which is very important when dealing with real-world data.

Noise and randomness in biological systems have often been treated as an unwelcome byproduct. Patrick Celka and co-workers identify different noise sources and their impact on dynamical systems. This contribution discusses the concept of randomness and how to best access the information one wants to retrieve. Different time series analysis techniques are presented. The applications govern speech enhancement, evoked potentials, cardiovascular system, and brain–machine interface.

The chapter of Ursula Gather and co-workers is dedicated to robust filtering procedures for signal extraction from noisy time series. The authors present various filter techniques with their specific properties and extensions in order to process noisy data or data contaminated with outliers. They point to the variety of different approaches and compare the advantages and disadvantages. By means of simulated data they demonstrate the different conceptual properties.

Dealing with bivariate time series analysis techniques, the chapter of Michael Rosenblum and co-workers is dedicated to the phenomenon of phase synchronization and the detection of coupling in nonlinear dynamical systems. The authors discuss the usage of model-based and nonmodel-based techniques and introduce novel ideas to detect weak interactions between two processes, together with the corresponding strength and direction of interactions. They illustrate their analysis techniques by application to data characterizing the cardiorespiratory interaction.

An approach to detect directional coupling between oscillatory systems from short time series based on empirical modeling of their phase dynamics is introduced by Dmitry Smirnov and Boris Petrovich Bezruchko. This time series analysis technique is utilized to analyze electroencephalography recordings with the purpose of epileptic focus localization and climatic data representing the dynamics of the North Atlantic Oscillation and El Niño/Southern Oscillation processes.

Phase synchronization analysis of brain signals, for instance intracranial electroencephalography data recorded from epilepsy patients, has come into the focus of neuroscience research. Mario Chavez and co-workers suggest a data-driven time series analysis technique to select the important contents in a signal with multiple frequencies, the empirical mode decomposition. They summarize this concept and demonstrate its applicability to model systems and apply it to the analysis of human epilepsy data.

For cases where the definition of the phase used by common approaches is impossible, Mamen Romano and co-workers present a way to detect and quantify phase synchronization using the concept of recurrences. Furthermore, to test for phase synchronization, an algorithm to generate surrogate time series based on recurrences is discussed. An application to fixational eye movement data complements the results for model systems.

Theoden I. Netoff and co-workers dedicated their work to infer coupling and interaction in weakly coupled systems, especially in the presence of noise and nonlinearity. To this end, they applied several analysis techniques to model data and to data obtained from an electronic circuit. They explored advantages and disadvantages of the methods in specific cases. The conclusion of their chapter is that nonlinear methods are more sensitive to detect coupling under ideal conditions. However, in the presence of noise, linear techniques are more robust.

Dealing with multivariate systems, the chapter of Manfred Deistler is dedicated to state space and autoregressive moving average models. He summarizes the basic ideas about state space models and autoregressive moving average models including external influence. He focuses on the mathematics and discusses approaches to parameter estimation. Lower dimensional parameterizations of these state space models are described to cope with high-dimensional time series.

David S. Stoffer and Myron J. Katzoff introduce an extension to spatio-temporal state space models. They concentrate on the concept of spatially constrained state-space models presenting ideas and mathematical aspects. Their application is dedicated to real-time disease surveillance by analyzing weekly influenza and pneumonia mortality collected in the northeastern United States that is essential in helping to detect the presence of a disease outbreak and in supporting the characterization of that outbreak by public health officials.

Graphical models are introduced in the chapter by Michael Eichler. He introduces the mathematical basis for a graphical representation of the interaction schemes obtained by multivariate analysis techniques. Moreover, the inference in these graphs is discussed and illustrated by means of model systems. Novel multivariate analysis techniques that allow distinction not only of direct and indirect interactions but also of the direction of interactions leading to such graphs are summarized and applied to neurophysiological and fMRI data.

The directed transfer function allows detection of directed influences in multivariate systems. Katarzyna J. Blinowska and Maciej Kamiñski introduce the directed transfer function, extend the concept to nonstationary data, and discuss approaches to decide its statistical significance. In their application, they analyze human electroencephalography data using the directed transfer function. They complement this work by comparisons of different multivariate analysis techniques.

Luiz A. Baccalá and co-workers are working on a multivariate analysis technique called partial directed coherence. Besides several applications of this technique, one of the challenges when applying this technique to real-world data is that a significance level is mandatory. Several approaches to evaluate statistical significance in practice are presented and discussed in their chapter. Moreover, they compare their technique to other techniques suggested for a similar purpose. The techniques are applied to electroencephalography data during and immediately before an epileptic seizure.

Another multivariate analysis technique to detect the directions of interactions between processes is discussed by Mingzhou Ding and co-workers. Bivariate Granger causality and conditional Granger causality are presented with particular emphasis on their spectral representations. Following a discussion of the theoretical properties and characteristics, the time series analysis technique is applied to model systems and to multichannel local field potentials recorded from monkeys performing a visuomotor task.

Pedro A. Valdés-Sosa and co-workers focus in their chapter on multivariate autoregressive models (MAR) based on a Bayesian formulation that combines several components of different types of penalizations as well as spatial *a priori* covariance matrices. This approach is shown to be practical by simulations and an application to concurrent EEG and fMRI time series gathered in order to analyze the origin of resting brain rhythms.

Ranging from univariate to multivariate analysis techniques, ranging from applications of physics to life sciences, covering an exceptionally broad spectrum of topics, beginners, experts as well as practitioners in linear and nonlinear time series analysis who seek to understand the actual developments will take advantage of this handbook.

2 Nonlinear Analysis of Time Series Data

Henry D. I. Abarbanel and Ulrich Parlitz

Nonlinear dynamical systems pose challenges in the analysis of observed time series. The required time-domain methods require more care than linear frequency-domain techniques, yet they are mature enough to answer important questions about the system producing the time series data. We review a set of standard methods for this analysis with an eye toward how they may be used in a practical sense and with a critical focus on their limitations. The key question in any such analysis is what aspect of the physical or biological system is of importance.

2.1 Introduction

Nonlinear dynamics plays an essential role in the behavior of physical and biological systems actually observed in experiments. Chaotic oscillations of moons orbiting heavy planets as well as action potential generation by neurons arise from nonlinear processes in those settings. This means one must step beyond the classical set of time series tools, such as Fourier analysis, utilized widely in the extraction of information from observed time series. Indeed, Fourier analysis is precisely suited for the simplification of linear time invariant dynamics. This method transforms and simplifies such dynamics from differential equations to algebraic problems since the transform kernel $e^{i\omega t}$ is the eigenfunction of the time translation operator. However, even the presence of a quadratic term in the dynamical variable leads to a convolution of the Fourier transform of that variable with itself, thus significantly complicating the analysis rather than simplifying it.

The methods for analyzing time series from nonlinear systems have thus been developed in time domain. We review here some methods in the analysis of such time series concentrating on those which have proven valuable over time and accepting that this chapter will thus miss recent developments which may prove valuable as they are critically used.

Our discussion will start with the embedding methods utilized to reconstruct a "proxy" phase space (or state space) for the observed system based on the geometric theorem of Whitney and brought to nonlinear dynamics by Takens [1] and the Santa Cruz "dynamics collective" [2] around 1980. Within that framework we

Handbook of Time Series Analysis. Björn Schelter, Matthias Winterhalder, Jens Timmer

ISBN: 3-527-40623-9

will address how to determine the key quantities within the embedding process: time delays and dimensions [3–5]. This in itself gives substantial clues to the dynamical system leading to the measurements. To classify that system we require some invariants of the dynamics, and we discuss dimensions and Lyapunov exponents. The latter also give us insight into the predictability of the system. From there, we discuss the job of predicting within the reconstructed phase space. At that stage we turn to estimating the parameters in models of the system producing the time series measurements.

Through this chapter we use an example from the Laboratory of Al Selverston at University of California, San Diego (UCSD) [6]. These are measurements of the cross membrane voltage in an isolated neuron of a small circuit, the pyloric central pattern generator of crustaceans. This neuron, called LP, when in the intact circuit produces quite regular voltage bursts which are coordinated with bursts of two other circuit neurons leading to an important three-phase functional outcome for the crustacean digestive system. While model equations of motion of the Hodgkin–Huxley form are known for this neuron [7], tests for the quality of those models relied in the past on visual, subjective aspects of the time series of voltage. The analysis here is both illustrative of how one uses the tools of nonlinear time series analysis and has important implications for the understanding of the entire neural circuit.

2.2 Unfolding the Data: Embedding Theorem in Practice

We will primarily focus on the usual and simplest case of time series measurements of a single signal $s(t)$. If more than a single measurement is available, there are additional questions one may ask and answer. The signal is observed with some accuracy, usually specified by an estimate of the "noise" level associated with interference of the measurement by other processes. The signal is also measured in discrete time, starting at an initial time t_0 and then typically at a uniform time interval τ_s we call the sampling time. $s(t)$ is thus the set of N measurements $s(t_0 + n\tau_s)$, $n = 1, 2, \ldots, N$.

The dynamical system from which the signal comes is usually unknown in detail. In the case of the LP neuron $s(t)$ is the membrane voltage ranging from about -70 mV to $+50$ mV, and while one has conductance-based Hodgkin–Huxley models for the dynamics [8], one does not know how many dynamical variables are needed nor does one know with any specificity the many parameters which enter such models. We are certain, however, that there is more than one dynamical variable and the system state space is not one dimensional even though the measurement is. To describe the state of the system we need more than amplitude and phase which is where linear analyses dwell.

The first task is to ask how many variables we will need to describe the system. If the dynamical system has a state space trajectory lying on an attractor of dimension d_A, then our observation is the projection of the multidimensional

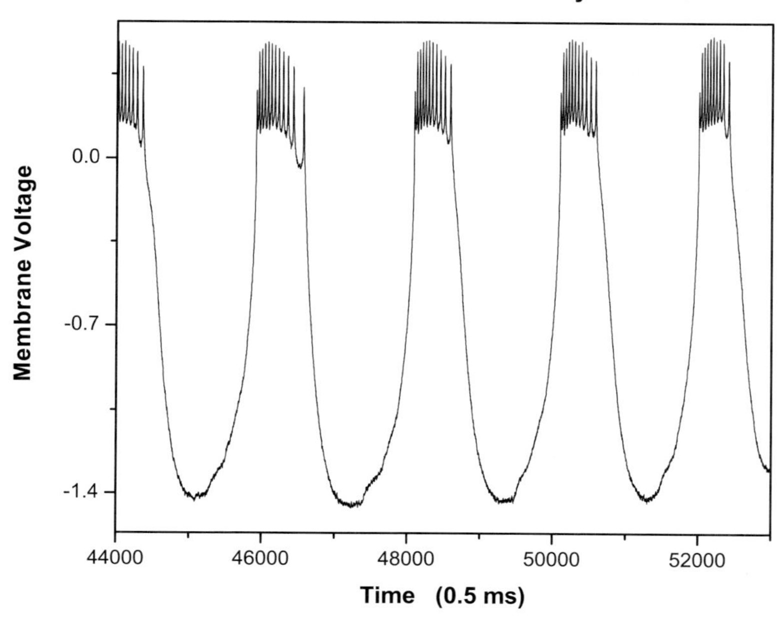

Fig. 2.1: Membrane voltage across the cell membrane of an isolated LP neuron from the crustacean pyloric central pattern generator. The amplitude on the y-axis is in scaled, arbitrary units. The time series is shown as a solid line, but the voltages were measured at 2 kHz or $\tau_s = 0.5$ ms. Altogether 200 000 data points or 100 s of data were recorded.

orbit in a space of integer dimension larger than d_A onto the measurement axis where we observe $s(t)$. If the dynamical system producing $s(t)$ is autonomous, then the orbit does not intersect itself in a high enough dimensional space capturing all the dynamical variables. In a space of integer dimension D a set of points with dimension d_A intersects itself in a set of points of dimension $d_A + d_A - D$. If D is large enough, this is negative, indicating no intersections at all. This tells us that if $D > 2d_A$, we are guaranteed that the space we use to describe the dynamics will have unfolded the projection made by the measurement. This is a sufficient condition. It could be that a dimension smaller than this unfolds the measurement projection, but we need another tool to determine that [3, 9–16].

It was probably David Ruelle's idea in the late 1970s that coordinates for the space of dimension D could be made out of the observations and their time delays. Takens proved a theorem [1] implying that the observed variable and any independent set of $D - 1$ other variables made from $s(t)$ would be acceptable coordinates for this space. The simplest set of variables, though not always the very best, is taken from the measurements themselves. One seeks a D-dimensional vector made from $s(t)$ and its time delays by forming

$$\begin{aligned} y(t = t_0 + n\tau_s) \\ &= [s(t_0 + n\tau_s), s(t_0 + (n - T)\tau_s), \ldots, s(t_0 + (n - (D - 1)T)\tau_s)] \,. \end{aligned} \quad (2.1)$$

This D-dimensional vector is composed of the observation $s(t_0 + n\tau_s)$ and the $j = 1, 2, \ldots, D-1$ earlier observations at $t_0 + (n-jT)\tau_s$. If $T = 1$, the components are selected at each sampling time.

To use this vector as a "proxy" for the degrees of freedom actually specifying the state of the system (unknown to us, of course) we need to determine values for D and T. To simplify the notation we will drop the initial time t_0 and the sampling time τ_s and write $s(n) = s(t_0 + n\tau_s)$ as well as

$$y(n) = [s(n), s(n - T), \ldots, s(n - (D - 1)T)] \,. \quad (2.2)$$

How do we know that the sampling time τ_s is small enough to capture significant variations of the dynamical signal $s(t)$? If we know nothing about the source of the observations $s(t)$, we cannot answer this question with any certainty. We will indicate how one can test this, but that comes in a moment. If we know that the source of the signal is an oscillating neuron, then we might know that the typical time scale of neural activity is in milliseconds, so if τ_s is 1 s, we probably have undersampled data. If τ_s is 1 μs, the data are probably oversampled. One always prefers the latter situation as selecting a subset of the data to describe that the system can be reliable. For now, let us assume that the system is properly sampled or possibly slightly oversampled.

2.2.1 Choosing T: Average Mutual Information

The goal of replacing the original signal $s(n)$ with a vector $y(n)$ is to provide independent coordinates in a D-dimensional space to replace the unknown coordinates of the observed system. The components of the vector $y(n)$ should thus be independent looks at the system itself, so all of the needed dynamical variations in the system are captured. If the time delay between the components $s(n - jT)$ and $s(n - (j - 1)T)$ is too small for some T, then the components are not really independent and we require a larger T. If T is too large, then the two measurements $s(n-jT)$ and $s(n-(j-1)T)$ are so far apart in time that the typical instabilities of nonlinear systems will render them essentially uncorrelated. We need some criterion which retains the connection between these measurements yet does not make them essentially identical.

While it is easy to evaluate the linear autocorrelation between measurements as a function of T, the usual criterion of seeking a zero in that quantity only leads to a value of T where the measurements are linearly independent. The dynamical interest of this is rather small. A much more motivated criterion, though harder to evaluate, was suggested by Fraser and Swinney in 1986: evaluate the average mutual information between measurements at time n and time $n - T$;

look for the first minimum in this quantity. This tells us when the two measurements are nonlinearly relatively independent, and this may provide a useful choice for T [17–21].

To evaluate the average mutual information, we need the distribution of the measurements $s(n)$ over the time series. This means we need to bin the amplitudes $s(n)$, $n = 1, 2, \ldots, N$, into a normalized histogram using the whole data set. This gives the frequency of occurrence $P(s(n))$. We also need to do the same for the time-delayed data $s(n-T)$, and we need the normalized histogram of the joint occurrence of $s(n)$ and $s(n-T)$ to find $P(s(n), s(n-T))$. The average mutual information

$$I(T) = \sum_{s(n), s(n-T)} P(s(n), s(n-T)) \log_2 \left[\frac{P(s(n), s(n-T))}{P(s(n))\, P(s(n-T))} \right] \tag{2.3}$$

tells us *in bits* how much, on average over the whole time series or the attractor, we know about the measurement at time n from the measurement at time $n - T$. $I(T) \geqslant 0$, and it acts as a nonlinear correlation function. The sums are over the binned values of the observations. Now the theorem of Takens indicates that (almost)[1] *any* value of T is acceptable, if the data are of infinite precision. Well, that is not likely, so how we choose T is bound to be somewhat arbitrary. In practice, as the goal of this handbook, we recommend that one find the value of T for which $I(T)$ has its first minimum and then evaluate all subsequent quantities we discuss for T, $T \pm 1$, $T \pm 2$, and perhaps $T \pm 3$. If the conclusions from that set of five calculations with different T are the same, then in a practical sense the selection of T is acceptable. Choosing different T is equivalent to selecting different coordinate systems, connected by an unknown nonlinear transformation, in which to view the unfolding of the observations. If the quantities of interest are expected to be independent of the coordinate system, which is usually an important criterion, then this is a simple practical test of that.

Let us look at our LP neuron data now. In Fig. 2.1 we present a selection of the data of scaled membrane voltage from an LP neuron isolated from all other electrophysiological or neurochemical input. The sampling time was $\tau_s = 0.5$ ms. Figure 2.2 shows the average mutual information evaluated using all 200 000 data points. There is a very shallow minimum near $T = 10$, corresponding to 5 ms in time. Note that the data are a collection of spikes riding on top of a slow, large amplitude variation of the membrane potential with a period about 1 s. The T selected by the first minimum of $I(T)$ reflects the variation of the spikes at about 30 Hz.

In the literature there are often suggestions that one should use the first zero of the autocorrelation function of the measured time series as a good choice for the time $T\tau_s$ to use in constructing the data vector $\mathbf{y}(n)$. In the case of the

[1] Some values of the delay time T may lead to a nonfaithful representation of the dynamics that is not equivalent to the original system. For example, a closed orbit is mapped to a point if T equals exactly the period of the oscillation.

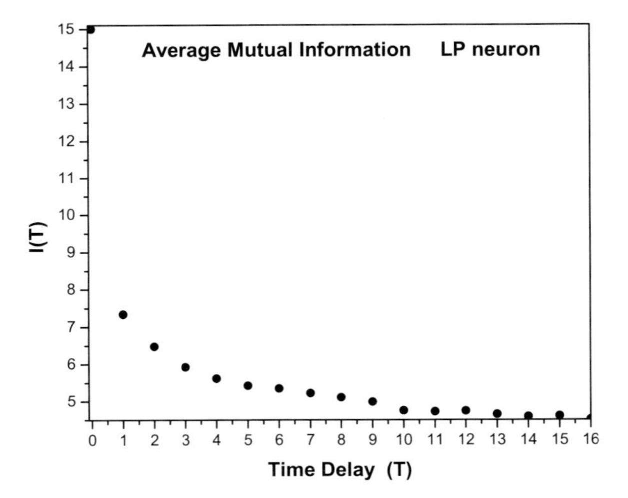

Fig. 2.2: The average mutual information $I(T)$ for the LP neuron membrane voltage time series shown in Fig. 2.1. $I(T)$ has a minimum in the neighborhood of $T = 10$; $T\tau_s \approx 5$ ms. It is a shallow minimum.

isolated LP neuron the Fourier power spectrum of the time series is shown in Fig. 2.3. Its Fourier transform is the autocorrelation function which shows a first zero crossing at 245 ms. This large number reflects the large amplitude oscillations near 1 Hz and washes out the dynamical structure of the spiking activity at each burst. That structure is reflected in the average mutual information choice of $T\tau_s \approx 5$ ms.

Procedure 2.1 (Average mutual information procedure). *From the amplitude range of the observations* $s(n)$ *form* B *bins. Record the frequency with which each bin is occupied by the values of* $s(n)$*. Normalize the frequency of occurrence by the total number of data. This normalized histogram is* $P(s(n))$*. Vary* B *to assure yourself that the amplitudes are properly sampled.*

Do precisely the same with the observations $s(n-T)$*. The corresponding distribution* $P(s(n-T))$ *should be the same as* $P(s(n))$ *if your data are stationary-independent of the origin of time-indicating autonomous oscillations of the signal source.*

From the amplitude range of the observations $s(n)$ *and* $s(n-T)$ *form* B^2 *bins. Record the frequency with which each bin is jointly occupied by the values of* $s(n)$ *and* $s(n-T)$*. Normalize the frequency of occurrence by the total number of data. This normalized histogram is* $P(s(n), s(n-T))$*. Vary* B *to assure yourself that the amplitudes are properly sampled.*

By summing over the bins evaluate

$$I(T) = \sum_{s(n),s(n-T)} P(s(n), s(n-T)) \log_2 \left[\frac{P(s(n), s(n-T))}{P(s(n))\, P(s(n-T))} \right]. \tag{2.4}$$

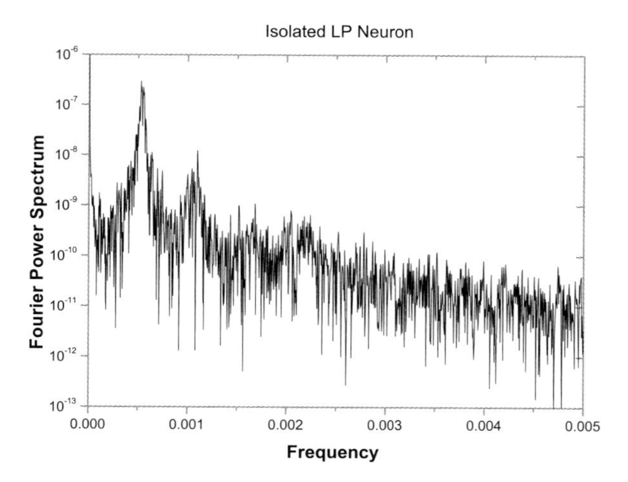

Fig. 2.3: The Fourier power spectrum for the LP neuron membrane voltage time series shown in Fig. 2.1. The frequency is given in units of $1/(0.5\,\text{ms}) = 2000\,\text{Hz}$. The peak in the power spectrum is at about 1 Hz corresponding to the nearly periodic low frequency oscillations of the isolated LP neuron. The higher frequency oscillations are irregular and show no sharp peaks. The Fourier transform of the power spectrum is the autocorrelation function which has its first zero crossing at approximately 245 ms. This is much longer than the first minimum of the average mutual information near 5 ms and reflects the large amplitude oscillations near 1 Hz in the original time series. A time delay this large will average out the important higher frequency spiking structure in the data. This should be a warning about the use of linear autocorrelation in nonlinear analysis.

The astute reader will note that we did not apply, as suggested in papers by Fraser [17, 18], the full machinery of information theory to the importance of having D components in the data vector $\mathbf{y}(n)$. Instead, we evaluated the importance of pairs of components on average over the data. The data requirements are daunting for the former, while our recommendation addresses the question of independence of pairs of elements of $\mathbf{y}(n)$ over the data set.

Figure 2.4 shows a two-dimensional delay embedding using a delay of $T = 9$. The high frequency spikes within the bursts are properly revealed whereas the chosen delay time is too small to unfold the slow dynamics between the bursts resulting in a reconstruction stretching along the diagonal. To obtain an optimal reconstruction of the slow dynamics we have to increase T but then the fast dynamics will be overfolded in a very complex manner. There are two characteristic time scales involved and in a two-dimensional delay embedding we have to make a decision whether we want to resolve the slow or the fast dynamics, because only a single delay time can be adjusted. This is different in higher dimensional delay embedding where we may use different delay times (mixed delays) for different components of the delay vector (e.g., $\mathbf{y}(n) = [s(n), s(n - T_1), s(n - T_2)]$) [22].

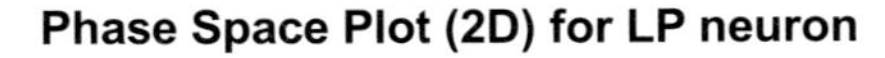

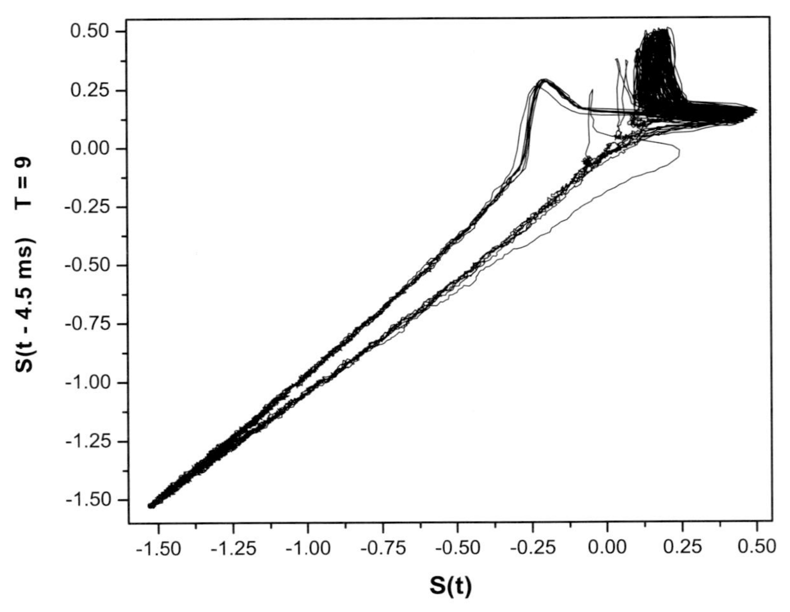

Fig. 2.4: The orbit of the LP neuron dynamics seen in two dimensions: $D = 2$. It is clear that the orbit is not yet fully unfolded; however, much of the state space structure is revealed even in this low dimension. These data will represent chaotic oscillations of the LP neuron—more on that as we go along—and this display even in two dimensions suggests strongly that it is phase space structures associated with the spiking on top of the slow oscillations of the membrane potential which lead to this.

An alternative for a two-dimensional representation that unfolds both time scales simultaneously is provided by the Hilbert transform $\mathcal{H}(s)$ of the time series. The Hilbert transform of a signal can be computed by shifting the phase of each Fourier component of the original signal by $\frac{\pi}{2}$. To understand why this is advantageous for an embedding we shall have a brief look at harmonic oscillations. A sinusoidal signal $s(t) = \sin(\omega t)$ is optimally embedded in the form of a circle if a delay of $T = \frac{1}{4}\frac{2\pi}{\omega}$ is used resulting in $s(t-T) = \sin(\omega t - \frac{\pi}{2}) = -\cos(\omega t)$, i.e., a phase shift of $\frac{\pi}{2}$. With the Hilbert transform we apply this optimal delay to each Fourier component separately and obtain an unfolding on all time scales when plotting $\mathcal{H}(s)$ versus s as shown in Fig. 2.5. When viewed in the complex plane this representation is also known as *analytic signal* and unfortunately there is no higher dimensional extension which would be necessary to reconstruct chaotic dynamics without intersections of trajectories.

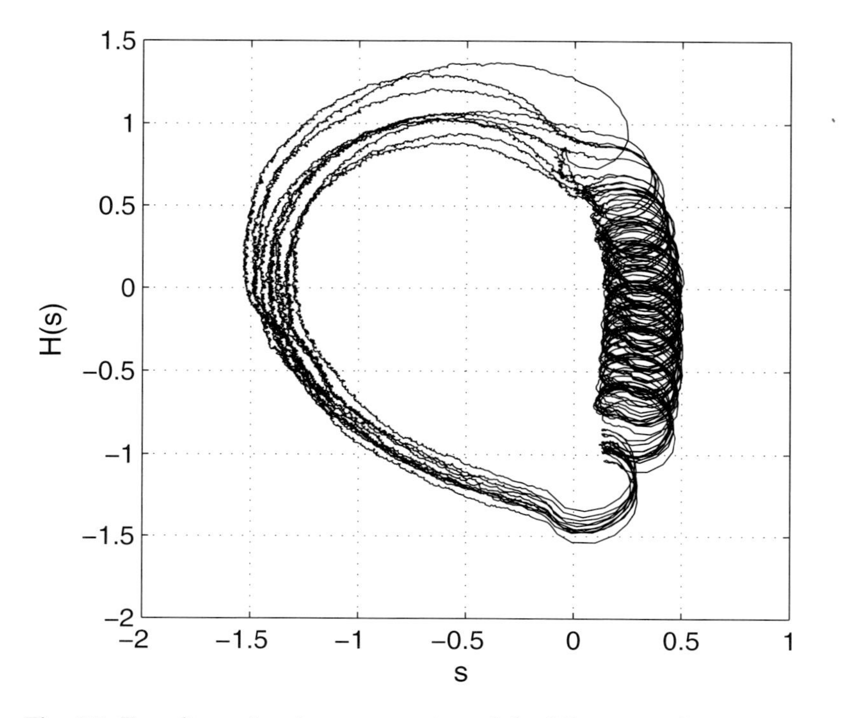

Fig. 2.5: Two-dimensional representation of the LP neuron dynamics in a plane spanned by the membrane potential signal s and its Hilbert transform $\mathcal{H}(s)$. Shown are only the first 20 000 samples of the time series.

2.2.2 Choosing D: False Nearest Neighbors

Global Embedding Dimension

The next question about the data vector

$$y(n) = [s(n), s(n-T), \ldots, s(n-(D-1)T)] \tag{2.5}$$

we need to address is the value of the integer "embedding dimension" D. Here is where Whitney's and Takens' results come into play. The key idea is that as we enlarge the dimension D of the vector $y(n)$ we eliminate step by step the intersections of orbits on the system attractor arising from our projection during the measurement process. If this is the case, then there might well be a global dimension allowing us to unfold a particular data set with particular coordinates as entries in $y(n)$ at a dimension less than the sufficient dimension of the Whitney/Takens geometric result.

To examine this we need the notion of crossing of trajectories, and this we realize in the close analogy of neighbors in the state space which are a result of the dynamics—true neighbors—and neighbors in the state space which are a result of

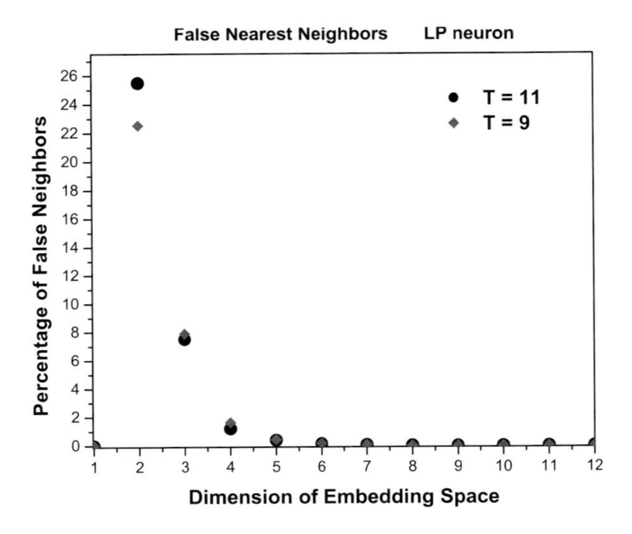

Fig. 2.6: False nearest neighbors as a function of the integer embedding dimension $D = 1, 2, \ldots$ for the observations of the membrane voltage of the isolated LP neuron. We evaluate this quantity for both $T = 9$ and $T = 11$ to determine whether our conclusion that $D = 4$ or $D = 5$ is adequate to unfold the attractor. Looking ahead we will show that the fractal dimension of the attractor here is about 3.15, so that a *sufficient* unfolding dimension, according to the Whitney/Takens geometric results, would be 7.

the projection during measurement—false neighbors [23]. If we select an embedding dimension D, then it is a matter of an order $N \log(N)$ search among all the points $\mathbf{y}(n)$ in that space to determine the nearest neighbor to a point $\mathbf{y}(k)$. If this nearest neighbor is not a close neighbor in dimension $D + 1$, then its "neighborliness" to $\mathbf{y}(k)$ is the result of a projection from a higher dimensional space. This is a false nearest neighbor, and we wish to eliminate all of them. We accomplish this elimination of the false nearest neighbors by systematically examining the nearest neighbors in dimension D and their "neighborliness" in dimension $D+1$ for $D = 1, 2, \ldots$ until there are no false nearest neighbors remaining. We call this integer dimension d_E.

Applying this idea to the LP neuron data, we evaluate the percentage of false nearest neighbors in dimensions $D = 1, 2, \ldots$ for $T = 9$ and $T = 11$. This is an example of a result, namely the smallest dimension where false nearest neighbors are absent, which we expect to be independent of the choice of T. Figure 2.6 shows this result. From this calculation we see that the conclusion that $d_E = 4$ or $d_E = 5$ would be a good embedding dimension is not dependent on the value of T in this small range, and, thus we can have confidence in this.

Procedure 2.2. *Global False Nearest Neighbor Procedure*

For dimension $\mathbf{D} = 1, 2, \ldots$, *form data vectors in the integer dimension* $\mathbf{D}$*:*

$$\mathbf{y}(\mathbf{n}) = [\mathbf{s}(\mathbf{n}), \mathbf{s}(\mathbf{n} - \mathbf{T}), \ldots, \mathbf{s}(\mathbf{n} - (\mathbf{D} - 1)\mathbf{T})] . \tag{2.6}$$

Using a search routine based on forming a "kd-tree" in dimension $\mathbf{D}$*, find the nearest neighbor of each point,* $\mathbf{y}(\mathbf{k})$*,* $k = 1, 2, \ldots, N - T$*, in dimension* $\mathbf{D}$*. By adding the component* $\mathbf{s}(\mathbf{n} - \mathbf{DT})$ *to the* $\mathbf{D}$*-dimensional vector, determine if the nearest neighbors in dimension* $\mathbf{D}$ *remain near neighbors in dimension* $\mathbf{D} + 1$*. "Near" implies a notion of distance, and, while any would do, we use the standard Euclidian distance.*

As a function of $\mathbf{D}$ *determine the number of false nearest neighbors—those which do not remain neighbors when seen in dimension* $\mathbf{D} + 1$*. This number will decrease as* $\mathbf{D}$ *increases, absent "noise." When the percentage of false nearest neighbors falls below some threshold, say 1 %, the embedding dimension* $\mathbf{d}_E$ *has been found. Further increasing the integer dimension of the embedding space does not further eliminate trajectory crossings.*

The threshold of 1 % for selecting an embedding dimension is clearly a useful, but not mathematically rigorous, choice. It is little more than a recognition that the accuracy of the measured data at the few percent level is what one often faces in "clean" observations. "Noise" seen as contamination of measurements by inaccuracies in the sensing devices or signals input from unwanted sources is formally of infinite dimension and 100 % false nearest neighbors would appear if noise alone were the signal.

2.2.2.1 Local or Dynamical Dimension

The embedding dimension we just selected is a global and average indicator of the number of coordinates needed to unfold the actual data $s(t)$ [24].

The global integer embedding dimension estimate tells us a minimum dimension d_E in which we can place (embed) the signal from our source. This dimension can be larger than the number of degrees of freedom in the dynamics underlying the signal $s(t)$. Suppose that *locally* the dynamics happened to be a two-dimensional map $(x_n, y_n) \rightarrow (x_{n+1}, y_{n+1})$ but the global structure of the dynamics placed this on the surface of an ordinary torus. To embed the points of the whole data set now lying on a torus, we would have to select $d_E = 3$; however, if we wish to determine equations of motion (or a map) to describe the dynamics, we really need only the local dimension of 2. This local dimension $d_L \leqslant d_E$, and is important when we wish to evaluate the Lyapunov exponent, as we do below, to characterize the dynamical system producing $s(t)$.

To determine d_L we need to move beyond the geometry in the embedding theorem and ask a *local* question about the data in dimension d_E where we know there are no unwanted intersections of the orbit associated with $s(t)$ and itself. The notion is that data vectors of dimension $d \leqslant d_E$,

$$y_d(n) = [s(n), s(n - T), \ldots, s(n - (d - 1)T)] , \tag{2.7}$$

will map without ambiguity locally into other vectors of dimension $d \leqslant d_E$. We can test for this by forming a d-dimensional local map

$$y_d(n + 1) = \mathbf{M}(y_d(n)) , \tag{2.8}$$

and asking whether this map accounts for the behavior of the actual data in $d \leqslant d_E$. For d too small, it will not. For $d = d_E$, it will. If for some intermediate d, the map is accurate, this is an indication of a lower dimensional dynamics than d_E needed to globally unfold the data.

To answer this question select a data vector $y(k)$ in d_E. Select N_B nearest neighbors in phase space to $y(k)$: $y^{(r)}(k)$; $r = 0, 1, 2, \ldots, N_B$, $y^{(0)}(k) = y(k)$. In d_E these are all true neighbors, but their actual time labels may or may not be near the time k. Choose in various ways a d-dimensional subspace of vectors $y_d^{(r)}(k)$. There are $\binom{d_E}{d}$ ways to do this and all are worth examining. This set of points maps in time into another set of points $y_d(k+1;r)$ near $y(k+1)$, and we expect that the map $\mathbf{M}(y)$ local to $y(k)$ will act as

$$y_d(k+1;r) = \mathbf{M}\left(y_d^{(r)}(k)\right) . \tag{2.9}$$

For other locations on the attractor the map will be different.

The map $\mathbf{M}(y)$ is unknown to us, but by choosing a parametric form for it, we can use the data to determine the parameters and ask, as a function of d, how well this map performs in describing the data. One easy way to parameterize the map is to make a Taylor expansion of it in d-dimensional space, and then we could determine the Taylor coefficients using a least-squares method. Write

$$\begin{aligned} y_d(k+1;r) &= \mathbf{M}(k; y_d^{(r)}(k)) \\ &= \mathbf{A}(k) + \mathbf{B}(k) \cdot y_d^{(r)}(k) + \mathbf{C}(k) \cdot y_d^{(r)}(k) y_d^{(r)}(k) + \cdots , \end{aligned} \tag{2.10}$$

and determine the local parameters, namely the vector $\mathbf{A}(k)$ and the tensors $\mathbf{B}(k)$, $\mathbf{C}(k), \ldots$, by minimizing

$$\sum_{r=0}^{N_B} |y_d(k+1;r) - \mathbf{A}(k) - \mathbf{B}(k) \cdot y_d^{(r)}(k) - \mathbf{C}(k) \cdot y_d^{(r)}(k) y_d^{(r)}(k) + \cdots|^2 . \tag{2.11}$$

The quality of this fit to the data will vary with the dimension $d \leqslant d_E$. The residuals after the least-squares fit will be large for small d and decrease as d increases. For the *local dimension* d_L, it will become independent of d. This is a suggested choice for the dimension of the dynamics.

To use this method of "local false nearest neighbors" with confidence, one should select d_E equal to and then larger than that indicated with global false nearest neighbors, and one should explore different choices for the vectors $y_d(k)$ in dimension d. If d_L is the local dynamical dimension, then its value will be independent of these variations.

The application of this procedure to our isolated LP data is shown in Fig. 2.7.

Procedure 2.3. *Local Dimension Determination Procedure*

For dimension $\mathbf{d_E} = 1, 2, \ldots$, *form data vectors in the integer dimension* $\mathbf{d_E}$:

$$y(n) = [s(n), s(n-T), \ldots, s(n-(d_E-1)T)] . \tag{2.12}$$

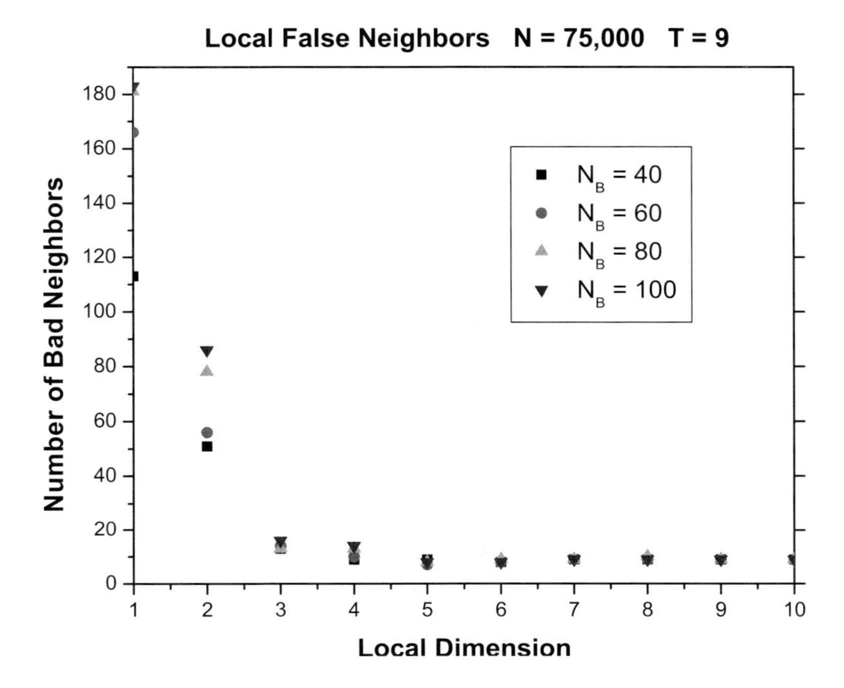

Fig. 2.7: The percentage of bad predictions (large residual in a least-squares fit) as a function of d arising from a local d-dimensional map for the data from the isolated LP neuron. We also vary the number of nearest neighbors N_B forming a neighborhood in d-dimensional space. These neighborhoods are mapped into one another in time by a local parametric map with parameters determined from the observed data. We use $N_B = 40$, 60, 80, and 100. It is clear that at $d_L = 4$, the quality of the predictions made by a local map becomes independent of d and of N_B. $d_E = 10$ was chosen.

Choose a subspace of dimension $\mathbf{d} = 1, 2, \ldots, \mathbf{d}_E$. *There are* $\binom{\mathbf{d}_E}{\mathbf{d}}$ *such subspaces. One plausible choice would be to form the sample covariance matrix in dimension* $\mathbf{d}_E$ *for data local to any time point* k *and select the* $\mathbf{d} = 1, 2, \ldots, \mathbf{d}_E$ *largest principal components.*

In dimension $\mathbf{d}$ *make a* local *map from a neighborhood of nearest neighbor points around* $\mathbf{y}(k)$ *into the points around* $\mathbf{y}(\mathbf{k} + \mathbf{K})$ *to which they go in* $\mathbf{K}$ *time steps. As a function of* $\mathbf{d}$ *evaluate the residual of a least-squares fit locally to the parameters in the map from time* k *to time* $k + K$. *When this residual representing the quality of the local fit in dimension* $\mathbf{d}$ *to the data is larger than some threshold, we have a bad prediction. Average the number of bad predictions for a fixed* $\mathbf{K}$ *over the data set. When this average number of "bad" predictions becomes independent of* d, *this indicates a good local dimension* $\mathbf{d}_L \leqslant \mathbf{d}_E$ *has been found.*

Repeat the procedure varying the number of nearest neighbors used in the neigh-

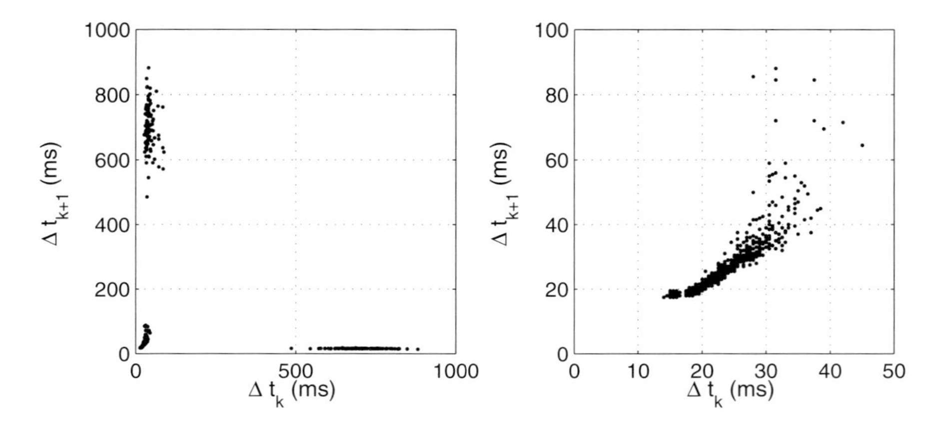

Fig. 2.8: Interspike intervals of the LP neuron dynamics shown in Fig. 2.1

borhood-to-neighborhood local map. We use $\mathbf{N}_B = 40, 60, 80,$ *and* 100*, but this is not dictated by any rigorous prescription.*

There are a number of parameters here including those in the local map and the horizon $\mathbf{K}$ *to which one uses the local map to predict ahead for comparison with the data. One must explore combinations of these parameters to arrive at consistency for the local dimension.*

2.2.3 Interspike Intervals

For spiking signals such as the neuron dynamics considered here another approach for state space reconstruction exists which is based on the time intervals $\Delta t_k = t_{k+1} - t_k$ between consecutive spikes occurring at times t_k and t_{k+1}. These *interspike intervals* (ISI) contain all relevant dynamical information and can also be used for delay embedding [25]. Figure 2.8 shows a two-dimensional ISI delay embedding of the neuron dynamics. Small interspike intervals (also shown enlarged) correspond to spikes within a burst and show a very regular pattern that can approximately be described by a one-dimensional function $\Delta t_{k+1} \approx g(\Delta t_k)$ as will be done in Section 2.6.1.

2.3 Where are We?

In the spirit of a handbook we should pause now and examine what the readers will have accomplished with their hard won data $s(t)$. So far we have indicated algorithms that do *not* analyze any data. We have only presented a series of steps to identify the space in which the data should be analyzed. We have taken observations of a single variable $s(t)$ from a multidimensional system and identified a multidimensional vector for that data

$$\mathbf{y}(t = t_0 + n\tau_s) = \mathbf{y}(n) = [s(n), s(n-T), s(n-2T), \ldots, s(n-(d_E-1)T)] , \quad (2.13)$$

where the integers T and d_E are a time delay that exposes the data on the system attractor and d_E is the smallest global dimension which eliminates trajectory overlaps associated with the projection of the multidimensional system to the single $s(t = t_0 + n\tau_s)$ axis. We have also identified a *local* dimension $d_L \leqslant d_E$ for the actual dynamics of the source producing the observations.

We want to move on to the extraction of other information from the original data, now properly formatted. To do this we must select a few of the myriad of questions about a dynamical system we might wish to ask. One question which we do *not* know how to answer is the reconstruction of the dynamical equations underlying the data. There are many attempts to do this, and they all eventually make a guess about the functional form of the equations and then determine the numerical parameters contained in that conjecture. If one knows a great deal about the equations, this may be successful, and we will discuss that in our last section.

Another question of interest is whether we can use the observed data, now properly formatted in a multidimensional state space with data vectors $y(n)$, to predict the future behavior of the observed variable $s(t)$. The answer to this is "yes" and one can also use the data to determine a prediction horizon for this. With nonlinear sources, prediction may be limited if chaotic oscillations are present. We discuss these ideas in our next two sections.

Of course, this does not address many interesting issues, and for that we apologize. Of particular interest is the idea of learning the characteristics of a communication channel with nonlinear elements in it and then using this information to correct the distortion of a signal propagating through that channel [26]. This is called "channel equalization" and has some significant practical applications to extending the range or enlarging the effective bandwidth of many channels.

2.4 Lyapunov Exponents: Prediction, Classification, and Chaos

The attractor of the dynamical systems producing $s(t)$ is contained in dimension d_E which assures that there is no residual overlap of trajectories from the projection to one dimension: $s(t)$. To characterize the attractor we can call on many different notions of dimension of the set of points $y(n)$. Each is an invariant of the dynamical system in the sense that a smooth coordinate change from those used in $y(n)$, including that involved in changing T, leaves these characteristic numbers unchanged. The invariance comes from the fact that each dimension estimate is a local property of the point set comprising the attractor, and smooth changes of coordinates do not alter this local property while they might change the global appearance of the attractor. These various dimensions are covered in many books, and each is interesting.

We here focus on a dynamical invariant of the attractor that also allows an estimate of dimension. The central issue is the stability of an orbit such as $y(n)$

under perturbations to points on the trajectory. This is a familiar question associated with the stability of fixed points or limit cycles as studied in classical fields such as fluid dynamics. If one has a fixed point x_0 of a dynamical system in d dimension $\mathbf{x}(t) = [x_1(t), x_2(t), \ldots, x_d(t)]$ with $x(t)$ satisfying

$$\frac{d\mathbf{x}(t)}{dt} = \mathbf{G}\big(\mathbf{x}(t)\big) , \tag{2.14}$$

so $\mathbf{G}(x_0) = 0$, then it is important to ask if state space points $x_0 + \Delta\mathbf{x}(t)$ are stable in the sense they remain near or return to x_0. Unstable points, where $\Delta\mathbf{x}(t)$ grows large, are not realized in observations of a dynamical system.

In the case of a fixed point, a Taylor series expansion assuming $\Delta\mathbf{x}(t)$ remains "small" leads to the analysis of the linearized dynamics

$$\frac{d\Delta\mathbf{x}(t)}{dt} = \mathbf{DG}(\mathbf{x}_0)\Delta\mathbf{x}(t) , \tag{2.15}$$

with

$$\mathbf{DG}(\mathbf{x})_{ab} = \frac{\partial G_a(\mathbf{x})}{\partial x_b}, \quad a, b = 1, 2, \ldots, d . \tag{2.16}$$

The fixed point is linearly stable (nonlinear stability is more subtle) if the d eigenvalues of the matrix $\mathbf{DG}_{ab}(\mathbf{x}_0)$ have zero or negative real part. If the dynamics is Hamiltonian, quite unlikely in the real world, then eigenvalues reflect the symplectic symmetry of the system and lie on the imaginary axis in complex conjugate pairs when the system is stable.

To determine the stability properties of the time-dependent orbit $\mathbf{y}(n)$ we need to go beyond the classical eigenvalue analysis of a matrix with fixed elements $\mathbf{DG}(\mathbf{x}_0)$. The needed analysis was made by Oseledec in 1968 [27]. To use his result we note that the continuous time dynamics $\frac{d\mathbf{x}(t)}{dt} = \mathbf{G}\big(\mathbf{x}(t)\big)$ is sampled by our observations every τ_s and we can replace it by a discrete time-one map in dimension $d_L \leqslant d_E$: $\mathbf{y}(n+1) = \mathbf{F}\big(\mathbf{y}(n)\big)$. We do not have an explicit form of $\mathbf{F}(\mathbf{x})$ in most cases. As above one should examine the $\binom{d_E}{d_L}$ choices of d_L-dimensional subspaces one has.

The discrete time map can be linearized about a solution $\mathbf{y}(n)$ by replacing $\mathbf{y}(n)$ by a nearby orbit $\mathbf{y}(n) + \Delta\mathbf{y}(n)$ leading to

$$\Delta\mathbf{y}(n+1) = \mathbf{DF}\big(\mathbf{y}(n)\big)\Delta\mathbf{y}(n) , \tag{2.17}$$

and iterating this L times leads to

$$\begin{aligned} \Delta\mathbf{y}(n+L) &= \mathbf{DF}^L\big(\mathbf{y}(n)\big)\Delta\mathbf{y}(n) , \\ \mathbf{DF}^L(\mathbf{y}(n)) &= \mathbf{DF}\big(\mathbf{y}(n+L-1)\big)\mathbf{DF}(\mathbf{y}(n+L-2)) \cdots \mathbf{DF}\big(\mathbf{y}(n)\big) . \end{aligned} \tag{2.18}$$

Oseledec forms the orthogonal matrix composed of $\mathbf{DF}^L(\mathbf{y}(n))$ and its transpose $[\mathbf{DF}^L(\mathbf{y}(n))]^T$,

$$\mathbf{OSL}\big(\mathbf{y}(n)\big) = \{[\mathbf{DF}^L(\mathbf{x})]^T \mathbf{DF}^L(\mathbf{x})\}^{\frac{1}{2L}} , \tag{2.19}$$

and proves that the limit as $L \to \infty$ exists and is independent of the d_E-dimensional $\mathbf{x}$ lying in the basin of attraction of the attractor defined by the orbit $\mathbf{y}(n)$. The logarithm of the eigenvalues of this limiting matrix are the Lyapunov exponents, and we write them as ordered $\lambda_1 \geqslant \lambda_2 \geqslant \cdots \geqslant \lambda_d$. They are also shown to be independent of the coordinate systems used to define the state space if these coordinates are connected by smooth transformations. The λ_a $a = 1, 2, \ldots, d$, are thus invariant characteristics of the dynamical system producing $s(t)$. If the dynamics is that of a flow with continuous time, for which our time-one map is a discrete approximation, then one of λ_a is zero. In any case, $\sum_{a=0}^{d} \lambda_a < 0$ for systems with dissipation, and the sum is zero for Hamiltonian systems.

If any of λ_a is positive, then the trajectory $\mathbf{y}(n) + \Delta\mathbf{y}(n)$ diverges from the original orbit, but since the attractors we encounter in real time series are compact, the new orbit does not diverge to spatial infinity. Instead it also visits points on the attractor though in an order quite different from that of the original orbit $\mathbf{y}(n)$. When one (or more) of λ_a are positive, the orbit is very sensitive to perturbations; in particular, it is sensitive to changes in the initial condition, and the resulting sensitive orbit we call chaotic.

λ_1 tells us how line segments in the state space (also the "proxy" state space of vectors $\mathbf{y}(n)$) increase ($\lambda_1 > 0$) or decrease ($\lambda_1 < 0$) or remain fixed ($\lambda_1 = 0$). $\lambda_1 + \lambda_2$ determines the same for areas in the state space, and $\sum_{a=0}^{d} \lambda_a$, the same for d-dimensional volumes. If some $\lambda_a > 0$, some subspaces have growing volumes while the whole space has a shrinking volume, $\sum_{a=0}^{d} \lambda_a < 0$. Somewhere in between dimension 1 and dimension d is a volume in a dimension which need not be integer, which neither grows nor shrinks. This is called the Lyapunov dimension D_L and turns out to be one of the many commonly defined fractional dimensions associated with an attractor. It is given by

$$D_L = K + \frac{\sum_{a=1}^{K} \lambda_a}{|\lambda_{K+1}|} , \tag{2.20}$$

where $\sum_{a=1}^{K} \lambda_a > 0$ and $\sum_{a=1}^{K+1} \lambda_a < 0$.

To actually evaluate λ_a and D_L, we need to estimate the $d \times d$ Jacobian $\mathbf{DF}(\mathbf{y}(n))$ along the proxy orbit $\mathbf{y}(n)$. We need information on the state space which probes the d dimensions so that we can fill the d^2 elements of $\mathbf{DF}(\bullet)$. One can evaluate these matrix elements by making *local* maps from N_B points $\mathbf{y}^{(r)}(n)$, $r = 0, 1, \ldots, N_B$, in the neighborhood of $\mathbf{y}(n) = \mathbf{y}^{(0)}(n)$. The points $\mathbf{y}^{(r)}(n)$ go, in one time step, into the points $\mathbf{y}(n+1; r)$ and the local map takes the form

$$\mathbf{y}(n+1; r) = \sum_{m=1}^{M} c(m, n)\phi_m\left(\mathbf{y}^{(r)}(n)\right) , \tag{2.21}$$

with $\phi_m(\mathbf{x})$ some basis set in the state space whose choice is up to us. Polynomials associated with a Taylor series often work well. The coefficients $c(m, n)$ are determined by minimizing the least-squares difference

$$\sum_{r=0}^{N_B} |\mathbf{y}(n+1;r) - \sum_{m=1}^{M} c(m,n)\phi_m(\mathbf{y}^{(r)}(n))|^2. \tag{2.22}$$

Once the coefficients are known, the estimate of the matrix elements $\mathbf{DF}(\mathbf{y}(n))_{ab}$ is given as

$$\mathbf{DF}(\mathbf{y}(n))_{ab} = \sum_{m=1}^{M} c(m,n)\frac{\partial \phi_{ma}(\mathbf{x})}{\partial x_b}|_{\mathbf{x}=\mathbf{y}(n)} . \tag{2.23}$$

To determine $\mathbf{DF}(\mathbf{y}(n))$ accurately one cannot always just use a local linear estimate of the neighborhood-to-neighborhood map as the curvature in the orbits on the attractor may lead to errors. Usually, if one is using polynomials locally, retaining quadratic and cubic terms in the Taylor expansion will be enough to accurately estimate the linear coefficient $\mathbf{DF}(\mathbf{y}(n))$.

With knowledge of the Jacobian matrices $\mathbf{DF}(\mathbf{y}(n))$ one can form and diagonalize the Oseledec matrix $\mathbf{OSL}(\mathbf{y}(n))$. It is important to use care in this as $\mathbf{DF}(\mathbf{y}(n))$ is very ill conditioned, because it contains both exponentially increasing and decreasing elements. A standard method to avoid large roundoff errors (or even over- and underflow) consists in a factorization of $\mathbf{DF}(\mathbf{y}(n))$ by means of repeated QR-decompositions [3, 28].

Lyapunov exponents have been the subject of many investigations and various techniques have been developed. We invite the reader to examine the various methods [28–49].

For our isolated LP neuron data set we evaluated the Lyapunov exponents in an embedding dimension $d_E = 5$ and a local dimension $d_L = 4$. The local dimension is quite important here because for $d_L = 5$, we would find five Lyapunov exponents, one of which would be false. Another way to determine if any Lyapunov exponents are false would be to evaluate the exponents both forward and backward along the time series and eliminate those which did not change sign [42]. This is always a very useful exercise giving further confidence in the choice of local dimension.

In Fig. 2.9 we show the Lyapunov exponents for the isolated LP neuron as a function of L the number of time steps along the attractor from a starting point $\mathbf{y}(n)$. These are the four ($d_L = 4$) eigenvalues of $\mathbf{OSL}(\mathbf{y}(n))$ as a function of L. The sum of the exponents is negative, as it must be. One of the exponents (λ_2) is consistent with zero, indicating that the dynamics of the isolated LP neuron is described by four differential equations. The Lyapunov dimension determined by these λ_a values is $D_L = 3.15$. λ_a are invariant characteristics of the isolated LP neuron dynamics, and any model of the LP neuron must reproduce these values. In checking it, be sure to sample the output of the model dynamics at $\tau_s = 0.5$ ms, as λ_a have a dimension of inverse time.

The Lyapunov exponents tell us one more important piece of information. If the perturbation of the orbit $\mathbf{y}(n)$ has initial length Δ_0, then since line segments

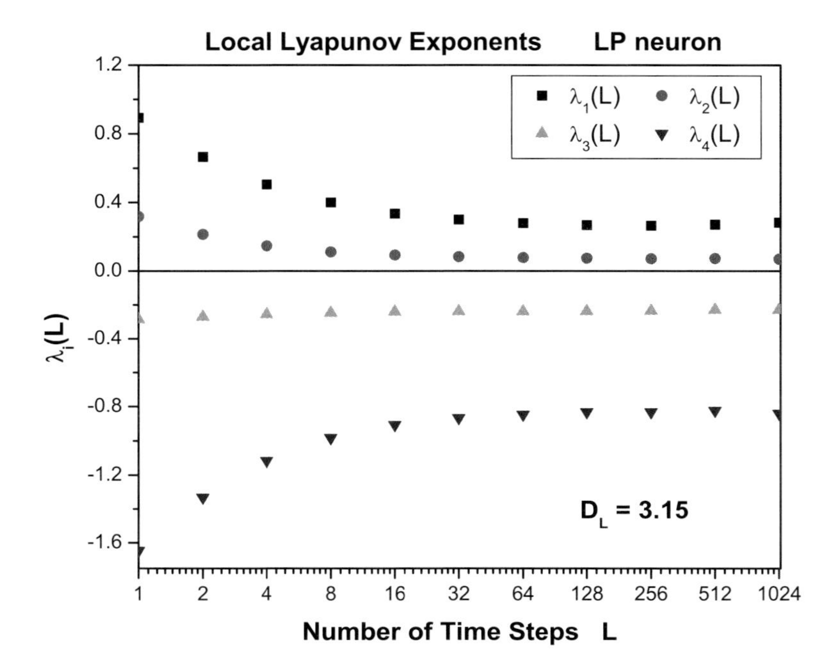

Fig. 2.9: Lyapunov exponents for the isolated LP time series in Fig. 2.1. These are evaluated as a function of the number of time steps L in the Oseledec matrix $\mathbf{OSL}(\mathbf{y}(n))$. The matrix is estimated in embedding dimension $d_E = 5$ where no trajectory crossings occur as indicated by the false nearest neighbor calculation. $d_L = 4$ exponents are determined as indicated by the evaluation of the local dynamical dimension for this data set. We see one positive exponent, one exponent consistent with zero, and two negative exponents. Their sum is negative, and the Lyapunov dimension of the attractor, seen in a two-dimensional projection in Fig. 2.4, is about 3.15. The presence of a zero exponent indicates that the underlying dynamics of this isolated neuron is described by differential equations: actually, $d_L = 4$ differential equations.

on the attractor grow as $\exp(\lambda_1 n \tau_s)$, this initial perturbation will grow to the size of the attractor R_A in a time

$$n\tau_s = \frac{1}{\lambda_1} \log\left[\frac{R_A}{\Delta_0}\right]. \tag{2.24}$$

R_A can be estimated by the range of the data $s(t)$ and Δ_0 is up to you. When a perturbation has grown to the size of the attractor itself, one has lost the ability to predict (more on that in a moment) and one commonly refers to $\frac{1}{\lambda_1}$ as the time horizon for predictability, in units of steps in τ_s.

Procedure 2.4. *Lyapunov Exponents Determination Procedure*

Starting with vectors in the global unfolding space of dimension $\mathbf{d}_E$ *select among the* $\binom{d_E}{d_L}$ $\mathbf{d}_L$*-dimensional subspaces and construct local maps from* $\mathbf{N}_B$ *neighbors of* $\mathbf{y}(\mathbf{n})$ *to* $\mathbf{N}_B$ *neighbors of* $\mathbf{y}(\mathbf{n}+1)$*. The linear term in this map as evaluated on the orbit* $\mathbf{y}(\mathbf{n})$ *yields the local Jacobian matrix* $\mathbf{DF_{ab}}(\mathbf{x})$*.*

Form the Oseledec matrix $\mathbf{OSL}(\mathbf{y}(\mathbf{n})) = \{[\mathbf{DF^L}(\mathbf{x})]^T\mathbf{DF^L}(\mathbf{x})\}^{\frac{1}{2L}}$ *and evaluate its eigenvalues as a function of the number of time steps* $\mathbf{L}$ *along the orbit. These eigenvalues should converge for large* $\mathbf{L}$ *to constants* λ_a*,* $a = 1, 2, \ldots, d_L$*. Use a careful method, such as the recursive QR decomposition, for dealing with the very ill-conditioned matrix* $\mathbf{OSL}(\mathbf{y}(\mathbf{n}))$*.*

Check this calculation by estimating λ_a *arising from perturbations at many places along the orbit. For* $\mathbf{L}$ *large enough the values of* λ_a *should be the same. The variation should decrease as a fractional power of* $\frac{1}{L}$*.*

$\sum_{a=1}^{d_L} < 0$*.* λ_1 *indicates the prediction horizon for the dynamics. The presence of one zero exponent indicates that the underlying dynamics is governed by differential equations:* $\mathbf{d}_L$ *of them.* λ_a *are characteristic of the dynamics. An estimate of the fractal dimension of the attractor is the Lyapunov dimension* $\mathbf{D_L}$*.*

2.5 Predicting

One goal of time series analysis is learning enough about the underlying dynamics from observations to be able to make predictions about the behavior in the future [3, 50]. Using information about the trajectories in d_E-dimensional state space, one can do this for nonlinear dynamical systems as well.

Before describing how to do this, certain caveats are in order.

- If only $s(n)$ was observed and nothing else is known about the underlying dynamical system, one will only be able to predict future values of $s(n)$ given a new value $s(n')$ at $t' = t_0 + n'\tau_s$.
- If one also knows the relation of the variable $s(n)$ to another dynamical variable of the underlying system, say, $v(n) = g(s(n))$, then one can predict $v(n)$ as well.
- If one has observed several variables $s_i(n)$, then one can use these in a mixed mode, if desired, to establish data vectors $\mathbf{y}(n)$, or one may use the simultaneously observed values of $s_i(n)$ and $s_j(n)$, say, to establish a relationship $s_j(n) = f_j(\mathbf{y}_i(t) = [s_i(n), s_i(n-T), \ldots, s_i(n-(d_E-1)T)])$. In other words, the state space need be reconstructed or unfolded only once for all dynamical variables [51].
- If one changes the underlying parameters, temperature, voltage, current, …, under which the data $s(t)$ were observed, the whole procedure of unfolding, … must be repeated, unless one has a physical or biophysical knowledge of the changes in the orbit as a function of these parameters. This is especially

important to remember when bifurcations, topological changes in the nature of the attractor, occur when these external parameters are varied.

With these items in mind, one can use the vectors in d_E-dimensional space to predict the orbit, approximately up to time $\frac{1}{\lambda_1}$. The idea is that one takes a set of observed data and within the d_E-dimensional unfolding dimension, constructs a d_L-dimensional map

$$\mathbf{y}(n+1;r) = \mathbf{F}(\mathbf{y}^{(r)}(n)) , \tag{2.25}$$

for the $r = 0, 1, \ldots, N_B$ neighbors of each orbit point $\mathbf{y}(n) = \mathbf{y}^{(0)}(n)$ which goes into the orbit point $\mathbf{y}(n+1;0) = \mathbf{y}(n)$. This map may be realized in local polynomials or in other basis functions

$$\mathbf{y}(n+1;r) = \sum_{m=1}^{M} c(m,n)\phi_m(\mathbf{y}^{(r)}(n)) , \tag{2.26}$$

as above. Once the basis functions are chosen, determine the coefficients $c(m,n)$ by the least-squares minimization. One can use other distance metrics, of course.

Now suppose a new data point $s(m)$ is measured. We would like to know what points will follow it in time: $s(m+1)$, $s(m+2)$, ... From the new data point, form a data vector in dimension d_E,

$$\mathbf{y}_{\text{new}}(m) = [s(m), s(m-T), \ldots, s(m-(d_E-1)T)] , \tag{2.27}$$

and in dimension d_E look for its nearest neighbor among the members of the original data set. Suppose it is $\mathbf{y}(q)$. Associated with $\mathbf{y}(q)$ is a local map in dimension d_L which carries $\mathbf{y}(q) \to \mathbf{y}(q+1) = \mathbf{F}_q(\mathbf{y}(q))$. It also carries the N_B nearest neighbors of $\mathbf{y}(q)$ one step forward in time.

Now we use the map $\mathbf{F}_q(\mathbf{x})$ (in dimension d_L) to map forward $\mathbf{y}_{\text{new}}(m)$. There are many ways to do this each utilizing information about the nearest neighbors in different roles. The simplest prediction would be

$$\mathbf{y}_{\text{new}}(m+1) = \mathbf{F}_q(\mathbf{y}_{\text{new}}(m)) , \tag{2.28}$$

and another might be

$$\mathbf{y}_{\text{new}}(m+1) = \sum_{r=0}^{N_B} w_r \mathbf{F}_q(\mathbf{y}_{\text{new}}^{(r)}(m)) , \quad \sum_{r=0}^{N_B} w_r = 1 , \tag{2.29}$$

which weights the mapping of all its nearest neighbors as well [26].

Now we are ready to map this point one step further into the future: find the nearest neighbor of $\mathbf{y}_{\text{new}}(m+1)$ in dimension d_E among all the members of the original data set. Suppose it is $\mathbf{y}(k)$; then the simplest map projecting this one step into the future would be

$$\mathbf{y}_{\text{new}}(m+2) = \mathbf{F}_k(\mathbf{y}_{\text{new}}(m+1)) = \mathbf{F}_k\Big(\mathbf{F}_q(\mathbf{y}_{\text{new}}(m))\Big) . \tag{2.30}$$

From this the procedure should be clear, and a temporal sequence $y_{new}(m)$, $y_{new}(m+1)$, ... is created by this.

The first component of the data vectors $y_{new}(\bullet)$ is precisely the predictions for the observations $s(m+1)$, $s(m+2)$, ... following the new observation $s(m)$.

Attending to the caveats noted at the outset to this section, we can predict forward to $s(m+K)$ with $K\tau_s \approx \frac{1}{\lambda_1}$.

Using our standard example of data from the isolated LP neuron, we have used 41 000 data points to learn local maps, and then made predictions five steps ahead and then 75 steps ahead for data points at time 53 000 to 55 000. In these predictions we used $d_L = 4$, $d_E = 5$, and $T = 9$. A quadratic polynomial was used for the local maps. In Fig. 2.10 we show the observed data and the predictions $K = 5$ steps ahead and in Fig. 2.11, the same for $K = 75$ steps ahead. The prediction horizon for these data is $1/\lambda_1 \approx 4$ steps, and we see that at $K = 5$ the predictions are rather good, while at $K = 75$ they become quite inaccurate for the rapidly varying spikes in the data. However, even at $K = 75$ the slowly varying part of the data, outside the spiking region, is well predicted (not shown in the figures).

Returning to the selection of the time delay in our embedding vectors $y(n)$ indicates that the choice $T = 9$ or 10 suggested by average mutual information considerations is required to properly sample and then predict the rapid variations in the data. Had we accepted the value of $T = 425$ from linear correlation considerations, we would have completely lost the ability to say anything about these rapid variations which are a distinct and important feature of the dynamics.

Procedure 2.5. *State Space Prediction Procedure*

Start with the data vectors $\mathbf{y}(\mathbf{n})$ in dimension $\mathbf{d}_E$ and the local maps in dimension

$$\mathbf{d}_L, \quad \mathbf{y}(\mathbf{n}+1) = \mathbf{F}_{\mathbf{n}}(\mathbf{y}(\mathbf{n}))\,. \tag{2.31}$$

Observe a new data point $\mathbf{s}(\mathbf{m})$ and form a new data vector in $\mathbf{d}_E$,

$$y_{new}(\mathbf{m}) = [\mathbf{s}(\mathbf{m}), \mathbf{s}(\mathbf{m}-\mathbf{T}), \ldots, \mathbf{s}(\mathbf{m}-(\mathbf{d}_E-1)\mathbf{T})]\,. \tag{2.32}$$

Locate its nearest neighbor in dimension $\mathbf{d}_E$ among the members of the original data set. Suppose it is $\mathbf{y}(\mathbf{k})$. Use the local map in dimension $\mathbf{d}_L$ to move the new vector forward in time by one step

$$y_{new}(\mathbf{m}+1) = \mathbf{F}_{\mathbf{k}}(y_{new}(\mathbf{m}))\,. \tag{2.33}$$

Continue until you have moved ahead approximately $\frac{1}{\tau_s \lambda_1}$ steps which is the approximate prediction horizon.

More details about the local modeling approach outlined above can be found in the chapter on local modeling in this handbook [26].

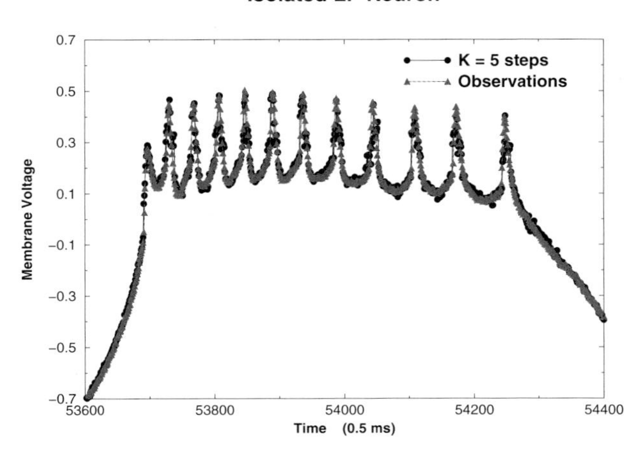

Fig. 2.10: Predictions using proxy state space methods on the data from an isolated LP neuron. Data are in Fig. 2.1. We used 41 000 data points in $d_E = 5$ to make local maps in $d_L = 4$ and selected $T = 9$. Predictions were made five time steps ahead for all points between 53 000 and 55 000. The prediction horizon $1/\lambda_1$ is approximately four steps for these data, and the predictions at five steps are still quite accurate.

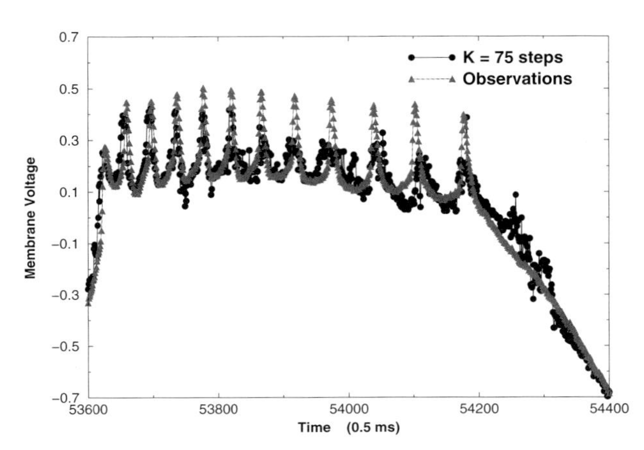

Fig. 2.11: Predictions using proxy state space methods on the data from an isolated LP neuron. Data are in Fig. 2.1. We used 41 000 data points in $d_E = 5$ to make local maps in $d_L = 4$ and selected $T = 9$. Predictions were made 75 time steps ahead for all points between 53 000 and 55 000. The prediction horizon $1/\lambda_1$ is approximately four steps for these data, and the predictions at 75 steps are no longer accurate for the rapidly varying spikes in the data, yet remain accurate for the slowly varying part of the trajectory in state space.

2.6 Modeling

Having analyzed the data in the previous sections we now want to derive models describing the underlying dynamical process. Depending on our pre-knowledge and potential applications we may distinguish basically two cases.

(A) We possess a model that was derived from first principles and the only remaining task is to find proper parameter values so that the model dynamics and the given time series fit together.

(B) (Almost) Nothing is known about the underlying dynamics except for the observed data.

Unfortunately, case (B) is (much) more common than case (A). If nothing is known but the available data, the only choice we have is so-called black-box modeling. There, a model is built from some pool of mathematical functions and then its parameters are estimated by comparing the model output with the data. The crucial step is to select a "good model" from the plethora of possible alternatives and this delicate topic is treated in depth in disciplines like machine learning or statistical learning theory [52, 53]. We shall discuss a few aspects of model selection in the following but this is certainly only the tip of the iceberg of this important field of applied mathematics and computer science. If a successful black-box model is found which reproduces the time series and its typical features, this may be (very) useful for forecasting or control of a process. However, in general it does not reveal any insight into the physics (or biology, ...) of the data source.[2] If, on the other hand, dynamical equations for the process are known which generated the observed time series, we usually still have to determine (in some cases very many) free parameters in that model. Furthermore, our observation may have (slightly) distorted the observable and this has in some cases to be taken into account in terms of some (unknown!) measurement function.

2.6.1 Modeling Interspike Intervals

To start with a simple example for black-box modeling we consider the interspike intervals of the given neuron time series shown in Fig. 2.8. As already mentioned in Section 2.2.3 this diagram suggests an approximate description in terms of some function $\Delta t_{k+1} = g(\Delta t_k)$. The goal of (black-box) modeling is to find a function g that fulfills this task in some optimal sense (to be specified). In particular, the function should have good generalization properties: not only the given time series has to be mapped correctly but also new data from the same source (not yet seen when learning the model). To achieve this (ambitious) goal overfitting has to be avoided, i.e., the model must not incorporate features of the particular realization of the given (finite!) time series. The performance of a good

[2] Historically, cases are known where data-driven modeling resulted in fundamental equations of physics like Kepler's laws, but this is certainly exceptional.

model does not depend on statistical features like random choice of initial conditions on a (chaotic) attractor or purely stochastic components (noise) that will *not* be the same for different measurements from a fixed (stationary) source or process. So the model has to be flexible enough to describe the data but should not be too complex, because then it will start to model also realization-dependent features. To determine some suitable level of complexity one may employ complexity measures and balance them with the prediction error. Another practically useful method is cross validation. The given data set is split into two parts: a learning or training set and an independent test set. The training set is used to specify the model (including parameters), and this model is then applied to the test set to evaluated its generalization properties and potential overfitting. To illustrate this point we shall fit a polynomial to the interspike intervals within a burst shown in Fig. 2.8 (enlargement on the right-hand side). These data points are randomly split into two halves, training and test set. A polynomial of degree m is fit to the training set and then used to map the test data points. For both sets of points mean squared errors are computed, called E_{train} and E_{test}. In Fig. 2.12 both errors are plotted versus m. Increasing the degree m renders our model more complex and results in a monotonically decreasing error E_{train} of the training set (dashed line in Fig. 2.12). For small m the error E_{test} of the test set also decreases but at $m = 3$ it starts to increase again, because for too complex polynomials overfitting sets in and the performance of the model on the test set deteriorates. To obtain a representative and robust evaluation of the model performance the time series has been split randomly several times in different training and test sets and the given errors are mean values of the corresponding training and test errors.

2.6.2 Modeling the Observed Membrane Voltage Time Series

We shall now consider the more ambitious task of modeling the temporal evolution of the amplitude of the observed membrane voltage time series shown in Fig. 2.1 without any additional knowledge about the underlying dynamics. As a first step we need a high-dimensional reconstruction of the state space dynamics unfolding both time scales. Mixed delays combining short and long time lags in a multidimensional state vector are a possible solution but not the best. It turned out that much better results can be achieved if the modeling problem is split into two parts in terms of two coupled models for the slow and the fast dynamics, respectively. To train these models we first separate both time scales by decomposing the given time series $s(t)$ into the sum $s_s(t) + s_f(t)$ of a slow and a fast signal using a linear low-pass filter. The task of the model describing the slow dynamics is to predict $s_s(t + \Delta t)$ ($\Delta t = 0.5\,\text{ms}$) from a reconstructed state $[s_s(t), s_s(t - T_s), s_f(t)]$ with $T_s = 200\Delta t$. Numerical simulations showed that such a three-dimensional reconstruction is sufficient for the slow dynamics. In contrast, the fast dynamics requires higher dimensional embed-

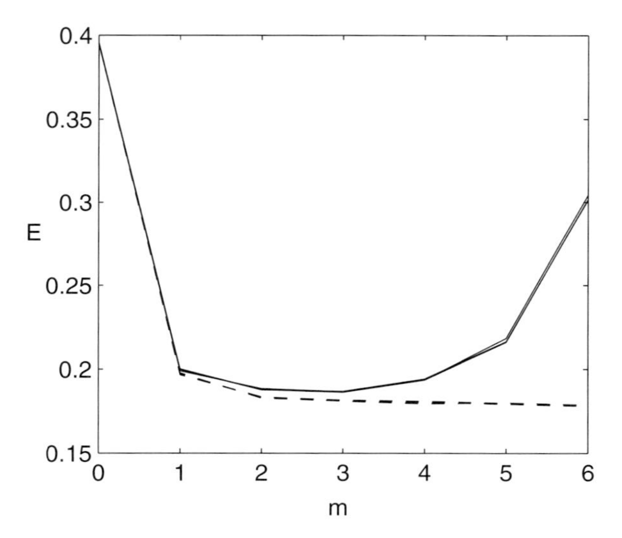

Fig. 2.12: Polynomial approximations of the interspike interval dynamics within bursts shown in Fig. 2.8. Computed are the mean squared errors of training data (dashed line) and test data (solid line) in dependence on the polynomial degree m. The training error decreases monotonically whereas the test error starts to increase once overfitting occurs.

dings. Good results have been obtained with a six-dimensional reconstruction $[s_f(t), s_f(t - T_f), \ldots, s_f(t - 4T_f), s_s(t)]$ ($T_f = 10\Delta t$) used as input for a second model which predicts the fast component $s_f(t + \Delta t)$. Note that both models are bi-directionally coupled due to the common elements $s_s(t)$ and $s_f(t)$ in both reconstructed states.

Having chosen suitable input spaces we can now proceed to solve the resulting function approximation tasks. Here we need a pool of possible functions in which we shall then search for good candidates to be included in our model. Motivated by Taylor expansions one might use multidimensional polynomial approximations but they typically suffer from a strong tendency to oscillate and diverge between and outside the given data points (and the resulting models possess rather poor generalization properties). An alternative avoiding these difficulties is linear superpositions of radial basis functions g_m,

$$s(t) = \sum_{m=1}^{M} c_m g_m(t) + e(t,) \tag{2.34}$$

where c_m denote model coefficients (parameters) and $e(t)$ ($t = 1, \ldots, N$) are modeling errors due to noise and model imperfections. To determine the model coefficients c_m the cost function

$$G = \sum_{t=1}^{N} e^2(t) + \mu \sum_{m=1}^{M} c_m^2 \tag{2.35}$$

is minimized (least-squares fit) which also includes a regularization term to control model complexity. Increasing the regularization parameter μ decreases complexity (as an additional countermeasure against overfitting and to cope with ill-posed least-squares problems).

Our pool consists of Gaussian basis functions $g_m(\mathbf{y}) = \exp(-\|\mathbf{y} - \mathbf{z}\|^2/\sigma_m^2)$ centered at the reconstructed (input) states with different sizes σ_m.

When extending the model, the cost function (2.35) is computed for all candidates from the pool and the function providing the largest cost reduction is chosen. Of course, when repeating this procedure again and again the model will grow and eventually become too complex. To avoid such overfitting a stopping criterion is needed. Like for the polynomial fit of the interspike intervals presented in the previous section we could split our data into training and test set and stop model extensions as soon as the test error starts to grow (compare Fig. 2.12). With this approach, however, only part of the data are used for determining the model. We might have obtained a better model if we would have used all data for training but then no data are left for detecting overfitting. A solution for this problem is leave-one-out cross validation (also called delete-1 cross validation). Only a single data point is used for testing and the remaining $N-1$ points are available for training the model. Of course, to obtain a statistically significant evaluation this procedure has to be repeated with many test points. To avoid extreme computational costs elegant numerical algorithms have been devised that basically exploit methods from linear algebra which are applicable because the model is linear in its parameters (for details see [54]). Similar to the test error in Fig. 2.12 the mean leave-one-out cross validation error[3] decreases first and begins to increase as soon as overfitting sets in. This increase is used as an indicator to stop the modeling process.

Figure 2.13 shows a result obtained with such a black-box modeling approach. The model describing the slow dynamics consists of 426 radial basis functions and the fast oscillations (spikes) are approximated using 972 Gaussians. Both models are coupled in terms of their input (delay) vectors and are iterated for 15 000 steps.

So far we trained and tested the model for a single time step. If the further future evolution is of interest, one has to apply the model iteratively using its output for generating the input delay vector of the next step. Unfortunately, models with very good single-step performance may provide poor results (or even diverge) when used iteratively. The best way to cope with this problem of error propagation is to use the multistep prediction error as cost function [52, 55]. This can significantly improve the performance of iterated predictions. The major dif-

[3] Applying the leave-one-out test in this way is also called PRESS statistics (Predicted REsidual Sums of Squares).

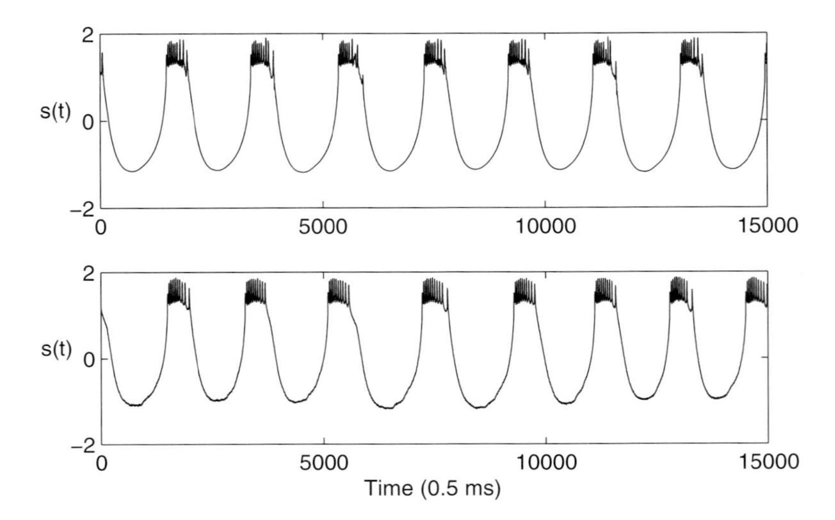

Fig. 2.13: Black-box modeling of the membrane voltage shown in Fig. 2.1 using two coupled radial basis models. Shown are 15 000 steps of the free running (iterated) model (top) and a similar section of the measured data (bottom) (courtesy of J. Dittmar).

ficulty with this approach is the fact that due to the iteration the cost function is no longer quadratic in the parameters and nonlinear minimization methods have to be employed to determine the optimal parameter set with high computational costs and the danger of getting stuck in local minima. And also the model structure derived with respect to the one-step error may turn out suboptimal when evaluated by means of the multistep prediction error.

Procedure 2.6. *Modeling Procedure*

Choose a suitable multidimensional representation of the dynamics. This is usually a delay embedding with proper delay times but additional processing (like separation of time scales) may be necessary. Note that this first step is crucial for all subsequent computations. Any dynamical information in the data which is not properly "translated" into input (state) vectors is lost!

Once the input state is specified select a model architecture and a pool of functions to be built into the model. Model architectures can be any kind of networks or just linear superpositions of basis functions. The latter model structure has the advantage to be linear in the parameters, which simplifies subsequent computations.

The next step is to decide which functions of the pool should enter the model. Using forward selection one includes in each step the function from the pool which reduces the error (cost) function most. This greedy strategy may result in suboptimal models. More effective but also more expensive is backward selection starting from a model containing all functions from the pool and deleting in each step the term which is most irrelevant. Of course, both selection methods can be combined.

For all term selection strategies a stopping criterion is required to avoid too complex models and overfitting. This can be implemented by monitoring the performance of the model applied to an independent test data set from the same data source (cross validation). Choose a model size (complexity) which minimizes the test error.

Try to improve the model by minimizing multistep prediction errors where the model is applied iteratively.

2.6.3 ODE Modeling

The modeling approaches considered so far were discrete in time. Sometimes, an ODE model is required or of advantage, because there it is easier to incorporate pre-knowledge in terms of (fundamental) laws of physics, for example. In the case of neuron dynamics many models have been suggested that are written as polynomial vector fields (ODEs) [56] and are basically extensions or simplified versions of the famous Hodgkin–Huxley model. Periodic spiking can, for example, be generated by a two-dimensional ODE of the following form:

$$\begin{aligned} \dot{v} &= p_1 + p_2 v + p_3 v^2 + p_4 v^3 - x \\ \dot{x} &= -p_5 x + p_6 v + p_7 v^2 \, . \end{aligned} \tag{2.36}$$

To determine the parameters p_i we use a shooting method, i.e., initial conditions $\big(v(0), x(0)\big)$ and parameters are varied until the resulting time series $v(t)$ of the model matches best the observed signal, for example in terms of a mean squared error. Figure 2.14 shows such a shooting fit for a spike sequence occurring in a burst of the experimental data. The parameter values obtained are $[0.1603, -0.3262, -0.2773, 0.8268, 3.5471, 2.9028, 0.5211]$.

To fit longer segments of the given time series more sophisticated models are required and the approximation task has to be split into several parts which are solved simultaneously using so-called multiple shooting techniques [57, 58]. Like other methods for fitting ODEs to data (e.g., minimization of synchronization errors [59]) this approach fails if no suitable ODE is chosen and therefore no parameter solution exists. Therefore, it remains an interesting problem to devise a combination of (multiple) shooting with some term selection method for generating appropriate ODE model structures.

2.7 Conclusion

This chapter has concentrated on methods which work for time series from both linear and nonlinear sources. It consists in finding the correct space in which to work, and then using properties of the points in that space to answer interesting questions about the source of the time series. If the signal is from a source which obeys linear dynamics globally, then the Fourier-based methods developed over many decades are likely to serve better for prediction, parameter estimation, ... If, however, the dynamics of the source is nonlinear globally over the state space

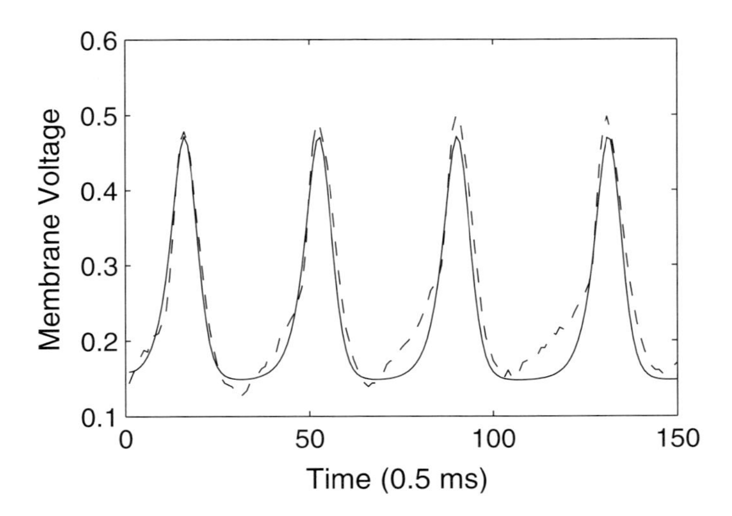

Fig. 2.14: Shooting fit of ODE (2.36) to a spike sequence of the experimental time series (dashed line).

of the system, then the time domain methods outlined in this chapter will be appropriate to use.

There are several ways to tell if the source is described by globally linear dynamics. One is to examine the Fourier power spectrum. If it is composed only of sharp lines which represent incommensurate frequencies, then the source is likely to be globally linear. Depending on the choice of coordinate system, there could be beat frequencies among the incommensurate fundamental frequencies. A globally linear source can only have periodic oscillations, and fixed points—DC signals. Another approach is to use the methods of this chapter and evaluate the Lyapunov exponents. If the system is stable, and it must be if one is observing it and signals are not moving to very large values where nonlinear saturation must apply, then the Lyapunov exponents must be zero, associated with stable oscillations, or negative, associated with fixed points. If there is a positive Lyapunov exponent, it cannot be a globally linear source that is being observed.

Once the state space in which to work has been established, then there are numerous questions one may ask of the system, and we have touched on only a few of them like forecasting and modeling the data. The reader is now equipped to answer those relating to his own interest.

Acknowledgements

This work of HA was partially funded by the U.S. Department of Energy, Office of Basic Energy Sciences, Division of Engineering and Geosciences, under grants no. DE-FG03-90ER14138 and no. DE-FG03-96ER14592; by a grant from the National Science Foundation, NSF PHY0097134, and by a grant from the National

Institutes of Health, NIH R01 NS40110-01A2. HA is also partially supported by the NSF sponsored Center for Theoretical Biological Physics at UCSD. HA also thanks Lev Tsimring, Nikolai Rulkov, Matt Kennel, Misha Sushchik, Lou Pecora, and Misha Rabinovich for discussions about this material over many years.

UP thanks C. Merkwirth, J. Wichard, J. Bröcker, A. Hornstein, D. Engster, and J. Dittmar for many interesting discussions on data analysis and modeling and the Institute for Nonlinear Science for support and kind hospitality during his stay at UCSD in 2002 and 2003.

References

[1] F. Takens. In D. A. Rand and L.-S. Young, editors, *Dynamical Systems and Turbulence*. Springer, Berlin, 1981.

[2] N. H. Packard, J. P. Crutchfield, J. D. Farmer, and R. S. Shaw. *Phys. Rev. Lett.*, 45:712, 1980.

[3] H. D. I. Abarbanel. *Analysis of Observed Chaotic Data*. Springer, New York, 1996.

[4] H. D. I. Abarbanel, R. Brown, J. J. Sidorowich, and L. S. Tsimring. *Rev. Mod. Phys.*, 65:1331, 1993.

[5] H. Kantz and T. Schreiber. *Nonlinear Time Series Analysis*. Cambridge University Press, Cambridge, 1997.

[6] H. D. I. Abarbanel, R. Huerta, M. I. Rabinovich, N. F. Rulkov, P. F. Rowat, and A. I. Selverston. *Neural Comput.*, 8:1567, 1996.

[7] D. Johnston and S. M.-S. Wu. *Foundations of Cellular Neurophysiology*. MIT Press, Cambridge, MA and London, 1997.

[8] M. Falcke, R. Huerta, M. I. Rabinovich, H. D. I. Abarbanel, R. C. Elson, and A. I. Selverston. *Biol. Cybern.*, 82:517, 2000.

[9] D. S. Broomhead and G. P. King. *Physica D*, 20:217, 1986.

[10] M. Casdagli, S. Eubank, J. D. Farmer, and J. Gibson. *Physica D*, 51:52, 1991.

[11] J. F. Gibson, J. D. Farmer, M. Casdagli, and S. Eubank. *Physica D*, 57:1, 1992.

[12] P. S. Landa and M. G. Rosenblum. *Physica D*, 48:232, 1991.

[13] M. Palus, , and I. Dvorak. *Physica D*, 55:221, 1992.

[14] T. Sauer and J. A. Yorke. *Int. J. Bif. Chaos*, 3:737, 1993.

[15] T. Sauer, J. A. Yorke, and M. Casdagli. *J. Stat. Phys.*, 65:579, 1991.

[16] J. Stark, D. S. Broomhead, M. E. Davies, and J. Huke. In *Proceedings of the 2nd World Congress of Nonlinear Analysts*, 1996.

[17] A. M. Fraser. *Physica D*, 34:391, 1989.

[18] A. M. Fraser. *IEEE Trans. Info. Theory*, 35:245, 1989.

[19] A. M. Fraser and H. L. Swinney. *Phys. Rev. A*, 33:1134, 1986.

[20] W. Liebert and H. G. Schuster. *Phys. Lett. A*, 142:107, 1989.

[21] J. M. Martinerie, A. M. Albano, A. I. Mees, and P. E. Rapp. *Phys. Rev. A*, 45: 7058, 1992.

[22] K. Judd and A. Mees. *Physica D*, 120:273, 1998.

[23] M. B. Kennel, R. Brown, and H. D. I. Abarbanel. *Phys. Rev. A*, 45:3403, 1992.

[24] H. D. I. Abarbanel and M. B. Kennel. *Phys. Rev. E*, 47:3057, 1993.

[25] T. Sauer. *Phys. Rev. Lett.*, 72:3811, 1994.

[26] D. Engster and U. Parlitz. In B. Schelter, M. Winterhalder, and J. Timmer, editors, *Handbook of Time Series Analysis*. VCH-Wiley, Weinheim, Germany, 2006.

[27] V. I. Oseledec. *Trans. Moscow Math. Soc.*, 19:197, 1968.

[28] K. Geist, U. Parlitz, and W. Lauterborn. *Prog. Theor. Phys.*, 83:875, 1990.

[29] H. D. I. Abarbanel, R. Brown, and M. B. Kennel. *Int. J. Mod. Phys. B*, 5:1347, 1991.

[30] G. Benettin, L. Galgani, A. Giorgilli, and J.-M. Strelcyn. *Meccanica*, 15:21, 1980.

[31] K. Briggs. *Phys. Lett. A*, 151:27, 1990.

[32] R. Brown, P. Bryant, and H. D. I. Abarbanel. *Phys. Rev. A*, 43:2787, 1991.

[33] P. Bryant, R. Brown, and H. D. I. Abarbanel. *Phys. Rev. Lett.*, 65:1523, 1990.

[34] M. Dämmig and F. Mitschke. *Phys. Lett. A*, 178:385, 1993.

[35] J.-P. Eckmann and D. Ruelle. *Rev. Mod. Phys.*, 57:617, 1985.

[36] J.-P. Eckmann, S. O. Kamphorst, D. Ruelle, and S. Ciliberto. *Phys. Rev. A*, 34: 4971, 1986.

[37] J. Fell and P. Beckmann. *Phys. Lett. A*, 190:172, 1994.

[38] R. Gencay and W. D. Dechert. *Physica D*, 59:142, 1992.

[39] J. Holzfuss and W. Lauterborn. *Phys. Rev. A*, 39:2146, 1989.

[40] H. Kantz. *Phys. Lett. A*, 185:77, 1994.

[41] J. Kurths and H. Herzel. *Physica D*, 25:165, 1987.

[42] U. Parlitz. *Int. J. Bif. Chaos*, 2:155, 1992.

[43] M. T. Rosenstein, J. J. Collins, and C. J. de Luca. *Physica D*, 65:117, 1993.

[44] M. Sano and Y. Sawada. *Phys. Rev. Lett.*, 55:1082, 1985.

[45] S. Sato, M. Sano, and Y. Sawada. *Prog. Theor. Phys.*, 77:1, 1987.

[46] I. Shimada and T. Nagashima. *Prog. Theor. Phys.*, 61:1605, 1979.

[47] R. Stoop and J. Parisi. *Physica D*, 50:89, 1991.

[48] A. Wolf, J. B. Swift, L. Swinney, and J. A. Vastano. *Physica D*, 16:285, 1985.

[49] X. Zeng, R. Eykholt, and R. A. Pielke. *Phys. Rev. Lett.*, 66:3229, 1991.

[50] J. D. Farmer and J. J. Sidorowich. *Phys. Rev. Lett.*, 59:845, 1987.

[51] H. D. I. Abarbanel, T. A. Carroll, L. Pecora, L. Tsimring, and J. J. Sidorowich. *Phys. Rev. E*, 49:1840, 1994.

[52] T. Hastie, R. Tishirani, and F. Friedman. *The Elements of Statistical Learning*. Springer, New York, 2001.

[53] V. N. Vapnik. *The Nature of Statistical Learning Theory*. Springer, New York, 2000.

[54] X. Hong, S. Chen, and P. M. Sharkey. *Int. J. Neural Syst.*, 14:27, 2004.

[55] L. Jaeger and H. Kantz. *Chaos*, 6:440, 1996.

[56] H. R. Wilson. *Spikes, Decisions and Actions*. Oxford University Press, Oxford, 1999.

[57] E. Baake, M. Baake, H. Bock, and K. Briggs. *Phys. Rev. A*, 45:5524, 1992.

[58] J. Timmer, H. Rust, W. Horbelt, and H. Voss. *Phys. Lett. A*, 274:123, 2000.

[59] U. Parlitz, L. Junge, and L. Kocarev. *Phys. Rev. E*, 54:6253, 1996.

3 Local and Cluster Weighted Modeling for Time Series Prediction

David Engster and Ulrich Parlitz

Local models are amongst the most precise methods for time series prediction. This chapter describes the basic parameters of local modeling and how these affect the model output. The choice of these parameters is crucial for the accuracy and stability of the model and an optimization procedure is described which often leads to good parameter values. To show the efficiency of this procedure, several artificial and real data sets are predicted using local models in conjunction with the optimization procedure. As an alternative to strict local modeling we discuss cluster weighted modeling, a modeling procedure first introduced by Gershenfeld et al., which combines a density estimation of the input data with a functional relationship to the output data. This leads to a number of local clusters, each containing its own model for describing the observed data. The parameters are optimized using an expectation-maximization (EM) algorithm, leading to a local optimum in parameter space.

3.1 Introduction

Given a data set of N pairs of points

$$\mathbf{\Omega} = \{(\mathbf{x}_1, y_1), (\mathbf{x}_2, y_2), \ldots, (\mathbf{x}_N, y_N)\}, \tag{3.1}$$

with vector inputs $\mathbf{x}_i \in \mathbb{R}^d$ and corresponding scalar outputs $y_i \in \mathbb{R}$ of an unknown system, the nonlinear modeling problem is to find an estimate $\hat{y}$ of the system output for a new vector input $\mathbf{q} \notin \mathbf{\Omega}$, which is often simply called the *query*.

A different and perhaps more familiar approach arises from the statistical viewpoint where one tries to find a good approximation for the *regression* $E[Y \mid \mathbf{X}]$. Here the pairs $(\mathbf{x}_i, y_i)$ are seen as realizations of the random variables $\mathbf{X}$ and Y, where Y and $\mathbf{X}$ are drawn from an unknown joint probability P. The regression $E[Y \mid \mathbf{X}]$ is the random variable which gives the conditional expectation $m(\mathbf{x}) \equiv E[Y \mid \mathbf{X} = \mathbf{x}]$. It is the best approximation for the output values y_i in a least-squares sense [1].

Handbook of Time Series Analysis. Björn Schelter, Matthias Winterhalder, Jens Timmer

ISBN: 3-527-40623-9

3.1.1 Time Series Prediction

We now want to specialize the described modeling problem to the case of time series prediction, where given a series $s_1, \ldots, s_N$, $s_i \in \mathbb{R}$, the model should be able to predict the next p time steps $s_{N+1}, \ldots, s_{N+p}$. We assume that the time series was generated by a nonlinear dynamical system with a deterministic time evolution. In the case of chaotic systems, even the exact knowledge of the underlying system does not allow the prediction of an arbitrary number of time steps due to the sensitivity on the initial conditions, i.e., the *prediction horizon* is limited. Additionally, if the time series is measured in an experiment it will always be corrupted by some measurement noise.

The input vectors $\mathbf{x}_i \in \mathbb{R}^d$ for the modeling algorithm can be obtained by reconstructing the attractor of the underlying dynamical system. Takens theorem [2] says that this can be accomplished by using a delay embedding of the time series with proper dimension d and delay τ, leading to the input vectors

$$\mathbf{x}_t = (s_t, s_{t-\tau}, \ldots, s_{t-(d-1)\tau}), \tag{3.2}$$

with $t = (d-1)\tau + 1, \ldots, N$. It is also possible to choose a *nonuniform* embedding, which instead of the fixed delay τ allows varying delays τ_i, $i = 1, \ldots, d-1$, between the components of the input vector [3].

To predict one step ahead, the corresponding output is given by $y_t = s_{t+1}$. For a further prediction of the next p steps, one can add the model output $\hat{s}_{N+1}$ to the given time series and repeat the modeling procedure until $\hat{s}_{N+p}$ is obtained, leading to an *iterated prediction*. However, if one is only interested in the model output $\hat{s}_{N+p}$, it is possible to do a *direct prediction* by using $y_t = s_{t+p}$ for the corresponding outputs. With iterated prediction, the errors of the model output accumulate, whereas for direct prediction the system output becomes more complex and is therefore more difficult to model correctly, especially for chaotic systems. There has been much discussion regarding whether iterated or direct prediction is the better choice [4]. This question cannot be answered in general, as it depends on the complexity of the system and the step size p. However, for chaotic systems iterated prediction has often shown to be superior in practice [5].

3.1.2 Cross Prediction

A more general case is the *cross prediction* of a time series, where one or more input time series $s^{(1)}_{1\ldots N}, \ldots, s^{(n)}_{1\ldots N}$ are given and one output time series $u_{1\ldots N}$ has to be predicted. The previous case of time series prediction can be seen as a special case of cross prediction, where the output time series is simply the input time series shifted p steps into the future. In the more general form with several different input time series, the construction of the input vector becomes more complicated. For every given input time series, a delay embedding must be performed. The delay vectors can then be concatenated to form the input vectors

$$\mathbf{x}_t = \left(s^{(1)}_t, s^{(1)}_{t-\tau_1}, \ldots, s^{(1)}_{t-(d_1-1)\tau_1}, s^{(2)}_t, s^{(2)}_{t-\tau_2}, \ldots, s^{(n)}_{t-(d_n-1)\tau_n}\right) \tag{3.3}$$

for the modeling algorithm.

However, even if the output of the dynamical system is completely determined by the input time series, in some cases the modeling problem becomes much easier if past values of the output time series are included in the input vector, effectively introducing a feedback into the modeling procedure. This can lead to an improvement of the prediction, but may lead to stability problems if the model is iterated over several time steps since the errors in the prediction accumulate. A practical example of such a cross prediction with feedback is shown in Section 3.4.3 with the modeling of friction phenomena.

3.1.3 Bias, Variance, Overfitting

For finding the mapping $y_i = f(\mathbf{x}_i)$ between dependent and independent variables, one has to consider that the model should not only be able to describe the given realization, but ideally also every other realization which is drawn from the joint probability $P(y, \mathbf{x})$. Even if one finds a perfect approximation for the regression $E[y \,|\, \mathbf{x}]$ *for one particular realization*, this does not in general lead to a model which will perform well on new data sets. In other words, the model should have the ability to *generalize* with respect to new data.

Given a realization $\mathbf{\Omega} = \{(\mathbf{x}_1, y_1), \ldots, (\mathbf{x}_n, y_n)\}$ of the data-generating process, the model based on this particular realization is written as $f(\mathbf{x}; \mathbf{\Omega})$. The expectation value of the squared error, given this realization, can be split into two parts

$$E[(y - f(\mathbf{x}; \mathbf{\Omega}))^2 \mid \mathbf{x}, \mathbf{\Omega}] = \underbrace{E[(y - E[y|\mathbf{x}])^2 \mid \mathbf{x}, \mathbf{\Omega}]}_{\text{variance}_y} + \underbrace{(f(\mathbf{x}; \mathbf{\Omega}) - E[y|\mathbf{x}])^2}_{\text{model error}}, \quad (3.4)$$

where the expectation is taken with respect to the joint probability P. The first term $E[(y - E[y|\mathbf{x}])^2 \mid \mathbf{x}, \mathbf{\Omega}]$ is the variance of y for a given $\mathbf{x}$ and is independent of the realization $\mathbf{\Omega}$ and the model $f(\mathbf{x})$. Therefore, the variance is a lower bound on the expectation value of the squared error, although it is of course through interpolation always possible to get a zero squared error for one particular realization. However, a model which simply interpolates the data will on other realizations lead to a larger squared error than the regression $E[y|\mathbf{x}]$, as it also tries to model realization-dependent features. This effect is called *overfitting* and can be avoided by introducing a bias which limits the variance of the model.

To see the connection between bias and variance, one has to examine the second term $\left(f(\mathbf{x}; \mathbf{\Omega}) - E[y \mid \mathbf{x}]\right)^2$, which describes the actual model error. It may be that for the particular realization $\mathbf{\Omega}$ our model perfectly approximates the regression $E[y|\mathbf{x}]$. However, the model might vary strongly depending on the given realization, or it might on average over all possible realizations be a bad approximator for the regression, making the model $f(\mathbf{x}; \mathbf{\Omega})$ an unreliable predictor of y. Since we want to have a model which has the ability to generalize, we must look at the expectation value of $\left(f(\mathbf{x}; \mathbf{\Omega}) - E[y \mid \mathbf{x}]\right)^2$ over all possible realizations, in the following denoted by $E_\Omega[\cdot]$. This term can again be split into two parts [1], the squared bias and the variance

$$E_\Omega[(f(\mathbf{x};\Omega) - E[y\,|\,\mathbf{x}])^2] = \underbrace{(E_\Omega[f(\mathbf{x};\Omega)] - E[y\,|\,\mathbf{x}])^2}_{\text{bias}^2} + \underbrace{E_\Omega[(f(\mathbf{x};\Omega) - E_\Omega[f(\mathbf{x};\Omega)])^2]}_{\text{variance}_f} . \quad (3.5)$$

The bias is the expectation value for the deviation between model output and the regression over all possible realizations. Therefore, a model with high bias will give similar results for different realizations, whereas a model with low bias and high variance can lead to very different model outputs and has a greater chance of overfitting. If the bias is zero we obtain $E_\Omega[f(\mathbf{x};\Omega)] = E[y\,|\,\mathbf{x}]$, i.e., our model is *on average* equal to the regression. However, from this we cannot conclude that *for one particular realization* the model $f(\mathbf{x};\Omega)$ is a good approximation for the regression $E[y\,|\,\mathbf{x}]$. A low bias typically comes with a large variance, making the model unreliable and leading to overfitting and therefore to an increase of the model error. This fact is known as the *bias variance dilemma* [1], which states that it is often not possible to have a low bias and a low variance at the same time. Instead, one has to find a good trade-off between these two.

3.2 Local Modeling

Given the modeling problem defined in the introduction, the most common procedure for getting an estimate for a new input vector $\mathbf{q}$ is to first fit a parametric function $f(\mathbf{x},\theta)$ on the data set Ω, where θ is a set of parameters which has to be optimized, e.g., with a maximum likelihood approach. After fitting the function $f(\mathbf{x},\theta)$, an estimate for $\mathbf{q} \notin \Omega$ can be obtained by evaluating $f(\mathbf{q},\theta)$. This procedure is also known as *global parametric modeling*, since a parametric function is fitted to the whole data set before the model can be queried.

In contrast to these global models, pure local models delay any computation until queried with the new vector input $\mathbf{q}$. A small neighborhood of $\mathbf{q}$ is located in the training set and a simple model using only the points lying in this neighborhood is constructed [6]. In statistical learning theory, local models are also referred to as *lazy learners* [7].

As the model is constructed in a neighborhood of the query $\mathbf{q}$, local modeling falls in the category of nonparametric regression, where no kind of functional form is preconditioned for the whole model. The data set is an unseparable part of the model construction and the quality of the resulting model highly depends on it. In contrast, in parametric regression the model $f(\mathbf{x},\theta)$ has a fixed functional form and the data points are only used to calculate or train the model parameters. After training, the resulting model can be separated from the data set and written down in closed form. Therefore, the model has a fixed functional form, which is particularly useful if this functional form is assumed or even known beforehand, e.g., by a physical theory where the parameters may also have a meaningful interpretation. In this case, parametric models are also more efficient than nonparametric ones, since they need less data for obtaining an accurate model that describes the data. However, if there does not exist any *a priori* knowledge, the functional form used may be unable to describe the data-generating process and the model will fail completely.

The neighborhood of the query in which the local model is constructed can be chosen in two different ways. The most common choice is to locate the k nearest neighbors $x_{nn_1}, \ldots, x_{nn_k}$ of x, i.e., the k points in the data set which have the smallest distance to the query point according to some arbitrary metric $\|\cdot\|$ (usually Euclidean). This type of neighborhood is also known as *fixed mass,* because the number of nearest neighbors remains constant. Alternatively, one can search for all points lying in some fixed neighborhood of the query point (*fixed size*) so that the actual number of neighbors varies. The fixed-mass neighborhood is easier to handle, since it varies its size according to the density of points and empty neighborhoods cannot occur.

The problem of finding nearest neighbors is very well studied and there are numerous algorithms for this task [8–10]. We use an algorithm called *ATRIA,* which relies on a binary search tree built in a preprocessing stage [11]. This algorithm is particularly effective when the points are close to a low-dimensional manifold, even when the actual dimension of the input space is large. Therefore, it is very well suited for the case where most of the data lie on a low-dimensional attractor.

3.2.1 Validation

As already described in Section 3.1.3, it is usually not possible to generate a model which offers low bias and low variance at the same time. The most common procedure for finding a good trade-off between bias and variance lies in the training of the model by using *cross validation*. Here the data set is split into two parts, the

- training set, used for training, and the
- test set, used for validating the model.

The usual iterative procedure is to switch between training and validation using the training and test data, respectively. With further training, the model error on the training set will usually monotonically decrease as the model is able to describe more and more features of the training data. At some point however, the model begins to overfit on the training data and the error on the test data will then begin to increase. Therefore, the minimum of the test error yields the optimal set of parameters and leads to a model which has still the ability to generalize. For comparing the performance between different models, another test data set is necessary which is only used to calculate one final error measure after the model training is completely finished. In this case, the test set for the cross validation is sometimes referred to as the *validation set* to distinguish it from the test data set for the model comparison.

The drawback of cross validation is the reduced number of points available for training. Therefore, the possibility remains that a better model could have been obtained without cross validation [12]. To minimize this possibility, the size of the test set should be chosen as small as possible. This leads to an "extreme" form of the cross validation, the *leave-one-out cross validation* (LOOCV), where the test set is reduced to one single test point. Of course, one has to repeat this validation procedure with enough different test points to get a good estimation of the actual model error. Local models are very well suited for LOOCV, as they are lazy learners which wait with the actual model calculations until they are queried. To implement LOOCV, they simply have to exclude the test point from their set of possible nearest neighbors.

Error Measures

The most common choice for calculating the model error is the mean squared error (MSE)

$$\mathrm{MSE}_1 = \frac{1}{|T_{\mathrm{ref}}|} \sum_{t \in T_{\mathrm{ref}}} \left(y_t - f^t(\mathbf{x}_t) \right)^2 , \tag{3.6}$$

where $|T_{\mathrm{ref}}|$ is the number of test points and $f^t(\mathbf{x})$ is the model which was constructed without the point $\mathbf{x}_t$.

For time series prediction, the MSE_1 gives the model error for predicting one step ahead in the future, but it is often desirable to have a model which predicts several steps p. This can be done by using *iterated prediction*, where the model is used p times successively. One has to consider, however, that the model error accumulates during the prediction. Otherwise, when the model is solely validated using the above MSE_1, one will mostly obtain models which are good for one-step but inferior for iterated prediction. Therefore, the MSE should be extended to average the error over p successive steps

$$\mathrm{MSE}_p = \frac{1}{p|T_{\mathrm{ref}}|} \sum_{t \in T_{\mathrm{ref}}} \left[\left(s_{t+1} - f^t(\mathbf{x}_t) \right)^2 + \sum_{i=1}^{p-1} \left(s_{t+i+1} - f^{t+i}(\hat{\mathbf{x}}_{t+i}) \right)^2 \right] . \tag{3.7}$$

The first point $\mathbf{x}_t$ is taken from the data set, whereas all further predictions depend on previous model outputs $\hat{\mathbf{x}}_{t+i}$.

If the time series is densely sampled, one has to take into account that the nearest neighbors of a test point $\mathbf{x}_t$ will mostly be points which are also close in time, i.e., points which are directly before or after this point on the same trajectory in phase space. Therefore, it is necessary to exclude not only the test point $\mathbf{x}_t$, but a whole segment of points lying in a certain interval $[t-c, t+c]$. For the new parameter c the average return time of the system can be used.

The model excluding these indices is denoted by $f^{t\pm c}(\mathbf{x}_t)$. Furthermore, it is good practice to normalize the model error with the variance of the time series. The normalized mean squared error (NMSE) over p steps is then given by

$$\mathrm{NMSE}_{p,c} = \frac{N}{p|T_{\mathrm{ref}}| \sum_{t=1}^{N}(s^t - \bar{s})^2} \sum_{t \in T_{\mathrm{ref}}} \left[\left(s_{t+1} - f^{t\pm c}(\mathbf{x}_t)\right)^2 + \sum_{i=1}^{p-1} \left(s_{t+i+1} - f^{(t+i)\pm c}(\hat{\mathbf{x}}_{t+i})\right)^2 \right]. \tag{3.8}$$

3.2.2 Local Polynomial Models

Local models use only a neighborhood of the query $\mathbf{q}$ to calculate the model output. Since the neighborhood is usually small, the actual model used should not be too complex. A good choice is to implement a polynomial model with low degree m, where the coefficients are calculated using the well-known least-squares method. In the following, we choose a fixed-mass neighborhood with k nearest neighbors.

One drawback of these simple local models is that they do not produce continuous output, because shifting the query point results in points suddenly entering or leaving the neighborhood. To smooth the model output, one can apply some kind of weights on the nearest neighbors, so that farther neighbors have a lesser effect on the output than the ones lying nearer to the query point.

To apply a weighted least-squares method, we define

$$\mathbf{X} = \begin{pmatrix} 1 & M(\mathbf{x}_{nn_1})_1^m \\ \vdots & \vdots \\ 1 & M(\mathbf{x}_{nn_k})_1^m \end{pmatrix}, \tag{3.9}$$

where $M(\mathbf{x})_1^m$ denotes all monomials of $\mathbf{x} \in \mathbb{R}^d$ with degree $1 \leqslant i \leqslant m$. The output vector is given by $\mathbf{y} = [y_{nn_1}, \dots, y_{nn_k}]^T$ and the coefficient vector by $\mathbf{v} = [v_1, \dots, v_l]^T$ with $l = |M(\mathbf{x})_1^m| + 1$. Additionally, we introduce the weight matrix $\mathbf{W} = \mathrm{diag}\{w_1, \dots, w_k\}$. The weighted sum of squared errors, which is to be minimized is now given by

$$P(\mathbf{v}) = (\mathbf{y} - \mathbf{X}\mathbf{v})^T \mathbf{W}^T \mathbf{W} (\mathbf{y} - \mathbf{X}\mathbf{v}). \tag{3.10}$$

Setting the gradient of this function to zero leads to the solution for the coefficient vector. With $\mathbf{X}_W = \mathbf{W} \cdot \mathbf{X}$ and $\mathbf{y}_W = \mathbf{W} \cdot \mathbf{y}$, we get

$$\mathbf{v} = (\mathbf{X}_W^T \mathbf{X}_W)^{-1} \mathbf{X}_W^T \mathbf{y}_W = (\mathbf{X}_W)^\dagger \mathbf{y}_W, \tag{3.11}$$

where $(\mathbf{X}_W)^\dagger$ denotes the pseudo inverse of $\mathbf{X}_W$, which can be calculated by using singular value decomposition [13] (see also Section 3.2.6).

3.2.3 Local Averaging Models

Setting the degree of the polynomial model to zero gives the local averaging model, where all input vectors are eliminated from the matrix $\mathbf{X}$. The coefficient vector can now be written as

$$\nu = (\mathbf{1}_k^T \mathbf{w}^2 \mathbf{1}_k)^{-1} \mathbf{1}_k^T \mathbf{w}^2 \mathbf{y} \tag{3.12}$$

$$= \frac{\sum_{i=1}^{k} w_i^2 y_{nn(i)}}{\sum_{i=1}^{k} w_i^2} \tag{3.13}$$

$$= \hat{y} \, ,$$

i.e., a weighted average of the output values of the k nearest neighbors. Although this seems to be an overly simplistic approach, this model can produce quite remarkable results [14]. It has several advantages over more complex models. Most important of all, local averaging models are always stable, as the model output is bounded by the output values of the nearest neighbors.

Furthermore, these models are very fast as they require almost no computation besides nearest-neighbor searching. Another advantage, especially when dealing with small data sets, is the ability of local averaging models to work with very small neighborhoods, as even one nearest neighbor is enough to produce a stable model output.

3.2.4 Locally Linear Models

Choosing a degree of $m = 1$ gives the locally linear model, where a weighted linear regression is performed on the output values of the nearest neighbors. The model output is now given by

$$\hat{y} = \langle [1 \ \mathbf{q}^T], \boldsymbol{\nu} \rangle \, . \tag{3.14}$$

In many cases, especially when many data points are available, the locally linear model gives far better results. However, to guarantee a stable model one usually has to perform some kind of regularization, which will be discussed in Section 3.2.6. Locally linear models are also computationally more expensive since they require a least-squares optimization of the coefficients.

3.2.5 Parameters of Local Modeling

Number of Nearest Neighbors

The number of nearest neighbors k is the most crucial parameter, as it directly affects the bias and variance of the resulting local model. A small number of nearest neighbors lead to a model with high variance and low bias. In the extreme case, a local averaging model with one nearest neighbor simply interpolates the outputs of the nearest neighbors of the data points. Conversely, a large number of neighbors lead to a model with high bias and low variance and in the extreme case to a very simple global model.

Weight Function

A good choice for weighting the nearest neighbors is functions of the form

$$w_n(r) = (1 - r^n)^n, \quad r = \frac{d_i}{d_{max}}, \tag{3.15}$$

where $d_{max} = |\mathbf{x}_k - \mathbf{q}|$ is the distance to the furthest nearest neighbor and $d_i = |\mathbf{x}_i - \mathbf{q}|$ is the distance to the nearest neighbor with index $i < k$. Depending on the exponent n, different kinds of weight functions can be obtained: with $n = 0$ no kind of weighting is applied, whereas $n = 1$ leads to linear weighting. Choosing $n = 2$ leads to biquadratic, $n = 3$ to tricubic weight functions. It is obvious that the type of weight function and the number of nearest neighbors are connected: choosing a high exponent n effectively reduces the number of nearest neighbors which affect the model output. However, the main motivation for using a weight function is to smooth the model output. Its effect on the accuracy of the model is mostly not very high, as long as any kind of weighting is done. Therefore, it is usually sufficient to choose n between 0 and 3.

Distance Metric

The kind of distance metric used has a strong influence on the neighborhood of the query point. Although the Euclidean metric will mostly be a good choice, other metrics can sometimes significantly improve model accuracy. By using the *diagonally weighted euclidian distance*

$$d_{dwe}(\mathbf{x}, \mathbf{q})^2 = \sum_{i=1}^{d} \lambda_i^2 (x_i - q_i)^2 = (\mathbf{x} - \mathbf{q})^T \boldsymbol{\Lambda}^2 (\mathbf{x} - \mathbf{q}), \tag{3.16}$$

where $\boldsymbol{\Lambda} = \mathrm{diag}(\boldsymbol{\lambda})$, $\boldsymbol{\lambda} \in \mathbb{R}^d$, one can specify which components of the input vectors should be more relevant when searching for nearest neighbors and which components should be more or less dropped. Unfortunately, one does not usually know beforehand which components are vital for modeling the data set and which are irrelevant or corrupted by noise. However, one can use an algorithm which uses the leave-one-out cross-validation error for optimizing the metric parameters (see also Section 3.2.7).

In the case of time series prediction, the input vectors are delay vectors in the form

$$\mathbf{x}_t = (s_t, s_{t-\tau}, \ldots, s_{t-(d-1)\tau}). \tag{3.17}$$

It is questionable why certain components of the input vectors should be favored, as a certain value s_i exists in different delay vectors at different positions. Nevertheless, in some cases the prediction can be improved by applying a special case of the diagonally weighted distance, the *exponentially weighted distance*

$$d_{\exp}(\mathbf{x}, \mathbf{q}) = \left(\sum_{i=1}^{d} \lambda^{i-1} (x_i - q_i)^2 \right)^{\frac{1}{2}} . \tag{3.18}$$

In the case of delay vectors, this method favors those components of $\mathbf{x}$ which are closer in time to the prediction. Furthermore, only one parameter has to be optimized and the standard euclidian metric can still be obtained by setting $\lambda = 1$. Therefore, an optimization procedure which optimizes λ can only improve the prediction accuracy compared to the Euclidean metric.

3.2.6 Regularization

Given enough data points, locally linear models are usually more precise than locally averaging models. One problem though lies in the calculation of the inverse of the matrix product $\mathbf{X}_W^\mathsf{T} \mathbf{X}_W$, which leads to unstable models if the resulting matrix is ill-conditioned. Therefore, some kind of regularization method must be applied. The two most common choices are the *principal component regression* (PCR) and the *Ridge regression* (RR), which will be described in the following section.

Principal Component and Ridge Regression

The basic principle of this method relies on the singular value decomposition of $\mathbf{X}_W$, which is given by

$$\mathbf{X}_W = \mathbf{U}\mathbf{S}\mathbf{V}^\mathsf{T} , \tag{3.19}$$

where $\mathbf{U} \in \mathbb{R}^{k \times k}$ and $\mathbf{V} \in \mathbb{R}^{l \times l}$ are orthonormal, and $\mathbf{S} = \operatorname{diag}(\sigma_1, \ldots, \sigma_p)$ is a diagonal matrix with the singular values $\sigma_1 \geqslant \sigma_2 \geqslant \cdots \geqslant \sigma_p \geqslant 0$ with $p = \min\{k, l\}$. The pseudoinverse of $\mathbf{X}_W$ can now be written as

$$\mathbf{X}_W^\dagger = \mathbf{V}\mathbf{S}^+\mathbf{U}^\mathsf{T} , \tag{3.20}$$

where $\mathbf{S}^+ = \operatorname{diag}(1/\sigma_1, \ldots, 1/\sigma_r, 0, \ldots, 0)$ and $r = \operatorname{rank}(\mathbf{X}_W)$ [15], i.e., we set $1/\sigma_i$ to zero if $\sigma_i = 0$. In practice, however, these singular values are usually not exactly zero. The matrix $\mathbf{X}_W$ is not singular but ill-conditioned. The PCR works by first setting these small singular values to zero and then calculating $\mathbf{S}^+$ as just noted. For this procedure it is crucial that the nearest neighbors are centered around the origin by subtracting the mean. This also simplifies the calculation of the locally linear model since the constant is now given by the weighted average

$\bar{y}_w = \sum_{i=1}^{k} w_i^2 y_{nn(i)} / \sum_{i=1}^{k} w_i^2$ of the images of the nearest neighbors. The column of 1's in the design matrix (3.9) can therefore be omitted [5].

However, there exist different possibilities in how to decide whether a singular value is so small that it should be dropped. The easiest way is the *truncated* PCR (TPCR), where a fixed number of the smallest singular values is automatically set to zero, without looking at their actual values. Alternatively, in *principal component threshold regression* (PCTR) every singular value below some previously defined threshold $\sigma_{\min}$ is dropped. This procedure can be further generalized by applying weights to the inverse singular values, which leads to a PCTR with *soft thresholding*. The model output for the locally linear model can then be written as

$$\hat{y} = \bar{y}_w + \sum_{i=1}^{p} \langle (\mathbf{q} - \bar{\mathbf{x}})^{\mathsf{T}}, \mathbf{v}_i \rangle \left(\frac{f(\sigma_i)}{\sigma_i} \right) \langle \mathbf{u}_i^{\mathsf{T}}, \mathbf{y}_W \rangle , \tag{3.21}$$

where in general any kind of weight function $f(\sigma)$ can be used. McNames [5] has suggested to use a modified biquadratic weight function

$$f(\sigma) = \begin{cases} 0 & s_{\min} > \sigma , \\ \left(1 - \left(\frac{s_{\max} - \sigma}{s_{\max} - s_{\min}} \right)^2 \right)^2 & s_{\min} \leqslant \sigma < s_{\max} , \\ 1 & s_{\max} \leqslant \sigma , \end{cases} \tag{3.22}$$

where $s_{\min}$ and $s_{\max}$ are given by

$$s_{\min} \equiv s_c(1 - s_w) \tag{3.23}$$

$$s_{\max} \equiv s_c(1 + s_w) . \tag{3.24}$$

The parameters s_c and s_w define the center and width of the threshold in which the singular values are weighted down to zero. Singular values above $s_{\max}$ remain unchanged, whereas those smaller than $s_{\min}$ are set to zero. With $s_w = 0$ we get $s_{\min} = s_{\max} = s_c$ and therefore a hard threshold at s_c.

Another good choice for the weight function is given by

$$f(\sigma) = \frac{\sigma^2}{\mu^2 + \sigma^2} , \tag{3.25}$$

so that for $\sigma \gg \mu$ we get $f(\sigma) \approx 1$, and for $\sigma \to 0$ the weight function becomes zero. The parameter $\mu \geqslant 0$ therefore defines the degree of regularization. This particular weight function leads to a special case of a regularization procedure known as *Ridge regression* or *Tikhonov regularization* [15, 16]. Here, the cost function (3.10) is modified by adding a penalty term which penalizes large values in the coefficient vector, leading to

$$P(\mathbf{v})_{RR} = (\mathbf{y} - \mathbf{X}\mathbf{v})^{\mathsf{T}} \mathbf{W}^{\mathsf{T}} \mathbf{W} (\mathbf{y} - \mathbf{X}\mathbf{v}) + \mathbf{v}^{\mathsf{T}} \mathbf{R}^{\mathsf{T}} \mathbf{R} \mathbf{v} . \tag{3.26}$$

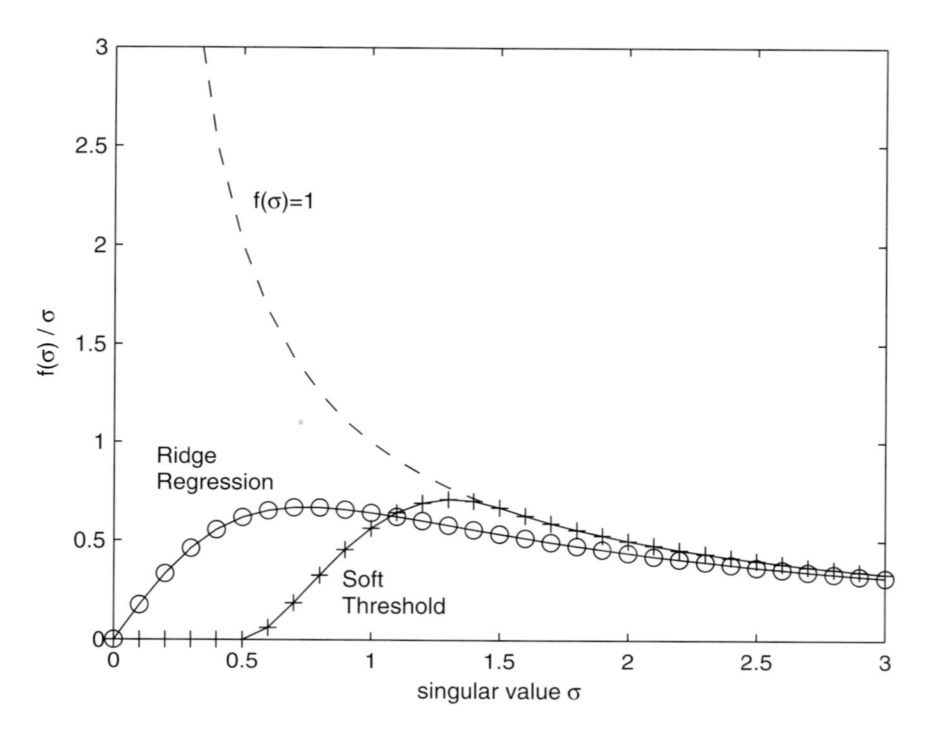

Fig. 3.1: Example for regularization with Ridge regression and TPCR with soft threshold. The dashed line shows the inverse $1/\sigma$, which goes to infinity for $\sigma \to 0$. The circles show the regularized singular value with Ridge regression and $\mu = 0.75$, while the crosses show TPCR with soft threshold and $s_c = 1$, $s_w = 0.5$.

The diagonal *Ridge matrix* $\mathbf{R} \equiv \mathrm{diag}(r_1, \ldots, r_l)$ weights the different components of the coefficient vector. The solution for $\mathbf{v}$ can now be written as

$$\mathbf{v} = (\mathbf{X}_W^T \mathbf{X}_W + \mathbf{R}^T \mathbf{R})^{-1} \mathbf{X}_W^T \mathbf{y}_W \,. \tag{3.27}$$

Therefore, the modified cost function (3.26) is equivalent to adding the values $r_1^2, \ldots, r_l^2$ to the diagonal of $\mathbf{X}_W^T \mathbf{X}_W$. A simple (and popular) choice for the Ridge matrix is $\mathbf{R} = \mu^2 \mathbf{I}$, i.e., all components of $\mathbf{v}$ are weighted with the same factor μ^2. The solution (Eq. 3.27) can now be easily obtained by using the singular value decomposition $\mathbf{X}_W = \mathbf{U}\mathbf{S}\mathbf{V}^T$ and this leads to

$$\mathbf{v} = \sum_{i=1}^{k} \frac{\sigma_i}{\sigma_i^2 + \mu^2} \langle \mathbf{u}_i^T, \mathbf{y}_W \rangle \mathbf{v}_i \tag{3.28}$$

and therefore to the above-mentioned weight function (3.25).

An illustration of both regularization techniques can be seen in Fig. 3.1. While TPCR has the advantage that it can locally adapt to the dimensionality of the data, Ridge Regression in its general form (3.26) can put different regularization parameters on each component through the regularization matrix $\mathbf{R}$. Both methods can produce good results; however, it has been shown that in the case of time series prediction of chaotic systems, principal component regression with thresholding is superior to Ridge regression [5].

Local Projection

Another possibility for regularization is to reduce the dimensionality of the points found in the neighborhood of the query before performing the least-squares optimization. This can be done by performing a principal component analysis (PCA) on the nearest neighbors and then projecting them onto the subspace which covers most of the variance of the data [17, 18]. Given the following matrix

$$\mathbf{A} = \begin{pmatrix} \mathbf{x}_{nn_1}^{\mathsf{T}} - \bar{\mathbf{x}}_{nn}^{\mathsf{T}} \\ \cdots\cdots\cdots\cdots \\ \mathbf{x}_{nn_k}^{\mathsf{T}} - \bar{\mathbf{x}}_{nn}^{\mathsf{T}} \end{pmatrix} \tag{3.29}$$

containing the centered nearest neighbors of the query, the eigenvalues and eigenvectors of the empirical covariance matrix $\mathbf{C} = \mathbf{A}^{\mathsf{T}} \cdot \mathbf{A}$ are calculated. The eigenvalues correspond to the variance in the direction given by the corresponding eigenvalue. Keeping only the first r eigenvectors with eigenvalues above some given threshold σ, we can define through these remaining eigenvectors a lower dimensional subspace which covers most of the variance of the data. The nearest neighbors projected into this subspace are given by $\tilde{\mathbf{A}} = \mathbf{A} \cdot \mathbf{P}_r$, with the projection matrix given by

$$\mathbf{P}_r = (\mathbf{v}_1 \cdots \mathbf{v}_r)\,, \tag{3.30}$$

consisting of the eigenvectors corresponding to the first r largest variances. This also effectively removes noise present in the data, given that the noise is small so that it only contributes a small amount to the variance. The coefficients for the local model can then be calculated in this lower dimensional subspace.

This procedure is very closely related to TPCR. In fact, for locally linear models it is equivalent, given that the nearest neighbors are centered around their mean, since the design matrix $\mathbf{X}$ from Equation (3.9) is then equal to $\mathbf{A}$. It follows that $\tilde{\mathbf{A}} = \mathbf{A} \cdot \mathbf{P}_r = \mathbf{U}_r \cdot \mathbf{S}_r$, where $\mathbf{U}_r$ and $\mathbf{S}_r$ denote the matrices $\mathbf{U}$ and $\mathbf{S}$ from the SVD in Equation (3.20), but reduced to the r largest singular values. The pseudo inverse of $\tilde{\mathbf{A}}$ is then given by

$$\tilde{\mathbf{A}}^{\dagger} = (\tilde{\mathbf{A}}^{\mathsf{T}}\tilde{\mathbf{A}})^{-1}\tilde{\mathbf{A}}^{\mathsf{T}} = \mathbf{S}_r^{-1}\mathbf{U}_r^{\mathsf{T}}\,. \tag{3.31}$$

Given the query $\mathbf{q}$, we obtain for the model output

$$\hat{y} = \mathbf{q} \cdot \mathbf{P}_r \cdot \tilde{\mathbf{A}}^{\dagger}\mathbf{y} = \mathbf{q} \cdot \mathbf{P}_r \cdot \mathbf{S}_r^{-1}\mathbf{U}_r^{\mathsf{T}}\mathbf{y}\,, \tag{3.32}$$

which is equivalent to TPCR since $\mathbf{P}_r = (\mathbf{v}_1 \cdots \mathbf{v}_r) = \mathbf{V}_r$.

For locally quadratic models or local models with other model types like radial basis function networks, the local projection and TPCR are no longer equivalent. TPCR with soft thresholding introduced in the previous section is more flexible and can often lead to better results than TPCR without soft thresholding or local projection.

3.2.7 Optimization of Local Models

Several different parameters have to be set for local modeling. Most of these parameters deal with the neighborhood of the query point: the kind of metric used for calculating the distances between the query point and its neighbors, the number of nearest neighbors k and the weight function applied. The other parameters deal with the model used in the neighborhood: one has to choose between locally averaging or locally linear models, and for the latter, one has to choose a regularization method. The regularization itself has additional parameters associated, which have a large influence on the stability and accuracy of the model, especially in the case where the model is iterated over several steps.

Although all these parameters have a more or less intuitive appeal, it is difficult to find good values based on simple "trial and error." Furthermore, these parameters are not independent of each other: the distance metric and weight function directly affect the form and size of the neighborhood which is primarily controlled by the number of nearest neighbors. On the other hand, changing the type of model or the regularization parameters often demands other forms of neighborhoods.

Good parameter values can be found by applying an optimization algorithm using the leave-one-out cross-validation error. Although local models allow an efficient calculation of this error, it is still a time-consuming task, especially for large data sets combined with multiple-step prediction. Moreover, gradient-based optimization algorithms are mostly not applicable, as only the regularization and metric parameters allow the calculation of a gradient.

One popular approach for such an optimization problem is to use genetic algorithms [19], as they do not need a gradient and are able to deal with integer and floating point parameters at the same time. They are well suited for optimizing embedding parameters, especially when a nonuniform embedding is used [20].

Genetic algorithms start with a randomly chosen population of parameter vectors which can contain the delays of the embedding as well as the number of nearest neighbors or any other parameter for local modeling. This population is then "evolved" by using different types of inheritance, mutation, and selection operators. The algorithm stops after a certain number of iterations.

However, it is not advisable to optimize all parameters at once with a genetic algorithm, as the initial population and the number of iterations have to be very large for the algorithm to converge. This may be due to the fact that the parameters are not of equal importance. Therefore, we first use a genetic algorithm to optimize only the delays for a nonuniform embedding and the number of nearest neighbors, since these are the most crucial parameters for a good model performance. During this optimizations step, the other parameters are held constant; we used biquadratic weights with an Euclidean distance, and for locally linear models the regularization procedure given by Eq. (3.22) with $s_c = 1 \times 10^{-4}$ and $s_w = 0.6$.

After this primary optimization step, the other parameters are optimized using a simple type of cyclic optimization, where all parameters are successively optimized with an exhaustive search in the case of integer parameters and with a semiglobal line search for floating point parameters [5]. Although one has a good value for the number of nearest neighbors, it should be included in this second optimization step since it is the most crucial parameter.

Because of local minima, this optimization procedure will usually not lead to the global minimum in parameter space, but nevertheless it will usually improve the prediction accuracy compared to manually chosen parameters.

3.3 Cluster Weighted Modeling

Cluster Weighted Modeling (in the following denoted as CWM), an algorithm first introduced by Gershenfeld et al. [21], lies between the local and the global modeling approach. It is global in the sense that the model has to be trained beforehand with the whole data set, i.e., it lacks the flexibility of lazy learning. But it is also local in the sense that usually only the points lying in a neighborhood of the query point are crucial for the model output.

CWM essentially tries to estimate the joint density $p(\mathbf{x}, y)$, since this density allows us to compute derived quantities such as the conditional forecast $\langle y \mid \mathbf{x} \rangle$ for new query points. To estimate the density, a Gaussian mixture model is used, which factors the density over distinct clusters $c_m, m = 1, \ldots, M$. But where conventional Gaussian mixture models only estimate the quantity $p(\mathbf{x})$, CWM includes an additional output term to capture the functional dependence of the output values y_i on the input vectors $\mathbf{x}_i$ as part of the density estimation.

Therefore, the density estimator is written as

$$\begin{aligned} p(y, \mathbf{x}) &= \sum_{m=1}^{M} p(y, \mathbf{x}, c_m) \\ &= \sum_{m=1}^{M} p(y, \mathbf{x} \mid c_m) p(c_m) \\ &= \sum_{m=1}^{M} p(y \mid \mathbf{x}, c_m) p(\mathbf{x} \mid c_m) p(c_m) , \end{aligned} \tag{3.33}$$

where the three terms are

- weights $p(c_m)$, $m = 1, \dots, M$,
- input domains $p(\mathbf{x}|c_m)$,
- output terms $p(y|\mathbf{x}, c_m)$.

The weights $p(c_m)$ are real values which describe the fraction of the data the cluster c_m explains. The input domains are given by multivariate Gaussians

$$p(\mathbf{x} \mid c_m) = \frac{|\mathbf{C}_m^{-1}|^{\frac{1}{2}}}{(2\pi)^{\frac{d}{2}}} \exp\left(-(\mathbf{x} - \boldsymbol{\mu}_m)^T \cdot \mathbf{C}_m^{-1} \cdot (\mathbf{x} - \boldsymbol{\mu}_m)/2\right) , \tag{3.34}$$

with $\boldsymbol{\mu}_m$ the means and $\mathbf{C}_m$ the covariance matrices of the Gaussians. The input domains are used to capture the density of the input vectors $\mathbf{x}$ in phase space. When dealing with high-dimensional spaces, it is advisable to reduce these input domains to separable Gaussians with single variances in each dimension $\sigma_{m,i}^2$ instead of using the full covariance matrix, so that the input term becomes

$$p(\mathbf{x} \mid c_m) = \prod_{i=1}^{d} \frac{1}{\sqrt{2\pi\sigma_{m,i}^2}} \exp(-(x_i - \mu_{m,i})^2/2\sigma_{m,i}^2) . \tag{3.35}$$

The output terms are also given by Gaussians

$$p(y \mid \mathbf{x}, c_m) = \frac{1}{\sqrt{2\pi\sigma_{m,y}^2}} \exp(-[y - f(\mathbf{x}, \boldsymbol{\beta}_m)]^2/2\sigma_{m,y}^2) , \tag{3.36}$$

but the means of these Gaussians are now given by parametric functions $f(\mathbf{x}, \boldsymbol{\beta}_m)$, where $\boldsymbol{\beta}_m$ denotes the parameters for the cluster c_m. These functions, which are often called local models but which will be in the following denoted as *cluster functions* to avoid confusion with the local models in the previous section, are usually chosen to be fairly simple. For an easy optimization of the cluster function parameters, it is necessary to use a linear parameterized model

$$f(\mathbf{x}, \boldsymbol{\beta}_m) = \sum_{i=1}^{I} \beta_{m,i} f_i(\mathbf{x}) , \tag{3.37}$$

where $f_i(\mathbf{x})$ are some suitable basis functions (usually monomials). Next to the number of clusters M, the number I and type f_i of the basis functions directly determine the complexity of the resulting CWM and hence control bias and variance.

The reason for choosing the output term in the above way becomes clear when we look at the model output, i.e., the conditional forecast which we can obtain by integrating the output values with respect to the conditional density $p(y \mid \mathbf{x})$,

$$\begin{aligned}\langle y \mid \mathbf{x}\rangle &= \int y\, p(y \mid \mathbf{x})\, dy = \int y \frac{p(y,\mathbf{x})}{p(\mathbf{x})}\, dy \\ &= \frac{\sum_{m=1}^{M} \int y\, p(y \mid \mathbf{x}, c_m)\, dy\, p(\mathbf{x} \mid c_m) p(c_m)}{\sum_{m=1}^{M} p(\mathbf{x} \mid c_m) p(c_m)} \\ &= \frac{\sum_{m=1}^{M} f(\mathbf{x}, \boldsymbol{\beta}_m) p(\mathbf{x} \mid c_m) p(c_m)}{\sum_{m=1}^{M} p(\mathbf{x} \mid c_m) p(c_m)} .\end{aligned} \tag{3.38}$$

The model output of the CWM is therefore given by a weighted average of the cluster functions $f(\mathbf{x}, \boldsymbol{\beta}_m)$. The Gaussians, which are given by the input domains $p(\mathbf{x} \mid c_m)$, control the interpolation of the cluster functions and therefore do not serve as approximators like in conventional radial basis function networks.

One is now confronted with the problem of finding good values for

- the weights $p(c_m)$,
- the means $\boldsymbol{\mu}_m$ and variances $\boldsymbol{\sigma}^2_{m,x}$ of the input domains,
- the variances of the output terms $\sigma^2_{m,y}$, and
- the parameters of the cluster functions $\boldsymbol{\beta}_m$.

The task of parameter optimization is done using an *expectation-maximization* (EM) algorithm.

3.3.1 The EM Algorithm

The EM algorithm is an iterative maximum-likelihood estimator and was first introduced by Dempster et al. [22]. Since it has proved to be a successful optimization strategy for conventional Gaussian mixture models (GMM) [23], it is reasonable to use it also for the related cluster weighted models. The EM algorithm is typically used when one is confronted with *incomplete data* or when the likelihood function involves *latent variables*. However, the distinction of these two cases is more a matter of interpretation, since we can always think of latent variables as data which we could not observe, therefore leading to incomplete data. In our case, the observation consists of the input and output data points $\{\mathbf{x}_i, y_i\}_{i=1}^{N}$, but we do not know which clusters in our model ansatz are "responsible" for which points.

The basic strategy of EM is to start with an initial guess for the unknown data and calculate the expectation of the likelihood for the *complete* data, where the expectation is taken with respect to the computed conditional distribution of the unobserved data; this is called the *expectation step* (E-step). Afterwards, we compute a new estimation of the unobserved data by maximizing the likelihood of the complete data; this is called the *maximization step* (M-step). More intuitively, one assumes in the E-step that the current estimate for the cluster parameters is correct, whereas in the M-step the cluster parameters are reestimated based on the distribution of the data. The EM algorithm alternates between these two steps until some stopping criterion is met, which may be a convergence criterion or the detection of overfitting through means like cross validation.

The algorithm is guaranteed to converge, but since it is basically a hill-climbing approach it may only reach a local maximum of the likelihood. For the case of CWM, this is usually not a problem since the large number of parameters which have to be estimated allow many realizations which will show approximately the same performance. However, the maximum likelihood estimate for a large number of parameters will usually lead to an overfitting of the resulting CWM, which forces us to keep the number of parameters reasonably low either by limiting the number of clusters or by using a stopping criterion like cross validation for the number of EM iterations.

Expectation Step

The EM optimization starts by first initializing the parameters. One would usually start with uniform weights $p(c_m) = 1/M$ and all variances and function parameters equal to 1. The center positions can be chosen randomly, e.g., by picking M random points out of the input training data $\mathbf{x}_i, i = 1, \ldots, N$.

In the E-step it is assumed that the given parameters are correct and on this assumption the posterior distribution for each cluster is calculated, which is given by

$$\begin{aligned} p(c_m \mid y, \mathbf{x}) &= \frac{p(y, \mathbf{x}, c_m)}{p(y, \mathbf{x})} = \frac{p(y \mid \mathbf{x}, c_m) p(\mathbf{x} \mid c_m) p(c_m)}{\sum_{l=1}^{M} p(y, \mathbf{x}, c_l)} \\ &= \frac{p(y \mid \mathbf{x}, c_m) p(\mathbf{x} \mid c_m) p(c_m)}{\sum_{l=1}^{M} p(y \mid \mathbf{x}, c_l) p(\mathbf{x} \mid c_l) p(c_l)} . \end{aligned} \tag{3.39}$$

This distribution relates each cluster to each data point. Looking at the resulting fraction, one can see that the posterior is the ratio between one and all clusters predicting one specific point.

Maximization Step

In the M-step, one assumes that the distribution of the data is correct and now calculates the weights

$$\begin{aligned} p(c_m)^{\text{new}} &= \int p(y, \mathbf{x}, c_m) \, dy \, d\mathbf{x} = \int p(c_m \mid y, \mathbf{x}) p(y, \mathbf{x}) \, dy \, d\mathbf{x} \\ &\approx \frac{1}{N} \sum_{n=1}^{N} p(c_m \mid y_n, \mathbf{x}_n) . \end{aligned} \tag{3.40}$$

Given these new weights, we can now update the cluster positions and variances. In principle, the new cluster means are given by calculating the expectation of $\mathbf{x}$ with respect to the conditional density $p(\mathbf{x} \mid c_m)$. However, we also want to position the clusters with respect to how well they predict the target values y; therefore, we also have to integrate over y, leading to

$$\begin{aligned}\boldsymbol{\mu}_m^{\text{new}} &= \int \mathbf{x}\, p(\mathbf{x} \mid c_m)\, d\mathbf{x} = \int \mathbf{x}\, p(y, \mathbf{x} \mid c_m)\, dy\, d\mathbf{x} \\ &= \int \mathbf{x} \frac{p(c_m \mid y, \mathbf{x})}{p(c_m)} p(y, \mathbf{x})\, dy\, d\mathbf{x} \\ &\approx \frac{1}{Np(c_m)} \sum_{n=1}^{N} \mathbf{x}_n p(c_m \mid y_n, \mathbf{x}_n) = \frac{\sum_{n=1}^{N} \mathbf{x}_n\, p(c_m \mid y_n, \mathbf{x}_n)}{\sum_{n=1}^{N} p(c_m \mid y_n, \mathbf{x}_n)} .\end{aligned} \tag{3.41}$$

This can be written in a more condensed form by defining the *cluster weighted expectation* of a function $\phi(\mathbf{x})$ as

$$\langle \phi(\mathbf{x}) \rangle_m \equiv \frac{\sum_{n=1}^{N} \phi(\mathbf{x}_n) p(c_m \mid y_n, \mathbf{x}_n)}{\sum_{n=1}^{N} p(c_m \mid y_n, \mathbf{x}_n)} , \tag{3.42}$$

so that the new cluster means are given by $\langle \mathbf{x} \rangle_m$. In the same way, the new variances can be written as

$$\sigma_{m,i}^{2,\text{new}} = \langle (x_i - \mu_{m,i})^2 \rangle_m . \tag{3.43}$$

Cluster Function Parameters

For updating the parameters $\boldsymbol{\beta}_m$ of the cluster functions (3.37), we maximize for each cluster c_m the log-likelihood with respect to $\boldsymbol{\beta}_m$, i.e., we must solve

$$\frac{\delta}{\delta \boldsymbol{\beta}_m} \log \prod_{n=1}^{N} p(y_n, \mathbf{x}_n) = 0 . \tag{3.44}$$

Using the cluster weighted density (3.33) and the chosen output term (3.36) we get

$$\begin{aligned} 0 &= \sum_{n=1}^{N} \frac{1}{p(y_n, \mathbf{x}_n)} p(y_n, \mathbf{x}_n, c_m) \frac{y_n - f(\mathbf{x}, \boldsymbol{\beta}_m)}{\sigma_{m,y}^2} \cdot \frac{\delta f(\mathbf{x}_n, \boldsymbol{\beta}_m)}{\delta \boldsymbol{\beta}_m} \\ &= \frac{1}{N\, p(c_m)} \sum_{n=1}^{N} p(c_m \mid y_n, \mathbf{x}_n)[y_n - f(\mathbf{x}_n, \boldsymbol{\beta}_m)] \cdot \frac{\delta f(\mathbf{x}_n, \boldsymbol{\beta}_m)}{\delta \boldsymbol{\beta}_m} \\ &= \left\langle [y - f(\mathbf{x}, \boldsymbol{\beta}_m)] \cdot \frac{\delta f(\mathbf{x}, \boldsymbol{\beta}_m)}{\delta \boldsymbol{\beta}_m} \right\rangle_m , \end{aligned} \tag{3.45}$$

where we again use the definition (3.42) for the cluster weighted expectation. Since we use linear parameterized cluster models (3.37), we obtain

$$\begin{aligned} 0 &= \langle [y - f(\mathbf{x}, \boldsymbol{\beta}_m)]\, f_j(\mathbf{x}) \rangle_m \\ &= \langle y\, f_j(\mathbf{x}) \rangle_m - \sum_{i=1}^{I} \beta_{m,i} \, \langle f_j(\mathbf{x})\, f_i(\mathbf{x}) \rangle_m \,. \end{aligned} \tag{3.46}$$

For each cluster c_m, we define the matrix $\mathbf{B}_{ji,m} = \langle f_j(\mathbf{x}) f_i(\mathbf{x}) \rangle_m$ and the vector $\mathbf{a}_{j,m} = \langle y f_j(\mathbf{x}) \rangle_m$, leading to the following simple update for the new cluster model parameters

$$\boldsymbol{\beta}_m^{\text{new}} = \mathbf{B}_m^{-1} \cdot \mathbf{a}_m \,. \tag{3.47}$$

As in the case of the linear local models, a regularization procedure should be used to deal with singular or ill-conditioned matrices $\mathbf{B}_m$. In our examples, we used a singular value decomposition with truncated principal components (see Section 3.2.6).

The updated output variances are now given by

$$\sigma_{m,y}^{2,\text{new}} = \langle [y - f(\mathbf{x}, \boldsymbol{\beta}_m)]^2 \rangle_m \,. \tag{3.48}$$

3.4 Examples

3.4.1 Noise Reduction

A possible application for the cross prediction introduced in Section 3.1.2 is the reduction of measurement noise from a deterministic dynamical system [24]. For this purpose, a noiseless time series from this dynamical system is necessary and a second time series is generated by corrupting the noiseless data with additive white noise. Afterward, a model is trained to predict from this noisy time series the noiseless one. This model can then be used as a tool for noise reduction on before unseen noisy data from the same dynamical system, given that the statistical properties of the noise are similar.

In our example, we want to reduce noise from the Rössler system [25]. For training, we generate a time series with 30 000 points and add white noise with an SNR of 10 dB. The embedding parameters obtained while training the local model are also used for the the cluster weighted model.

Through the prediction the SNR could be raised to 18 dB. An example for the prediction of the local model is shown in Fig. 3.2. The attractors reconstructed by an 3D embedding of the original and the predicted test data can be seen in Fig. 3.3.

3.4.2 Signal Through Chaotic Channel

Closely related to the previous example, where we subtracted measurement noise from a dynamical system, we now want to reconstruct a signal which is sent

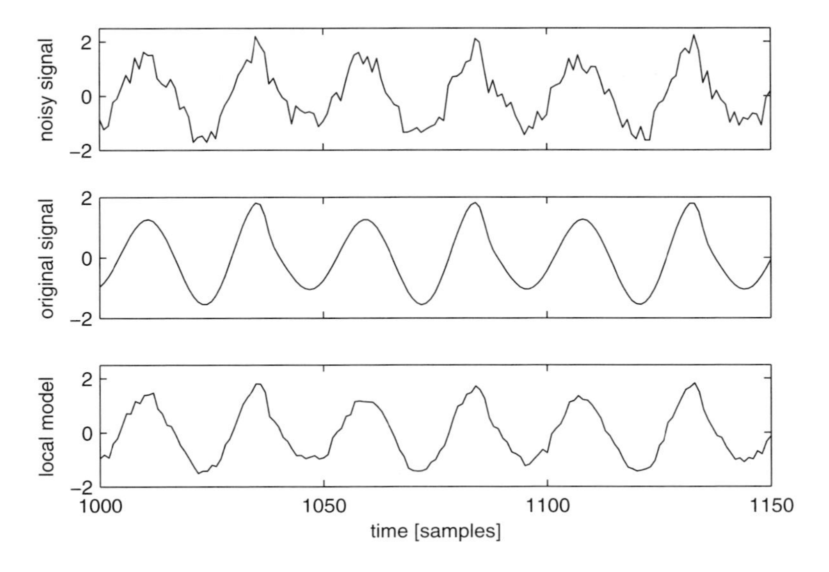

Fig. 3.2: Noisy input (top), original signal (middle), and local model prediction (bottom). The CWM prediction looks almost identical.

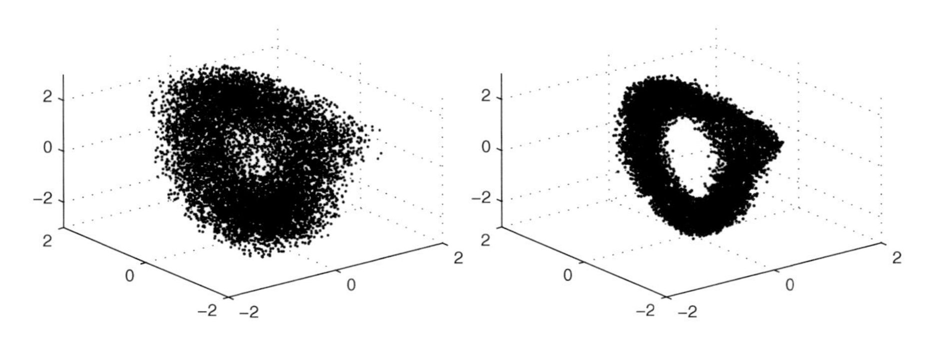

Fig. 3.3: Original noisy attractor (left) and local model test data prediction (right).

through a chaotic dynamical system, where the signal can be seen as a special case of dynamical noise. In our numerical example, a music wave file is taken as the signal and the Lorenz system as the chaotic system. The signal is added to the first ODE of the Lorenz system, while the variable y is taken as the output (see Fig. 3.4).

We now want to construct a model which is able to predict the original signal given the output, without providing any *a priori* knowledge of the underlying dynamical system. Like the previous example, this is the case of a cross prediction without feedback. In the following example, the model is trained using 30 000 point pairs, consisting of the original signal and the output of the chaotic system.

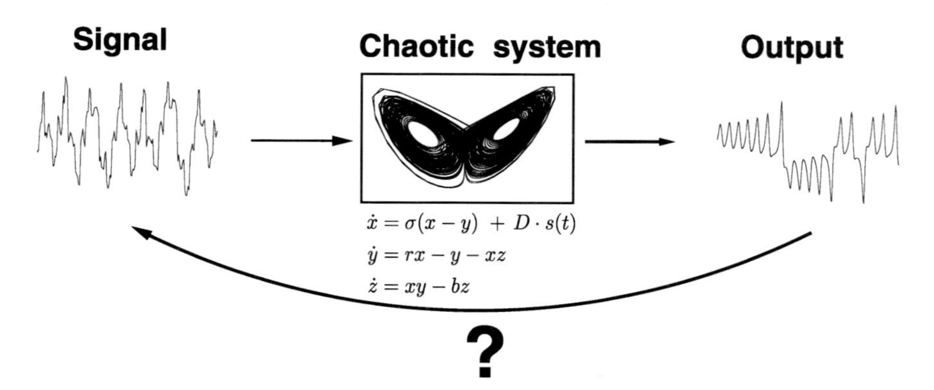

Fig. 3.4: Illustration of a chaotic channel.

First, we train a locally linear model which also yields good embedding parameters, which are also used for training a cluster weighted model with locally quadratic functions. Both models are tested on 10 000 test data points. The locally linear model has a NMSE of 11.5 %, while the cluster weighted model (200 clusters) performed slightly better with a NMSE of 9.8 %. The latter result can be seen in Fig. 3.5. Although the NMSE is quite large, which can also be seen in the plot of the residuals, the model still shows a good reconstruction of the basic signal properties.

3.4.3 Friction Modeling

Friction is a very complex and nonlinear phenomenon, comprising various regimes and behavioral facets. While there exist numerous analytical approaches for describing different aspects of friction phenomena, a model which could explain all aspects of friction is still missing. In practical control applications where a high accuracy is demanded, the highly nonlinear dependence of the friction force on displacement is one of the main problems. Black-box models, which do not depend on any *a priori* physical knowledge, can help us deal with this problem.

Experimental friction data, obtained from an experimental setup done by Al-Bender, Lampaert, and Tjahjowidodo at the University of Leuven [26], are used to train local models as well as cluster weighted models. The data consist of the (desired) displacement $\mathsf{P}(t)$ for the model input and the friction force $\mathsf{F}(t)$ (to be applied) for the model output. Therefore, we again have a cross prediction from $\mathsf{P}(t)$ to $\mathsf{F}(t)$, but in this case the accuracy of the modeling can be greatly improved by adding past values $\mathsf{F}(t - \delta)$ of the friction force to the input vector, introducing a feedback into the modeling procedure. The training data set consisted of 90 000 data points and the models were tested on 20 000 points. Here, the models are freely iterated over the complete test data set, i.e., while the position

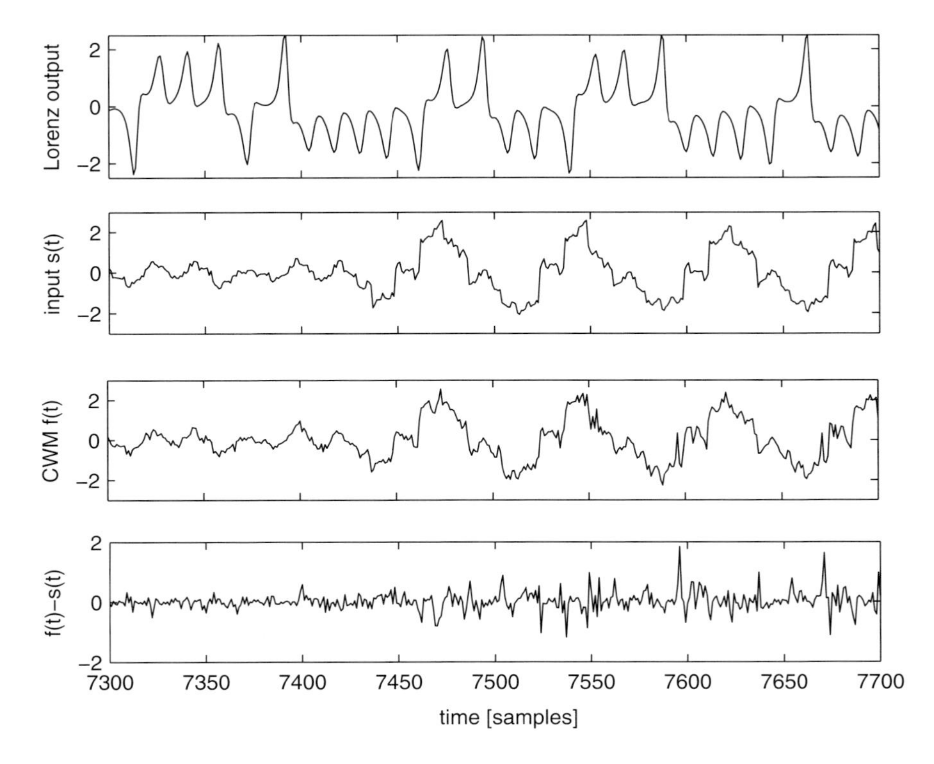

Fig. 3.5: Prediction of test data for a music signal using a cluster weighted model. The first two plots show the output signal from the Lorenz system and the original input signal. The lower two plots show the CWM prediction and the residuals. The result from the local model looks almost the same.

values in the input vector are always exact, the friction force is always estimated (except for the starting value, which is also exact).

Like in the previous examples, we first trained the locally linear model to obtain good embedding parameters. In this case, we obtained the 5D embedding vector

$$\mathbf{x}(t) = \big(P(t), P(t-16), P(t-66), P(t-67), F(t-19)\big)\,, \tag{3.49}$$

therefore consisting of four position values and one past force value. It is important to note that the optimal delay for the past force value (in this case $\delta = 19$) can only be obtained through an optimization which depends on the multistep prediction error. Since the time series is very densely sampled, the optimization on the one-step prediction error would yield an "optimal" value for the delay of $\delta = 1$, with the model simply repeating the last force value. Of course, such a model will lead to bad prediction results when freely iterated over the test data set.

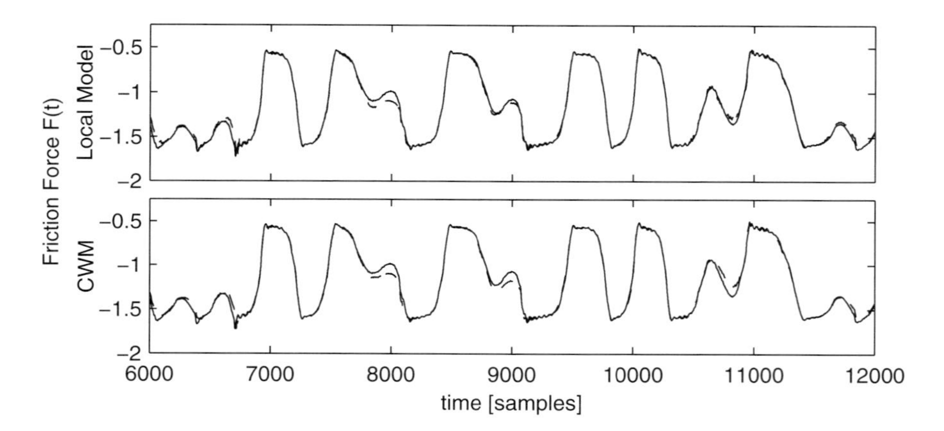

Fig. 3.6: Local model (top) and CWM prediction (bottom) for a section of friction test data; predictions are given by the dashed lines.

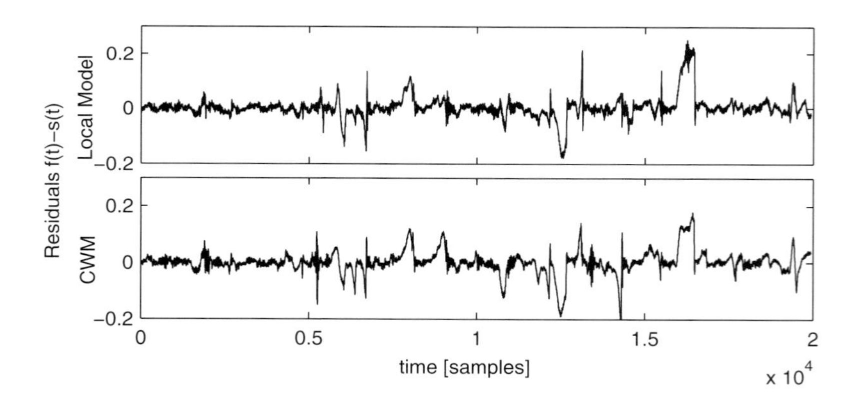

Fig. 3.7: Residuals of local model (top) and CWM prediction (bottom) for the complete friction test data.

Another important effect of the multistep prediction error is the better stability of the final model during iteration over several steps. In fact, as our tests show, the last position value $P(t-67)$ is crucial for the stability of the local model, though it may first seem redundant since it is almost equal to the previous one as they are only separated by a delay of 1. However, even with this additional position value, the cluster weighted model could not produce stable results when iterated over the test data, since it tends to oscillate with a period given by the delay of the past force value. Although one can enforce stability by simply clipping the model output with the minimum and maximum value of the given output data from the training set, the model error gets quite large. Although it is possible to dampen the oscillations through filtering, the filter introduces new parameters (order, cutoff frequency) which somehow have to be optimized.

Our approach for solving this problem is to use not one, but three different CWMs, each having a slightly different delay for the past force value (in this case we used 17, 19, and 21). This is called a *model ensemble*, and it is well known that such ensembles can often lead to better predictions than each single model in this ensemble could provide [27], although in our case we are more interested in stability features. The additional position value $P(t - 67)$ was now omitted, as it was no longer necessary for stability and led in this case to slightly worse prediction results. Since every model has a different delay for the force value, each model will tend to oscillate with different periods. When predicting the test data, we first calculate the three model outputs for each point and simply take the median, i.e., in this case the output of the model lying between the other two. The median is fed back to all three models, practically dampening beginning oscillations. Of course, this procedure can be extended to an arbitrary amount of models.

The local model (260 neighbors, linear weight function, Euclidean distance, TPCR with soft threshold and $s_c = 3 \times 10^{-3}$ and $s_w = 0.67$) yields a NMSE of 1.01 % over the 20 000 test data points. The CWM ensemble, where each CWM used quadratic functions and 600 clusters, has almost the same performance with a NMSE of 1.05 % (see Fig. 3.6). This ensemble error is lower than each of the single model outputs (though only slightly). The residuals for both models can be seen in Fig. 3.7.

3.5 Conclusion

As our examples show, cluster weighted modeling (CWM) can yield similar performance as local modeling. For pure cross prediction without feedback, CWM is very easy to use since besides the choice of the cluster functions there is only one parameter to choose, namely the number of clusters. If one seeks primarily a compact model, a low number of clusters with linear or even constant models is the obvious choice. However, from our experience, if the data-generating process is reasonably complex, a CWM with a low number of clusters can usually not compete with local models in terms of accuracy. Therefore, if high accuracy is important, the number of clusters must be chosen high enough (in our examples several hundred) and quadratic cluster functions are preferable since they often perform better than linear ones. However, since the number of parameters for such a CWM becomes very large, especially in high-dimensional spaces, a large number of data points for training must be available. Additionally, cross validation or some other means for preventing overfitting is crucial for training such CWMs. In practice, CWMs with a large number of clusters will often begin to overfit after only a few iterations of the EM algorithm.

For prediction with feedback like in the friction modeling example, CWM has the same problem as almost all other modeling techniques which rely on the minimization of the one-step prediction error: they often perform badly if iterated over several steps. In the case of friction modeling we were not able to generate a CWM which could compete with local models in terms of accuracy, and at the same time be stable when applied iteratively for predicting the complete test data set. The usage of an ensemble of CWMs proved to be a good solution for this problem, but at the cost of a more complicated modeling procedure and higher computational investment.

Local models have the advantage that they can be explicitly trained on the multistep prediction error, making them particularly suitable for prediction with feedback. They are very flexible and can be immediately used without the need for a training procedure. However, to get very accurate results, one has to optimize the different model parameters like the number of nearest neighbors, metric, weighting, and regularization. One advantage of CWMs is that they provide a density estimation for the joint probability, from which the model uncertainty can be estimated. Since clusters are only put in regions of the phase space which contain data points, CWMs also work well in high-dimensional spaces, and the clusters can also be used to obtain global properties like dimension estimates [28].

Acknowledgements

The authors thank F. Al-Bender, V. Lampaert, and T. Tjahjowidodo from the Catholic University Leuven for providing us the experimental friction data. The authors also thank the DFG Graduiertenkolleg "Identification in Mathematical Models: Synergy of Stochastic and Numerical Methods" and the Volkswagenstiftung (grant no. 1/76938) for financial support.

References

[1] S. Geman, E. Bienenstock, and R. Doursat. *Neural Comput.*, 4:1, 1992.

[2] F. Takens. In D. A. Rand and L. S. Young, editors, *Dynamical Systems and Turbulence*, page 336. Springer, New York, 1981.

[3] K. Judd and A. Mees. *Physica D*, 120:273, 1998.

[4] K.-S. Chan and H. Tong. *Chaos, a Statistical Perspective*. Springer, New York, 2001.

[5] J. McNames. PhD thesis, Stanford University, 1999.

[6] J. D. Farmer and J. J. Sidorowich. *Phys. Rev. Lett.*, 59:845, 1987.

[7] C. G. Atkeson, A. W. Moore, and S. Schaal. *Artif. Intell. Rev.*, 11:11, 1997.

[8] J. L. Bentley. *Commun. ACM*, 18:509, 1975.

[9] J. H. Friedman, J. L. Bentley, and R. A. Finkel. *ACM Trans. Math. Softw.*, 3: 209, 1977.

[10] J. McNames. *IEEE Trans. Pattern Anal. Mach. Intell.*, 23:964, 2001.

[11] C. Merkwirth, U. Parlitz, and W. Lauterborn. *Phys. Rev. E*, 62:2089, 2000.

[12] C. Schaffer. *Mach. Learn.*, 10:153, 1993.

[13] W. H. Press, B. P. Flannery, S. A. Teukolsky, and W. T. Vetterling. *Numerical Recipes in C: The Art of Scientific Computing*. Cambridge University Press, Cambridge, 1992.

[14] J. McNames, J. A. K. Suykens, and J. Vandewalle. *Int. J. Bif. Chaos*, 9:1485, 1999.

[15] G. H. Golub and C. F. van Loan. *Matrix Comput.* The Johns Hopkins University Press, Baltimore, MA, 1996.

[16] A. A. Björck. *Numerical Methods for Least Squares Problems*. SIAM, Philadelphia, 1996.

[17] I. T. Jolliffe. *Principal Component Analysis*. Springer, New York, 1986.

[18] D. S. Broomhead and G. P. King. *Physica D*, 20:217, 1986.

[19] D. E. Goldberg. *Genetic Algorithms in Search, Optimization and Machine Learning*. Addison-Wesley, Reading, MA, 1989.

[20] V. Babovic and D. R. Fuhrman. Technical report 0401-2. Technical report, D2K, 2001.

[21] N. Gershenfeld, B. Schoner, and E. Metois. *Nature*, 397:329, 1999.

[22] A. Dempster, N. Laird, and D. Rubin. *J. R. Stat. Soc.*, 39:1, 1977.

[23] G. McLachland and D. Peel. *Finite Mixture Models*. John Wiley and Sons, New York, 2000.

[24] S. A. Billings and K. L. Lee. *Int. J. Bif. Chaos*, 14:1037, 2002.

[25] O. E. Roessler. *Phys. Lett. A*, 57:397, 1976.

[26] U. Parlitz, A. Hornstein, D. Engster, F. Al-Bender, and V. Lampaert. *Chaos*, 14:420, 2004.

[27] T. Hastie, R. Tibshirani, and J. Friedmann. *The Elements of Statistical Learning*. Springer, New York, 2001.

[28] B. Schoner and N. Gershenfeld. In A. Mees, editor, *Nonlinear Dynamics and Statistics*. Birkhäuser, Boston, 2001.

4 Deterministic and Probabilistic Forecasting in Reconstructed State Spaces

Holger Kantz and Eckehard Olbrich

A typical time series analysis task is to extract knowledge from the past in order to make predictions about the future. Such an endeavor relies on the presence of correlations in time. We present concepts, methods, and algorithms for this task. Special emphasis is laid on nonlinear stochastic processes, probabilistic predictions, and their verification. Whereas in processes with a rather strong deterministic component one is used to predict the most probable future value together with some uncertainty (error bar), in strongly random processes it is more useful to forecast the probability that the future observation will fall inside a certain range of values. Such a range of values in applications often relates to an "event," so that we also recall the statistical theory of classification and classification errors.

4.1 Introduction

Prediction of the future is a ubiquitous desire of mankind. Whereas ancient cultures might have used rather obscure techniques (e.g., oracles), we try nowadays to make our forecasts on the basis of objective facts. Evidently, we are in a very comfortable situation if the phenomenon which we like to forecast follows some deterministic time evolution, and if moreover we have a full understanding of the process, and finally are able to determine with sufficient accuracy the current state of the system. In weather forecasting, one is close to this situation: the dynamics of the atmosphere is deterministic and its physics is well understood. Hence, numerical weather prediction schemes are rather accurate models of what happens in nature, and feeding them with current measurements of the relevant inputs yields rather reliable forecasts. The remaining uncertainty is mainly due to the lack of input data, in particular over the oceans, and partly due to insufficient knowledge of some parameterized processes, such as the microphysics inside clouds where extremely complicated processes at the phase transition of water droplets and water vapor occur. If the knowledge about the dynamics of a process is insufficient or if it is impossible to measure those observables which are needed to define the actual state of the system, then time series approaches

Handbook of Time Series Analysis. Björn Schelter, Matthias Winterhalder, Jens Timmer

ISBN: 3-527-40623-9

might be a way out. Nonetheless, time series approaches are similar in spirit as predictions using models: The time series data have to be used for two purposes. We need to extract rules for the time evolution, i.e., an equivalent to the equations of motion of a system in its state space, and we need to identify the current state of the system on which the dynamics acts. Since in this reduced setting many phenomena which in principle are deterministic might appear stochastic, a weaker form of prediction will naturally arise: probabilistic predictions, where we cannot reliably give a precise value for a quantity at some time in the future, but only a probability that the value will be inside some interval. Such probabilistic prediction schemes are still very useful, but they require more sophisticated concepts for their verification than deterministic predictions: Probabilistic predictions do not allow us to compute a prediction error.

In this chapter we start with the concept of state space reconstruction which was introduced about 25 years ago for the analysis of time series data for low-dimensional deterministic systems. Prediction schemes on different technical levels are then rather straightforwardly derived from the concept of determinism. As a step beyond standard results, we will discuss already here a state-dependent uncertainty of predictions. In other words, we will interpret additional structure in data in a way to "predict the error" of a specific prediction. The need to do so is evident: As an example, we are all aware of weather conditions where a forecast about rain on the next days is highly likely to be true, and that there are other weather conditions where such a forecast is very uncertain. In fact, based on the technique of ensemble forecasts, weather forecasts are nowadays often annotated by labels such as "certain" or "less certain," and here we will outline how to achieve similar information from time series data.

We will then argue in more detail why deterministic phenomena are rare when we deal with time series data. The much more appropriate model class which should be represented by our prediction consists of nonlinear stochastic processes. As a detour, we will first recall a method to reconstruct Fokker–Planck equations from data. This is a conceptually very interesting approach, which suffers, however, from two shortcomings: First of all, the process must be Markovian, and secondly one must be able to record all state variables of the system. In cases where one or both requirements are violated, a continuous state Markov chain, i.e., a model in discrete time, may be a useful approximation.

In the following we will ignore the everyday experience and assume that the process underlying our time series data is stationary. We hence use the hypothesis that process parameters in the future will be identical to those in the past, and that, more precisely, (a) all conditional probabilities do not explicitly depend on time, and (b) that the process is recurrent. We can then assume that (a) what we extract from the past is a good characterization of the future and that (b) there exist similarities in the past which allow us to extrapolate from the presence into the future.

4.2 Determinism and Embedding

Let us assume that the time evolution of the system which we observe can be described by deterministic equations of motion. As examples where this is certainly true we can list many physical laboratory experiments, such as (nonlinear) electric resonance circuits and mechanical devices, but also the almost nondissipative motion in our planetary system. Formally, such dynamics are described by a set of first-order differential equations,

$$\dot{\mathbf{x}} = \mathbf{f}(\mathbf{x}), \mathbf{x} \in \Gamma \subset \mathbb{R}^D, \quad \mathbf{f}\colon \mathbb{R}^D \to \mathbb{R}^D \text{ Lipschitz continuous,} \tag{4.1}$$

which together with an initial condition $\mathbf{x}(0) = \mathbf{x}_0$ uniquely determine the trajector $\mathbf{x}(t)$ for all times $t \geqslant 0$.

A time series $\{s_n\}, n = 1, \ldots, N$, is obtained by applying a measurement function $s = h(\mathbf{x})\colon \Gamma \to \mathbb{R}$ to the trajectory $\mathbf{x}(t)$ at equidistant times $t_n = n\Delta$. Δ is called the sampling interval and $1/\Delta$ is the sampling frequency, giving rise to a finite number of measurements N. The sampling interval has to be adopted to the time scales involved in the dynamics, i.e., it must not be too large. On the other hand, too small Δ (called oversampling) is a waste of resources and leads to time series which can be compressed by down-sampling. Experience shows that for irregular (chaotic) fluctuations one should use about 20 to 50 sample points per typical oscillation period. Of course, if more than a single measurement device is used, one might record simultaneously the values of different measurement functions, which is advantageous. In the following we will follow the folklore and assume the worst case of a single observable, but the extension to multivariate data (multichannel measurements) will be outlined as well.

Having introduced the notion of sampling, we are dealing with a dynamical system in discrete time. The situation of Eq. 4.1 can be formally cast into discrete time: denoting by $\mathbf{F}(\mathbf{x})$ the time-Δ map of the flow, it gives rise to the iteration of the map $\mathbf{x}((n+1)\Delta] = \mathbf{F}[\mathbf{x}(n\Delta))$. Predictions in this situation would mean to propose a value $\hat{\mathbf{x}}$ which is as close as possible to the yet unknown state $\mathbf{x}((n+1)\Delta)$. Knowing $\mathbf{F}$ and $\mathbf{x}(n\Delta)$ we can evidently make the perfect prediction $\hat{\mathbf{x}} = \mathbf{F}(\mathbf{x}(n\Delta))$. Having only the observations $s_n, s_{n-1}, \ldots$ available modifies the situation in several respects: First, the goal of the prediction can only be a proposed value for s_{n+1}, not for $\mathbf{x}((n+1)\Delta)$, since $\mathbf{x}((n+1)\Delta)$ will remain unknown even after the future measurements and hence a prediction for $\mathbf{x}((n+1)\Delta)$ could not be verified.[1] We know that $s_{n+1} = h\big(\mathbf{x}((n+1)\Delta)\big)$; hence we can hope to be able to infer the current state $\mathbf{x}(n\Delta)$ from our measurements and to find from the past data a rule how to propagate this state one time step into the future. The above-introduced structure is in principle sufficient to provide the

[1] Note, however, the conceptually very interesting approach outlined in [1] which in principle enables one to reconstruct model equations in the unobserved phase space from time series measurements (assuming knowledge of $h(\mathbf{x})$), at the cost of solving a highly demanding nonlinear minimization problem.

mathematical tools for this; however, it is very useful to introduce an additional concept, namely the concept of the attractor.

In most dynamical systems describing physically relevant situations, a single initial condition does not create a trajectory which explores the full phase space. Instead, "typical" (i.e., apart from a set of Lebesgue-measure zero) initial conditions lead to trajectories $\mathbf{x}(t)$ which asymptotically settle down on an invariant set $\mathcal{A} \subset \Gamma$, whose dimensionality D_0 can be (much) smaller than D. A drastic physical example is the Rayleigh–Bénard experiment, where some liquid is confined by two plates and heated from below but cooled from above: For moderate temperature differences, convection rolls can be observed, which are a kind of low-dimensional collective behavior of the more than 10^{23} degrees of freedom in the system (i.e., $D \approx 10^{23}$, $D_0 = O(10)$).

The important result is stated by the Takens theorem [2] and its more recent formulation [3]: Given a dynamical system $\dot{\mathbf{x}} = \mathbf{f}(\mathbf{x})$ in a phase space $\Gamma \subset \mathbb{R}^D$, a measurement function $h\colon \mathbb{R}^D \to \mathbb{R}$, and a sampling interval Δ. Let the trajectory $\mathbf{x}(t)$ be confined to an $\mathbf{f}$-invariant set $\mathcal{A} \subset \Gamma$, with the box-counting dimension D_0. Denote the scalar measurements obtained through the sampling by $s_n := h(\mathbf{x}(t = n\Delta))$. Consider the delay embedding space spanned by delay vectors $\mathbf{s}_n = (s_n, s_{n-\tau}, s_{n-2\tau}, \ldots, s_{n-(m-1)\tau})$ for some positive integers m and τ. If $m > 2D_0$, then there exists a unique smooth map from $\mathcal{A}$ into the delay embedding space, which is invertible and has nonzero derivative on the image of $\mathcal{A}$ in $\mathbb{R}^m$. $\mathcal{A}$ is then said to be immersed in $\mathbb{R}^m$. This holds for prevalent h, generic $\mathbf{f}$, almost all Δ, and every $\tau \in \mathbb{N}$.

Hence, the m-dimensional delay embedding space is equivalent to the original unobserved phase space of the dynamical system, since in particular the dynamics of $\mathbf{s}$ is deterministic: Denote the projection by the measurement and the subsequent embedding by i; then the following commutative diagram exists:

$$\begin{array}{ccc} \mathbf{s}_n & \stackrel{\mathbf{G}}{\longmapsto} & \mathbf{s}_{n+1} \\ \uparrow i & & \uparrow i \\ \mathbf{x}(t = n\Delta) & \stackrel{\mathbf{F}}{\longmapsto} & \mathbf{x}(t = (n+1)\Delta) \end{array} \tag{4.2}$$

where $\mathbf{F}$ denotes the time-Δ map of the flow $\mathbf{f}$. For a deeper discussion of this theorem and also of the choice of the time lag τ see, e.g., [4, 5]. If we do not know D_0 or not even D, we have to infer the smallest number m which gives rise to an embedding empirically, and again there are several concepts available [4, 5]. Since this review is concerned with predictions, we will forget about these considerations and treat m and τ just as parameters in the prediction schemes which have to be tuned for optimal predictions.

The scheme (4.2) now gives the guideline for predictions: Find a "good" integer m so that $\mathbf{s}_n$ is equivalent to the unobserved state vector $\mathbf{x}(n\Delta)$ and apply the dynamics $\mathbf{G}$ to it. Since the embedding vector $\mathbf{s}_{n+1}$ one time step ahead can be found by copying all but the very first component, s_{n+1}, from $\mathbf{s}_n$, the unknown part of $\mathbf{G}$ is a scalar function $g(\mathbf{s}_n) = s_{n+1}$. This function has to be "learned" from the data.

Indeed, the time series $\{s_1, \ldots, s_n\}$ contains many training pairs $(\mathbf{s}_k, s_{k+1})$ from which we can extract g. If we assume that g is differentiable, then we can write

$$s_{n+1} = s_{k+1} + \nabla g(\mathbf{s}_n)(\mathbf{s}_n - \mathbf{s}_k) + O(|\mathbf{s}_n - \mathbf{s}_k|^2) . \tag{4.3}$$

Following Farmer and Sidorowich [6], we can use this relation to construct a locally constant (or zeroth-order) predictor in the following way:

$$\hat{s}_{n+1} = \frac{\sum_{k:\, |\mathbf{s}_k - \mathbf{s}_n| < \epsilon} s_{k+1}}{\sum_{k:\, |\mathbf{s}_k - \mathbf{s}_n| < \epsilon} 1} , \tag{4.4}$$

where $\epsilon \ll 1$ is another parameter. Differentiability of g is only required for a formal control of the remainders, whereas the expression itself yields reasonable predictions if g is continuous, and can even be applied if g is noncontinuous on a set of measure zero.

A better approximation of the local dynamics can be achieved if linear corrections with ∇g are taken into account. For that purpose one again considers all past states $\mathbf{s}_k$ in the ϵ-neighborhood $\mathcal{U}_n$ of $\mathbf{s}_n$ and introduces the averages $\bar{s}_{n-l} := \langle s_{k-l} \rangle_{\mathcal{U}_n}$ (the average $\bar{s}_{n+1}$ then simply being the predictor (4.4)). The set of relations $s_{k+1} = \bar{s}_{n+1} + \sum_{i=1}^{m} a_i(n)(s_{k-i+1} - \bar{s}_{n-i+1}) + \sigma_k$ for $k\colon |\mathbf{s}_k - \mathbf{s}_n| < \epsilon$ gives rise to a minimization problem for $\sum_k \sigma_k^2$, which yields the coefficients $a_i(n)$ of $\nabla g(\bar{\mathbf{s}}_n)$. Hence the locally linear (or first-order) predictor reads

$$\hat{s}_{n+1} = \bar{s}_{n+1} + \sum_{i=1}^{m} a_i(n)(s_{n-i+1} - \bar{s}_{n-1+1}) . \tag{4.5}$$

Let us emphasize that the construction of the coefficients a_i is repeated for every prediction, so that these are explicitly n-dependent. This is reflected by the name *locally* linear predictor. Both prediction schemes have been widely used in the literature. Moreover, there are many interesting issues related to predictions and modeling, see, e.g., [1, 7, 8]. The function $g(\mathbf{s})$ is a scalar field on $\mathbb{R}^m$. If it is sufficiently smooth, one can of course also approximate it by a single global function $\hat{g}$ which can be fitted to the data. Then one has to minimize the mean-squared prediction error

$$\bar{e}^2 := \sum_{n=m}^{N-1} \left(s_{n+1} - \hat{g}(\mathbf{s}_n)\right)^2 , \tag{4.6}$$

where the minimization is done with respect to the parameters contained in $\hat{g}$. If parameters enter linearly in $\hat{g}$, i.e., if $\hat{g}$ is a linear combination of parameter free terms, then this minimization is straightforward. In fact, the locally constant and

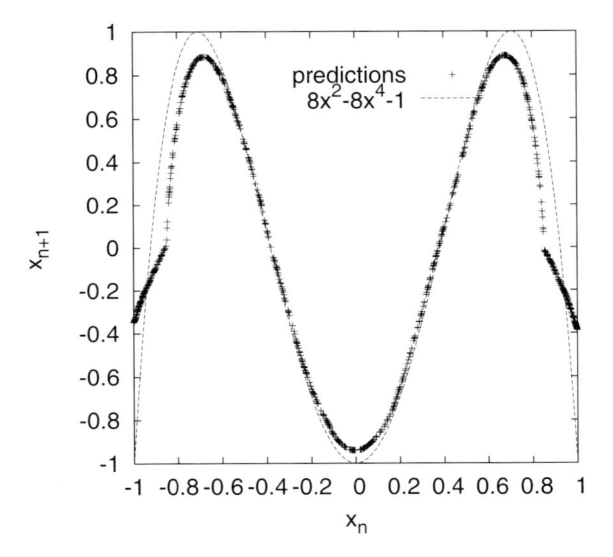

Fig. 4.1: The predicted values $\hat{s}_{n+1}$ obtained by the locally constant predictor (4.4) for a time series of the second iterate of the logistic equation, $s_{n+1} = f(s_n)$, $f(s_n) = -1 + 8s_n^2 - 8s_n^4$, together with the graph of $f(s)$. The systematic deviation between predictions and graph is a result of the smoothing due to the finite ϵ in the predictor (for didactic purposes, we used the rather big value $\epsilon = 0.15$).

the local linear models result from this minimization if one assumes $\hat{g}$ to be constant or linear in $\mathbf{s}_n$, respectively, and restricts the sum over the training pairs, n, to a neighborhood of the actual delay vector. Examples of global nonlinear functions are multivariate polynomials [9] on $\mathbf{s}$ or so-called radial basis functions [10]. In feed-forward neural networks, which also establish a nonlinear input–output relation [11], parameters enter in a nonlinear way in $\hat{g}$, so that sophisticated minimization schemes (e.g., error back-propagation) are needed, such that training neural networks is an art of its own.

The minimization of the mean-squared prediction error leads to an unbiased estimate of the dynamics only if the training pairs are uncorrelated, which for time series data is evidently violated but which usually (for sufficiently large data sets) is not a practical problem. If errors are assumed to be Gaussian, then together with the independence this minimization leads to the optimal predictor in the framework of the maximum likelihood principle.

We want to finish the section on deterministic data by a discussion of the origin of prediction errors and of their size distribution. Evident systematic errors might be introduced if the data do not fulfill the assumptions made by the predictor about the dynamics. This concerns in an evident way a too small embedding dimension m, but also smoothness or continuity. Moreover, an important issue is overfitting or generalization: The predictor contains the information of those situations on which the predictor has been optimized. For a check of whether the predictor really describes the dynamics or whether it just describes the past data one has to apply it to a test set, i.e., to a set of pairs $(s_{k+1}, \mathbf{s}_k)$ which have not been used to fit the coefficients in the predictor. If the data set is small, this is usually done by *(complete) cross-validation* or *leave one out statistics*: When performing a prediction for s_{n+1}, one excludes the training pair $(s_{n+1}, \mathbf{s}_n)$ from the database (and usually also all those training pairs in the future which are correlated with s_{n+1}). A prediction error $\langle(\hat{s}_{n+1} - s_{n+1})^2\rangle$ computed this way is called *out-of-sample error* and is the only quantity which faithfully tells how well the predictor will perform on future data (stationarity provided).

Even with great care and in favorable situations there will remain two sources of errors: systematic errors which are introduced by a lack of flexibility of the model of g which is (implicitly or explicitly) established by the predictor, and statistical errors because of all kinds of noises. As an example of the systematic errors, think of the seemingly parameter-free model (4.4): In the language of statistics, this is a kernel estimator with a bandwidth ϵ, i.e., the true structures in g are smoothed out by a length scale ϵ. Hence, if ϵ is large compared to the structure in g, such a predictor will generate systematic errors. In addition, for points at the boundary of the attractor, neighboring points are systematically located on the inside, so that the prediction is systematically biased. In Fig. 4.1 we illustrate this for data generated by the logistic equation. Unfortunately, the size of ϵ is limited from below by the time series length N and the embedding dimension m: Inside the m-dimensional ϵ-neighborhood of $\mathbf{s}_n$ we must find at least one neighboring point in order to have an estimate of $\hat{s}_{n+1}$, so that the mean nearest-neighbor distance is a lower bound for ϵ.

Statistical errors are usually introduced through measurement noises on the recorded data. On the one hand, this uncertainty causes some uncertainty about the current state (which is represented by the noisy delay vector), on the other hand also the future observation will be noisy, the noise part not being predictable. Hence, in the best case the root-mean-squared prediction error is the standard deviation of this noise. For linear processes, the average prediction error should be independent of the state $\mathbf{s}_n$. For nonlinear processes, however, deficiencies of the predictor as well as the amplification of noise on the delay vector might depend on the state and hence cause prediction errors whose magnitude depends on the state. Chaotic systems contain directions in state space which are expanding, i.e., trajectories originating from nearby points diverge exponentially fast with probability one. This stretching or instability of solutions causes an amplification of every uncertainty about the current state of a system. However, the stretching rates can depend significantly on the state vector. Irrespective of whether this is really the origin of prediction errors, the magnitude of the expected error can be easily predicted. The more accurate but also more computation intense way is to first compute the prediction errors on a large training set (possibly by complete cross-validation) and storing them. Then performing the actual predictions, one can search for neighbors in the training set and postulate that the actual error will be bound by the maximum of the errors made for predictions on the neighboring points. This is what we call *predicted error amplitude* in Fig. 4.2. As we see, it is indeed a fair estimate of the upper bound of the individual prediction errors on the test set. A simpler, slightly less accurate approach (because it cannot incorporate systematic errors and hence reflects only errors due to noise) can be easily derived within the framework of the locally constant predictor: In the locally constant predictor, the prediction is the mean of the images of neighboring state vectors. The expected error is naturally restricted by the width of the distribution of these images, i.e., the true future value should be inside the range of these images, which hence could be given as an error interval. If the input data are noisy, so that this amplifies the uncertainty, the standard deviation

$$\hat{\sigma}_n = \sqrt{\langle (s_{k+1} - \hat{s}_{n+1})^2 \rangle_{\mathcal{U}_n}} \tag{4.7}$$

should be characteristic of the distribution of state-dependent prediction errors. That the actual prediction errors are actually much smaller, when determinism is strong, is related to the fact that the deterministic part of the misprediction is related to the difference between the mean value of the neighbors and the actual value from which the prediction starts, amplified by the local stretching rate. Since this distance often is tiny, prediction errors for numerically generated deterministic data without additive noise are usually much smaller. In Fig. 4.2 the prediction of the error amplitude is illustrated for experimental data from an NMR laser which are representing low-dimensional dynamics contaminated by a few percent of noise. As this example shows, the magnitude of the prediction error can be correctly predicted in this way, i.e., one can easily equip every prediction with an error bar. In the example shown the error amplitudes can evidently vary by almost two orders of magnitude, so that the advance knowledge of this magnitude is a valuable additional information.

Often, one wants to predict more than one time step ahead. This could be done by iterating predictions, i.e., by using the prediction $\hat{s}_{n+1}$, either to construct a new delay vector $\hat{\mathbf{s}}_{n+1}$, which is the input for the prediction $\hat{s}_{n+2}$, and so on, or to use a predictor which in a single step makes a prediction for the time $n+r$. The latter is easily done by replacing the training pairs $(\mathbf{s}_k, s_{k+1})$ by training pairs $(\mathbf{s}_k, s_{k+r})$. In both cases, the prediction error will grow exponentially in r (if the data represent a deterministic chaotic process), but whereas in the latter case it will saturate at the standard deviation of the data, for iterated predictions it should saturate at about $\sqrt{2}$ times the standard deviation. The reason is that the images s_{k+r} for big r will cover the whole range of s, so that the single jump predictor for large r will just produce the mean value of s as output (leading to a root-mean-squared prediction error identical to the standard deviation), whereas the iterated one-step predictions will smoothly pass over to modeling (if the

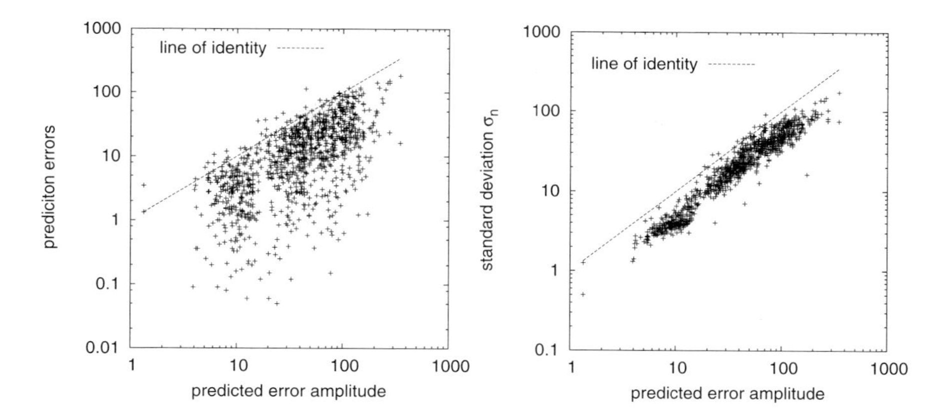

Fig. 4.2: Left panel: a set of 1000 prediction errors versus the individually predicted error amplitude. Right panel: plotting the predicted error amplitudes versus the standard deviations $\hat{\sigma}_n$ shows that the latter offer a slightly less accurate but much simpler prediction of the error amplitudes. Data: experimental data from an NMR laser discussed in detail in [5].

predictor is good), i.e., the predicted trajectory will continue to fluctuate as the true process, but for large r these fluctuations will lose their correlation with the true time series. However, if one wants to use global nonlinear functions $g(\mathbf{s})$ for predictions (e.g., multivariate polynomials or neural networks), one should use iterated one-step predictions: The map $\mathbf{s}_n \mapsto s_{n+1}$ is much less nonlinear than the map $\mathbf{s}_n \mapsto s_{n+r}$, $r > 1$, so that a faithful global fitting of the latter requires a much more flexible function g and hence many more coefficients to be fitted.

If more than one observable is recorded, i.e., if we work with multi-channel recordings, then the only modification needed lies in the definition of the state vectors $\mathbf{s}_n$: One then uses multivariate delay vectors, i.e., one combines all recorded observables taken at successive time steps to delay vectors. The total number of elements in these vectors should then replace the integer m in the Takens theorem and hence should be larger than twice the dimension of the invariant set. A numerical toolbox for the analysis of data in terms of determinism, including the two nonparametric predictors (4.4) and (4.5), is freely available as the TISEAN package [12].

4.3 Stochastic Processes

Although a deterministic relationship between the current state and the future is most desirable, in many situations it does not exist. Even in cases where it exists it is often not explorable. As an example, it is intuitively clear that a local wind speed measurement is such a poor representation of the wind field in the three-dimensional space that the determinism of the Navier–Stokes equations cannot be exploited for forecasting the local wind speed. More formally, even if the dynamics in its state space lives on a finite-dimensional attractor, the attractor dimension is often much too large for a reconstruction of the state vectors by delay vectors. This has two reasons: Even without chaos, the embedding of dynamics on a high-dimensional set poses practical problems, since neighboring vectors have a large average distance. More precisely, if one distributes N vectors in a D-dimensional hypercube of unit length, then the average inter-point-distance is about $N^{-1/D}$. An even more severe problem is introduced by chaos: The instability and irregularity of chaos has the consequence that the invariant set in the time delay embedding space has a much more complicated structure than the same set in the original state space. This is related to the lack of correlations between time series elements which are far apart in time. Both effects together lead to the observation that high-dimensional chaotic dynamics can hardly be identified as being deterministic by time series tools applied to time series of tractable length and with realistic noise levels [13].

Hence, it is often plausible that the dynamics underlying a given time series is generated by a stochastic process. If we assume some dominant deterministic feedback loops acting on some relevant variables and represent all other variables by white noise and damping, then the mathematical formulation of the equations of motion is a vector-valued stochastic differential equation

$$dx_i = f_i(\mathbf{x})\,dt + \sum G_{ij}(\mathbf{x})\,dW_j\,, \quad i = 1, \ldots, D \tag{4.8}$$

(Ito stochastic calculus assumed) where $G_{ij}(\mathbf{x})$ is a $D \times D$ tensor which determines the amplitudes and correlations of the noise inputs represented by dW_i, the differentials of the Wiener process [14]. Such an equation can be converted into a Fokker–Planck equation which describes the time evolution of the phase space density,

$$\dot{\rho}(\mathbf{x}) = -\partial_i(\mathbf{D}_i^{(1)}(\mathbf{x}) + \mathbf{D}_{ij}^{(2)}(\mathbf{x})\partial_j)\rho(\mathbf{x})\,. \tag{4.9}$$

The drift terms are simply $\mathbf{D}_i^{(1)}(\mathbf{x}) = f_i(\mathbf{x})$; the diffusion terms read $\mathbf{D}_{ij}^{(2)}(\mathbf{x}) = G_{ik}(\mathbf{x})G_{kj}(\mathbf{x})$ (summing over multiply occurring indices).

In data analysis, the task is now to reconstruct the drift field $\mathbf{D}_i^{(1)}(\mathbf{x})$ and the diffusion tensor $\mathbf{D}_{ij}^{(2)}(\mathbf{x})$ from time series data. This can only be done if either the state space variables $\mathbf{x}$ are directly recorded or they can be obtained from the observed quantities by a simple transformation. In other words, the structure of the Fokker–Planck equation cannot be converted into any kind of delay embedding space.

We will therefore assume that the recorded time series contain the simultaneous measurement of a multicomponent state vector of a system. Then the first test of whether this hypothesis is reasonable is a test for the Markov property: If indeed a measured vector represents the state of a Markov process, then a property called the Chapman–Kolmogorov equation is fulfilled,

$$P(\mathbf{x}(t_2) \mid \mathbf{x}(t_1)) = \int P(\mathbf{x}(t_2) \mid \mathbf{x}(t'))P(\mathbf{x}(t') \mid \mathbf{x}(t_1))\, d\mathbf{x}(t') \quad \forall\, t_1 < t' < t_2\,, \tag{4.10}$$

where $P(\mathbf{x}(t_2) \mid \mathbf{x}(t_1))$ denotes the conditional probability to observe $\mathbf{x}(t_2)$ at time t_2 provided to have measured $\mathbf{x}(t_1)$ at time t_1. If the Markov property is thus established, the drift and diffusion terms can be estimated by the following conditional averages (provided Δ is sufficiently small):

$$\mathbf{D}_i^{(1)}(\mathbf{x}) \approx \frac{1}{\Delta}\langle x_i\,(t+\Delta) - x_i\,(t)\rangle|_{\mathbf{x}(t)=\mathbf{x}} \tag{4.11}$$

$$\mathbf{D}_{ij}^{(2)}(\mathbf{x}) \approx \frac{1}{2\Delta}\Big(\big\langle\big(x_i(t+\Delta) - x_i(t)\big)\big(x_j(t+\Delta) - x_j(t)\big)\big\rangle_{\mathbf{x}(t)=\mathbf{x}} - \Delta^2\mathbf{D}_i^{(1)}\mathbf{D}_j^{(1)}\Big)\,. \tag{4.12}$$

In a sequence of pioneering publications Friedrich and Peinke [15, 16] have applied this modeling technique to several data sets. Among them is highway traffic, where the independent observables are the flux of cars and their speed, which could nicely be described by a Fokker–Planck equation [17]. Another concerns an erratic metal cutting process [18]. The limitation of this conceptually very nice method is not only given by the requirement to deal with multivariate data which represent the state vectors of the system, but also by the attempt to adopt a model in continuous time to data with discrete sampling. If the sampling rate is too coarse, a proper estimation of the drift and diffusion terms suffers from systematic errors of the order of the square of the sampling interval.

In order to generate a forecast from this model, one would integrate the Fokker–Planck equation forward in time, starting with an initial density which is a δ-peak located at the most recent measurement. The prediction which minimizes the root-mean-square (rms) prediction errors is the (time-dependent) mean of the evolving probability density. An alternative to integrating the Fokker–Planck equation (which is a partial differential equation) would be to represent the temporally evolving density by a finite sample of trajectories which themselves are solutions of the corresponding Langevin equation. For this purpose, one has to select a suitable integration scheme for stochastic differential equations. A very simple one is the Euler integrator, where the noise amplitude has to be rescaled with the square root of the step width

$$\mathbf{x}(t+\Delta) \approx \mathbf{x}(t) + \Delta\Big(\mathbf{f}(\mathbf{x}(t)) + \frac{1}{\sqrt{\Delta}}\mathbf{G}(\mathbf{x})\xi\Big)\,. \tag{4.13}$$

In many time series applications, a Fokker–Planck model is out of reach, most often, because the time series is univariate. In such cases, and also, if the sampling interval is large compared to the internal time scales of the process, a time discrete modeling which allows for a more than one-step memory is more useful. If the number of possible states is finite (i.e., if the time series is a symbol sequence with a finite number of different possible symbols), then such a model is known as Markov chain. The relation between all possible states (inputs) and all possible states (output) is fully specified by a Markov matrix (a_{ij}) with the following properties:

$$0 \leqslant a_{ij} \leqslant 1 \quad \forall i, j \tag{4.14}$$

$$\sum_j a_{ij} = 1 \quad \forall i\,. \tag{4.15}$$

The entry a_{ij} of this matrix then denotes the transition probability from state i to state j. If the order of the Markov chain is $m > 1$, i.e., if the current state is encoded by a sequence of m symbols rather than by a single symbol, then the set of all possible sequences of m symbols has to be enumerated, and the transition probabilities in between all states $j \to k$ which do not fulfill the condition that state k can be gained from j by chopping the first symbol and by appending a last symbol has to be zero. Evidently, a sufficiently long symbol sequence can easily be converted into a correspondingly large set of input–output states, which can be used to determine the matrix coefficients a_{ij} by simple counting with proper normalization.

For real-valued time series, this concept can be easily generalized [19, 20]: If the state of the stochastic process is fully defined by the sequence of the m past measurements, then the probability for finding a given value s' in the following measurement is most sharply defined, i.e., then no better knowledge of the future state is possible than given by the conditional probability $p(s' \mid s_n, s_{n-1}, \ldots, s_{n-m+1})$. Hence the full process is characterized by all possible conditional probabilities of this kind. This is a continuous state Markov chain of order m.

The conditional probability densities (cpdf) $p(s' \mid s_n, s_{n-1}, \ldots, s_{n-m+1})$ can be estimated from time series data under the assumption that they are continuous under the condition $\mathbf{s}_n = (s_n, s_{n-1}, \ldots, s_{n-m+1})$, i.e., that the probability $p(s' \mid \mathbf{s}_n)$ remains almost unchanged if one changes $\mathbf{s}_n$ slightly. Then one proceeds as in deterministic cases: For given $\mathbf{s}_n$ one determines the set $\mathcal{U}_n = \{\mathbf{s}_k : |\mathbf{s}_k - \mathbf{s}_n| \leqslant \epsilon\}$ with $k < n$ (for causality). Following the above assumption, the known "futures" s_{k+1} are distributed according to the unknown $p(s' \mid \mathbf{s}_n)$ which can hence be estimated from this finite sample of $\{s_{k+1}\}$ (e.g., as a histogram). The value $\hat{s}_{n+1}$ which minimizes the error with respect to the true future s_{n+1} in the root-mean-squared sense is again the mean of this distribution, which can be estimated directly (i.e., without estimating the cpdf) from the set of $\{s_{k+1}\}$ as

$$\hat{s}_{n+1} = \frac{1}{|\mathcal{U}_n|} \sum_{k:\, \mathbf{s}_k \in \mathcal{U}_n} s_{k+1}\,. \tag{4.16}$$

As a consequence, when we use the locally constant predictor, then we need not think about whether the process is deterministic or stochastic, the algorithm is equally well justified. Also in the stochastic case one will optimize the rms prediction error with respect to m, i.e., one will vary the embedding dimension. We should add that even if the observed process in its true state space is Markovian, it is usually not Markovian when reconstructed through some observable. Therefore, in theory in most situations no finite perfect m exists, but in practice (finite data set, measurement noise) some not too big m is optimal. For very large m, a fixed diameter neighborhood $\mathcal{U}$ will be typically empty, so that no conditional pdf can be estimated, and a nonempty neighborhood must have such a big diameter that the continuity assumption of the cpdf can no longer justify that the set of $\{s_{k+1}\}$ should be distributed according to $p(s' \mid s_n, s_{n-1}, \ldots, s_{n-m+1})$, i.e., we end up with a systematic misestimation of this cpdf. However, such problems will inevitably worsen the out-of-sample prediction error, so that m can be safely optimized by simple variation and calculation of this error.

The Langevin dynamics (4.8) has, through the state-dependent tensor $g(\mathbf{x})$, a state-dependent noise amplitude. Translated into the Markov chain model, this means that the variances of the cpdf can well depend on the state $\mathbf{s}_n$. In fact, as in the deterministic case, the standard deviation of the cpdf is an estimate of the prediction error to be expected. This is illustrated in Fig. 4.3 for the prediction of wind speed measurements by Markov chains. We see a good correlation between the actual differences $|\hat{s}_n - s_n|$ and the standard deviation of the corresponding cpdf. As an alternative, one could easily determine the value δs_{n+1}, so that the true future observation is inside an interval $\hat{s}_{n+1} \pm \delta s_{n+1}$ with, e.g., 90 % probability.

Having determined a probability distribution rather than a single value allows us to perform various other predictions. As an evident example, one can determine the probability that the next measurement will be above or below a certain threshold or that it will be outside some interval. One could also determine the median instead of the mean, or the most probable value, which minimizes other measures of error.

At the end of this section we want to mention a special case of time discrete stochastic models, the linear Gaussian models. They are most generally defined as a linear stochastic process in a vector-valued state space,

$$\mathbf{x}_{n+1} = A\,\mathbf{x}_n + \boldsymbol{\nu}_n \tag{4.17}$$

$$s_{n+1} = C\mathbf{x}_{n+1} + \eta_{n+1} \tag{4.18}$$

with $\boldsymbol{\nu}_n$ and η_n denoting dynamical and measurement noise, respectively, and $\mathbf{x}_n$ a hidden state. In its state space the process is Markovian. However, as a step closer to realistic situations, one assumes that a time series represents, as before, an observable which is a linear function of the state variables, including some measurement noise. Also in this purely linear setting, the recorded time series does generally no longer represent a Markov process. Two alternative approaches for the analysis and modeling of such data are fully worked out: Either one tries to infer the model in its state space. This is outlined in the literature under the

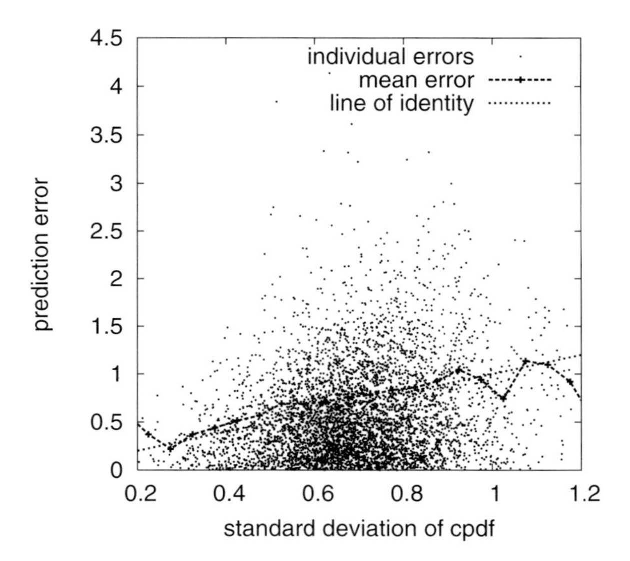

Fig. 4.3: The individual forecast errors obtained from 4000 predictions of surface wind speeds by Markov chains of order 20, and their mean values obtained on vertical stripes, versus the standard deviations of the corresponding cpdfs. As one sees, these standard deviations yield a faithful estimate of the magnitude of the errors to be expected.

keyword *state space models* [21], where the essential tool is the Kalman filter, and which is also related to *hidden Markov models*.

The alternative is modeling the dynamics of the observable s directly. Then the class of models is known as moving average MA(m) and as autoregressive AR(n) models, which generally are more efficient in their combination to ARMA(n, m) models. The MA part is non-Markovian and takes the loss of information about the current state by projection from the state space to the reals into account. For simplicity, we will restrict our discussion here to AR models. An AR(n) model generates the future observation s_{n+1} as a linear combination of the last n time series elements (assuming zero mean of the data set), plus some Gaussian distributed random number

$$s_{n+1} = \sum_{k=1}^{n} a_k s_{n-k+1} + \xi_n \,, \tag{4.19}$$

where ξ_n are independent Gaussian random numbers with zero mean and unit variance. Stability of the output (s_n should remain finite for $n \to \infty$) imposes some constraints on the coefficients a_k [22], which are not necessarily fulfilled if the coefficients are obtained from a fit to data [23]. However, for predictions stability is not an issue, since the model is iterated only once or a few time steps. So for a given time series one finds a_k by writing down the mean-squared error obtained by the predictor $\hat{s}_{n+1} = \sum_{k=1}^{n} a_k s_{n-k+1}$ and minimizes the latter with respect to the parameters a_k. This leads to a set of coupled equations which are linear in a_k and hence can be solved by matrix inversion. So, $\hat{s}_{n+1}$, the deterministic part of the model, is the prediction, and due to the model structure, the true observation (provided the fitted coefficients are a fair estimate of the coefficients used to generate the data) follows a Gaussian distribution with unit variance around this predicted value. So the prediction error is independent of the state of the process. If this is not the case (a cross-correlation analysis between the residuals $\hat{\xi}_n = \hat{s}_{n+1} - s_{n+1}$ and s_{n+1} is recommended), then either the order of the model is not matched or the real process is more complex than linear. From Eq. (4.19) it becomes evident that an AR(m) model is a special case of a continuous state Markov chain of order m, the cpdfs being Gaussians of unit variance centered around the deterministic $s_{n+1} = \sum_{i=1}^{m} a_i s_{n-i+1}$.

What we wrote about multichannel measurements and about more-than-one-step-ahead predictions in the deterministic case applies here as well.

4.4 Events and Classification Error

Another typical prediction task is the prediction of events. In this case, the prediction itself is a classification: either the precise value of the observable is such that we say that the event happens, or it is different and no event is happening. As an illustration, one could record a river level and speak about a flood if the river level exceeds the height of the levees, or we can record a human EEG and speak about an epileptic seizure if the EEG shows a certain signature. So even if as inputs we use some real-valued variables, the prediction is a yes/no classification. This has several implications and difficulties.

The standard approach to optimize predictions is to minimize the prediction error, as we did before. Here, an obvious definition of a prediction error does not exist, and instead one speaks about the classification error. Predicting an event to happen, there could be no event taking place, which is a false alarm. On the other hand, events can take place if they have not been predicted. If events are as rare as they are in fact in many applications (think of earthquakes), a kind of prediction error will be reasonably low if our prediction scheme just promises no event to happen at all. This is evidently a worthless prediction scheme. Instead, our predictions are good if we have a high hit rate (number of hits divided by the total number of events) at a low false alarm rate (number of false alarms divided by the number of nonevents). As a benchmark, an algorithm which generates predictions randomly without any knowledge about the reality with a given rate of predicted events will cause the hit rate to be the same as the false alarm rate. This problem is also sometimes discussed as the issue of sensitivity versus specificity [24]: The predictor should sense that something is coming up (sensitivity), but its prediction should be specific, i.e., it should only predict if the real thing is coming. The statistical tool for the analysis of these properties is the ROC (receiver operating characteristic) statistics [25] and will be outlined below.

If the events to be predicted are rare, also the ROC statistics does not report the success of predictions fairly. As an example, think again of an earthquake: If one were able to predict a major earthquake for a given day, but it takes place a day earlier or later, we would call this a quite precise prediction. In terms of hit rate and false alarms, this one day shift would cause one false alarm and one missed hit, the same as if the prediction had been off by a month. However, if we think that this prediction was targeting a city to be evacuated, a few-day misprediction might already cause a disaster, which illustrates that the tolerance in time which determines whether the prediction is a good or a bad hit is not just governed by the phenomenon to be predicted (dramatic earthquakes in large cities have a return period of centuries, so that even a year of misprediction would not be bad), but also by how we (can) make use of the prediction. And exactly this is the reason why currently no better error measure than the classification error is used.

In the following we will assume that the time series reports the value of some continuous observable s_n, and that an event takes place when its value is inside some specific interval. As an example, think of the water level of a river, where a flood occurs when it exceeds a certain threshold, $s_n > c$.

If a phenomenon has a low-dimensional deterministic time evolution, the best strategy for the prediction of an event would be to predict the real-valued observable $\hat{s}_{n+1}$ and to convert this value into whether the event takes place or not. If the rms prediction error of this prediction is low, then also the events should be fairly well predicted. Nonetheless, the latter prediction will be characterized by the hit rate and the false alarms rate.

In many more situations, the phenomenon will appear stochastic. In this case one should not use the predicted value $\hat{s}_{n+1}$, since for principal reasons the true value s_{n+1} will deviate from the prediction, and this deviation might be large. Therefore, one should study the predicted distribution for the future values and estimate how probable it is that the future value will fulfill the criterion for the event to take place. If the stochastic time series model is an AR model, then the distribution of the future is a Gaussian with known (state-independent) width around the predicted value $\hat{s}_{n+1}$ (compare Eq. (4.19)). If the stochastic time series model is a continuous state Markov chain, then the distribution of the future value depends on the current state in all its details, but it can be estimated from the data in generalization of Eq. (4.16),

$$p(\mathcal{C}) := \int_{\mathcal{C}} p(s \mid \mathbf{s}_n) ds \approx \frac{1}{|\mathcal{U}_\epsilon(\mathbf{s}_n)|} \sum_{k:\, \mathbf{s}_k \in \mathcal{U}_\epsilon(\mathbf{s}_n)} \Xi_{\mathcal{C}}(s_{k+1}), \tag{4.20}$$

where $\Xi_{\mathcal{C}}(x)$ is the index function of the set $\mathcal{C}$, i.e., $\Xi_{\mathcal{C}}(x) = 1$ if $x \in \mathcal{C}$ and 0 else, and $\mathcal{U}_\epsilon(\mathbf{s}_n)$ is the set of ϵ-neighbors of the conditioning vector $\mathbf{s}_n$, i.e., $\mathbf{s}_k \in \mathcal{U}(\mathbf{s}_n)$: $|\mathbf{s}_k - \mathbf{s}_n| < \epsilon$ and $|\mathcal{U}_\epsilon(\mathbf{s}_n)|$ denotes the number of elements in this set. If we choose for $\mathcal{C}$ the bins of a histogram, we can thus obtain an estimate of $p(s \mid \mathbf{s}_n)$. Under the aspect of events, one would split the range of the observable s into a set $\mathcal{C}$ and its complement, defining an event e to take place at time n if $s_n \in \mathcal{C}$. Then Eq. (4.20) applies again and we have

$$\begin{aligned} \hat{p}_{n+1} &= \text{prob(event at time } n+1) = p(e \mid \mathbf{s}_n) \\ &= \frac{\text{number of images } s_{k+1} \text{in } \mathcal{C}}{|\mathcal{U}_\epsilon(\mathbf{s}_n)|} . \end{aligned} \tag{4.21}$$

In summary, probabilistic prediction of events supplies an estimated time dependent probability $0 \leqslant \hat{p}_{n+1} \leqslant 1$ for the event to take place at time $n+1$. Before we ask for the performance of this prediction scheme, we have to think about its validation: How can we verify that the predicted probabilities are meaningful if the future observation to be made will give a yes/no answer? The self-consistency check which is employed in weather predictions is called reliability test. It consists in the construction of suitable subsamples. The first one performs a large number of predictions. Then for all possible values $0 \leqslant r \leqslant 1$ of the predicted probability one constructs a sample $S_r = \{k\colon \hat{p}_{k+1} \in [r, r+\delta r]\}$. For each of these samples one can compute the number of events, i.e., the number n_r of situations where $s_{k+1} \in \mathcal{C}$. If we denote by N_r the total size of the sample S_r, then $f_r = n_r/N_r$ is a number between 0 and 1. Now consistency requires that $f_r \approx r$ within the statistical errors. A systematic deviation of $f_r \approx r$ indicates a systematic misestimation of the predicted probabilities and hence some bias in the algorithm and is therefore a starting point for improvements. However, even if this test is successful, it does not say anything about the performance as a predictor. As an extreme but evident example consider a predictor which gives a time-independent probability for the event to come. Then there is only one nonempty subsample S_r which contains all time series elements. The value for f_r in this case is just the average rate of events. So if the constant prediction is exactly this value, then the predictor passes the test, otherwise not. If in Germany the maximum daily temperature remains below 0 °C on 30 days per year on average, then the prediction that tomorrow the maximum temperature will be below zero with a probability of $\frac{30}{365}$ is correct and consistent, but this is evidently not the optimal prediction since it ignores the existence of the seasons. Hence, passing the reliability test is a necessary property of a probabilistic predictor, but it does not tell anything about its optimality.

As we mentioned before, predicting events is a kind of classification task. In fact, in classical statistics classification is a well-defined problem: One assumes that the system has some well-defined state, and tries to figure out in which state the system is based on observations. When doing predictions we want to predict the state in the future, which requires more sophisticated data analysis (or might even be impossible), but in terms of assessing the performance it is the same statistical problem. And this problem is the issue of the Bayesian equality: From the history we can collect all events, and we can therefore study the conditional probability that a particular state vector $\mathbf{s}$ has been observed given the following event. However, for predictions we need the opposite, the conditional probability of an event to follow given a certain state vector. These two conditional probabilities are related by the Bayesian theorem which follows straightforwardly from multiplying each of the two conditional probabilities by a marginal probability in order to arrive at the same joint probability

$$P(e \mid \mathbf{s}) = P(\mathbf{s} \mid e)\frac{p(e)}{p(\mathbf{s})} . \tag{4.22}$$

The hit rate of our predictor is related to the knowledge of $P(\mathbf{s} \mid e)$, whereas the false alarm rate is related to $(1 - P(e \mid \mathbf{s}))p(\mathbf{s})/(1 - p(e))$. What Eq. (4.22) shows is simply that there is no easy relation between hit rate and false alarm rate, but that instead these are two independent quantities characterizing the combination of predictor and process. Finding the optimal predictor means to find the optimal representation of the state $\mathbf{s}$ such that the conditioning explores all available information about $p(e)$.

So the statistics to study is the false alarm rate versus the hit rate. The probabilistic predictions of an event do not yet generate an alarm at all—above we just outlined how to predict the state-dependent probability of the event to happen. So the predicted probability has to be converted into an alarm. This can be done by introducing a threshold p_c for the predicted probability: If $\hat{p}_{n+1} > p_c$ we predict the event to follow; otherwise the absence of the event is predicted. Depending on the numerical value of p_c, we thus generate more or less warnings. Clearly, a natural threshold value p_c might be such that the total number of warnings is identical to the total number of events, but in the end p_c is really an adjustable parameter which can be used to adapt the prediction scheme to one's needs. Evidently, a low value of p_c leads not only to a good hit rate, but also to a large false alarms rate, whereas for high p_c both rates are low. So depending on whether one wants to avoid false alarms or whether one cannot accept missed hits one can adjust the sensitivity of the predictor through p_c. If the predictor were insensitive to the current state of the system and hence alarms were given without correlation to the true future, the hit rate would be identical to the false alarms rate, which is the benchmark for a null predictor. Only if the hit rate is larger than the false alarms rate the predictor is useful.

The wind speed of surface wind can fluctuate tremendously, as it is quantified by a quantity called turbulence intensity [26]. Strong wind gusts, i.e., the increase of the wind speed within a short time interval, can cause considerable damage, since humans and machinery have no time to adjust themselves against it. Gusts are quantified by the increase of the wind speed in m/s inside a time interval of some s, which is a kind of acceleration (turbulent wind speed is not differentiable), with no commonly used definition. Depending on their magnitude g and the turbulence intensity of the current weather condition, we are discussing here events which occur about 10–500 times a day, in a data set of about 700 000 measurements per day (8 Hz sampling rate). The prediction of such gusts by Markov chains is studied in [27], including the consistency check by reliability tests and the analysis of the performance by the ROC statistics. Statistical analysis of wind data shows that as a simplified model, an AR(1) model with multiplicative noise is not too bad. Apart from the evident nonstationarity, it describes the decay of the autocorrelations and the magnitude of the fluctuations properly, if we choose the following coefficients:

$$v_{n+1} = 0.95v_n + 0.1 \cdot (|v_n| + 0.5) \cdot \xi_n \,, \quad s_n = |v_n| \tag{4.23}$$

(in fact, the observed quantity s_n in our time series is the modulus of the horizontal wind speed $\mathbf{v}$). However, this model cannot correctly generate the higher order statistics, and, more importantly, it cannot reproduce the observed nonstationarities. In the Markov chain approach, the nonstationarity is assumed to be caused by the variation of hidden parameters (such as weather conditions or the main wind direction), which correspond to different parts of a higher dimensional extended state space of the model. The method does not require to investigate these hidden parameters in more detail; one just optimizes the order m of the chain such that there is no detectable memory ignored. The latter can be studied by entropy analysis and is a topic of ongoing work. For our data, a Markov chain of order $m \approx 20$ is the best compromise between memory and statistical accessibility of the conditional pdfs which one has to extract from the past data.

In Fig. 4.4 we show a ROC plot for the prediction of turbulent wind gusts. In this case, we use bivariate input data, namely the wind speeds recorded at 20 m and at 30 m above the ground, in order to predict the gusts at a 20 m altitude. The alarm of such a gust is given on first computing the state (=time)-dependent probability of a gust to come and then to issue the alarm if this probability exceeds p_c. The parameter p_c is altered from 1 to 0 along the curves from the lower-left to the upper-right corner and hence tunes the rate of total alarms given. In this example, the algorithm performs much better than expected. As a specific feature, the better the predictability the stronger the gust to be predicted (the larger g). The ongoing work [28] shows that this is a typical feature in situations where the event is defined by a jump in the signal (which is also the case with stock market crashes or freak waves in the ocean), and part of its origin lies in the fact that the event is not independent of the last observation, since it is defined by the difference of some future values and the current observation.

4.5 Conclusions

In this contribution we tried to present a comprehensive view of data-driven predictions. Let us recall here that the main obstacle in practice is given by nonstationarities in the phenomenon to be predicted, or in insufficient historic data. Both applies in particular when we try to predict rare events. In cases where data-driven methods alone are insufficient, one might wish to incorporate addi-

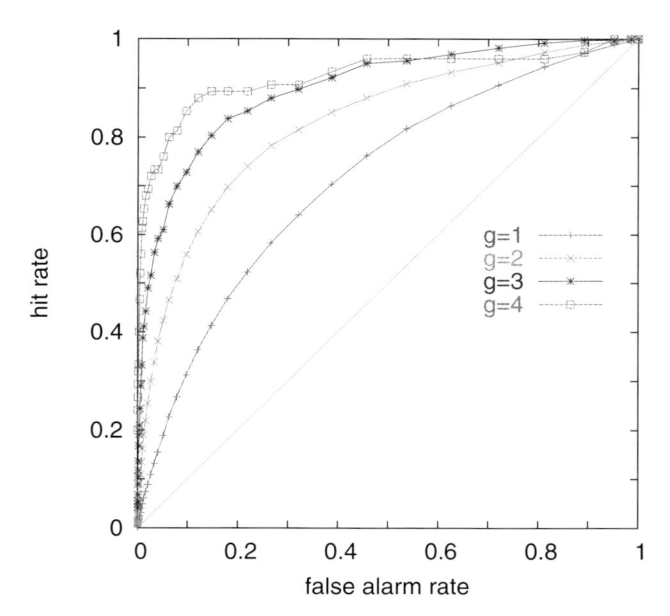

Fig. 4.4: The ROC statistics for forecast of turbulent wind gusts. The different curves show results for different classes of events, namely gusts with different amplitude g. The prediction scheme is explained in the text.

tional knowledge about the studied phenomenon into the time series methods, such as symmetries, principal bounds to the observable, or long-range correlations. It remains a challenge to combine time series methods with such additional constraints, where one possibility is offered by Bayesian statistics and its prior probabilities. Also, one should stress that working successfully with data is not just an issue of methods but also of experience.

References

[1] H. U. Voss, J. Timmer, and J. Kurths. *Int. J. Bif. Chaos 1*, 4:1905, 2004.

[2] F. Takens. *Detecting Strange Attractors in Turbulence*, volume 898 of *Lecture Notes in Math.* Springer, New York, 1981.

[3] T. Sauer, J. Yorke, and M. Casdagli. *J. Stat. Phys.*, 65:579, 1991.

[4] H. D. I. Abarbanel. *Analysis of Observed Chaotic Data.* Springer, New York, 1996.

[5] H. Kantz and T. Schreiber. *Nonlinear Time Series Analysis.* Cambridge University Press, Cambridge, 1997.

[6] J. D. Farmer and J. J. Sidorowich. *Phys. Rev. Lett.*, 59:845, 1987.

[7] M. Casdagli and S. Eubank. *Nonlinear Modeling and Forecasting*. Santa Fe Institute Studies in the Science of Complexity, Proc. Vol. XII. Addison-Wesley, Reading, MA, 1992.

[8] M. Casdagli. *J. R. Stat. Soc. B*, 54:303, 1991.

[9] R. P. Kapsch, H. Kantz, R. Hegger, and M. Diestelhorst. *Intl. J. Bif. Chaos*, 11: 1019, 2001.

[10] D. Broomhead and D. Lowe. *Complex Syst.*, 2:321, 1988.

[11] S. Haykin. *Neural Networks: A Comprehensive Foundation*. Prentice-Hall, Englewood Cliffs, NJ, 2nd edition, 1999.

[12] R. Hegger, H. Kantz, and T. Schreiber. *Chaos*, 9:413–435, 1999. URL `http://www.mpipks-dresden.mpg.de/~tisean`. The TISEAN software package can be downloaded from this url.

[13] H. Kantz and E. Olbrich. *Phys. Lett. A*, 232:63, 1997.

[14] L. Arnold. *Stochastic Differential Equations: Theory and Applications*. Wiley, New York, 1974.

[15] R. Friedrich and J. Peinke. *Phys. Rev. Lett.*, 78:863, 1997.

[16] S. Siegert, R. Friedrich, and J. Peinke. *Phys. Lett. A*, 243:275, 1998.

[17] S. Kriso, J. Peinke, R. Friedrich, and P. Wagner. *Phys. Lett. A*, 299:287, 2002.

[18] J. Gradisek, S. Siegert, R. Friedrich, and I. Grabec. *Phys. Rev. E*, 62:3146, 2000.

[19] F. Paparella, A. Provenzale, L. Smith, C. Taricco, and R. Vio. *Phys. Lett. A*, 235:233, 1997.

[20] M. Ragwitz and H. Kantz. *Phys. Rev. E*, 65:056201, 2002.

[21] P. J. Brockwell and R. A. Davis. *Introduction to Time Series and Forecasting*. Springer, New York, 1996.

[22] G. E. P. Box and G. M. Jenkins. *Time Series Analysis*. Holden-Day, San Francisco, 1976.

[23] W. Press et al. *Numerical Recipes*. Cambridge University Press, Cambridge, 1992.

[24] M. S. Pepe. *The Statistical Evaluation of Medical Tests for Classification and Prediction*. Oxford University Press, Oxford, 2003.

[25] K. Fukunaga. *Introduction to Statistical Pattern Recognition*. Academic Press, San Diego, 1990.

[26] T. Burton, D. Sharpe, N. Jenkins, and E. Bossanyi. *Wind Energy Handbook*. Wiley, New York, 2001.

[27] Holger Kantz, Detlef Holstein, Mario Ragwitz, and Nikolay K. Vitanov. *Physica A*, 342:315, 2004.

[28] S. Hallerberg, E. G. Altmann, D. Holstein, and H. Kantz. *Precursors of Extreme Events*. Preprint, 2006. URL `http://arxiv.org/abs/physics/0604167`.

5 Dealing with Randomness in Biosignals

Patrick Celka, Rolf Vetter, Elly Gysels, and Trevor J. Hine

"If we know the laws of change, we can precalculate in regard to it, and freedom of action thereupon becomes possible. Changes are in the imperceptible tendencies to divergence that, when they have reached a certain point, become visible and bring about transformations."

I Ching, Ta Chuan/The Great Treatise.

Biosignals originate from complex biological tissues' own dynamics and exchange energy with their environment. From an inside biological tissue viewpoint, biosignals intrinsically contain high- dimensional deterministic dynamics superimposed to random fluctuations from their environment. From the outside-tissue view, this may be the other way round. The term *random* conveys the idea of high *uncertainty* and results from the separation of the two worlds, outside and inside tissue, which is purely arbitrary. Thus randomness is a question of viewpoint. This chapter makes clear the concept of randomness and how to best access the information we want to retrieve. Different signal processing methodologies for performing this task are presented: from linear to nonlinear techniques. Applications to real-life signals are provided such as processing electrocardiograms, electroencephalograms, and speech signals.

5.1 Introduction

This chapter aims to introduce the concepts of noise and randomness in systems and signals together with their potentially still controversial origins and use by biological systems. In this regard, we show the first results of the use of randomness by the human visual system. The chapter goes into the details of some signal and system techniques for reducing the effect of undesirable interference with the signals of interest. We also present results and comparison of these techniques on some specific biosignals. The conclusions we can draw from the application of these techniques is that more understanding of the sources and nature of these perturbations, random or not, is required in future scientific and engineering disciplines.

Handbook of Time Series Analysis. Björn Schelter, Matthias Winterhalder, Jens Timmer

ISBN: 3-527-40623-9

5.1.1 Determinism: Does It Exist?

Since the 18th century with Isaac Newton, Gottfried Wilhelm Leibniz, David Hume, and Immanuel Kant to some extent, determinism got a royal position both in the emerging modern sciences and philosophy. Objects of all kinds, small or big, have a position and velocity at a given time, which entirely determines the future of these two quantities by the law of classical dynamical mechanics. Determinism was then synonymous with causality. Some disturbing problems with unexpected peculiar planet orbits were attributed to the hand of God. The 19th century, with the discovery by the mathematician Henry Poincaré and the meteorologist Edward Lorenz of nonperiodical still bounded trajectories as solutions of low-dimensional nonlinear differential equations, completely changed the deterministic picture. These solutions were referred to resulting from deterministic chaos. Further in the 20th century, determinism fell apart with the quantum mechanics description of the world, initiated by Albert Einstein, and Max Planck, and further finalized by Werner Heisenberg, Niels Bohr, Erwin Schrödinger, and Wolfgang Pauli. In quantum mechanics, probability theory plays a major role in describing the subatomic level of matter. Still, the probability field associated with a quanta is deterministically computed either from the Heisenberg matrix or Schrödinger wave formalism. Heisenberg's uncertainty principle puts a definitive end to a fully deterministic view of our universe, both at the micro- and macrolevels, as we shall discuss further in this section. The concept of determinism is nowadays quite unclear and comes with attributes such as weak, strong, or effective.

5.1.2 Randomness: An Illusion?

Classical statistical mechanics describes phenomena *en masse* from which individual component shows erratic behavior. Mathematicians have developed a tool to deal with erratic and unpredictable phenomena known as statistics. Statistics is based on probability theory developed in the 17th century by Blaise Pascal and in the 18th century by Pierre–Simon Laplace, and fully established in its modern form by Andrey Nikolaevich Kolmogorov by mid of the 20th century. There is not just one theory of probability which already shed light on its controversial application to physics and other fields of science [1]. From this perspective, statistics is a way to model the macroscopic evolution of high-dimensional systems where a complete picture of the initial conditions of the system under study cannot be obtained. On a smaller scale, quantum systems already have their inherent indeterminacy by the introduction of the probability wavefunction ψ. One of the dramatic consequences of the introduction of this probability function and quantum operators has been established by Heisenberg: the discovery of the principle that bears his name and known as the uncertainty principle. Combining quantum mechanics of low-dimensional systems and high-dimensional statistics gives us quantum statistics which aims at describing large scale quantum objects. From this, we can deduce that the randomness which appears as a fundamental property of small scale objects and also impacts larger scale objects may lead to the discovery of large scale quantum objects manifesting macroscopic randomness.

We can also infer that chaos in low-dimensional quantum mechanical systems [2] can produce macroscopic effects from small scale quantum fluctuations due to either a nonlinear dynamic properties of the overall system, the so-called Poincaré resonance occurring in Poincaré systems as studied by the Brussels–Austin group [3, 4], or to the Heisenberg uncertainties which prevent us grasping the initial conditions of a system in the state space. Similar macroscopic collective behavior can be obtained from time-varying linear systems at the edge of instability where small microscopic effects could have large macroscopic consequences by a back and forth crossing of the stability region during a small amount of time. Last but not least, at the microscopic level the fourth uncertainty relation in a nonrelativistic view, i.e., $\delta t \delta E \geqslant \hbar$, prevents us understanding what is happening actually at the instant we perform a measurement, i.e., the collapse of the wavefunction. During a small instant δt, the system under study can be in any state and can even violate the laws of conservation of energy and momentum. Indeed, recent results have shown that the classical interpretation of quantum mechanics is at fault when interpreting negative probabilities [5].

Probabilities thus appear from three different sides: (1) as a fundamental property of small or large scale quantum objects, (2) as our lack of means for initial-condition grasping of the macroscopic objects, and (3) as a result of nonlinear dynamics under specific ranges of the system's parameters. From a pragmatic point of view, Edwin Jaynes says: *What we consider to be fully half of probability theory as it is needed in current applications—the principle of assigning probabilities by logical analysis of incomplete information—is not at all present in the Kolmogorov system* [6]. When agreeing with Jaynes, probability theory is a practical way to cope with our ignorance, and not as a necessity due to inherent randomness in nature. This perspective leads to inference theory of *en masse* phenomena which is of practical use for scientists and engineers. However, the intrinsic indeterminism of quantum objects, including our consciousness, leads to a singular view of our world where each single event, leading to conscious or unconscious perception, has to be taken into account. Unfortunately, still no available mathematical theory can handle this view and it thus has to be left aside for the technical part of this chapter.

A natural source of randomness is the α particle emission from the spontaneous disintegration of an unstable radioactive element. We can model how a large number of these elements will behave, but are completely unable to predict when one element will actually produce an α particle. Another example is how brain neurons behave under visual stimuli. We can model the global spatio-temporal behavior to some extent, but are totally unable to predict when and which individual neuron will fire.

For these reasons, probability theory plays a major role in modeling macroscopic phenomena or large ensembles of microscopic ones, but remains silent about predicting individual events. The roots of randomness have to be searched for in the deep structure of matter and its connection with the *process of measurement* and the fourth uncertainty relation $\delta t \delta E \geqslant h$.

5.1.3 Randomness and Noise

We would like to clearly differentiate between the concepts of noise and randomness by starting with the following assertions.

1. Noise is relative to your *knowledge*: i.e., people speaking other languages can be considered as noise in our language.

2. Noise is relative to the type of system and sensors which belong to that system: i.e., anything that the system cannot *interpret* is considered as noise.

3. Noise is relative to the *inside–outside* picture of the system: i.e., where you are in the system. This is directly linked with the concept of semiopen[1] and closed system. In an open system, the input energy can be turned into a useful signal if the inside system can *interpret* the messages or can adapt and learn to do it. By contrast, in a closed system the outside energy will always be foreign to the inside and in that sense not *knowledgeable*: i.e., noise. To give a practical example: the 50 Hz power line electromagnetic field is considered as noise for an electrocardiogram electrode, but is not from the viewpoint of the power line.

What we have shown here is that the concept of noise is essentially relative to where you are measuring, with what sensors and to your knowledge. We thus see that noise as we have introduced has little to do with randomness. Noise can then be defined as the signal that your system's sensors cannot interpret. From a system's point of view, noise is a signal that interferes with its natural behavior in terms of dynamic invariant.

Noise can be a nuisance or potentially a rich source of material from which the system can benefit in terms of efficiency (see point 3). In particular, this latter effect will be described in Section 5.2.2. Noise as a nuisance is the particular focus of the rest of the chapter, and is introduced in Section 5.3. We will describe some techniques which can reduce the nuisance for the further analysis of the system's natural properties. In a natural environment such as those of biological systems, the nuisance is more often nonstationary. We will thus focus in Section 5.4 on some techniques which can cope with the nonstationary nature of the signals.

The chapter is organized as follows. Section 5.2 browses the concepts of randomness in biological systems, Section 5.3 presents strategies developed by scientists and engineers to cope with randomness in signals and systems, Section 5.4

[1] A semiopen system is one that allows energy and/or information to come inside or outside with a given transfer ratio, like a semitransparent mirror.

elaborates on different techniques for reducing the effect of randomness, and Section 5.5 presents different applications of these techniques to biosignals. From the experience we obtain from the real-life applications, Section 5.6 tends to raise issues linked with the use of probabilities in analyzing signals from living systems and concludes the chapter.

5.2 How Do Biological Systems Cope with or Use Randomness?

5.2.1 Uncertainty Principle in Biology

As pointed out in the introduction, the uncertainty principles as discovered by Heisenberg and further developed by Bohr and de Broglie [7], have puzzled scientists for years, and their interpretations are still controversial and a matter of renewed research [8]. Basically, the uncertainties can be understood as either originating from the measurement process that disturbs the system under study and prevents measurements of conjugated variables, or as originating in the statistical interpretation of the wavefunction ψ of conjugated variables, namely the spread of ψ. Note that the second interpretation is valid before and after the measurement, and do not need the measurement to exist.

Biologists and geneticists, while studying small scale systems such as DNA molecules, and their role in the creation and maintenance of life, have been looking at *chance phenomena* in further details [9, 10]. Actually, it is thought that the *chance phenomena* in biology as manifested, for instance, by the random but still useful appearance of the four acid basis—T-G-C-A—in DNA, are somehow linked with quantum mechanical effects and the uncertainty principle [9, 11, 12]. At the larger scale of the neuron, single photon experiments in the visual system of vertebrate and invertebrates paved the way to quantum effects in the visual pathways [13]. In this situation, the uncertainty principle in biology is viewed as the engine for a creative process. Thus in this case the quantum probability wave has the potential for increasing the negentropy. By contrast randomness by itself generates entropy. Thus, the interplay of *chance phenomena* and the quantum wave probability function has the *potentia*, à la Descartes, for creation in the sense of increase of negentropy.

The physicist H. Stapp claims that the *act of measurement* as perceived by the person's consciousness is what he refers to a *Heisenberg event* [14]. An Heisenberg event is nothing less than the actualization of a large scale quantum structure. When this happens in the brain neuronal network this leads to a *discrete* conscious event. The suite of these discrete conscious events carries the process of knowledgeability. Randomness appears completely alien in this picture, because a Heisenberg event brings *instantaneous* knowledge which immediately collides with the unknownness of randomness. Note that the instantaneous nature of the

Heisenberg event contradicts the fourth uncertainty principle because this would require an infinite amount of energy.

Thus we conclude that quantum theory applied to biology has the *potentia* to bring innovative ideas about life processes, and that moreover quantum constrained randomness plays the crucial role of an engine for knowledge.

5.2.2 Stochastic Resonance in Biology

In the previous section, we have discovered that randomness at a nanoscopic scale can in fact be useful for biological systems. At a macroscopic scale, a phenomena called stochastic resonance (SR) has been shown to have very useful effect on the performance and robustness of biological systems such as the electrodetection in the paddlefish [15]. SR was discovered in the 1980s and modeled in the context of nonlinear systems, either static or dynamic. SR has also been recently discovered in quantum mechanical systems and seems to be quite common in both classical and quantum systems in the presence of bistability [16]. Biological applications of SR are now vast and a review can be found in [17]. SR is a global effect that improves the signal detection sensitivity of nonlinear systems by using random internal or external excitation signals, see Section 5.4.1 for a definition of dynamic noise or what is called here internal noise: the resonance parameter is the noise level of the input signal. SR has also been shown to play an important role in phase synchronization between nonlinear dynamic systems [18]. In biological neurons, the neurotransmitters release *quanta* of information by nerve terminals in order to activate the opening and closing of ion channel gates. This process appears to be random, and provides the postsynaptic potential with additional variability which is further used to improve signal communication in neuronal assemblies. The randomness provides some additional richness to the pure deterministic or randomless neuron. The nature of this randomness is still nevertheless an open question and may rely on quantum effects [19] as proposed earlier in this section.

While the previously described SR effect appears at a low level such as a single neuron, SR effects have also been reported in macroscopic systems such as neuronal assemblies and the retinal cells [20]. At the highest level of psychophysic perception experiments, SR has also been shown to have some impact [17]. In fact, recent evidence has come to light implicating a random process in increasing sensitivity to relative motion perception. There exists an illusion—the jitter aftereffect (JAE)—where adaptation takes place to a visual dynamic random noise pattern like the "snow" on an untuned television [21–23]. After such adaptation, the visual system becomes hypersensitive to motion. The result of this is the JAE: movement of the image on the retina due to the ever present small fixational eye movements is revealed and the unadapted parts' image seems to move relative to the background. Movement of the retinal image due to these fixational eye movements is normally never seen. This adaptation process is not related to the

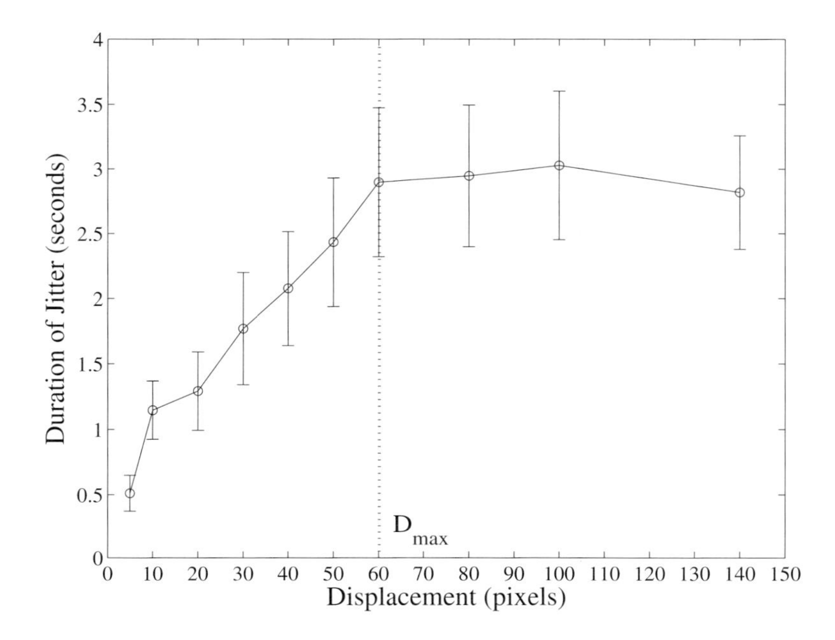

Fig. 5.1: JAE as a function of random movement displacement during adaptation, measured in pixels. The randomly textured background pattern is displaced 32 times a second. D_{max} is the point at which no motion signal exists and the viewer sees random noise. Error bars are ±1 SEM for eight observers (Hine and Dunn, private communication).

presence of a motion signal in the adaptation noise pattern, but conversely, is determined by the presence of random, uncorrelated noise. Hine and Dunn (submitted) have recently completed psychophysical experiments demonstrating this and results are given in Fig. 5.1. Here, the postadaptation sensitivity as measured by the duration of the JAE increases as a function of the level of perceived incoherence in the adapting signal. This perceived incoherence is related to the D, the amplitude of the displacement of the randomly textured background from frame-to-frame. For $D < D_{max}$ (equivalent to about 0.8 deg of visual angle, dotted line on Fig. 5.1), as D is increased the background looks as though it is moving randomly around like a "jitterbug" with increasing energy and incoherence. At $D > D_{max}$, no motion is seen at all, only incoherent noise, and at this point the JAE both peaks and plateaus.

SR thus seems to be a well-spread phenomenon across disciplines boundaries and at every scales of time and space. The noise-induced order of SR is quite counter-intuitive but certainly makes a great use of randomness in nonlinear systems. This phenomenon is undoubtedly one promising landmark for future search for the use of randomness in nature, and eventually the understanding of micro- to macro properties of matter and life.

5.3 How Do Scientists and Engineers Cope with Randomness and Noise?

According to our definition of noise from Section 5.1, when designing instruments, engineers and scientists have to decide on what is knowledgeable and what is not, and on the type of information they want to access. Note that the instrument we are talking about here can be a physical probe or measuring device, but it can also be an algorithm. In doing so, we define the inside–outside parts of our instruments, and also define what we can call the signal or information space E_o and the noise space E_e, such that the observed signal $\mathbf{x}$ is in the space $E = E_o \cap E_e$ and is composed of the signal component $\mathbf{s}$ and the noise component $\mathbf{e}$. The signal space E_o contains the information $\mathbf{s}$ we want to access, retrieve, compress, crypt, or transmit, and the noise space E_e contains the unknown part of the signal $\mathbf{x}$. We still have to define what is the kind of space we are talking about. The space is, generally speaking, an abstract parameter domain where the signal *or* noise information is embedded. Indeed, if we know what we do not know, i.e., the noise, we know the complementary part which is the information. The most well-known spaces are: time, frequency, position, and momentum. But there are many other possible spaces such as the state space of a dynamical system, the color space of an image, the auto-regressive moving-average space of a time series model, the spin or energy space of a particle, the principal component space, the wavelet space, etc. Figure 5.2 shows a schematic view of the two different spaces E_o and E_e. An observed signal $\mathbf{x}$ belongs to the space $E_o \cap E_e$. Its position inside that space determines the amount of information and noise: i.e., the so-called signal-to-noise ratio (SNR). The SNR can be computed by the ratio of the shortest path length between $\mathbf{x}$ and $\mathbf{e}$, and the shortest path length between $\mathbf{x}$ and $\mathbf{s}$. The task of extracting the information is thus to move $\mathbf{x}$ along a given path P such that it eventually reaches the boundary $\partial(E_o \cap E_e)_e$. If $\mathbf{s}$ belongs to $\partial(E_o \cap E_e)_e$, then we have completely reduced the noise. Sometimes, the signal of interest lies in $E_o \setminus E_e$ and we will never be able to reduce the noise to zero. Different noise reduction techniques will be represented by different paths leading to different locations more or less close to $\partial(E_o \cap E_e)_e$ and thus different performances in terms of noise reduction.

It is a fundamental aspect of the design of the instrument to define the most appropriate space E for performing the information extraction. For instance, in speech processing, the most common space is the log-transformed auto-regressive space known as linear predictive coefficients while others exist (see Section 5.5.3). In heart rate variability studies, the frequency space is well known to be useful in characterizing the so-called sympathovagal balance while others also exist (see Section 5.5.1). When designing our instrument, we are sometimes limited to the type of space we can access, and thus we may use space transformations T to map our signals into the appropriate space $\tilde{E}$. More advanced techniques use a subspace decomposition operation H_k, with $k = 1, \ldots, M$, of

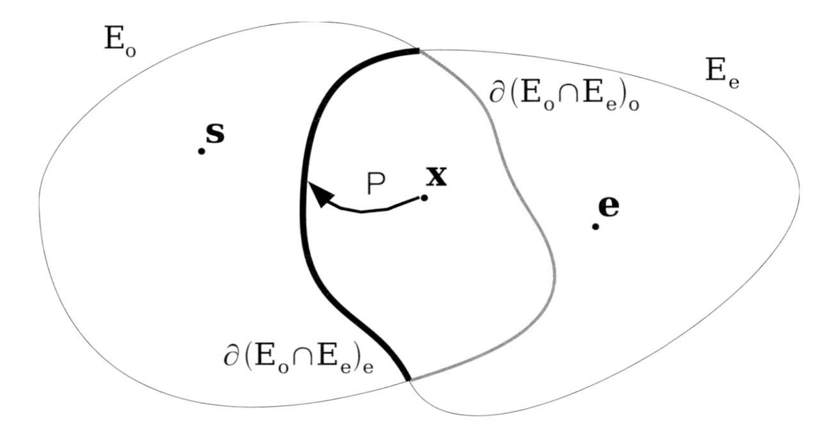

Fig. 5.2: Schematic view of the two spaces E_o and E_e

the original space $E = \cup_{k=1}^{M} E_k$ before applying specific transformations T_k into the space $\tilde{E}_k$. Typical examples of such a subspace decomposition is a filter bank when E is the frequency space. We will address different subspace decompositions in Section 5.4.1 and 5.4.2. Once the signal $\mathbf{x}$ is decomposed by H_k and transformed by T_k, we obtain $\tilde{\mathbf{x}}_k$, we actually perform the noise reduction method of our choice by applying a linear or nonlinear projector $\tilde{P}_k$ onto the subspace $\tilde{E}_{c_k}$ which is supposed to move the signal $\mathbf{x}_k$ closer to the boundary $\partial\left((E_o)_k \cap (E_e)_k\right)_e$; more precisely move the signal $\tilde{\mathbf{x}}_k$ closer to the boundary $\partial\left((\tilde{E}_0)_k \cap (\tilde{E}_e)_k\right)_e$. Finally, we perform the inverse transform T_k^{-1} and inverse subspace decomposition G_k, which reconstruct an estimate of the information signal $\hat{\mathbf{s}} \approx \mathbf{s}$. The full scheme is illustrated in Fig. 5.3 and summarized by the following set of equations:

$$\tilde{\mathbf{x}}_k = (T_k \circ H_k)[\mathbf{x}] \tag{5.1}$$

$$\tilde{\mathbf{x}}_{c_k} = \tilde{P}_k[\tilde{\mathbf{x}}_k] \tag{5.2}$$

$$\hat{\mathbf{s}}_k = (G_k \circ T_k^{-1})[\tilde{\mathbf{x}}_{c_k}] \tag{5.3}$$

with the perfect reconstruction property $H_k \circ G_k = 1$ ($\circ$ denotes the operator composition rule). In Fig. 5.3, the space $\tilde{E}_{c_k}$ or E_{c_k} is a copy of the space $\tilde{E}_k$ or E_k which is meant for the "cleaned" space, which is in fact a subspace of $\tilde{E}_k$ or E_k respectively. We usually assume that the noise is additive to the information signal, such that we have the following model:

$$\mathbf{x} = \mathbf{s} + \mathbf{e}, \tag{5.4}$$

where $\mathbf{e}$ can be a noise source originating from inside the system, the so-called internal noise or dynamical noise, or a noise source originating from outside the

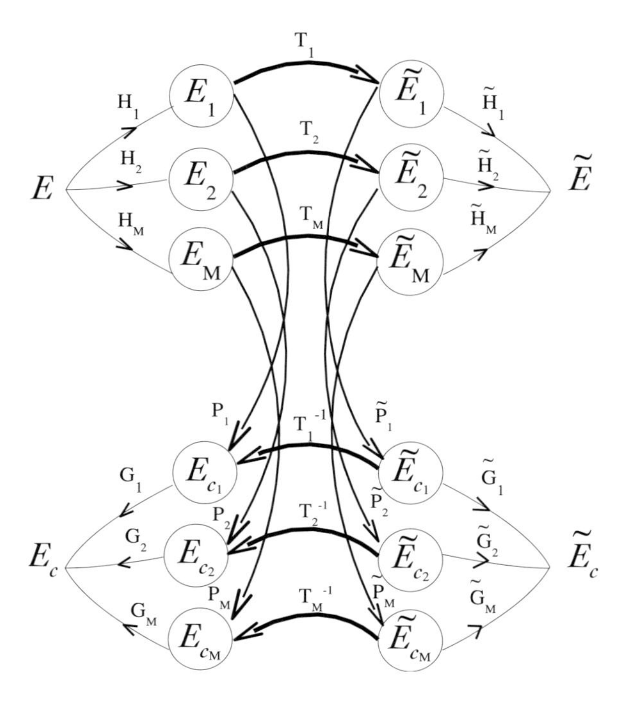

Fig. 5.3: The generic scheme of noise reduction techniques using subspace projection.

system under inspection, the so-called measurement noise (see Section 5.4.1). In this situation the reconstructed signal is computed as

$$\hat{\mathbf{s}} = \sum_{k=1}^{M} \hat{\mathbf{s}}_k \,. \tag{5.5}$$

Finally, we would like to point out that the transformation T and T^{-1} have to have some *smoothness* properties to guarantee the preservation of topological properties of the space E when mapped into $\tilde{E}$, such as the intrinsic *dynamical invariants* of the system (see Section 5.4.2 for more details).

In the following sections, we will deal only with time series signals and thus the primary parameter space is the time either continuous $t \in \mathbb{R}$ or discrete $t_k = k/F_s$ where F_s is the sampling frequency, and use the notation $\mathbf{x}(k) = \mathbf{x}(k/F_s)$. While numerous subspace mapping H_k and transformations T do exist, we will concentrate in the chapter on principal component analysis for which we have $T = I$ and H_k are the band pass filters associated with each principal components' eigenvectors (see Section 5.4.1).

5.4 A Selection of Coping Approaches

5.4.1 Global State-Space Principal Component Analysis

Principal component analysis (PCA) has been widely used by scientists and engineers to analyze and extract features from N_T multidimensional random signals $\mathbf{x}(k) \in \mathbb{R}^{n_s}$ [24] where $k = 1, \ldots, N_T$. PCA has been developed independently in early 20th century by the mathematicians Karl Pearson and Harold Hotelling. PCA enables us to linearly expand the n_s-dimensional space spanned by some eigenvectors $\mathbf{\Phi}_k$ by the sole use of the set of linearly correlated multidimensional signals. These eigenvectors can be further sorted by their decreasing length. Most of the time it is assumed that the largest eigenvectors contain most of the information, but this is a mistaken view as we will see below in this chapter. Moreover, this is done without any *a priori* knowledge on the data statistics, while it is known to be optimal for Gaussian distributed signals. The eigenvectors span the space $\mathbb{R}^{n_s}$ in the following way:

$$\mathbf{x}(k) = P_E \mathbf{p}(k) \tag{5.6}$$

with

$$P_E = [\mathbf{\Phi}_1 \cdots \mathbf{\Phi}_{n_s}], \tag{5.7}$$

where the rows of the matrix P_E contain the eigenvectors $\mathbf{\Phi}_k$. The vectors $\mathbf{p}(k)$ are called the principal components, which can be gathered into a matrix $P_C = [\mathbf{p}(1) \cdots \mathbf{p}(N_T)]$ (see Eq. (5.23)). When the vectors $\mathbf{\Phi}_k$ form an orthonormal basis in $\mathbb{R}^{n_s}$, the vectors $\mathbf{p}(k)$ are orthogonal to each other, which in a statistical sense means uncorrelated. The principal components are expressed as a linear transformation of the data

$$\mathbf{p}(k) = P_E^T \mathbf{x}(k), \tag{5.8}$$

where T is the transpose operation. PCA offers the great advantage to be model independent, i.e., data driven, as compared to Fourier- or Wavelet-based approaches which are both linear expansions of some predetermined eigenvectors. Indeed, in PCA, the eigenvectors $\mathbf{\Phi}_k$ are computed only from the knowledge of the data, while the Fourier or wavelet approaches impose a model for these eigenvectors: i.e., complex exponentials for Fourier and the so-called wavelets for Wavelet. We will show next how these transformations can be useful when one wants to separate the information from the noninformation bearing spaces spanned by the eigenvectors. Here one has to keep in mind that the determination of the information space is purely application dependent. Adding to that picture the information space is sometimes not known *a priori* and thus each eigenvector has the same *a priori* probability to be chosen for building the information space.

PCA can in turn be used for signals which are supposedly obtained by some measurement on a dynamical system S. In this situation the vectors $\mathbf{x}(k) \in \mathbb{R}^{n_s}$ represent the measurements on S. A signal as measured on S is a perturbed so-called observable. An observable is a smooth map from the manifold M of dimension m, where the system's trajectories lie, to the set of real numbers $\mathbb{R}$. In order to recover the underlying dynamic law from the measurements, we have to perform somehow an inverse mapping that brings us back to M or a $\mathbb{R}^m$ which is diffeomorphically equivalent. This inverse mapping is called an embedding and we recall hereafter some basic elements for its construction.

PCA can be seen as a uniform filter bank designed from the data: the eigenvectors $\mathbf{\Phi}_i$ being the coefficients of finite impulse response filter i in the ith subband. This can easily be seen from Eq. (5.23), where the trajectory matrix Z is multiplied by the projector matrix P_E which acts as a filter-bank convolution matrix. Principal component filter bank has been shown to be optimal at minimizing the mean square reconstruction error [25] and is thus of great interest for coding and noise reduction.

Embedding

In this section, we recall some basic assumptions about dynamical systems and state the main theorems concerning the embedding theory. Let M be a compact manifold of dimension m. A dynamical system on M is defined by a vector field generated by f in the continuous time case (usually, we call f the vector field). The dynamical system is then represented by

$$\mathbb{R} \to M\colon t \to \mathbf{s}(t) = f^t(\mathbf{s}_0), \tag{5.9}$$

where $\mathbf{s}_0 = \mathbf{s}(t_0)$ is the initial condition. The trajectory is noted $\{(t, \mathbf{s}(t))\}_{t=t_0}^{T}$ and the orbit $f^t(\mathbf{s}_0)$. The flow on M is defined as f^t, and the dynamics is often described by a set of differential equations $d\mathbf{s}(t)/dt = f(\mathbf{s}(t))$ and an initial condition.

As the dynamical system evolves on M, trajectories are trapped in a subset $\mathcal{A}$ of M which is often of smaller noninteger dimension $d \leqslant m$ (fractal set). Any point $\mathbf{s}$ on M that belongs to a trajectory of the dynamical system is called a *state*. The set $\mathcal{A}$ is called an *attracting set* for the dynamical system, or simply an attractor, under some conditions [26]. From a typical trajectory on M we define an observable as a smooth function $\nu\colon M \to \mathcal{R}^l\colon \mathbf{s} \to \nu(\mathbf{s})$. The function ν can be nonlinear.

An observable maps any point of the manifold M onto real values that may be acquired during experimental measurements. Observe the difference between an observable and measurements related to this observable. Essentially, the difference comes from the measurement process and all the perturbations linked with it. Usually $l \leqslant m$, and l denotes the number of available signals. In most of the situations $l = 1$, and forthcoming embedding theorems hold in that case. Let us assume $l = 1$. We have the following problem:

Problem 5.1. *For some dynamical system with a flow f^t, given an observable ν, a flow sampling time τ, and a corresponding time ordered set of points in $\mathbb{R}$, i.e., $\{\nu(\mathbf{s}), \nu(f^\tau(\mathbf{s})),$*

$\nu(f^{2\tau}(\mathbf{s})),\ldots\}$, *how could we have information about the original dynamical system from this observable.*

The time τ may be considered as a sampling time interval at which the data from the flow are observed. This flow sampling time is not specified at that time. We are now ready to define an embedding:

Definition 5.1. An embedding is a smooth map $\Phi\colon M \to U$ such that $\Phi(M) \subset U$ and Φ is a diffeomorphism between M and $\Phi(M)$.

The dimension n_s of the embedding space U is obviously greater than m. Due to the diffeomorphic property of the map Φ, the flow in the space U will be equivalent to the flow in M and all the properties of the flow f on M will be conserved in the reconstructed state space. The problem is finally to determine the map Φ and the dimension n_s. In the problem statement, problem 5.1, the time τ was not specified and can be arbitrarily chosen. This time interval was called the flow sampling time for obvious reasons. As we do not have access to this parameter, we cannot set a correspondence between the time series sampling time dt and τ. Moreover, as we will see in the next section, the existence of measurement noise forces us to introduce this delay. Recall that we have already a part of the map Φ: the observable ν. It remains to build a map h from $\mathbb{R}$ to U.

Definition 5.2. From a time ordered set of points $V = \big(x(0), x(1), \ldots\big)$ where $x(k) \in \mathbb{R}$, we define the function

$$h\colon V \to \mathbb{R}^{n_s}: x(k) \to \mathbf{x}(k) = \Big(x(k), x(k-J), x(k-2J), \ldots, x\big(k-(n_s-1)J\big)\Big)^T .$$

The time lag $J\,dt$ is known as the embedding delay. From the observable ν and the corresponding time series $\mathbf{x}$ introduced above, the embedding theorem is [27]:

Theorem 5.1. *Let M be a smooth compact manifold of dimension m, f a vector field on M, ν an observable $\nu\colon M \to \mathbb{R}$, and h the map $h\colon \mathbb{R} \to \mathbb{R}^{n_s}$ as defined in 5.2. The map $\Phi = h \circ \nu$ is an embedding of M into $\mathbb{R}^{n_s}$ if $n_s \geqslant 2m+1$.*

The embedding theorem provides us with the matrix of reconstructed *state-space* vectors

$$Z = [\mathbf{x}(1) \cdots \mathbf{x}(N_T)] \tag{5.10}$$

which is called a state-space trajectory matrix. If the time series is long enough, i.e., $N \gg J$, it is preferable to cut the embedded vector time series to the first $N_T = N - (n_s - 1)J$ samples. Practically, the observable ν should be measured. Distortions and noise due to measurement are thus introduced. Both of these perturbations are included in a *measurement noise* signal $e_m(t)$. The noise signal $e_m(t)$ is considered as a random perturbation whose samples are independent of the samples in $\mathbf{x}$: $E\{x(t)e_m(t')\} = 0\ \forall t, t'$. Note that an additional random perturbation $\mathbf{e}_d$ coming from the dynamical system itself may exist. It is called the *dynamical noise*. When we take into account all these effects we obtain from an observable v, and two random variable $(\mathbf{e}_d, e_m)$, the measurement map

$$u: M \times \mathbb{R} \to \mathbb{R}: (\mathbf{s}, e_m) \to u(\mathbf{s}, e_m) = v(\mathbf{s}) + e_m, \tag{5.11}$$

where

$$\frac{d\mathbf{s}(t)}{dt} = f(\mathbf{s}(t)) + \mathbf{e}_d(t). \tag{5.12}$$

This measurement map naturally induces a perturbed time series $\mathbf{x}$. It turns out finally that the theoretical **Problem 1**, has been transformed into the following

Problem 5.2. *Given a time series* $\mathbf{x}$ *and a sampling time* dt, *how could we extract information about the original dynamical system from* $\mathbf{x}$?

A good review of how to find the embedding delay J can be found in [28].

State-Space Principal Component Analysis

For discrete time signals, global PCA (GPCA) transforms can be expressed from Eq. (5.8) as a linear weighted sum of linearly independent vectors $\mathbf{\Phi}_k$. We assume $J = 1$ throughout this section for the sake of simplicity but the case $J > 1$ can be handled with few careful mathematical steps. The core of PCA is based on a close inspection of the eigenvectors and associated eigenvalues computed from the $n_s \times n_s$ covariance matrix C_{zz} of the trajectory matrix Z. The trajectory matrix may be viewed as a cloud of points in $\mathcal{R}^{n_s}$ which may be approximated by an n_o-dimensional ellipsoid. The n_o principal axis of this ellipsoid is given by the eigenvectors of the matrix C_{zz} corresponding to the n_o largest eigenvalues where

$$C_{zz} = \frac{1}{N_T} ZZ^T \tag{5.13}$$

from which the eigenvalues and vectors are given by

$$C_{zz}\, \mathbf{\Phi}_i = \lambda_i \mathbf{\Phi}_i \quad i = 1, \ldots, n_s\,. \tag{5.14}$$

The eigenvalues are rank ordered $\lambda_1 \geqslant \cdots \lambda_{n_s}$, and $D_\lambda = \text{diag}(\lambda_1, \ldots, \lambda_{n_s})$ is a diagonal matrix. These eigenvalues and eigenvectors play a key role in this singular spectrum analysis (SSA) as we will discover later on. The matrix C_{zz} of dimension $n_s \times n_s$ is the covariance matrix of $\mathbf{x}(k)$, averaged over the entire trajectory. Eigenvectors of C_{zz} formed a basis in the space $\mathcal{R}^{n_s}$. The space $\mathcal{R}^{n_s}$ is split into two orthogonal spaces: $\mathcal{R}^{n_s} = \mathcal{R}^{n_o} \times \mathcal{R}^{n_e = n_s - n_o}$. The n_e-dimensional space E_e is considered as the *noise space* while the n_o-dimensional space E_o is denoted as the *signal space*[2], i.e.,

$$\begin{aligned} E_o &= \text{Span}\{\mathbf{\Phi}_1, \ldots, \mathbf{\Phi}_{n_o}\} \\ E_e &= \text{Span}\{\mathbf{\Phi}_{n_o+1}, \ldots, \mathbf{\Phi}_{n_s}\}. \end{aligned} \tag{5.15}$$

[2] This decomposition can be used for noise cleaning.

The trajectory matrix can be projected on E_o and E_e. Resulting vectors are called principal components (P_C) and noise components (N_C) respectively. This decomposition method is close to the *principal component analysis* (PCA) [29]. Indeed, if we note the projection operators on E_o respectively E_e by P_{E_o} and P_{E_e}, with

$$P_{E_o} = [\mathbf{\Phi}_1, \ldots, \mathbf{\Phi}_{n_o}] \tag{5.16}$$

$$P_{E_e} = [\mathbf{\Phi}_{n_o+1}, \ldots, \mathbf{\Phi}_{n_s}] \tag{5.17}$$

Eq. (5.14) can be rewritten as

$$C_{zz}[P_{E_o} P_{E_e}] = C_{zz} P_E = P_E D_\lambda . \tag{5.18}$$

Multiplying the lhs of Eq. (5.18) by P_E^T, using the orthogonality property $P_E P_E^T = I$ of the eigenvectors $\mathbf{\Phi}_i$, and the definition of C_{zz} we obtain

$$(Z^T P_{E_o})^T (Z^T P_{E_o}) = D_{\lambda_o} \tag{5.19}$$

$$(Z^T P_{E_e})^T (Z^T P_{E_e}) = D_{\lambda_e} \tag{5.20}$$

$$(Z^T P_{E_o})^T (Z^T P_{E_e}) = 0 \tag{5.21}$$

$$(Z^T P_{E_e})^T (Z^T P_{E_o}) = 0, \tag{5.22}$$

where $D_\lambda = D_{\lambda_o} + D_{\lambda_e}$ is the natural decomposition of the eigenvalue matrix with respect to E_o and E_e. Equations (5.19) and (5.20) express the fact that the components of the projected trajectories, i.e., the principal components P_C supposedly representing the signal space

$$P_C = P_{E_o}^T Z = [\mathbf{p}(1) \cdots \mathbf{p}(N_T)] \tag{5.23}$$

and the principal components N_C supposedly representing the noise space

$$N_C = P_{E_e}^T Z \tag{5.24}$$

are orthogonal. Equations (5.21) and (5.22) express the fact that the noise and feature spaces are mutually orthogonal. Equations (5.19)–(5.22) also show that λ_i is the mean square value of the projected trajectory on the eigenvector $\mathbf{\Phi}_i$. Therefore, the matrix D_λ effectively provides major information about the trajectory in the state space. Projecting the trajectory matrix onto the E_o space allows us to extract the deterministic part of the process. PCA decomposes the trajectory matrix in the *state-space* domain. Indeed, $\mathbf{\Phi}_i$ are main directions in the state space where the dynamical system trajectories spreads.

A *time* domain counterpart of this method can be applied. Instead of computing C_{zz}, we can directly perform a singular value decomposition of the trajectory matrix to obtain an equivalent space decomposition.[3] Let us introduce the matrix B_{zz}

$$B_{zz} = \frac{1}{N_T} Z^T Z = \frac{1}{N_T} \begin{bmatrix} \mathbf{x}(1)\mathbf{x}(1)^T & \mathbf{x}(1)\mathbf{x}(2)^T & \dots & \mathbf{x}(1)\mathbf{x}(N_T)^T \\ \mathbf{x}(2)\mathbf{x}(1)^T & \mathbf{x}(2)\mathbf{x}(2)^T & \dots & \mathbf{x}(2)\mathbf{x}(N_T)^T \\ \vdots & \vdots & \vdots & \vdots \\ \mathbf{x}(N_T)\mathbf{x}(1)^T & \dots & \dots & \mathbf{x}(N_T)\mathbf{x}(N_T)^T \end{bmatrix} . \quad (5.25)$$

The matrix B_{zz} is of dimension $N_T \times N_T$ and is named the *structure matrix*. B_{zz} has eigenvectors $\boldsymbol{\Psi}_i \in \mathcal{R}^{N_T}$ and corresponding eigenvalues σ_i

$$B_{zz}\, \boldsymbol{\Psi}_i = \sigma_i \boldsymbol{\Psi}_i \quad i = 1, \dots, N_T . \quad (5.26)$$

The matrix B_{zz} is highly degenerated because $\mathrm{rank}(B_{zz}) \leqslant N_T$. Two orthogonal spaces can be constructed

$$\begin{aligned} S_o &= \mathrm{Span}\{\boldsymbol{\Psi}_1, \dots, \boldsymbol{\Psi}_{n_o}\} \\ S_e &= \mathrm{Span}\{\boldsymbol{\Psi}_{n_o+1}, \dots, \boldsymbol{\Psi}_{n_s}\} . \end{aligned} \quad (5.27)$$

There is a relationship between the vectors $\boldsymbol{\Phi}_i$ and $\boldsymbol{\Psi}_i$ given by

$$Z^T \boldsymbol{\Phi}_i = \sigma_i \boldsymbol{\Psi}_i \quad \text{for} \quad i = 1, \dots, n_s \quad (5.28)$$

yielding the singular value decomposition (SVD) of the trajectory matrix Z

$$Z^T = P_S D_\sigma P_E^T , \quad (5.29)$$

where $D_\sigma = \mathrm{diag}(\sigma_1, \dots, \sigma_{n_s})$. The elements of D_σ are called the *singular values* of Z. The projection operators are given by

$$P_{S_o} = [\boldsymbol{\Psi}_1, \dots, \boldsymbol{\Psi}_{n_o}] \quad (5.30)$$

$$P_{S_e} = [\boldsymbol{\Psi}_{n_o+1}, \dots, \boldsymbol{\Psi}_{n_s}] \quad (5.31)$$

$$P_S = [P_{S_o} P_{S_o}] . \quad (5.32)$$

Using Eqs. (5.28) and (5.29) together with the projection operators P_{S_o} and P_{S_e}, we get similar relation as in the spatial analysis

$$(ZP_{S_o})^T (ZP_{S_o}) = D_{\sigma_o} \quad (5.33)$$

$$(ZP_{S_e})^T (ZP_{S_e}) = D_{\sigma_e} \quad (5.34)$$

$$(ZP_{S_o})^T (ZP_{S_e}) = 0 \quad (5.35)$$

$$(ZP_{S_e})^T (ZP_{S_o}) = 0 . \quad (5.36)$$

It is easily seen from Eqs. (5.18) and (5.29) that

$$D_\sigma^2 = D_\lambda . \quad (5.37)$$

[3] One can also use the Karhunen–Loève Transformation (KLT).

Equation (5.37) shows that the singular values of Z are simply the square roots of eigenvalues of C_{zz}. Starting with Eq. (5.29), and using the decomposition of the projection operators in the *signal and noise* spaces we obtain

$$Z = \hat{X} + X_n , \tag{5.38}$$

where

$$\hat{X} = P_{S_o} D_\sigma P_{E_o}^T \tag{5.39}$$

$$X_n = P_{S_e} D_\sigma P_{E_e}^T . \tag{5.40}$$

Moreover, using Eqs. (5.23), (5.24), (5.29), (5.33), and (5.34), we observe that the principal and noise components are also given by

$$P_C = P_{S_o} D_\sigma \tag{5.41}$$

$$N_C = P_{S_e} D_\sigma . \tag{5.42}$$

From Eqs. (5.29), (5.41), and (5.42), we observe that adding P_C to N_C allows us to reconstruct Z and thus the entire noisy signal $\mathbf{x}$. The decomposition Eq. (5.38) allows us, in principle, to distinguish between the deterministic $\hat{X}$ and random X_n parts of the measurement $\mathbf{x}$. Using this observation, we can expect that the first column of $\hat{X}$, noted $\hat{\mathbf{x}}$ will approach the time series $\mathbf{x}$ corresponding to the observable v. Equation (5.39) can be seen as a denoising procedure of the time series $\mathbf{x}$. This is partially true because the noise signal modifies all the singular values of Z, and thus, the matrix $\hat{X}$ is also contaminated by the noise. Several methods are available to estimate the noise contribution to the feature space [30], but we will not investigate this further.

If we are interested in the noise reduction ability of SSA, we must have a closer look at Eq. (5.39). In fact, as explained in the previous paragraph, the first column of the matrix $\hat{X}$ gives an approximation of $\mathbf{x}$. But the same is true for all the other columns apart from an inherent delay and a noise part. The noise contributions are expected to be independent of all the columns. Thus taking the mean value of all the delayed columns will result in a more accurate reconstruction of $\mathbf{x}$.

From a signal processing point of view, we can also interpret Eq. (5.23) as a linear finite impulse response filtering of the data contained in Z. The coefficients of the filters being the vectors $\mathbf{\Phi}_i$. The beauty of this approach relies on the fact that the filter coefficients are determined from the data set and not *a priori* prescribed by the user. From this point of view they are optimally determined. The next section will provide support for these considerations.

Choosing the Best Basis: Minimum Description Length

Constructing a model for the prediction of time series, system identification or noise reduction involves both the selection of a model class and the selection of a model within the selected model class. Successful selection of a model class appears to be a very difficult task without prior information about the time series. The selection of a model inside a class appears to be a more manageable problem. A parametric model, with a parameter vector $\Xi = [\xi_1, \xi_2, \dots, \xi_M]^T$, may be constructed under the assumption that there is a class of conditional probability functions $P(y \mid \Xi)$, each assigning a probability to any possible observed time series or sequence $y = [y(1), y(2), \dots, y(N)]^T$ of N sample points. The parameter vector Ξ is to be estimated to optimize the model. This can be done by maximizing $P(\mathbf{y}|\Xi)$ or its logarithm with respect to Ξ only if we ignore prior information about $P(\Xi)$, which is known as the maximum likelihood (ML) approach

$$\hat{\Xi} = \max_{\Xi}\{\ln[P(y \mid \Xi)]\} . \tag{5.43}$$

It can be shown that the maximum likelihood estimation criterion can also be expressed in coding theoretic terms [31]. For any parameter vector $\hat{\Xi}$, one can assign a binary code sequence to y that is uniquely decodable. The corresponding mean code length of this sequence, which is equal to $L(y \mid \Xi) = -\ln[P(y \mid \Xi)]$, is called its *entropy*. The minimal value of the entropy is attained for $\Xi = \hat{\Xi}$. Hence determining the ML estimate or finding the most efficient encoding of y in a binary code sequence are equivalent tasks. Up to now, the number of parameters k has been supposed to be known. Things are getting more complicated if it is not so, which is by far the most frequent situation. When applying the ML estimator consecutively for all increasing k values, one may end up with as many parameters as sample points, and a very short code sequence for y. But if a binary code sequence for y is constructed for some Ξ, this parameter vector has to be known at the decoder side for successful decoding. A more realistic encoding demands the parameters to be encoded themselves and added to the code sequence. In this case, the total code string length is

$$L(y, \Xi) = L(y \mid \Xi) + L(\Xi) . \tag{5.44}$$

The crucial point in this representation of the total code length is the balance between the code length for the data $L(y \mid \Xi)$ and the code length for the parameter vector $L(\Xi)$. For a rough description of the data with a parsimonious number of parameters, these latter are encoded with a moderate code length, whereas the data need a relatively long code length. However, describing data with a high number of parameters may request a short code length for the data but the price must be paid with a longer code for the parameters. From that point of view, it is reasonable to look for the parameters that minimize the total code length. This gives rise to the *minimum description length* (MDL) of data [32]. When minimizing Eq. (5.44), the general derivation of the MDL leads to

$$\hat{\Xi} = \min_{\Xi}\{-\ln[P(y \mid \Xi)] - \ln[P(\Xi)] - \sum_{j=1}^{M} \ln[\delta_j]\}, \tag{5.45}$$

where[4] $P(\Xi)$ is the probability distribution of the parameters, and δ_j is the precision on the jth parameter. The first term comes from $L(y \mid \Xi)$ (the data), while the other two terms come from $L(\Xi)$ (the parameters). The last term in Eq. (5.45) decreases if a coarser precision is used (larger δ_j) while the first term generally increases. An estimation of the precision coefficients δ_j can be found by solving [33]

$$\left(\frac{\partial^2 \mathbf{Q}}{\partial \Xi^2} \cdot \delta\right)_j = 1/\delta_j \quad \text{with} \quad \mathbf{Q} = L(y \mid \hat{\Xi}), \tag{5.46}$$

where $\delta^T = [\delta_1, \ldots, \delta_M]$. Equation (5.46) comes from the minimization of the log-likelihood function $L(y, \Xi)$ using a second-order approximation of $L(y \mid \Xi)$, i.e.,

$$L(y \mid \Xi) \leqslant L(y \mid \hat{\Xi}) + \delta^T \mathbf{Q} \delta / 2 \,. \tag{5.47}$$

One has to now decide which are the n_o principal eigenvalues. Or in other words, how could we separate the noise space from the feature space? We intend to use here Rissanen's MDL criteria to separate the feature from the noise space in an objective and unsupervised manner. Clearly, the unknown parameter vector is $\Xi = \{\lambda_1, \ldots, \lambda_{n_s}, \mathbf{\Phi}_i, \ldots, \mathbf{\Phi}_{n_s}, \sigma_e^2\}$, where σ_e^2 is the noise power. Assuming that we have the trajectory matrix Z, the likelihood function is

$$P(Z \mid \Xi) = P(\mathbf{x}(1), \ldots, \mathbf{x}(N_T) \mid \Xi) \tag{5.48}$$

$$= \prod_{i=1}^{N_T} \frac{e^{-\mathbf{x}(i)^T \Sigma_{zz}^{-1} \mathbf{x}(i)}}{(2\pi)^{n_s/2} \det(\Sigma_{zz})^{1/2}} \tag{5.49}$$

with Σ_{zz} being the covariance matrix of the vectors $\mathbf{x}(i)$, and C_{zz} defined as Eq. (5.13) being the estimated covariance matrix of the embedded data: $E[C_{zz}] = \Sigma_{zz}$. The trace term may disappear in the minimization of the MDL if we use the following C_{zz} estimator $(n_s - 1)/n_s C_{zz}$. We have assumed that the vectors $\mathbf{x}(i)$ are zero meaned. From Eq. (5.48), the log-likelihood function is given by

$$L(Z|\Xi) = -\ln[P(Z \mid \Xi)] \tag{5.50}$$

$$\approx -\frac{N_T}{2}\left(\ln[\det(\Sigma_{zz})] + \mathrm{tr}(\Sigma_{zz}^{-1} C_{zz})\right) \,. \tag{5.51}$$

It can be shown that maximizing the log-likelihood function provides a vector $\hat{\Xi}$, and following Eq. (5.45) we obtain the MDL criteria for our problem [33, 34]

[4] The requirement that the code length should be an integer is ignored.

$$\mathrm{MDL}(l) = -N_T(n_s - l)\ln\left[\frac{\prod_{i=l+1}^{n_s}\lambda_i^{1/(n_s-l)}}{\frac{1}{n_s-l}\sum\limits_{i=l+1}^{n_s}\lambda_i}\right] + \left(\tfrac{1}{2} + \ln[\gamma]\right)(l+1) - \sum_{j=1}^{l}\ln[\delta_j], \quad (5.52)$$

where l is the number of eigenvectors $\boldsymbol{\Phi}_i$ taken into account in the expansion. The code length is taken to be $\gamma = 32$ which corresponds to a floating point representation. The δ_j are computed with Eq. (5.46) where Q is the covariance matrix C_{zz}. Note that using $Q = C_{zz}$ we implicitly take into account the precision of only the singular values. One can also use a simplified version of Eq. (5.52) for large data sets

$$\mathrm{MDL}_S(l) = -\left(N_T(n_s - l)\right)\ln\left[\frac{\prod_{i=l+1}^{n_s}\lambda_i^{1/(n_s-l)}}{\frac{1}{n_s-l}\sum\limits_{i=l+1}^{n_s}\lambda_i}\right] + n_f(l)\ln[N_T]\,. \quad (5.53)$$

The number of free adjustable parameters is $n_f(l)$. As soon as we have obtained the decomposition Eq. (5.29), we can proceed to the analysis of D_λ. Adding eigenvectors by eigenvectors, we rebuild Z progressively. After l steps, we compute $\mathrm{MDL}(l)$. The result is finally a representation of various processes in z: deterministic + stochastic. The number of free parameters is computed by counting the number of parameters in Ξ and subtracting the number of constraints linked with the orthonormal conditions on the eigenvectors $\boldsymbol{\Phi}_i$. This leads to [35]

$$n_f(l) = (2n_s l + l + 1) - 2l - l(l-1) = l(2n_s - l) + 1\,. \quad (5.54)$$

Finally the model order n_o should satisfy

$$n_o = \min_l \mathrm{MDL}(l)\,. \quad (5.55)$$

The model order n_o is also related to the *statistical dimension* introduced by Vautard [35, 36], and it has been shown that there is no simple relation between n_o and the dimension of the dynamical system. Note that we refer to L_2 norm MDL criteria, Eqs. (5.52) and (5.53) and to the L_1 norm MDL if instead of D_λ we use D_σ.

We conclude this section with some remarks. We have developed a model order selection procedure based on Rissanen's criteria applied to PCA decomposition of measurements. This theory assumes the existence of a feature and a noise space: i.e., the dimension of the noise space is different from zero. Thus, applying this theory on noiseless data sets would lead to some order overdetermination or even to indeterminacies. To overcome this drawback, we can introduce artificially some noise as explained later. Singular spectrum analysis works preferably for quasiperiodic motions displaying strong frequency components.

5.4.2 Local State-Space Principal Component Analysis

Local principal component analysis (LPCA) has been often applied by the engineering community to perform noise reduction or bring further insights on the behavior of observed dynamical systems [37–39]. Herein, we focus on LPCA and its application to noise reduction. The basic idea of these approaches is to observe the data locally in a large n_s-dimensional space of delayed coordinates. Since noise is assumed to be random, it extends approximately in a uniform manner to all the directions of this space. In contrast, the dynamics of the deterministic system underlying the signal confines its trajectories to a lower dimensional subspace of dimension $n_0 < n_s$. Consequently, the eigenspace of the noisy signal is partitioned into a noise and a signal-plus-noise subspace and noise reduction is performed by projecting the noisy data onto the signal-plus-noise subspace. The main problem of these algorithms with respect to real-world applications is the optimal choice of the different parameters. Indeed, the number of parameter to be estimated is large and the parameter values depend generally on noise level, data length and the natures of signal and noise.

LPCA is based on the exploitation of the predictability of an observed process [28], that is, the estimation of the signal from its past history or samples. To infer the foundations of LPCA and its relation to noise reduction consider the following dynamical system:

$$s(k) = g\{s(k-1), \dots, s(k-m), e_d(k), \boldsymbol{\Theta}\}, \tag{5.56}$$

where g is an unknown, nonlinear function assumed to be smooth, m is the dimension of the dynamical manifold, $\boldsymbol{\Theta}$ is a vector containing the model parameters, and $e_d(k)$ is called the dynamical noise which is assumed to be white. It is important at this point to shed light on the distinction between *dynamical noise* $e_d(k)$ and *measurement noise* $e_m(k)$. While measurement noise represents a harmful alteration of the useful signal, dynamical noise is an inherent part of it. However, they may not be often distinguishable *a posteriori* by the analysis of the observed time series and thereby we will treat them often as one simple noise contribution denoted by $e(k)$. To introduce the proposed approach let us rewrite the dynamics of the signal given by Eq. (5.56) such that they are neither forward nor backward in time, i.e., in an implicit way

$$\tilde{g}\{\mathbf{s}(k), e_d(k), \tilde{\boldsymbol{\Theta}}\} = 0, \tag{5.57}$$

where $\mathbf{s}(k) = [s(k), \dots, s(k-1), \dots s(k-m)]^T$ is the vector of delayed coordinates as in definition 5.2 and $\tilde{\boldsymbol{\Theta}}$ is the parameter vector of this implicit representation. Assume that $\tilde{g}$ is a smooth function of the coordinates, i.e., it is at least piecewise differentiable. If we are faced with a system with no dynamical noise, the local linearization of $\tilde{g}$ in the vicinity of a given point $\bar{\mathbf{s}}_k$ leads to [28]

$$\{\mathbf{s}(k) - \bar{\mathbf{s}}^{(l)}\}^T \tilde{\boldsymbol{\Theta}}^{(l)} = 0 + O\{\|\mathbf{s}(k) - \bar{\mathbf{s}}^{(l)}\|^2\} \quad l = 1, \dots, L, \tag{5.58}$$

where $\tilde{\boldsymbol{\Theta}}^{(l)}$ is the parameter vector of the implicit linear model in the neighborhood $\mathcal{N}^l$ and

$$\bar{\mathbf{s}}^{(l)} = \frac{1}{|\mathcal{N}^l|} \sum_{s(k)\in\mathcal{N}^l} s(k) \tag{5.59}$$

is the center of mass of neighborhood $\mathcal{N}^l$ and $|\mathcal{N}^l|$ denotes the cardinality, namely the number of points of the neighborhood $\mathcal{N}^l$. The partitioning of the space of delayed coordinates in local neighborhoods $\mathcal{N}^l$ for $l = 1, \ldots, L$ is application and signal dependent. A methodology for speech enhancement is described in Section 5.5.3. Assume that this is done in such a way that the second right-hand side term of Eq. (5.58) can be neglected. Thus, Eq. (5.58) would be zero in the noiseless case. In contrast, for noisy time series a supplementary noise-related term is added and Eq. (5.58) becomes

$$\{\mathbf{x}(k) - \bar{\mathbf{x}}^{(l)}\}^T \boldsymbol{\Theta}^{(l)} = n(k) \quad \mathbf{x}(k) \in \mathcal{N}^l \tag{5.60}$$

and $\left(\mathbf{x}(k), \bar{\mathbf{x}}^{(l)}\right)$ replaces $\left(\mathbf{s}(k), \bar{\mathbf{s}}^{(l)}\right)$ in Eq. (5.59). The crucial idea of local projective algorithms is to use a delay coordinate vector of large dimension n_s whereas the dynamics of the underlying deterministic system confine the trajectories to a lower dimensional manifold of dimension $n_0 < n_s$. Consequently, there exists $n_s - n_0$ mutually independent vectors $\boldsymbol{\Phi}_j^{(l)}, j = 1, \ldots, n_s - n_0$, fulfilling Eqs. (5.58) and (5.60) for the noiseless and noisy case respectively. The noise-free attractor does not extend to the space spanned by these $n_s - n_0$ vectors, which constitutes the null space of the problem. In contrast, for noisy sequences this null space is not empty but contains contributions of the noise and, consequently, will be called the noise subspace.

For the sake of clarity we suppress subsequently the notation (l) and substitute $\mathbf{x}(k)$ for $\mathbf{x}(k) - \bar{\mathbf{x}}^{(l)}$. This implies that we are dealing locally with zero mean variables. However, the reader should keep in mind that the presented linear approach is only valid locally and that the direction $\boldsymbol{\Phi}^{(l)}$ depends on the position in the space of the delayed coordinates.

The core of the presented noise reduction algorithm is to identify the noise subspace and to remove the corresponding components from the noisy sequences. This can be achieved by seeking $n_s - n_0$ vectors $\boldsymbol{\Phi}_j, j = 1, \ldots, n_s - n_0$ such that the projection of the noisy data onto these vectors is minimum. For normalized vectors $\boldsymbol{\Phi}_j$, the projection of the data onto the noise subspace is

$$\sum_{j=1}^{n_s - n_0} \boldsymbol{\Phi}_j \{\boldsymbol{\Phi}_j^T \mathbf{x}(k)\} \tag{5.61}$$

which is required to have minimum norm. Taking into account that the sought vectors are orthonormal leads the following Lagrangian to be minimized [28]

$$\mathcal{L} = \sum_{s(k)\in\mathcal{N}^l} \left[\sum_{j=1}^{n_s - n_0} \boldsymbol{\Phi}_j \boldsymbol{\Phi}_j^T \mathbf{x}(k) \right]^2 - \sum_{j=1}^{n_s - n_0} \lambda_j \left\{ \boldsymbol{\Phi}_j^T \boldsymbol{\Phi}_j - 1 \right\}, \tag{5.62}$$

where $\lambda_j, j = 1, \ldots, n_s - n_0$ are the Lagrange multipliers. The minimization with respect to $\mathbf{\Phi}_j$ and λ_j can be done separately for each j yielding the following eigenvalue problem:

$$\mathbf{C}\mathbf{\Phi}_j = \lambda_j \mathbf{\Phi}_j \quad j = 1, \ldots, n_s, \tag{5.63}$$

where $\mathbf{C}$ is the $n_s \times n_s$ sample covariance matrix of $\mathbf{x}(k)$ within the neighborhood $\mathcal{N}^l$

$$\mathbf{C} = \frac{1}{|\mathcal{N}^l|} \sum_{\mathbf{x}(k) \in \mathcal{N}^l} \mathbf{x}(k)\mathbf{x}(k)^T. \tag{5.64}$$

Therefore, we end up with a classical eigenvalue problem of the local covariance matrix. The global minimum is given by the $n_s - n_0$ eigenvectors associated with the smallest eigenvalues. According to section, the 5.4.1 noise reduction is provided by replacing the noisy sequences by

$$\hat{\mathbf{s}}(k) = \sum_{j=1}^{n_0^{(l)}} \mathbf{\Phi}_j \{\mathbf{\Phi}_j^T \mathbf{x}(k)\}, \tag{5.65}$$

where $\mathbf{\Phi}_j, j = 1, \ldots, n_0^{(l)}$ are the eigenvectors associated with the largest eigenvalues of $\mathbf{C}$. Following the results of Section 5.4.1 one can implement this algorithm in an elegant manner by principal component analysis (PCA) of the local zero mean data vector $\mathbf{x}(k) = \mathbf{x}(k) - \bar{\mathbf{x}}^{(l)}, \forall \mathbf{x}(k) \in \mathcal{N}^l$. The residuals of this approximation are given by

$$\hat{\mathbf{e}}(k) = \sum_{j=n_0^{(l)}+1}^{n_s} p_j(k)\mathbf{\Phi}_j \quad p \leqslant n_s\,. \tag{5.66}$$

Equations (5.65) and (5.66) show clearly that the eigenspace of the noisy data is partitioned into a noise subspace determined by $\mathbf{\Phi}_j$ for $j = n_0^{(l)} + 1, \ldots, n_s$ and into a signal-plus-noise subspace determined by $\mathbf{\Phi}_j$ for $j = 1, \ldots, n_0^{(l)}$. Optimal noise reduction performance, i.e., minimal signal distortion and maximal noise reduction, can be attained only if $n_0^{(l)}$ and n_s are chosen optimally. Among the possible selection criteria, the MDL criterion, described in its general form by Eqs. (5.52) and (5.53), has been shown in multiple domains to be a consistent model order estimator especially for short time series [40]. MDL selects the model that produces the minimum code length for the given data. If we apply the general MDL selection criterion given by Eq. (5.52) to PCA and take into account eccentricity considerations of the local confidence ellipsoids, we obtain in the case of additive white Gaussian noise after some simplifications [41]

$$\mathrm{MDL}(n_0^{(l)}) = -\ln \left\{ \frac{\prod_{j=n_0^{(l)}+1}^{n_s} \hat{\lambda}_j^{\frac{1}{n_s-n_0^{(l)}}}}{\frac{1}{n_s-n_0^{(l)}} \sum_{j=n_0^{(l)}+1}^{n_s} \hat{\lambda}_j} \right\}^{(n_s-n_0^{(l)})N} + M \cdot \left(\frac{1}{2} + \ln[\gamma]\right) - \frac{M}{n_0^{(l)}} \sum_{j=1}^{n_0^{(l)}} \ln[\hat{\lambda}_j \sqrt{2/N}], \quad (5.67)$$

where $M = n_0^{(l)} n_s - n_0^{(l)2}/2 + n_0^{(l)}/2 + 1$ is the number of free parameters. The parameter γ determines the selectivity of MDL. Accordingly, $n_0{}^{(l)}$ is given by the minimum of $\mathrm{MDL}(n_0^{(l)})$. For $\gamma = 64$ one obtains a very parsimonious approach while $\gamma = 1$ provide a less restrictive selection.

A further important point in the design of a noise reduction algorithm lies in the adequate choice of the embedding dimension n_s and the neighborhood size $|N^l|$ for $k = 1, \ldots, L$, or, equivalently, the number of local regions L. In classical methods this is often done in an empirical manner. However, in a robust noise reduction algorithm n_s and L should also be selected through an objective criterion. Therefore, we apply the MDL criterion to this parameter estimation problem. The application of the MDL principle to estimate n_s and L requires the description of a family of competing models and density functions that we are considering. Since these are global parameters of the noise reduction algorithm we consider the residual error of an approximation of the data set with parameter values n_s and L

$$e_{(n_s,L)}(k) = x(k) - \hat{s}_{(n_s,L)}(k) \qquad \begin{cases} n_s = n_{s\,\min}, \ldots, n_{s\,\max} \\ L = L_{\min}, \ldots, L_{\max}. \end{cases}$$

The parameters n_s and L have to be chosen such that these residuals are most likely to the added noise, and more precisely to Gaussian white noise. We propose to base the likelihood on the singular values of the covariance matrix

$$\mathbf{C}_{ee} = \frac{1}{N_T} \sum_{t=0}^{N_T-1} \mathbf{e}(k)\mathbf{e}(k)^T, \quad (5.68)$$

where $\mathbf{e}(k) = [e_{(n_s,L)}(k), \ldots, e_{(n_s,L)}(k-1), \ldots e_{(n_s,L)}(k-l_e)]^T$ is the delayed embedding of the residual noise and l_e+1 is the embedding dimension. The value of l_e is not critical and an appropriate value is given by $l_e = n_{s\,\max}$. The choice of this model is mainly motivated by the fact that the singular value spectrum is a very efficient and salient data representation and as such constitutes a promising route for a maximum likelihood approach. More specifically one can directly apply Eq. (5.67) for the selection of n_s and L, that is, the optimal values of the embedding dimension n_s and the number of local regions L are given by the minimum of the MDL of Eq. (5.67) but using the eigenvalues of $\mathbf{C}_{ee}$.

5.5 Applications

5.5.1 Cardiovascular Signals: Observer of the Autonomic Cardiac Modulation

This section describes a typical biomedical application where PCA-based noise reduction is used on cardiovascular signals in order to allow a subsequent consistent extraction of the autonomic cardiac modulation. The development of noninvasive indicators of the beat-to-beat modulation of the autonomic cardiac outflow has been motivated by the growing evidence that autonomic reflex alterations play an important role in many pathophysiological situations. Classically, the spectral analysis of heart beat intervals or blood pressure is used to yield such an indicator [42]. However, while high frequency (HF) fluctuations (HF range: 0.15 Hz to 0.4 Hz) are generally recognized to reflect parasympathetic modulation, interpretation of the low frequency (LF) fluctuations (LF range: 0.04 Hz to 0.15 Hz) is more controversial [42]. In order to solve this controversy a method based on blind source separation (BSS) which separates LF fluctuations in heart rate (RR) and arterial blood pressure (ABP) into two independent signals can be applied [43]. However, this method requires simultaneous recordings of ECG and ABP, which may be cumbersome in clinical applications. An alternative method is based on blind source separation of short-term fluctuations of RR and QT (time interval between the bottom of the R wave and the end of the T wave in an ECG complex) time series, which requires only the recording of a surface ECG [44]. Due to the presence of stochastic influences on RR and QT such as measurement and quantification noise, BSS cannot be applied directly to RR and QT time series. Prior noise reduction is required to allow BSS to operate correctly.

Two antagonistic parts of the autonomic nervous system (ANS), i.e., the cardiac sympathetic (CSNA) and the parasympathetic (CPNA) activities, control the heart beat rhythm. Changes in the level of CSNA and CPNA influence functional heart properties through alterations of the respective electrophysiological subsystem [44, 45]. These alterations are then reflected on global ECG parameters such as RR and QT intervals.

Respiration (RE) acts also on ECG parameters through the autonomic nervous system (solid line) and through mechanical influences (dashed line). However, the latter cause only about 10% of the overall interaction between heart and respiration [46] and are neglected in this approach. ECG parameters can be seen as noisy mixtures of CSNA and CPNA. The noise consists of unknown stochastic influences on RR and QT, measurement and quantification noise. The latter represents important contributions in QT time series.

The task of an observer of the autonomic cardiac outflow consists in reconstructing hidden signals (CSNA, CPNA) using only accessible noisy mixtures of these signals (RR, QT). This is a problem of BSS which is solved by a two-step algorithm. First, a noise reduction is performed on the noisy mixtures. Then, the hidden source signals are reconstructed from the enhanced mixtures by a BSS method for temporally correlated sources. For the application of BSS, one has to assume the independence of the sources. Although it has been established that the sympathetic and parasympathetic activities are not globally independent [47], previous works using BSS [43], have shown that it is possible to reconstruct two independent components that are sensitive to CSNA and CPNA respectively.

5.5.1.1 Noise Reduction by Spatio-Temporal PCA

Since we assume that the model underlying the data is linear we focus here on GPCA. To simultaneously take advantage of the correlations existing between the observed noisy mixtures and temporal time correlations of the source signals we apply spatio-temporal GPCA. Thus, we consider the following n_s-dimensional vector, obtained by an embedding in the space of the delayed coordinates (see Section 5.4.1)

$$\mathbf{y}(k) = [y_1(k), \ldots, y_1(k-(n1_s-1)J), \ldots, y_r(k), \ldots y_r(k-(nr_s-1)J)]^T, \quad (5.69)$$

where $n_s = n1_s + \cdots + nr_s$ is the embedding dimension, and r is the number of spatial dimension. All the $y_j(k)$ constitute spatially distributed signals. Usually, we take $n1_s = \cdots = nr_s$. GPCA with associated MDL-based parameter selection is then applied on this multidimensional observation to perform noise reduction.

5.5.1.2 Blind Source Separation of Noisy Mixtures

Blind source separation (BSS) is now a well-known technique in the signal processing community [48, 49] (see further references therein). Its goal is to recover hidden source signals of which only observed mixtures are available. In this chapter we focus on instantaneous BSS which is based on the following model underlying the observed data:

$$\mathbf{y}(k) = \mathbf{x}(k) + \mathbf{e}(k) \quad (5.70)$$

$$\mathbf{x}(k) = \mathbf{A}\mathbf{s}(k) \quad k = 1, \ldots, N_T, \quad (5.71)$$

where $\mathbf{y}(k) = [y_1(k) \cdots y_r(k)]^T$ are r observed noisy linear mixtures of the r hidden source signals $\mathbf{s}(k) = [s_1(k) \cdots s_r(k)]^T$, $\mathbf{A}$ is the unknown mixing matrix, $\mathbf{e}(k) = [e_1(k) \cdots e_r(k)]^T$ is an additive noise vector, and N_T is the number of samples. The aim of BSS is the estimation of a de-mixing matrix $\hat{\mathbf{B}}$ such that $\hat{\mathbf{s}}(k) = \hat{\mathbf{B}}\mathbf{y}(k)$ constitutes a perfect reconstruction of the hidden source signals (up to a scaling factor and a permutation). Generally, this task can be achieved satisfyingly by BSS for vanishing noise levels. In contrast, increasing noise level may significantly degrade the blind reconstruction performance [49]. Thus, we see that high performance in noisy environments can only be obtained if the BSS algorithm is preceded by an efficient noise reduction system.

We further take advantage of the fact that the source signals in our given application are temporally correlated [43] and apply a method proposed in [48, 49].

This method requires that not only the instantaneous correlations but also the delayed correlations between the output signals vanish. This leads to a generalized eigenvalue problem [48].

The application of the BSS algorithm for noisy mixtures to the observed ECG parameters (RR, QT) provides two independent signals (u_1, u_2) which supposedly represent fluctuations of CSNA and CPNA. After rising the amplitude and permutation ambiguities inherent to BSS by using prior knowledge about cardiovascular signals [44], a quantitative marker of the sympathovagal balance can be based on the ratio

$$\hat{R} = \frac{\hat{\sigma}^2_{\text{CSNA}}}{\hat{\sigma}^2_{\text{CPNA}}}, \tag{5.72}$$

where $\hat{\sigma}^2_{\text{CSNA}}$ and $\hat{\sigma}^2_{\text{CPNA}}$ are the variance of the reconstructed sympathetic and parasympathetic activities.

5.5.1.3 Results

A tough task in the development of algorithms based on BSS for biomedical applications is their validation. Indeed, BSS techniques reconstruct generally hidden variables such as CSNA and CPNA, which are not accessible in humans. However, a validation procedure requires information about these hidden variables. An elegant way to circumvent this limitation consists in its application to subjects under experimental conditions known to elicit or inhibit sympathetic or parasympathetic response. This shows then clearly if the observer is able to highlight changes in the levels of CPNA and CSNA. Appropriate experimental protocols have been conducted on six free breathing subjects

1. Protocol Pe: Phenylephrine© ($0\,\mu\text{g} \cdot \text{kg}^{-1} \cdot \text{min}^{-1}$ to $1.5\,\mu\text{g} \cdot \text{kg}^{-1} \cdot \text{min}^{-1}$) was infused for 15 min in order to increase the mean arterial pressure by 10 mm Hg.

2. Protocol Ni: Nipride© ($0\,\mu\text{g} \cdot \text{kg}^{-1} \cdot \text{min}^{-1}$ to $1.5\,\mu\text{g} \cdot \text{kg}^{-1} \cdot \text{min}^{-1}$) was infused for 15 min to decrease the mean arterial pressure by 10 mm Hg.

Ni is known to induce a sympathetic stimulation and parasympathetic inhibition whereas Pe mainly has the opposite effect. After providing informed consent, we obtained from subjects 6 min recordings of surface ECG on a 486 Intel PC with an A/D board (Labmaster) at a sampling frequency of 500 Hz. This signal has then been oversampled at 1000 Hz and corresponding RR and QT interval time series have been extracted. The main goal of the oversampling was to enhance the accuracy of the QT detection algorithm. Finally, since this study takes into account only LF and HF components of the various signals, all of them have been re-sampled at 1 Hz and bandpass filtered (0.04 Hz to 0.4 Hz).

Results computed by the proposed observer without prior denoising ($\hat{R}$) and with prior denoising ($\hat{R}$d) are given in Table 5.1. To compare our method with a traditional approach we have also evaluated the FFT-based indicator of the sympathovagal balance $\hat{R}_{FFT}$ which consists in the ratio of LF to HF components of RR intervals [42].

Tab. 5.1: Indicators of sympathovagal balance for six subjects without ($\hat{R}$) and with ($\hat{R}d$) prior denoising or with an FFT-based indicator ($\hat{R}_{FFT}$).

	$\hat{R}$		$\hat{R}d$		$\hat{R}_{FFT}$	
Subject	Ni	Pe	Ni	Pe	Ni	Pe
1	0.01	0.38	1.22	0.92	3.64	0.59
2	19.0	49.0	1.23	0.60	0.79	1.03
3	5.67	2.33	1.53	0.73	2.99	0.65
4	99.0	0.01	1.39	0.74	8.46	0.89
5	0.01	0.02	1.33	0.64	1.36	0.32
6	0.02	0.01	1.48	0.79	4.78	0.75

We can remark that prior denoising is necessary, for the indicator without prior denoising $\hat{R}$ provides inconsistent results. In contrast, we note that $\hat{R}d$ allows the classification of subjects under different experimental conditions. Indeed, $\hat{R}d(\text{Ni})$ is always larger than 1 while $\hat{R}d(\text{Pe})$ is smaller than 1. The statistical reliability of the proposed observer is confirmed by analysis of variance tests (ANOVA) which provides $p = 5 \times 10^{-6}$. Therefore, since Ni is known to induce a sympathetic stimulation and parasympathetic inhibition whereas Pe has the opposite effect, results show that the proposed observer is able to shed light on changes in the level of the sympathovagal balance. The analysis of results for the traditional FFT-based indicator $\hat{R}_{FFT}$ shows that the discrimination of Ni and Pe is significant from the statistical point of view ($p = 0.025$). Nevertheless, this indicator does not provide a classification of subjects under different experimental conditions. Indeed, one cannot find a number κ satisfying $\hat{R}_{FFT}(\text{Ni}) > \kappa$ and $\hat{R}_{FFT}(\text{Pe}) < \kappa$ for all subjects.

5.5.2 Electroencephalogram: Spontaneous EEG and Evoked Potentials

Among the research fields where noise has a primary role is brain research, be it for clinical, physiological, or psychological purposes. In this section, we investigate the use of noise reduction techniques to spontaneous and evoked brain electrical responses. The first application is the potential use of brain electrical signals as captured by the surface electroencephalogram (EEG) for controlling a device. This paradigm is called brain machine interface. The second is the analysis of single trial visual evoked response potentials in cognitive/psychological and clinical neuroscience context.

Brain Machine Interface

Brain-machine interfaces (BMI) allow for communication and control of systems that do not depend on the brain's normal output channels of peripheral nerves and muscles [50]. A BMI enables a person to control an instrument, being software or hardware, by generating specific states of the brain, which leave their signature in the EEG. These states should be as independent as possible to facilitate the decision of the machine. Many BMIs use motor imagery paradigms. The mental tasks chosen in this study were imagination of repetitive self-paced left-hand movement ('L') and imagination of repetitive self-paced right-hand movement ('R'). The imagined hand movement was a flexion at the wrist causing the hand moving up and down. The third task was that activating the language center. This task, 'W,' consisted of generating words that begin with the same letter, freely chosen by the subject. The words were not spoken. All tasks were executed with opened eyes.

A BMI system is usually composed of three subsystems: (1) a preprocessor, (2) a feature extraction and selection, and finally (3) a classification stage which takes the final decision.

EEG signals were recorded with the 32-channel Biosemi ActiveTwo system©. The electrodes were placed on the scalp according to an extension of the 10–20 international electrode placement system. The ground electrode is replaced by two separate electrodes, located between C3 and Cz, and Cz and C4 respectively. The EEG signals were digitized at a sampling rate $F_s = 2048\,\text{Hz}$, subsampled to $F_s = 512\,\text{Hz}$ and stored for the offline analysis. Together with the EEG signals, we recorded the "task signal," indicating which task the subject is doing at every moment. Five naive (without training) subjects participated in the experiment.

The features computed from the EEG signals were frequency band power densities and spatially grouped and averaged synchronization measures between different EEG signals. The results show the three classes' discrimination obtained using a combination of three support vector machine (SVM) classifier with a linear kernel. The overall performance of the BMI is measured with the correct classification rate ($[\text{CR}] = \%$) and the rate of unknown response ($[\text{UR}] = \%$), referring to the case where the classifier cannot determine the class. The remaining percentage is the error rate ($[\text{ER}] = \%$). We refer the reader to [51, 52] for a detailed description of the methodology.

We applied GPCA and LPCA noise reduction to test whether they are able to remove irrelevant information from the EEG signals while keeping the relevant information the classifier is using to distinguish the different mental tasks. It is difficult to *a priori* define which part of the EEG signals actually corresponds to noise and which part contains the relevant information. Therefore, the noise reduction algorithms were evaluated by their improvement or deterioration of the CR. Table 5.2 reports the parameters used for the noise reduction methods.

Table 5.3 presents results with features computed after global PCA noise reduction. None of the denoising methods achieved a better average CR than the

Tab. 5.2: Parameters for GPCA and LPCA noise reduction and values used in our experiments.

Parameter/flag	Description	Value
mode_lag	$J = 1$ if 'n'; computed from correlation if 'y'	'y'–'n'
n_s	Embedding space dimension	20–10
mode_MDL	use of MDL for determining n_o if 'y'	'y'–'n'
n_o	The projection space dimension if mode_MDL='n'	4-5–10
mode_eig	use L_2 norm if 'y'; use L_1 norm if 'n'	'y'–'n'
$\nu = \lvert\mathcal{N}^l\rvert$	Default neighborhood size	50
L	Number of neighborhood	$0.1{*}N_T$

CR obtained without application of noise reduction. For subjects 2 and 4, however, sometimes an improvement of CR was detected. Originally, 51.11 % classification accuracy was obtained, the best denoising method yielded and average CR of 51.06 %. For subjects 3 and 5 no results better than those without noise reduction could be obtained.

Tab. 5.3: Results for GPCA noise reduction with the parameter set: (n_s, n_o, mode_-lag, mode_eig, mode_MDL).

Global PCA	Subj	1	2	3	4	5	Average
None	CR	**61.56**	40.27	**63.50**	39.49	50.71	**51.11**
	UR	6.19	4.06	7.20	8.02	6.95	6.49
20-4-n-n-y	CR	60.59	40.76	62.81	39.85	50.38	50.88
	UR	6.42	4.67	6.87	8.28	6.92	6.63
20-4-n-n-n	CR	60.20	39.78	62.82	40.33	50.53	50.73
	UR	6.54	5.17	6.47	8.34	7.14	6.73
20-4-y-n-n	CR	53.74	38.72	43.72	40.13	43.43	43.95
	UR	4.75	4.22	6.10	5.46	5.28	5.16
20-4-n-y-n	CR	60.20	39.78	62.82	40.33	50.53	50.73
	UR	6.51	5.17	6.47	8.34	7.14	6.72
20-4-y-y-y	CR	58.46	39.05	59.97	40.67	48.08	49.25
	UR	6.72	5.76	7.50	7.40	7.48	6.97
20-10-n-n-n	CR	60.62	41.16	63.00	40.19	50.35	51.06
	UR	6.39	4.67	6.78	8.34	6.89	6.61

Table 5.4 presents the average CR obtained with LPCA denoising which never outperforms that obtained without denoising. The best average was CR = 50.96 % as compared to 51.11 % originally. For subjects 2, 4 and 5, LPCA denoising could yield improved results. As compared to GPCA denoising, the best average CRs was slightly better: 51.06 % and 50.96 % for global and LPCA denoising respectively. The different settings of parameters considered did not affect the CR significantly. Using MDL to determine the dimension of the subspace yielded slightly better results than those obtained with a manually chosen n_o, i.e., without MDL.

Tab. 5.4: Results for LPCA denoising (ν, n_s, n_o, mode_MDL).

Local PCA	Subj	1	2	3	4	5	Average
None	CR	**61.56**	40.27	**63.50**	39.49	50.71	**51.11**
	UR	6.19	4.06	7.20	8.02	6.95	6.49
50-10-5-n	CR	60.34	41.66	62.53	39.94	49.45	50.79
	UR	6.70	5.30	7.40	8.06	8.29	7.15
50-10-5-y	CR	59.83	41.59	62.69	39.93	50.57	50.92
	UR	7.16	4.27	7.88	8.82	7.26	7.08
50-10-5-n,	CR	60.86	40.60	62.75	39.86	50.38	50.89
$J = 1$	UR	6.52	4.30	7.15	8.87	7.29	6.83
50-10-5-y,	CR	60.28	41.25	62.97	39.54	**50.79**	50.96
$J = 1$	UR	6.60	4.18	6.99	8.65	6.70	6.63
50-20-4-n	CR	60.11	**41.71**	59.37	40.09	50.29	50.31
	UR	6.45	5.29	7.86	7.93	7.36	6.98
50-20-4-y	CR	60.46	40.67	62.87	40.04	50.35	50.88
	UR	6.51	4.12	7.23	7.90	7.26	6.61

In these experiments, discriminating EEG during different mental tasks using power spectral densities and PLV, using noise reduction methods provided negligible increase or even decrease in classification performances. The reasons for this can be: (1) the "removed noise" contains actually some information in its power spectrum density, (2) the noise reduction techniques have "linearized" the signals (which is more unlikely with LPCA, and thus better performances) which could contain relevant phase information. In both cases, a need for a better understanding of the noise components in recorded electrical brain activity is necessary.

Visual Evoked Potentials

In this section, we present results of denoising on visual brain evoked potential (VEP) which is a subclass of evoked potentials (EP). A visual stimulus was presented to a subject from which the EEG was recorded. The EEG was recorded using the standard 10–20 electrode placement system and we have used 47 channels. The careful visual inspection of the signals was performed to discard any recording with a bad skin contact which produces heavily artifactual data. The signals were sampled at 500 Hz. The duration of a trial was 800 ms with 100 ms of prestimulus. NbTrial = 400 trials were recorded and further averaged to provide the averaged signal $s_A(n)$.

Neuro-scientists and -clinicians usually perform a large collection of visual stimulus triggered VEP response signals, and then perform a statistical averaging of the stimulus-locked VEP. In recent years, however, it has been pointed out that the statistical averaging can in fact deteriorate the single trial informational signal. Moreover, due to the natural nonstationarity of the signals and the experimental procedure, the statistical average should be applied. For this reason, methods for using single-trial EP have been developed. Due to the low signal-to-noise ratio of the single trials, noise reduction methods have been developed with more or less success [53–55]. These methods make use of a thresholding on wavelet coefficients to perform the actual denoising, and sometimes the manually selected set of *good* wavelet coefficients.

We have applied global and local PCA, together with a hard-threshold wavelet noise reduction method [56]. For the sake of completeness, we have also performed a Wiener filtering method in each subband as constructed from the wavelet filter bank used in the previous technique. The wavelet and Wiener noise removal approaches can be explained in the same framework as in Section 5.3(see also [57]). We have used the biorthonormal B-spline wavelet of order $(Nh = 4, Ng = 20)$ where Nh is the order of the analysis FIR filter and Ng the order of the synthesis FIR filter. The performance measure assumes that $s_A(n)$ is the noise-free signal and is defined as follows:

$$\text{NMSE} = 100 \frac{E[(s_A(n) - \hat{s}(n))^2]}{E[s_A^2(n)] + E[\hat{s}^2(n)]} \tag{5.73}$$

where $\hat{s}(n)$ is the noise reduced signal. NMSE tends to zero when $E[(s_A(n) - \hat{s}(n))^2]$ and thus the estimated signal $\hat{s}(n)$ is close to $s_A(n)$. We further make use of the many trials to improve the noise reduction by averaging the noise reduced single trials. So for instance, at the trial K, we apply the noise reduction methods on the K trials and then we average them to produce an average noise reduced VEP $\hat{s}^{(K)}(n)$. We can then use the following K-averaged measure:

$$\text{NMSE}^{(K)} = 100 \frac{E[(s_A(n) - \hat{s}^{(K)}(n))^2]}{E[s_A^2(n)] + E[(\hat{s}^{(K)})^2(n)]} \,. \tag{5.74}$$

The parameters used in each method are reported in Table 5.5 (see Sections 5.4.1 and 5.4.2 for details).

Tab. 5.5: Parameters used in our experiments for LPCA and GPCA.

Parameter/flag	Description	Value
J	Embedding lag	1
n_s	GPCA embedding space dimension	30
n_s	LPCA embedding space dimension n_s	30
$\nu = \lvert\mathcal{N}^l\rvert$	Default neighborhood size	50
L	Number of neighborhood	$0.1{*}N_T$

Figure 5.4 shows the statistics of the four different methods. The upper panel shows the $\text{NMSE}^{(K)}$ in function of the number of trials. As K increases, the $\text{NMSE}^{(K)}$ decreases to reach zero when $K = \text{NbTrial}$. A significant improvement is seen already after three trials. The wavelet denoising should be preferred for a

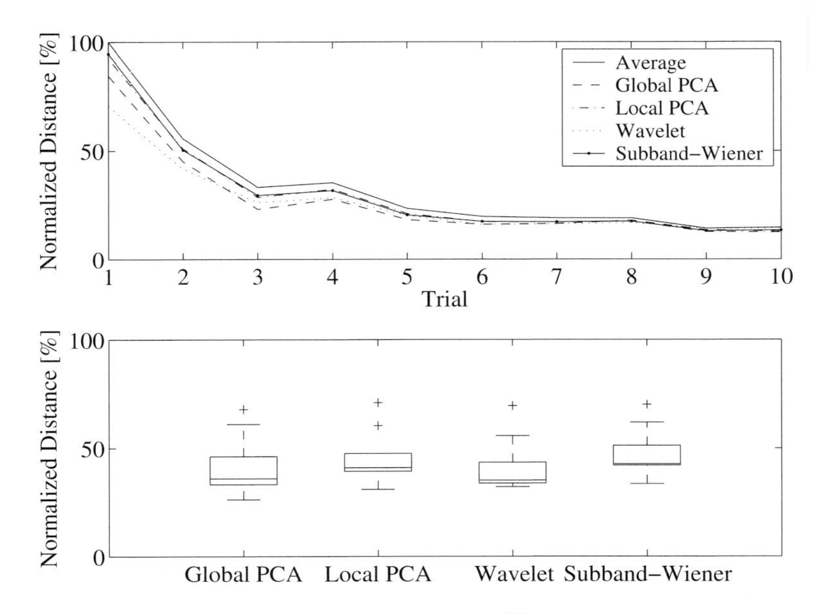

Fig. 5.4: Upper panel: the normalized $\text{NMSE}^{(K)}$ in function of the number of trials K for the four different techniques. Lower panel: a Boxcar display of the statistical results of the four different methods. The box has lines at the lower quartile, median, and upper quartile values with whiskers showing the rest of the data. The (+) indicates outliers data beyond the whiskers.

small number of trials, while GPCA performs better after three trials. Almost all methods perform well after six trials due to the averaging effect. The lower panel shows the quartiles of the nonaveraged NbTrial values of $\text{NMSE}^{(K)}$ for the four different methods.

Figure 5.5 shows the result on one particular channel (channel 21) for $K = 3$. The P100 (positive deflection at about 100 ms after stimuli) and N200 (negative deflection at about 200 ms after stimuli) waves are clearly extracted from the various techniques. The sharpness of the waves with noise removed is much more pronounced than with the total averaged waves where it is smeared out by the averaging process. The quality of the noise reduction method should be assessed with a clinical specialist within the framework of a specific application, e.g., inverse solution problem, time delay estimation, amplitude estimation, time-frequency content, or instantaneous phase extraction. We can also observe some severe deflections at about 400 ms, 500 ms, and 600 ms in the cleaned signals that do not show up in the total averaged signal. These deflections can be due to artifacts and should be dealt with according to artifact detection and removal methods [58].

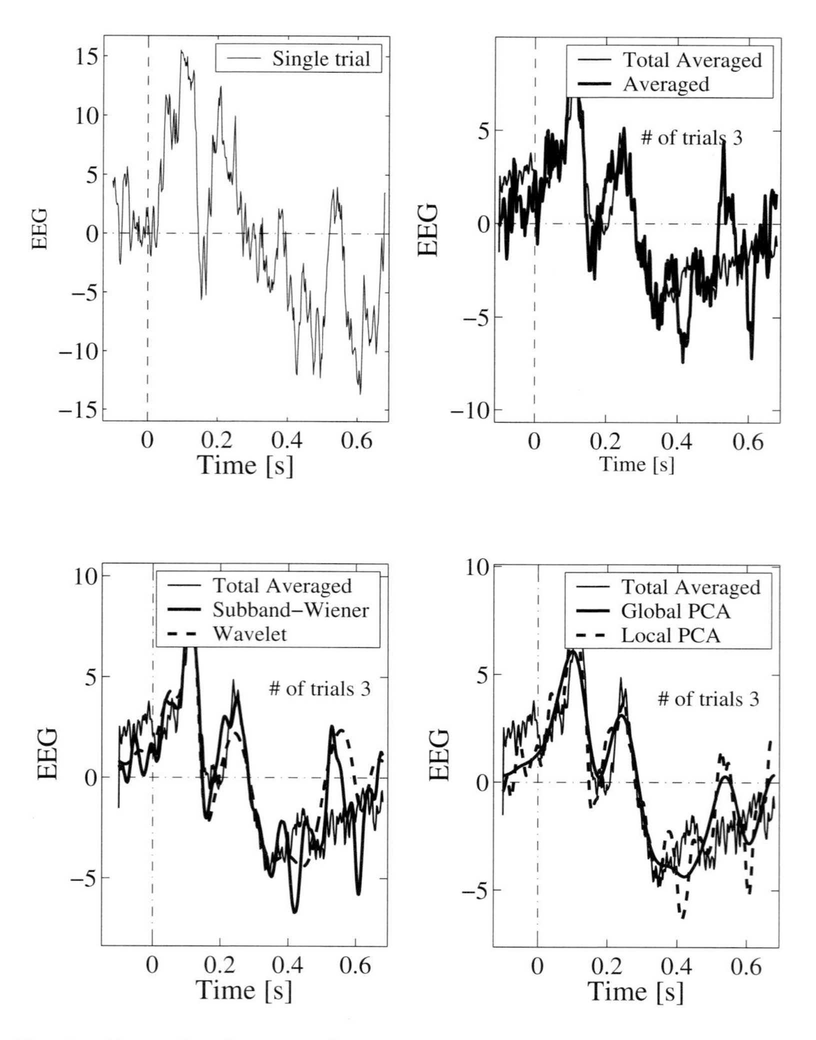

Fig. 5.5: Example of noise reduction using the four different techniques. Top left: single trial EVP. Top right: total 400 trial average with three trial average. Bottom left: Wavelet and subband Wiener techniques. Bottom right: GPCA and LPCA techniques.

5.5.3 Speech Enhancement

The performance of automatic speech processing systems degrades drastically in noisy environments. Therefore, several single channel enhancement algorithms using the discrete Fourier transform (DFT), such as subtractive-type approaches [59, 60] or Wiener filtering, have been developed. The major problem with most of these methods is that they suffer from a distortion called "musical noise." To reduce this distortion, the DFT can be replaced by the discrete cosine transform (DCT) [61] or the KLT [62]. The enhancement is obtained by nulling the noise subspace as explained in Section 5.4.1, with an additional optimal weighting of the signal-plus-noise subspace.

In this section we present a subspace approach for single channel speech enhancement and recognition in highly noisy environments based on the KLT, and implemented via PCA. This choice is motivated by the fact that the KLT provides an optimum compression of information, while the DFT and the DCT are suboptimal. The main problem in subspace approaches is the optimal choice of the different parameters. We present therefore an approach for the optimal subspace partition using MDL.

5.5.3.1 Proposed Subspace Approach

Consider a speech signal $s(k)$ corrupted by an additive stationary background noise $e(k)$ as in Eq. (5.4). Our noise reduction algorithm operates on a frame-by-frame basis and the general enhancement scheme is represented in Fig. 5.6. A very efficient and robust implementation of the subspace approach is provided by the GPCA of the n_s-dimensional vector $\mathbf{x}(k)$, obtained by an embedding in the space of the delayed coordinates. We have used $J = 1$ in this section. In speech GPCA-based noise reduction processing the n_0 components are generally weighted, as proposed by Ephraim et al. in [62]. From Eq. (5.23) and using the weighting matrix G_o we get

$$\hat{\mathbf{s}}(k)_{\text{Eph95}} = G_o P_{E_o} P_C \tag{5.75}$$

with $G_o = \text{diag}(\exp\{-\kappa\sigma_n^2/\lambda_j\})$ for $j = 1, \ldots, n_0$ and $\kappa = 5$. The parameters n_s and n_0 are generally chosen in such a way that the noise is essentially relegated to the residuals of the signal approximation given by Eq. (5.75).

5.5.3.2 Subspace Partitioning

The optimal design of a PCA-based noise reduction algorithm for speech enhancement is a difficult task. The parameters n_s and n_0 should be chosen in an optimal manner through appropriate selection rules. Furthermore, the use of a weighting matrix G_w in Eq. (5.75) introduces a considerable amount of speech distortion. Therefore, in order to simultaneously maximize noise reduction and minimize signal distortion, we present in this section an approach consisting in a partition of the eigenspace of the noisy data into three different subspaces (see Fig. 5.6).

1. A noise subspace which contains mainly noise contributions. These components are nulled during reconstruction.
2. A signal subspace which contains principal components $\mathbf{p}_j(k)$ with a high signal-to-noise ratio $\text{SNR}_j \gg 1$. Components of this subspace are not weighted

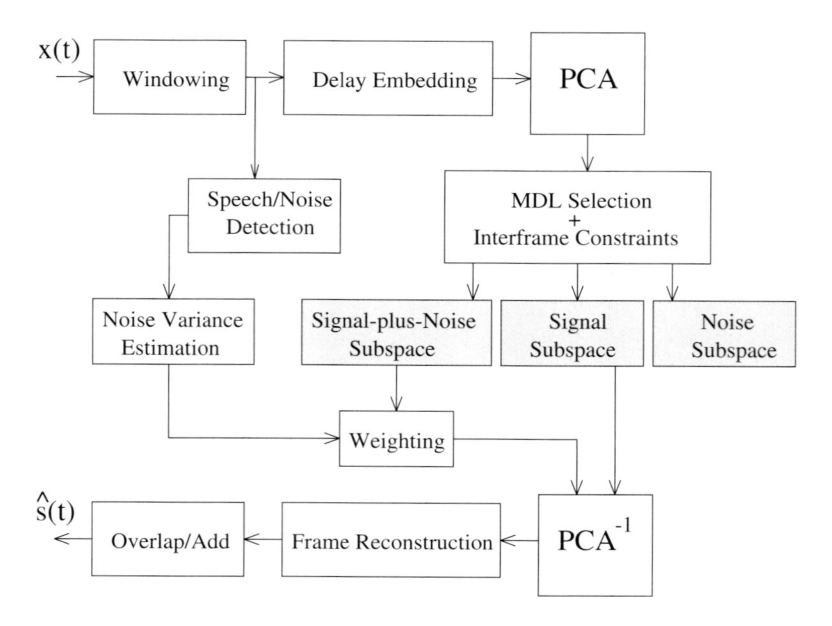

Fig. 5.6: The proposed enhancement algorithm.

since they contain mainly components from the original signal. This allows a minimization of the signal distortion.

3. A signal-plus-noise subspace which includes the components $\mathbf{p}_j(k)$, $\mathrm{SNR}_j \approx 1$. The estimation of its dimension can only be done with a high error probability. Consequently, principal components with $\mathrm{SNR}_j < 1$ may belong to it and a weighting is applied during reconstruction.

Using this new partition, the reconstructed signal is given by

$$\hat{\mathbf{s}}(k) = P_{E_o^1} P_C + G_o^2 P_{E_o^2} P_C, \tag{5.76}$$

where $P_{E_o^1}$ is the projection matrix corresponding to the n_{01} first eigenvectors and $P_{E_o^2}$ is the projection matrix corresponding to the $n_{02} - n_{01} + 1$ following eigenvectors, and $G_o^2 = \mathrm{diag}(\exp\{-\kappa\sigma_n^2/\lambda_j\})$ for $j = n_{01} + 1, \ldots, n_{02}$ and $\kappa = 5$. We note that the proposed approach requires the determination of the parameters n_{01} and n_{02}. The parameter n_{01} should provide a very parsimonious representation of the signal whereas n_{02} should also select components with $\mathrm{SNR}_j \approx 1$. The parameter γ determines the selectivity of MDL. Accordingly, n_{01} and n_{02} are given by the minimum of $\mathrm{MDL}(n_{0i})$ with $\gamma = 64$ and $\gamma = 1$, respectively.

A crucial point is the adequate choice of the embedding dimension n_s of the PCA. In this chapter we use a rule for the determination of n_s that has been proposed in the context of the singular spectrum analysis [63]. It is applicable if the useful signal is constituted of quasiperiodic contributions of a bandwidth Δf_x and is given by

$$n_s < \min\{1/\Delta f_x \quad (N_T/3 + 1)\}. \tag{5.77}$$

For speech signals, we found that an appropriate value for n_s is in the range from 40 to 80.

5.5.3.3 Results

For the performance evaluation, we have compared the following single channel enhancement algorithms: nonlinear spectral subtraction using the DFT (NSS) [60], subspace approach by Ephraim et al. using the KLT (Eph95) [62], proposed subspace approach (PCA–MDL). The testing database has been created by adding different types of background noises from the Noisex database to the clean speech signals, at SNRs ranging from -6 dB to $+\infty$ dB. The sampling frequency is 8 kHz. The frame size is $N_T = 400$ and we apply Hanning windowing with 50 % overlap. We have based our performance evaluation on the segmental SNR ([SNR] = dB), the Itakura–Saito distortion measure (IS), the observation of the spectrograms as well as informal listening tests. We have observed that generally subspace approaches based on the PCA (Eph95 and PCA–MDL) outperform linear and nonlinear subtractive-type methods using DFT. In particular, the use of a subspace approach significantly reduces the "musical noise."

Tab. 5.6: Segmental SNR and Itakura–Saito measure in the case of white Gaussian noise.

Noisy		Eph95		PCA–MDL	
SNR	IS	SNR	IS	SNR	IS
0	6.2	6.5	4.1	8.8	3.2
6	5.1	10.5	3.2	12.6	3.1
18	2.2	21.9	1.1	22	0.9

If we compare the subspace approaches, we can see in Table 5.6 that our method provides similar performance with respect to Eph95 for high input SNRs. However, it leads to a higher noise reduction and a lower signal distortion (smaller value of IS) for low SNRs. These results highlight the efficiency and consistency of the MDL-based subspace algorithm. Furthermore, this approach does not require parameter tuning based on empirical considerations. One important additional feature of our method is that it is highly efficient in detecting speech pauses, even under very noisy conditions. In order to be able to apply the MDL selection approach to colored noises, we have to modify the covariance matrix $\mathbf{C}$ of the noisy data by taking into account the covariance matrix of noise computed during speech pauses. This leads to the results presented in Table 5.7 for helicopter cockpit noise. We can see, that even in this case, our method provides good performance over subtractive-type algorithms.

Tab. 5.7: Segmental SNR and Itakura–Saito measure in the case of helicopter cockpit noise.

Noisy		NSS		PCA–MDL	
SNR	IS	SNR	IS	SNR	IS
0	3.1	5.2	3	6.7	2.4
6	2.1	10.1	1.9	10.9	1.1
18	0.5	20.2	0.4	20.5	0.3

We have applied our enhancement algorithm as a preprocessing stage to speech recognition in noise. We have used a speech recognizer which has been designed and trained on clean speech for the isolated digit recognition. The recognizer has been built up by the HTK HMM toolkit version 2.1. The features for speech recognition are the 12 MFCC and the energy, together with the first- and second-order derivatives of these 13 parameters. The training database is constituted of 400 recordings of seven digits. The general model for the isolated digit recognition consists of a model for silence between the digits (three emitting states). The testing database contains 50 sequences of seven digits with additive white Gaussian noise.

Tab. 5.8: Correctness of recognition in the case of white Gaussian noise.

Input SNR	Noisy %	NSS %	Eph95 %	PCA–MDL
−6	16	20	27	37
0	20	31	39	44
6	35	50	60	68

Table 5.8 gives the recognition results in terms of correctness for the compared algorithms. These results underline that our method allows an extraction of the relevant features of speech even under highly noisy conditions.

5.6 Conclusions

This chapter has presented the multifaceted world of randomness and noise, both from biological systems' and scientists' viewpoints. We have tried to emphasize the fact that these two viewpoints are still quite different, and make the two worlds quite separate. Future development in science and engineering should take this fact into account in order to move toward more efficient and sustainable system design. This step forward will be made possible by new understanding of the relationship we have with *nature*. Grasping the essence of randomness will be a key factor in this direction.

The time series analysis is the landscape for exercising statistical signal processing: modeling, system identification, prediction, and noise reduction. We have also seen that the statistical approach can be very efficient when a sufficient amount of data are available, some good *a priori* knowledge about the problem is available, and we aim at describing or analyzing en masse phenomena. However, when no or few *a priori* knowledge is at one's disposal, and the system is very complex, these methods tend to provide poor results. These two extreme cases have been exemplified by speech enhancement, visual evoked potential noise reduction and cardiovascular system analysis on one hand, and noise reduction in brain-machine interfaces on the other. Obviously, the last application is far more complex than the other ones and requires much further understanding of the brain function than what is actually available.

We have described a generic technique based on subspace decomposition and projection which allows for great flexibility, allowing us to deal with the most complex signals which are nonstationary and nonlinear.

References

[1] T. L. Fine. Theories of probability: An examination of foundations. *Academic Press*, 1973.

[2] M. Gutzwiller. Chaos in classical and quantum mechanics. 1990.

[3] T. Petrosky, I. Prigogine, and S. Tasaki. *Physica A*, 173:175, 1991.

[4] T. Petrosky and I. Prigogine. *Proc. Natl. Acad. Sci.*, 90:9393, 1993.

[5] S. Albeverio, R. Cianci, N. De Grande-De Kimpe, and A. Khrennikov. *Russian J. Math. Phys.*, 6:3, 1999.

[6] E. T. Jaynes. Probability theory: The logic of science. *Cambridge University Press*, 2003.

[7] L. de Broglie. *Les Incertitudes d'Heisenberg et l'Interprétation Probabiliste de la Mécanique Ondulatoire*. Gauthier-Villars, Paris, 1982.

[8] H. Martens and W. M. de Muynck. *J. Phys. A*, 25:4887, 1992.

[9] S. Hagan, S. Hameroff, and J. Tuszynski. *Phys. Rev. E*, 65:61901, 2002.

[10] W. Nagl, M. Rattemayer, and F. A. Popp. *Naturwissenschaften*, volume 572. 1981.

[11] S. Hameroff and R. Penrose. Toward a science of consciousness—the first tucson discussions and debates. *MIT Press*, 1996.

[12] J. C. Eccles. *Proc. R. Soc. Lond. [Biol]*, 227:411, 1986.

[13] W. Bialek. In H. Flyvbjerg, F. Jülicher, P. Ormos, and F. David, editors, *Physics of Biomolecules and Cells:Les Houches Session LXXV*, EDP Sciences. Springer, Berlin, 2002.

[14] H. Stapp. Mind, matter and quantum mechanics. *Springer*, 2004.

[15] J. J. Collins. *Nature*, 402:241, 1999.

[16] D. E. Makarov and N. Makri. *Phys. Rev. B*, R2257:52, 1995.

[17] F. Moss, L. M. Ward, and W. G. Sannita. *Clin. Neurophysiol.*, 115:267, 2004.

[18] V. Anishchenko, F. Moss, A. Neiman, and L. Schimansky-Geier. *Sov. Phys. Usp.*, 42:7, 1999.

[19] S. Hecht, S. Schlaer, and M. H. Pirenne. *J. Gen. Physiol.*, 25:819, 1942.

[20] P. Prokopowicz and P. Cooper. *Int. J. Comp. Vision*, 16:191, 1995.

[21] I. Murakami and P. Cavanagh. *Nature*, 395:798, 1998.

[22] I. Murakami and P. Cavanagh. *Vision Res.*, 41:173, 2001.

[23] T. J. Hine, A. Renneflott, and M. Chappell. In *25th Annual Meeting of the European Conference on Visual Perception, Glasgow*, volume 31, page 130, 2002.

[24] I. T. Jolliffe. *Principal Component Analysis*. Springer, Berlin, 1986.

[25] S. Akkarakaran and P. Vaidyanathan. *IEEE Trans. Inform. Theory*, 47:1003, 2001.

[26] J.-P. Eckmann and D. Ruelle. *Rev. Mod. Phys.*, 57:617, 1985.

[27] F. Takens. Detecting strange attractors in turbulence. *Lect. Notes Math.*, 1981.

[28] H. Kantz and T. Schreiber. Nonlinear time series analysis. *Cambridge University Press*, 1997.

[29] D. S. Broomhead and G. P. King. *Physica D*, 20:217, 1986.

[30] M. R. Allen and L. A. Smith. *Phys. Lett. A*, 234:419, 1997.

[31] J. Rissanen. *Encycl. Stat. Sci.*, 5:523, 1985.

[32] J. Rissanen. Stochastic complexity in statistical inquiry. 1989.

[33] K. Judd and A. Mees. *Physica D*, 82:426, 1995.

[34] M. Wax and T. Kailath. *IEEE Trans. Acoust. Speech Sig. Proc.*, 33:387, 1985.

[35] R. Vautard and M. Ghil. *Physica D*, 35:395, 1988.

[36] P. Celka and P. Colditz. *Med. Eng. Phys.*, 24:1, 2002.

[37] R. Cawley and G. Hsu. *Phys. Rev. A*, 46:3057, 1992.

[38] P. Grassberger, R. Hegger, H. Krantz, C. Schaffrath, and T. Schreiber. *Chaos*, 3:127, 1993.

[39] J. P. Saul, R. D. Berger, P. Albrecht, S. P. Stein, M. H. Chen, and R. J. Cohen. *Am. J. Physiol.*, 261:1231, 1991.

[40] R. Vetter, P. Celka, J.-M. Vesin, G. Thonet, E. Pruvot, M. Fromer, U. Scherrer, and L. Bernardi. *Ann. Biomed. Eng.*, 26:293, 1998.

[41] R. Vetter. *Extraction of efficient and characteristic features of multidimensional time series*. PhD thesis, EPFL, Lausanne, 1999.

[42] Heart rate variability—standards of measurement, physiological interpretation, and clinical use. Technical report, Task Force of the European Society of Cardiology, the North American Society of Pacing, and Electrophysiology, 1996. Circ..

[43] R. Vetter, J.-M. Vesin, P. Celka, and U. Scherrer. *IEEE Trans. Biomed. Eng.*, 46: 322, 1999.

[44] R. Vetter, N. Virag, J.-M. Vesin, P. Celka, and U. Scherrer. *IEEE Trans. Biomed. Eng.*, 47:578, 2000.

[45] M. R. Boyett, A. Clough, J. Dekanski, and A. V. Holden. *Computational Biology of the Heart*. Wiley, New York, 1997.

[46] H. Herzen and H. Seidel. *Modeling the dynamics of biological systems*. Springer, Berlin, 1995.

[47] M. Kollai and K. Koizumi. *J. Aut. Nerv. Syst.*, 1:33, 1979.

[48] L. Molgedey and H. G. Schuster. *Phys. Rev. Lett.*, 72:3634, 1994.

[49] A. Belouchari, K. Abed-Meriam, J. F. Cardoso, and E. Moulines. *IEEE Trans. Sig. Proc.*, 45:434, 1997.

[50] T. Vaughan, W. Heetderks, L. Trejo, M. Weinrich W. Rymer, M. Moore, A. Kbler, N. Birbaumer B. Dobkin, E. Donchin, E. Wolpaw, and J. Wolpaw. *IEEE Trans. Neur. Syst. Rehab. Eng.*, 2:94, 2003.

[51] E. Gysels and P. Celka. *IEEE Trans. Neur. Syst. Rehab. Eng.*, 12:406, 2004.

[52] E. Gysels, P. Renevey, and P. Celka. In *Proceedings of the Biosignal Interpretation Conference*, 2005. Tokyo, Japan.

[53] R. Q. Quiroga and H. Garcia. Single-trial event-related potentials with wavelet denoising. *Clin. Neurophysiol*, 114:376, 2003.

[54] A. Effern, K. Lehnertz, T. Schreiber, P. David T. Grunwald, and C. E. Elger. *Physica D*, 140:257, 2000.

[55] A. Effern, K. Lehnertz, T. Grunwald, P. David G. Fernández, and C. E. Elger. *Clin. Neurophysiol.*, 11:2255, 2000.

[56] S. Mallat. A wavelet tour of signal processing. *Academic Press*, 1998.

[57] S. P. Ghael, A. M. Sayeed, and R. G. Baraniuk. *Proc. SPIE Math. Imag.*, 3169: 389, 1997.

[58] M. Browne and T. R. Cutmore. *Clin. Neurophysiol.*, 113:1403, 2002.

[59] N. Virag. *IEEE Trans. Speech Audio Proc.*, 7:126, 1999.

[60] P. Lockwood and J. Boudy. *Speech Communic.*, 11:215, 1992.

[61] I. Y. Soon, S. N. Koh, and C. K. Yeo. *Speech Communic.*, 24:249, 1998.

[62] Y. Ephrahim and H. L. Van Trees. *IEEE Trans. Speech Audio Proc.*, 3:251, 1995.

[63] R. Vautard, P. Yiou, and M. Ghil. *Physica D*, 58:395, 1992.

6 Robust Detail-Preserving Signal Extraction

Ursula Gather, Roland Fried, and Vivian Lanius

We discuss robust filtering procedures for signal extraction from noisy time series. Particular attention is paid to the preservation of relevant signal details like abrupt shifts. Moving averages and running medians are widely used but have shortcomings when large spikes (outliers) or trends occur. Modifications such as modified trimmed means and linear median hybrid filters combine advantages of both approaches, but they do not completely overcome the difficulties. Better solutions can be based on robust regression techniques, which even work in real time because of increased computational power and faster algorithms. Reviewing the previous work we present filters for robust signal extraction and discuss their merits for preserving trends, abrupt shifts and local extremes as well as for the removal of outliers.

6.1 Introduction

Linear filters have long been the primary device for the extraction of a time-varying level (a "signal") from time series because of the profound theory of linear systems, computational ease, simple design, and optimal attenuation of additive Gaussian noise. However, they are neither suitable if there are sudden changes from one signal level to another nor in the case of impulsive noise generating strongly deviant outliers ("spikes") caused by measurement problems for instance. Change points are often the most important information and should be preserved, while at the same time a substantial amount of outliers should be resisted, since previous data cleaning is not possible in automatic application. Tukey [1] suggests standard median filters ("running medians") for these purposes, but these still have some shortcomings as we will point out in the following.

To fix notation, we assume a simple data-generating model. Let (y_t) be a time series, observed at discrete time points $t \in \mathbb{Z}$. Of course there will be only a finite number of measurements $y_1, \ldots, y_N$ available, but the main difference is that additional rules are needed for handling the endpoints. Extrapolation of the results from the first and last windows or adding the first and the last observed

Handbook of Time Series Analysis. Björn Schelter, Matthias Winterhalder, Jens Timmer

ISBN: 3-527-40623-9

values a sufficient number of times are possible ways of dealing with finite sets of data. We assume that the data are generated as

$$y_t = \mu_t + u_t + \nu_t, \quad t \in \mathbb{Z}, \tag{6.1}$$

where the sequence (μ_t) is the signal, while u_t is the "ordinary" observational noise with constant median zero and variance σ_t^2. Sporadic measurement problems are represented by the impulsive (spiky) noise ν_t from an outlier generating mechanism. It is zero most of the time, but can take very large absolute values occasionally.

The construction of filtering procedures is usually guided by some demands. One aim is to preserve certain signal characteristics, e.g., like linear or more generally monotonic trends and abrupt, long-term level shifts. Good noise attenuation is not enough to yield acceptable signal quality. Filters with optimal noise reduction could be derived under restrictions guaranteeing detailed preservation if we were willing to specify a distribution for the noise, or at least a suitably small family of distributions. However, knowledge about the noise distribution is often scarce, in particular when we are faced with measurement problems resulting in large, irrelevant spikes. Moreover, we are often confronted with other phenomena such as heteroscedasticity due to time-varying "environmental" conditions. Therefore we propagate robust filters, which perform reasonably well under a broad range of conditions and do not strongly rely on a completely specified model which is most likely misspecified.

To illustrate the previous arguments we give a simple example: If we just impose that a time invariant constant signal value $\mu_t = \mu$ is to be approximated and assume the observational noise to be independently Gaussian distributed, the most efficient method in terms of the error variance is the sample mean, i.e., the arithmetic average of all available observations. However, it is well known that the sample mean is not at all robust against deviations from normality. A simple measure of robustness is the finite-sample breakdown point of an estimator, which gives us the minimal fraction of deviant observations possibly making the estimate completely meaningless [2]. It is well known that a single outlier has an unbounded effect on the sample mean, resulting in a finite-sample breakdown point of $1/N$. A possible solution are M-estimators, which achieve some robustness and large efficiency within the so-called contamination neighborhoods of the Gaussian distribution Φ [3]. These neighborhoods contain all mixtures $(1-\epsilon)\Phi + \epsilon F$ with a constant $\epsilon \in (0, 1)$ and an arbitrary distribution F. However, there is a trade-off between efficiency and robustness: Designing the estimator for a larger neighborhood increases robustness, but reduces the efficiency at the Gaussian. The median finally is the Huber M-estimator with maximal asymptotic breakdown point 50 %, guaranteeing optimal protection among all reasonable location estimators: About half of the sample needs to be contaminated for the effects to become arbitrarily large. We will focus on methods with high breakdown points.

Instead of optimizing a single criterion, statistical procedures intended to deal with real-world data should behave well in many different aspects [4]. Common criteria in routine application are the existence of a unique solution, low computation time, the preservation of important signal details, high robustness against outliers, and satisfactory finite-sample efficiency under Gaussian or other prototype distributions. We restrict ourselves to filters fulfilling the first two demands, and compare candidate methods wrt the latter three properties.

In the general situation of a time-varying signal there are different approaches to filter construction: Recursive filters update the estimate for the previous time point including the information from the incoming observation. Exponentially weighted moving averages (EWMA) are perhaps the most common example. These filters are designed for sequential ("online") application, where one approximates the signal value at the most recent time point without delay. The resulting estimates are optimal wrt a weighted least-squares loss and very vulnerable to outliers. Robustifications based, e.g., on weighted least absolute deviations are possible, but computationally expensive and their statistical properties are difficult to analyze [5]. Recursive filters such as EWMA tend to follow changes like abrupt level shifts or monotonic trends with some delay since they only include past observations. A further major difficulty is to construct filters which preserve fine signal details like temporary shifts, while removing short sequences of irrelevant outliers.

Moving window techniques slide a time window through the series for local approximation of the signal from the data in the window. Moving averages and running medians are prominent representatives of such filters. However, moving averages and linear filters in general are not suitable for removing outliers and they always blur level shifts (also called "step changes" or "jumps"), see Fig. 6.1.

Moving window techniques can be designed for retrospective (fixed sample) or online (sequential) application. In retrospective application, a time delay does not cause a problem. Here one approximates the signal value in the center of the window, including both past and future observations in the calculations. In the online analysis one approximates the signal at the most recent time point, i.e., at the end of the window. To unify notation, we denote the time window used for the approximation of the signal value μ_t at time t by $y_{t-m}, \ldots, y_{t+\tilde{m}}$, where $\tilde{m} = m$ in the symmetric retrospective and $\tilde{m} = 0$ in the online situation. For determination of the window width $n = m + \tilde{m} + 1$ we need to choose a suitable value of m.

We discuss moving window techniques which allow us to preserve relevant signal details like level shifts and provide considerable robustness against deviations from the modeling assumptions, particularly against outlying spikes. For distinguishing between relevant temporary level shifts and irrelevant sequences of spikes we assume the latter to have shorter durations. The filter can be designed accordingly by choosing appropriate substructures and window widths. For more extensive reviews of (robust) nonlinear filters, see [6–10].

This chapter is organized as follows: Section 6.2 illustrates robust detail-preserving signal extraction using location-based filters like running medians. Section 6.3 proposes regression-based procedures which achieve large improvements in trend periods. Section 6.4 presents ideas for the modification and combination of the filters studied before. Section 6.5 draws some conclusions.

6.2 Filters Based on Local Constant Fits

Location-based filters apply a location estimator for the approximation of the signal value μ_t from $y_{t-m}, \ldots, y_{t+\tilde{m}}$. Such methods implicitly assume the signal to be almost constant within each time window, i.e., $\mu_{t-m} \approx \cdots \approx \mu_t \approx \cdots \approx \mu_{t+\tilde{m}}$ for all t. This assumption can be justified when choosing m small since the signal is assumed to vary slowly, but the cost is reduced smoothing. Generally, the window width needs to be chosen by a compromise between several aims: On the one hand, the assumption of a constant level within each window is less reasonable for large m. This causes problems particularly in the online situation as we then rely on a simple extrapolation. On the other hand, a large width stands for smaller variability, produces smoother estimates and increases robustness.

6.2.1 Standard Median Filters

Standard median filters, also called running medians, have been introduced by Tukey [1] and are perhaps the most prominent robust location-based filters. They approximate μ_t by the median of $y_{t-m}, \ldots, y_{t+\tilde{m}}$,

$$\mathrm{StM}(y_t) = \tilde{\mu}_t = \mathrm{med}(y_{t-m}, \ldots, y_{t+\tilde{m}}), \quad t \in \mathbb{Z}.$$

Like all filters based on "reasonable" location estimators, standard median filters are location and scale equivariant, meaning that adding a constant or multiplying by a constant changes the filter output in the same way. The quality of filters with these properties hence does not depend on the underlying measurement scale.

The asymptotic variance of the median is $1/[4nf^2(0)]$ if the noise has a density f with median zero. Accordingly, its asymptotic efficiency relatively to the mean is 63.7 % for the Gaussian, but 200 % for the Laplace distribution.

The finite-sample breakdown point of the median applied to n observations is $\lfloor (n+1)/2 \rfloor / n$, where $\lfloor c \rfloor$ represents the largest integer not larger than c. This means that at least half of the data needs to be shifted to completely change the estimate. This property can be used for designing running medians: To remove sequences of up to ℓ outliers and preserve level shifts with a duration of at least $\ell + 1$ observations, we can apply a running median with window width $n = 2\ell + 1$.

The exact fit point provides information on the preservation of relevant signal details and the removal of spikes under idealized conditions with no observational noise, i.e., $\sigma_t^2 \equiv 0$. Applied to a regression functional $T \colon \mathbb{R}^n \to \mathbb{R}^p$, the exact fit point corresponds to the smallest possible fraction of contamination which can cause T to deviate from a fit $\tilde{\gamma} \in \mathbb{R}^p$. Consider a sample $\mathbf{y}_n = \{(x_1, y_1), \ldots, (x_n, y_n)\}$ of n observations of a response y and a p-variate regressor x such that $y_i = \tilde{\gamma}' x_i$ for all $i = 1, \ldots, n$, and let $\mathbf{y}_{k,n}$ be a sample where k out of the n observations in $\mathbf{y}_n$ are replaced by arbitrary values. The exact fit point of T then becomes

$$\delta_n^*(T, \mathbf{y}_n) = \min_k \left\{ \frac{k}{n} \,\middle|\, \text{ there exists a sample } \mathbf{y}_{k,n} \quad \text{such that } T(\mathbf{y}_{k,n}) \neq \tilde{\gamma} \right\}.$$

The median, like all location estimators, regresses on a constant only, i.e., $p = 1$ and $x_i = 1, i = 1, \ldots, n$. Its exact fit point is equal to its finite-sample breakdown point. While the latter corresponds to the minimal number of spikes which can render the extracted value meaningless, the former yields the number of spikes a filter can remove completely in the absence of observational noise. A running median with width $n = 2\ell + 1$ can hence remove up to ℓ subsequent spikes completely if $\sigma_t^2 \equiv 0$. In retrospective application, it can preserve a level shift from one constant signal value to another exactly if it lasts at least for $\ell + 1$ observations, while in online application the shift is delayed by ℓ observations. Another notable property of the running median in retrospective application is that it recovers monotonic trends exactly under noise-free conditions.

The exact preservation of signal characteristics as described above applies only under idealized conditions. Nevertheless, the deviations can be expected to be small in the presence of little (as compared, e.g., to the height of shifts) observational noise since the median is Lipschitz continuous with constant 1: The median deviates at most by δ from μ_t if for all $i\, |u_{t+i}| < \delta$ and if not more than one of the following occurs in the window: at most ℓ spikes, a single level shift, or a monotonic trend. Lipschitz-continuous functionals are to be recommended in general since this property restricts the influence of minor changes in the data due to small observational noise or rounding [4].

Nevertheless, the performance of running medians becomes worse at monotonic changes (edges): It suppresses noise less efficiently there, and it shows a bias which is related to the noise power and the height of the edge. Further problems arise when more than one data pattern occurs in a single window: Running medians suffer from edge jitter, i.e., they move shifts toward preceding close-by spikes into the same direction. A shift during a monotonic trend can be preserved only if the shift and the trend point to the same direction. The shift gets blurred otherwise, and a single spike within a trend causes smearing [11–15]. Median filters with an adaptive window width have been suggested to reduce edge and plateau jitter caused by spikes close to edges [16–18]. The window width can be chosen using criteria such as the current signal slope [18], the length of detected outlier sequences [17, 19], or a variance decomposition assuming the noise variance to be stationary [20, 21].

When designing a filter we often want certain signals to pass the filter unperturbed. For a linear filter, such eigenfunctions can be characterized in the frequency domain by its passband and stopband. Signals which pass a nonlinear filter unchanged are called its roots and can be analyzed in the time domain. The roots of a running median with width $n = 2\ell + 1$ contain only edges of monotonic increase or decrease, separated by at least $\ell + 1$ constant values [22]. Thus, the roots of a running median are also roots of all running medians with smaller width. A running median reduces any time series within a finite number of repetitions to one of its roots.

Recursive medians are a simple variation of running medians, replacing the observations before time t by the already filtered values when calculating the output at t. A recursive median possesses the same set of roots as a running median with the same window width, but a time series may be filtered to different roots by the two filters. Recursive medians reduce every series to become a root in a single step; they provide better smoothing and they are more robust than running medians, but they distort edges more strongly [23, 24].

6.2.2 Modified Order Statistic Filters

Instead of the median, other order statistics (OS) can be applied for filtering as well. Switching to a higher or lower order statistic can improve the preservation of shifts [25]. More generally, OS-Filters, or L-filters, are based on linear combinations of order statistics [26]. Using a set of weights $w_1, \ldots, w_n$ summing up to 1, the filter output is calculated as

$$\mathrm{OS}(y_t) = \sum_{i=1}^{n} w_i y_{t(i)}, \tag{6.2}$$

where $y_{t(1)}, \ldots, y_{t(n)}$ are the ordered observations within the window. A suitable choice of the weights allows us to dampen noise with different tail behavior efficiently [26, 27]. Order statistic filters are location and scale equivariant. They preserve linear trends exactly in retrospective application and under noise-free conditions if the weights are chosen symmetric, $w_i = w_{n-i+1}$, $i = 1, \ldots, n$. As special cases we obtain the mean ($w_i = 1/n$ for $i = 1, \ldots, n$), the median ($w_{\ell+1} = 1$ and $w_i = 0$ otherwise for odd $n = 2\ell + 1$), the midpoint ($w_1 = w_n = 1/2$ and all other $w_i = 0$), and the α-trimmed means ($w_i = 1/(n - 2\lfloor \alpha n \rfloor)$ for $i = \lfloor \alpha n \rfloor + 1, \ldots, n - \lfloor \alpha n \rfloor$ and $w_i = 0$ otherwise). Order statistic filters with nearly minimal mean squared error (MSE) for a given error distribution can be designed using analytical approximations even in real-time application [28].

α-trimmed means (α-TM) have received considerable attention since they constitute a compromise between the mean ($\alpha = 0$) and the median ($\alpha = 0.5$). Often $\alpha \in [0.2, 0.275]$ is suggested to yield good efficiency for a broad family of distributions including the Gaussian [29, 30], i.e., we trim between 20 % and 27.5 % of the smallest and the largest observations. The price to be paid for increased efficiency close to the Gaussian as compared to the median is a smaller resistance to outliers: The breakdown point of an α-TM is asymptotically $2\alpha \times 100\,\%$.

Accordingly, an α-TM filter with $\alpha < 0.5$ smoothes a level shift to a ramp edge with $(1-\alpha)n$ observations [31–34]. More generally, running medians are the only order statistic filters which can preserve shifts exactly [29].

Order statistic filters with data-adaptive choice of the weights have been suggested to overcome this deficiency. They achieve considerable robustness against outliers and at the same time high efficiency under a broad range of conditions including time-varying, heterogeneous noise. Modified trimmed mean (MTM) filters are defined in analogy to trimmed means, but they choose the fraction of trimming α depending on the data in the current window. Observations which are further away than a distance q_t from the local median are trimmed and the average of the remaining observations is taken as filter output:

$$\begin{aligned} \mathrm{MTM}(y_t) &= \frac{1}{|I_t|} \sum_{i \in I_t} y_{t+i}\,, \\ I_t &= \{i = -m, \ldots, \tilde{m} : |y_{t+i} - \tilde{\mu}_t| \leqslant q_t\} \\ \tilde{\mu}_t &= \mathrm{med}(y_{t-m}, \ldots, y_{t+\tilde{m}}), \quad t \in \mathbb{Z}\,. \end{aligned} \tag{6.3}$$

Hence, MTM filters are a data-adaptive compromise between the running median ($q_t = 0$) and the moving average ($q_t = \infty$), compare also Fig. 6.1. An *a priori* choice of q_t can be based on the expected height of the shifts. A data-adaptive alternative can be formulated using a robust scale estimate like the local median absolute deviation about the median (MAD),

$$\tilde{\sigma}_t^M = c_n \cdot \mathrm{med}(|y_{t-m} - \tilde{\mu}_t|, \ldots, |y_{t+\tilde{m}} - \tilde{\mu}_t|).$$

Here, c_n is a correction factor depending on the window width n, usually chosen to achieve unbiasedness in the case of Gaussian noise. For n not very small we set $c_n = 1.483$. A reasonable range of choices is $q_t \in [2\tilde{\sigma}_t^M, 3\tilde{\sigma}_t^M]$, see [32, 35].

Double window modified trimmed mean (DWMTM)-filters are a variant of MTM-filters. They apply two windows with different widths. The median and the MAD are calculated from a short signal window with width $k < n$ to retain signal details. Then all observations deviating more than q_t from this median are trimmed from the larger window with width n, before the remaining values are averaged for better attenuation of observational noise. MTM filters can be seen as DWMTM filters with $k = n$. DWMTM-filters with adaptive choice of that factor, by which the local MAD is multiplied, have been suggested for removing signal-dependent noise [36].

Analyzing the breakdown and exact fit points shows that a DWMTM can remove up to $\lfloor k/2 \rfloor$ subsequent spikes from a constant signal under noise-free conditions. The smaller window width k should hence be chosen depending on the minimal duration of relevant signal details. Using a short inner window improves the preservation of shifts, see Fig. 6.1, but reduces the attenuation of noise. A DWMTM-filter can be tuned to be considerably more efficient for Gaussian noise and preserve large shifts better than a running median with the same n choosing k and q_t large enough [10, 35].

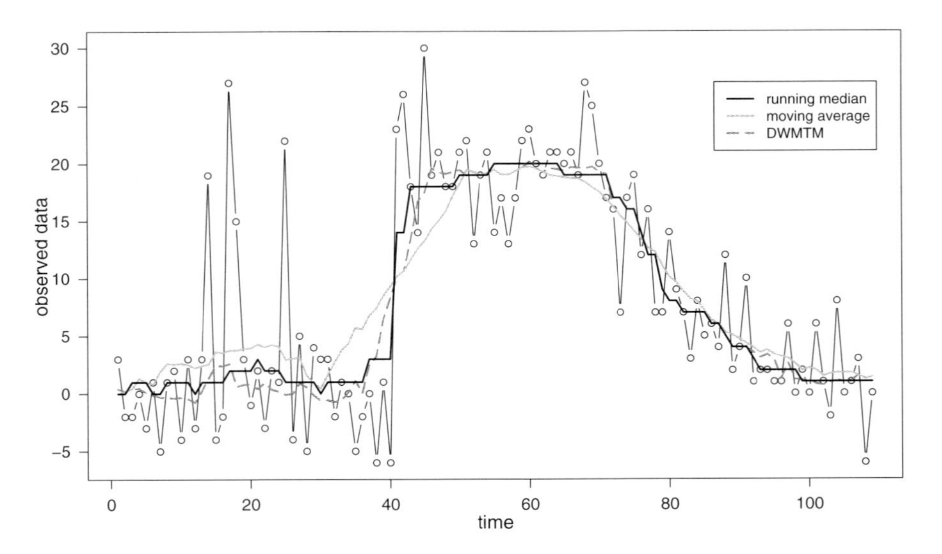

Fig. 6.1: Observed time series with outliers ∘, level shift and trend together with approximations of the signal by means of location-based filters: moving average, median and DWMTM-filter with $m = \tilde{m} = 10$, i.e., $n = 21$, and $k = 9$. In this example the median performs very well; however, it approximates the linear trend by a step function.

DWMTM-filters calculate the mean of a subsample which is chosen according to the distances from an initial estimate. This means a kind of hard thresholding and implies discontinuity. Soft thresholding with a smooth transition between acceptance and rejection can be obtained in the form of weighted averages with weights chosen according to the initial distances [16]. This idea will be explained in more detail at the end of the next section.

Further location estimators have been proposed for filtering. The Hodges–Lehman–Bickel (HLB) estimate of location is the median of averages of symmetrically placed order statistics,

$$\mathrm{HLB}_n(y_{t-m}, \ldots, y_{t+\tilde{m}}) = \mathrm{med}_{i=1,\ldots,\lfloor(n+1)/2\rfloor} \frac{y_{t(i)} + y_{t(n-i+1)}}{2} . \tag{6.4}$$

It is location and scale equivariant, preserves trends in the absence of noise, and has a breakdown point of 25 %. Nevertheless, the application of suitably trimmed means seems preferable [30].

6.2.3 Weighted Median Filters

The standard median and (modified) order statistic filters defined in the previous sections do not take into account the temporal distances between the target point t at which we estimate the signal and the observation times of the measurements included in the calculation. This causes problems if the implicit assumption of a locally constant signal within each window is not fulfilled. A remedy is to weight the observations according to their temporal distances [37], giving smaller weight to observations more distant from the target point.

We focus on weighted median (WM) filters: while the median minimizes the L_1-distance (the sum of the absolute deviations) to the data points, the weighted median of $y_{t-m}, \ldots, y_{t+\tilde{m}}$ for arbitrary positive real weights $w_{-m}, \ldots, w_{\tilde{m}}$ minimizes the weighted L_1-distance

$$\mathrm{WM}(y_t) = \arg\min_{\mu} \sum_{i=-m}^{\tilde{m}} w_i \cdot |y_{t+i} - \mu|. \tag{6.5}$$

Running medians correspond to uniform weights $w_i = 1$, $i = -m, \ldots, \tilde{m}$. WM filters have become popular because of their high flexibility: A running median necessarily applies a window of width $n = 2\ell + 1$ to preserve signal details of length $\ell + 1$ and to remove up to ℓ outlying spikes. Weighting of the observations allows us to use longer windows and thus yields a stronger noise reduction [23, 37].

Denoting the ordered observations in the window by $y_{t(1)} \leqslant \cdots \leqslant y_{t(n)}$ and the corresponding positive weights by $w_{(1)}, \ldots, w_{(n)}$, the weighted median corresponds to the kth-order statistic $\hat{\mu} = y_{t(k)}$, where

$$k = \max\left\{h\colon \sum_{i=h}^{n} w_{(i)} \geqslant \frac{1}{2}\sum_{i=1}^{n} w_i\right\}. \tag{6.6}$$

For example, the WM of 1, 2, 3, 9 with weights 0.1, 1.6, 1.4, and 0.5 is $y_{(3)} = 3$, since $0.5 + 1.4 \geqslant 3.6/2$. Generally, Eqs. (6.6) and (6.5) yield the same results. However, the whole interval $[y_{(k-1)}, y_{(k)}]$ solves Eq. (6.5) whenever $\sum_{i=k}^{n} w_{(i)} = \frac{1}{2}\sum_{i=1}^{n} w_i$. The solution $y_{(k-1)}$ would be obtained in Eq. (6.6) by summing from the bottom instead of from the top and taking the minimum instead of the maximum. This ambiguity can be solved as usual by choosing the center of the interval (the only choice which gives affine equivariance). For nonnegative integer valued weights $w_1, \ldots, w_n$, a simple equivalent representation of the weighted median of $y_{t-m}, \ldots, y_{t+\tilde{m}}$ is

$$\mathrm{WM}(y_t) = \mathrm{med}(w_{-m} \diamond y_{t-m}, \ldots, w_{\tilde{m}} \diamond y_{t+\tilde{m}}), \tag{6.7}$$

where $w \diamond y$ denotes the replication of y to obtain w identical copies.

Even though there is an infinite number of real weights, there is only a finite number of WM filters for a given window width. In particular, for every WM with arbitrary positive real weights there is an equivalent WM with integer weights [38]. Two weighted medians with respective weights $w_1, \ldots, w_n$ and $\tilde{w}_1, \ldots, \tilde{w}_n$ are called equivalent iff they give the same result for every sample. This is the case iff for every subset $I \subset \{1, \ldots, n\}$ of indices we have

$$\sum_{i \in I} w_i \geqslant 0.5 \sum_{i=1}^{n} w_i \iff \sum_{i \in I} \tilde{w}_i \geqslant 0.5 \sum_{i=1}^{n} \tilde{w}_i.$$

In particular, to get equivalence to the standard median it is crucial that the weights are balanced, such that no subset of less than $\lfloor (n+1)/2 \rfloor$ weights sums up to at least half the total mass. For an overview on the equivalence of WMs, see [39].

WM filters are unbiased for the mean in the case of symmetric noise. Formula for output central moments and the variance of WM filters can be found in [40], as well as an algorithm to obtain WM filters which minimize symmetrically distributed noise under the constraint that certain signal details are to be preserved under noise-free conditions. The optimal WM filter does not depend on the underlying error distribution, and it is optimal both under the MSE and under the mean absolute error (MAE) criterion. In the absence of structural constraints, the WM filter with minimal MAE and MSE for a given window width is the running median.

Root signal properties of general WM filters are much more difficult to derive than those of running medians [23]. Weighted median filters are basically low-pass filters, like the other filter classes treated here. The frequency response of selection type nonlinear filters like WMs can be analyzed by comparison with a linear filter having the sample selection probabilities as coefficients [41, 42]. Weighted median filters can be used for high-pass and band-pass filtering by allowing for negative weights [43] or by the linear combination of several weighted medians [44]. Ideas for robust periodograms and robust short-time Fourier transforms based on M-estimators in general and medians in particular can be found in [45]. For similarities between WM filters and linear filters with finite impulse response (FIR), see [23].

Weighting according to the temporal distances can of course also be applied to location estimators different from the median. DWMTM filters with additional weighting according to the temporal order of the observations can retain desired signal frequencies in addition to edge preservation and impulse suppression [46]. Again we can also apply the soft thresholding described at the end of subsection 6.2.2 [47]. Let $w_{-m}, \ldots, w_{\tilde{m}}$ be weights according to the temporal distances in the design space as before. Further, additional weights for the distances in the observation space are derived using an unimodal affinity function A, which is controlled by initial robust estimates of location μ and spread γ, e.g., the median and the MAD. Then the resulting weighted order statistic (WOS) affine FIR filter reads

$$\mathrm{WAF}(y_t) = \sum_{i=-m}^{\tilde{m}} w_i A_i^{\mu,\gamma} y_{t+i} \Big/ \sum_{i=-m}^{\tilde{m}} w_i A_i^{\mu,\gamma} . \tag{6.8}$$

Filters defined like this are data adaptive and location equivariant, and they can preserve trends and shifts exactly under noise-free conditions. For the preservation of shifts and the suppression of spikes the affinity function needs to decay sufficiently fast to zero.

Very general filter classes have been derived by linear combination of all order statistics with weighting according to both the temporal and the rank order [46, 48], but these are difficult to design except for multiplicative weights. Similarly, generalized Wilcoxon filters can be constructed combining linear rank statistics and temporal weighting, but they seem to be inferior to DWMTM filters both wrt edge preservation and noise attenuation [49].

6.3 Filters Based on Local Linear Fits

Location-based filters like those discussed before have difficulties in trend periods since the assumption of a local constant level is only appropriate when using very short time windows. These filters lose both efficiency and robustness in trend periods. Neither can they preserve arbitrary shifts during trends, nor can they remove spikes completely, not even under idealized conditions. Only DWMTMs can keep their good properties during trends if the inner window is sufficiently short [15, 50, 51].

It suggests itself that local linear fits are preferable to local constant fits [52], as they improve the approximation. In the context of time series filtering this means that we assume the data in a moving time window to be locally well approximated by a linear trend, $\mu_{t+i} = \mu_t + i\beta_t$, $i = -m, \ldots, \tilde{m}$. For estimation of the level μ_t and the slope β_t at time t we can apply robust linear regression to fit this local model, see also Fig. 6.2. In addition to the location and scale equivariance of location-based filters, a filter thus obtained offers invariance to (linear) trends [15] when using a regression-equivariant functional. This property guarantees that the quality of signal extraction does not depend on an underlying local linear trend. When varying the trend in the window, i.e., replacing $y_{t-m}, \ldots, y_{t+\tilde{m}}$ by $y_{t-m} - mc, \ldots, y_{t-1} - c, y_t, y_{t+1} + c, \ldots, y_{t+\tilde{m}} + \tilde{m}c$, the level estimate at time t remains the same, while the slope estimate increases by c.

6.3.1 Filters Based on Robust Regression

Contrary to the median for robust estimation of location, no generally accepted unique standard exists for robust linear regression. Comparisons of common robust regression techniques in the retrospective and in the online situation, respectively, can be found in [50, 53].

Like the median, standard L_1-regression minimizes the least absolute deviations (LAD)

$$(\hat{\mu}_t^{L1}, \hat{\beta}_t^{L1}) = \operatorname{argmin}\{(\mu, \beta): \sum_{i=-m}^{\tilde{m}} |y_{t+i} - \mu - \beta(t+i)|\}. \tag{6.9}$$

The hierarchical repeated median (RM) [54] at the target point t is

$$\begin{aligned} \hat{\beta}_t^{RM} &= \text{med}_{j=-m,\ldots,\tilde{m}}\left(\text{med}_{i\neq j}\,\frac{y_{t+i}-y_{t+j}}{i-j}\right), \\ \hat{\mu}_t^{RM} &= \text{med}\left(y_{t-m}+m\hat{\beta}_t^{RM},\ldots,y_{t+\tilde{m}}-\tilde{m}\hat{\beta}_t^{RM}\right). \end{aligned} \tag{6.10}$$

The RM firstly calculates a slope estimate $\hat{\beta}_t^{RM}$ by taking repeated medians of all pairwise slopes in the window, and then a level estimate $\hat{\mu}_t^{RM}$ as the median of the trend-corrected observations. It has turned out to outperform standard L_1-regression in most respects.

The Hampel–Rousseeuw least median of squares (LMS) [55, 56] minimizes the median of the squared distances,

$$(\hat{\mu}_t^{LMS}, \hat{\beta}_t^{LMS}) = \text{argmin}\{(\mu,\beta)\colon\ \text{med}_{i=-m,\ldots,\tilde{m}}[y_{t+i}-\mu-\beta(t+i)]^2\}. \tag{6.11}$$

A generalization is the least quantile of squares (LQS), replacing the median by another quantile.

Both the RM and the LMS possess the maximal breakdown point $\lfloor n/2 \rfloor / n$ for regression-equivariant estimators calculated from a sample of size n. This implies the same asymptotic 50 % breakdown point as for the standard median. The breakdown point of L_1-regression is smaller than this and asymptotically not larger than 25 % in the case of an equidistant design like in time series filtering [57].

For regression- and scale-equivariant functionals, the exact fit point is never smaller than the finite-sample breakdown point [58]. In the case of linear regression, an exact fit point of k/n means that whenever $y_{t+i} = \tilde{\mu} + \tilde{\beta} i$ fits at least $n-k$ of the n observations exactly, then the estimate becomes $(\tilde{\mu}, \tilde{\beta})$ whatever the other k observations are. The exact fit point of the LMS is $\lceil n/2 \rceil / n$, see [58], while for the RM it is $\lfloor n/2 \rfloor / n$, i.e., one less observation is needed to pull the fit away if the sample size is odd.

The RM and the LMS have the same breakdown point, but the LMS better resists many large outliers as even almost 50 % outliers of any size do not cause it to be strongly biased. Accordingly, it is able to preserve a level shift almost exactly in retrospective application. The strong negative bias of the corresponding scale estimate can be used to determine an LQS adaptively by comparison with the residual standard deviation [59]. To its disadvantages belongs its computational complexity of order n^2 [60], yielding computation times rapidly increasing with the window width. Besides, the LMS filter output is very wiggly since it is not continuous and its Gaussian efficiency is less than 25 % in small samples, and even decreasing in n.

In spite of the benefits of the LMS, the repeated median can be recommended both for retrospective and online applications [50, 53]. It offers almost the same Gaussian efficiency of about 65 % as the standard median, but independently of the underlying slope, Lipschitz continuity implying stability in the case of small changes in the data, and reasonable robustness as it resists well up to about 30 % outliers in a single window. It is computationally faster than the LMS, particularly so since a fast update algorithm is available which allows calculating the next filter value in linear time when moving the window forward [61]. Its main disadvantage consists probably in increased smoothing in the case of a level shift, see Fig. 6.2. A common phenomenon of local linear fits in online application is an overshoot of the new signal value after a shift [62]. The RM performs considerably better than the LMS in this respect. Further improvements retaining the robustness can be achieved by an adaptive choice of the window width based on residual sign tests [63].

Least trimmed squares (LTS) regression [64] can be seen as a modification of the LMS and has also been suggested for filtering purposes [65]. The LTS has better asymptotic properties than the LMS, especially a nonzero Gaussian efficiency, but it is computationally even more expensive and performs similar to the LMS in finite samples [53]. In the same way, no significant advantages of deepest regression [66] have been found as compared to the RM.

6.3.2 Modified Repeated Median Filters

In analogy to the modified trimmed mean filters defined in Section 6.2.2, we can fit a least-squares regression line, trimming or more generally down-weighting observations with large residuals in a preliminary robust regression step. This allows us to retain the breakdown point of the initial estimate when giving observations with huge residuals zero weight. Reweighted least squares (RLS) based on an initial LMS fit is popular for robust regression since it increases the Gaussian efficiency of the LMS considerably, but RLS can be unstable like the LMS because of its inherent lack of continuity.

Trimmed repeated median (TRM)-filters suggested in [51] use the RM in the initial step and apply least squares to the trimmed observations in a second step. A suitable trimming constant q_t can be obtained by estimating the variability about the RM regression line, e.g., by the MAD of the regression residuals [67].

Since TRM filters apply regression-equivariant functionals in both steps, they are not only location and scale equivariant, but like RM filters also trend invariant. Instabilities have not been observed, although RM filters are not Lipschitz-continuous because of the hard thresholding [51]. A TRM filter can be substantially more efficient than the RM with the same width n, depending on the amount of trimming. Choosing q_t as three (two) times the MAD yields, e.g., the Gaussian efficiency of 92 % (76 %). TRM filtering is computationally feasible since an update algorithm can be applied for the initial RM [61].

Double window filters with a shorter inner window width $k < n$ in the initial step improve the preservation of signal details, especially of abrupt shifts, see Fig. 6.2. The choice of k should depend on the length of outlier patches the filter should cope with: Up to $\lfloor k/2 \rfloor - 1$ outliers in the inner window can be resisted before the output can be completely wrong, according to the breakdown point of the initial RM. In practice, k should even be chosen about three times the length of outlier patches to be removed since one-third outliers can have a big, though limited influence on the RM. The benefits obtained in the case of a level shift increase with the difference $n - k$ between the outer and the inner window widths.

6.3.3 Weighted Repeated Median Filters

Application of a regression instead of a location estimator to the data in a moving window (implicitly) replaces the assumption of a locally constant level by that of a locally constant slope. Using ideas similar to those underlying weighted medians, we can weight the observations according to their temporal distances. Doing so we aim at increasing the window width of standard robust regression filters, without increasing the bias when the signal slope is time varying.

Weighted repeated median (WRM) filters and weighted L_1-filters for detail-preserving robust filtering are investigated in [68]. Weighting reduces the breakdown point of the repeated median, while it can increase that of L_1-regression when down-weighting observations far away from the target point t. The breakdown point can be further increased when confining to an approximative weighted L_1-solution: Starting from a high breakdown fit like the standard RM, we can iterate a finite number of steps between maximization wrt μ given β and vice versa. In the case of standard L_1-regression, this increases the breakdown point asymptotically to $1 - 1/\sqrt{2} \approx 0.293$.

The WRM in combination with the so-called Epanechnikov weights $w_i^{(1)} = 1 - [|i|/(m + 1)]^2$, $i = -m, \ldots, 0$, is well adapted for online application, while L_1-regression with weights $w_i^{(2)} = (1 + |i - t|)^{-1/2}$, $i = -m, \ldots, m$ performs even better in the retrospective case. Similar to weighted medians, the weighting in combination with the possible longer window widths increases considerably the Gaussian efficiency of these filters in the respective situation.

A simple WRM designed for preserving level shifts in retrospective application uses a shorter window for the initial slope estimation, applying uniform weights [51]

$$\begin{aligned} \mathrm{DWRM}(y_t) &= \mathrm{med}(y_{t-m} + m\hat{\beta}_t, \ldots, y_{t+m} - m\hat{\beta}_t) \\ \hat{\beta}_t &= \mathrm{med}_{i=-h,\ldots,h}\left(\mathrm{med}_{j=-h,\ldots,h,j\neq i} \frac{y_{t+i} - y_{t+j}}{i - j}\right). \end{aligned} \quad (6.12)$$

The DWRM slope is little affected until the shift intrudes into the inner window, resisting a shift almost as good as a standard median in the case of a constant signal if $h \ll m$. It is almost as efficient as the median with the same width $2m+1$ in the case of a constant signal, but it is trend invariant like all WRMs. Different from the double window filters presented in Section 6.3.2, the DWRM is Lipschitz continuous with constant $2h + 1$.

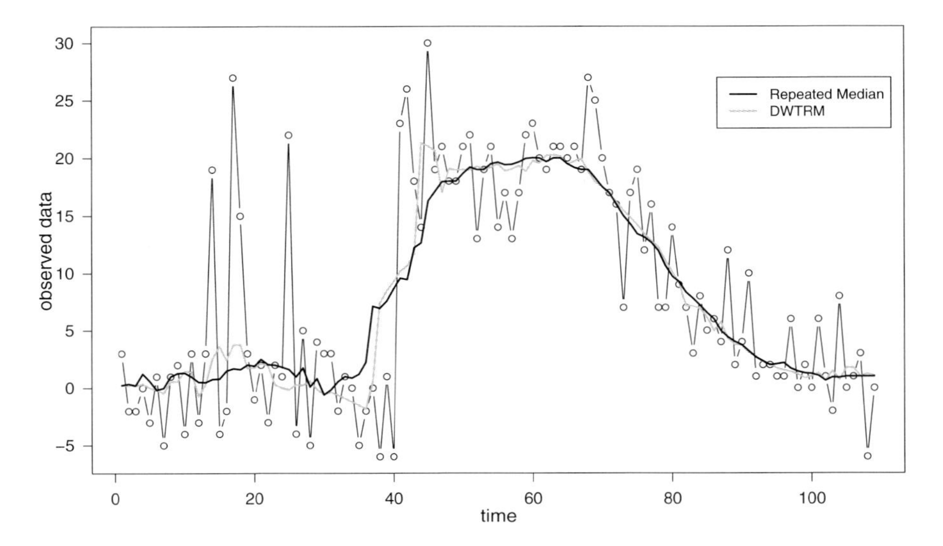

Fig. 6.2: Observed time series with outliers ∘, level shift and trend together with approximations of the signal by means of filters based on local linear fits: simple repeated median filter and DWTRM filter with $m = \tilde{m} = 10$, i.e., $n = 21$, and $k = 9$.

6.4 Modifications for Better Preservation of Shifts

A major disadvantage of the previous filters based on the repeated median is the smearing of level shifts, which is stronger than for median-based filters in the case of a constant signal. Double window filters reduce this effect, but they do not solve the problem completely. In the following we present some possibilities for further improvement.

6.4.1 Linear Median Hybrid Filters

Linear median hybrid (LMH) filters take the median value of linear subfilters $\Phi_1, \ldots, \Phi_M$ as the filter output [69–71]. When all subfilters give nonzero weight to only a finite number of observations, the resulting procedure is called linear median hybrid filter with finite impulse response, briefly FMH filter

$$\mathrm{FMH}(y_t) = \mathrm{med}\left[\Phi_1(y_t), \Phi_2(y_t), \ldots, \Phi_M(y_t)\right], \quad t \in \mathbb{Z}. \tag{6.13}$$

The linear subfilters used for preprocessing reduce the computational costs as compared to a running median with the same width, and they provide increased flexibility due to the many choices possible. They can be designed to track well polynomial trends of different degrees p.

Simple FMH filters are adapted to a constant signal ($p = 1$), using $M = 3$ subfilters, namely two one-sided moving averages and the current observation y_t

$$\Phi_1(y_t) = \frac{1}{m}\sum_{i=1}^{m} y_{t-i}, \quad \Phi_2(y_t) = y_t, \quad \Phi_3(y_t) = \frac{1}{m}\sum_{i=1}^{m} y_{t+i}. \tag{6.14}$$

Including the central observation as central subfilter allows us to preserve level shifts even better than running medians [71]. FMH filters are suitable only in retrospective application when using backward forecasting filters. We thus set $\tilde{m} = m$ in general.

Predictive FMH filters apply subfilters for one-sided extrapolation of linear trends ($p = 1$)

$$\mathrm{PFMH}(y_t) = \mathrm{med}\left[\Phi_F(y_t), y_t, \Phi_B(y_t)\right], \tag{6.15}$$

where $\Phi_F(y_t) = \sum_{i=1}^{m} h_i y_{t-i}$ and $\Phi_B(y_t) = \sum_{i=1}^{m} h_i y_{t+i}$. The minimal MSE predictions for a linear trend in the case of Gaussian noise under the restriction that the exact signal value is obtained in the deterministic situation without noise use the weights $h_i = \frac{4m-6i+2}{m(m-1)}$, $i = 1, \ldots, m$, see [70].

Combined FMH filters use predictions of different degrees,

$$\mathrm{CFMH}(y_t) = \mathrm{med}\left[\Phi_F(y_t), \Phi_1(y_t), y_t, \Phi_3(y_t), \Phi_B(y_t)\right], \tag{6.16}$$

where $\Phi_1(y_t)$, $\Phi_3(y_t)$, $\Phi_F(y_t)$, and $\Phi_B(y_t)$ are the subfilters for forward and backward extrapolation of a constant signal or a linear trend given above.

FMH filters have a smaller bias error at level shifts than running medians at the expense of a larger variance around the shift [71]. They do not suffer from edge jitter, but a spike—distant at most m time points from a shift—causes some smearing as the height of a shift and a constant signal value close to the shift change [69]. FMH filters recover linear trends in the absence of noise exactly, but only the PFMHs are trend invariant and thus can preserve shifts within trends as good as in constant periods.

However, PFMHs are neither very efficient for Gaussian noise nor very robust. All FMH filters dampen isolated outliers better than running medians [70], but already two outliers can affect them strongly [15]. The CFMH filters improve the Gaussian efficiency of PFMHs considerably when the signal is constant, becoming about as efficient as a simple FMH or a median with the same width [69]. However, this advantage gets lost with increasing signal slope. Every FMH filter is Lipschitz continuous with constant $\max|h_i^j|$, the maximal absolute weight given by one of the subfilters.

Different from running medians, FMH-filters create new values and can smooth oscillations between two measurements. Besides signals consisting only of local constant neighborhoods and edges, among the roots of FMH filters we find, e.g., triangular waves, which are not roots of running medians [69, 70]. Repeated filtering with increasing window widths helps us to overcome the typical triangular wave form of FMH-filtered time series.

Variations of FMH filters have been proposed. Recursive FMH filters apply the previously filtered values in the forward predictions. They provide better noise reduction than their nonrecursive counterparts and running medians, but they distort edges because of larger bias errors [71]. In-place growing FMH filters use a cascade of FMH filters of different widths [72],

$$z_t^{(0)} = y_t \tag{6.17}$$

$$z_t^{(j)} = \text{med}\left[\Phi_{lj}(y_t), z_t^{(j-1)}, \Phi_{rj}(y_t)\right] \tag{6.18}$$

with subfilters Φ_{lj} and Φ_{rj} of width increasing in j. These filters preserve shifts better than median and (recursive) FMH filters. Similar variants have been suggested for improved trend elaboration [73]. Finally, weights can be given to the linear subfilters. An optimal FIR–WOS hybrid filter under the MAE criterion can be found by an adaptive algorithm [23].

A general framework for adaptive order statistic, i.e., location-based filtering is developed in [74]. Similar to the hybrid filters discussed before, the idea is to use test statistics for selecting one of the location estimates obtained from different subwindows, or more generally, to obtain a weighted linear combination of all of them. A triple window median filter turned out to perform particularly well for the retrospective elimination of impulsive noise and edge preservation.

6.4.2 Repeated Median Hybrid Filters

To overcome the lack of robustness of FMH filters, we can construct hybrid filters with robust instead of linear subfilters [15]. We replace the half-window averages in the simple and the combined FMH by half-window medians, and use half-window repeated medians RM^F and RM^B for a linear trend

$$RM^F(y_t) = \text{med}(y_{t-m} + m\hat{\beta}_t^F, \ldots, y_{t-1} + \hat{\beta}_t^F), \tag{6.19}$$

$$\hat{\beta}_t^F = \text{med}_{i=-m,\ldots,-1}\left(\text{med}_{j=-m,\ldots,-1,j\neq i} \frac{y_{t+i} - y_{t+j}}{i-j}\right), \tag{6.20}$$

$$RM^B(y_t) = \text{med}(y_{t+1} - \hat{\beta}_t^B, \ldots, y_{t+m} - m\hat{\beta}_t^B), \tag{6.21}$$

$$\hat{\beta}_t^B = \text{med}_{i=1,\ldots,m}\left(\text{med}_{j=1,\ldots,m,j\neq i} \frac{y_{t+i} - y_{t+j}}{i-j}\right). \tag{6.22}$$

The resulting repeated median hybrid (RMH) filters are Lipschitz continuous with the same constant $2m+1$ as the RM. Fast update algorithms are available for the computation [15]. RMH filters are location and scale equivariant, but as for FMH filters only the predictive version is trend invariant. Replacing the central observation by the median of the whole window increases robustness and the Gaussian efficiency while also preserving shifts, but destroys the trend invariance, see Fig. 6.3.

RMH filters have the same nice properties wrt shift preservation as FMH filters, while improving upon them wrt the removal of spikes. RMH filters preserve

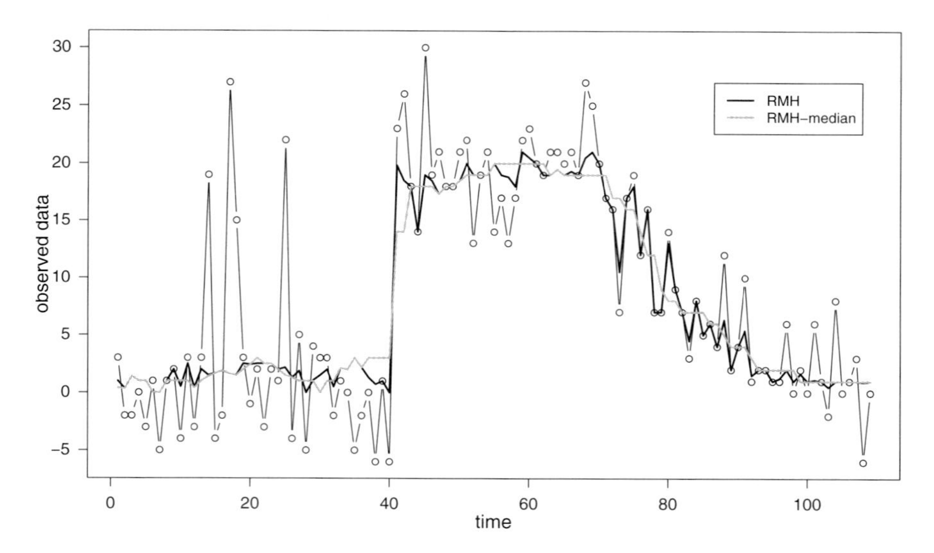

Fig. 6.3: Observed time series with outliers ○, level shift and trend together with approximations of the signal by means of median hybrid filters: RMH filter and robustified RMH filter (median replacing the central observation) with $m = \tilde{m} = 10$.

shifts better than the median even if the signal is constant. Only the predictive RMH preserves shifts irrespective of a trend because of its invariance, while the combined RMH has problems with shifts into the opposite direction of a trend, but less than the median.

The predictive RMH and the combined RMH have breakdown points $(\lfloor m/2 \rfloor + 1)/n$ and $(\lfloor m/2 \rfloor + 2)/n$, respectively, so that they guarantee some protection against up to five and six outliers, respectively, within $n = 21$ observations, while FMH filters do so only for a single outlier. The situation with a single long outlier patch starting right in the center of the window turns out to be a worst case situation for most of these hybrid methods, while several short patches have much smaller effects. In simulations, two outliers are found to damage the FMHs considerably, while the RMHs resist them substantially better. The combined RMH even resists about 25 % outliers when the signal is constant. However, the RMHs can be more affected by a patch of successive outliers than the standard RM.

With respect to the Gaussian efficiency, RMH filters are only slightly worse than the respective FMH filters. Again, the combined versions are more efficient than the predictive ones if the signal is constant, but this gain gets lost with increasing slope.

Summarizing, RMH filters are preferable to FMH filters since they provide the same benefits and are considerably more robust for the price of only a small loss in efficiency under the Gaussian. As compared to the standard RM, they attenuate Gaussian noise and long sequences of spikes less efficiently and are more variable, but they can preserve shifts and local extremes even better than the median when the signal is constant. The combined RMH improves the efficiency and the robustness of the predictive RMH in the case of a constant signal, but the latter preserves shifts irrespective of a trend.

6.4.3 Level Shift Detection

Instead of designing filters for improved shift preservation, we can incorporate rules for shift detection so that appropriate actions can be taken. Accordingly, an abundance of edge detection rules has been suggested. Some kind of low-pass filtering followed by differentiation is a common approach. FMH detectors combining several edge detection rules are outlined in [75]. However, detection rules based on differences which are optimized, e.g., for the Gaussian distribution can be adversely affected by deviations from this assumption, and in particular such nonrobust rules often confuse spikes with shifts.

Robust shift detection can be based on a comparison of two robust level estimates. Considerable robustness with only a small loss under the Gaussian can already be achieved when using an F-test comparing trimmed means and using a winsorized variance for standardization [76, 77]. A retrospective multilevel filter for edge detection and efficient suppression of different types of noise is suggested in [78]. If two half-sided median subfilters deviate largely, an edge is detected and the filter output is calculated as the median of these half-sided medians and the current observation. Alternatively, it is decided if the shift has happened at this or the previous time point depending on whether or not a shift had been detected before, and the filter output is chosen accordingly as one of the half-sided medians. If no shift is detected, the filter output is the average of the half-sided medians. Optionally, a preliminary impulse detection step can be added and the half-sided medians can be replaced by half-sided averages or mid-points to suppress noise with normal or light tails more efficiently. The deviation between the half-sided medians from which a shift is detected can be determined in a Bayesian framework by specifying the *a priori* probability of a shift. These filters assume a piecewise constant signal, but in simulations they perform better than running medians also during trends. Gradient estimates formed from differences of medians, or more generally trimmed means, have also been suggested as robust alternatives to ordinary means [79].

Edges can also be identified via detecting an increase of the local variability. Quasi-ranges $y_{t(n-i+1)} - y_{t(i)}$ are simple scale estimates, and the interquartile range is a usual robust standard. Double-window Hodges–Lehman–Bickel (HLB) and HLB median hybrid filters for improved edge preservation and noise suppression based on this principle are derived in [80]. The hybrid filters replace the HLB estimate by the median when a shift is detected. The double window filter trims all observations which are far from the median before calculating the HLB estimate. The HLB double window filters are found to provide better noise suppression than running medians, but they are outperformed by HLB hybrid filters. The interquartile range is also applied in [34]. The filter output is taken to be either the median or a trimmed mean with an adaptive amount of trimming, depending on whether a shift has been detected or not. A basic problem is the choice of the threshold for shift detection. A comparison of the neighbors of the median for edge detection is suggested in [81], i.e., $y_{t(m+2)} - y_{t(m)}$, where $n = 2m + 1$. If additionally the difference between the central observation and the median is large, it is concluded that a spike has occurred in addition to the shift, and a modified trimmed mean centered at a suitably chosen neighbor of the median is calculated to reduce edge shifting [81].

The empirical variance s_t^2 within the time window can also be applied for edge detection. Adaptive L filters (AL) can be constructed as a convex combination of the local mean and median for retrospective application, where $\tilde{m} = m$ [82]. The weight of the median increases with s_t^2 since a shift is regarded as more likely then,

$$AL(y_t) = w_t y_{t(m+1)} + (1 - w_t)\bar{y}_t \,, \tag{6.23}$$

where $w_t = \nu_t^2/(\sigma_u^2 + \nu_t^2)$, σ_u^2 is the noise variance, and $\nu_t^2 = s_t^2 - \sigma_u^2$ estimates the local variation of the signal. The noise variance is assumed to be constant and needs to be known or estimated from smooth signal regions. The lack of robustness of the empirical variance does not cause a problem since the filter output tends to the median as s_t^2 goes to infinity. The filter is unbiased in the case of symmetric noise and inherits the good properties of the median, namely edge preservation and removal of spikes, while offering larger efficiency under the Gaussian distribution. A modified version replacing the median by a weighted median for using larger windows is also suggested. The current observation y_t could be taken instead of the median for better edge preservation [83], but then the filter loses its robustness completely. Improvements are possible, e.g., by replacing the mean by an adaptively chosen trimmed mean based on the tails of the noise. For edge detection we can use $(s_t^2/\sigma_u^2)(s_t^2/\sigma_u^2 - 1)$. This quantity is close to zero if $s_t^2 \approx \sigma_u^2$, and largely positive if $s_t^2 \gg \sigma_u^2$, but there could occur again a confusion of spikes and shifts.

Further possibilities for edge detection are tests based on linear rank statistics, particularly the Wilcoxon and the median test [84]. The former is almost as effective as tests based on averages in the case of Gaussian noise, but it is more robust to deviations from the Gaussian assumption, while the latter performs well even in the presence of a substantial amount of impulses. The main disadvantage of these tests is probably that for short time windows a given significance level is difficult to obtain, because of the discretization due to using ranks. A comparative study shows that the rank-based tests and the tests comparing two robust level estimates outperform tests based on the local variability [85]. The latter seem interesting mainly for 2D-signal (image) processing since they do not need specification of a direction for the shift.

All the tests described above rely on shifts to arise from one constant signal value to another. Robust regression techniques can be applied to adapt these rules for being suitable during trends. We will just outline one such possibility, which is described in more detail in [86].

Robust shift detection within trends is made possible by a simple majority rule, applied to the repeated median residuals in the current window. A positive level shift is detected if more than half of the most recent RM residuals, or another appropriate large fraction of them, is larger than a multiple of a robust estimate of the variability, e.g., the MAD about the regression line. An analogous rule is used for negative shifts. Using twice the scale estimate for the threshold is a reasonable standard choice since small shifts are often irrelevant and can be accommodated otherwise. If we base the shift detection on the most recent $\lfloor n/2 \rfloor$ observations, requiring that at least half of them deviate widely from the regression fit, one quarter of outlying observations in the current window can have arbitrarily large effects. This means an indispensable loss of robustness when adding such a rule, but a shift can still be detected with a short delay if almost a quarter of the observations after the shift are outliers.

When a shift is detected, suitable actions need to be taken and the procedure be restarted. Shortening the window minimizes the blurring of edges [87]. For restarting we typically need to specify the time point at which the shift has happened. A simple possibility is to use the first time point at which a signal was triggered, or the first time point at which we found a large deviating residual when applying the above majority rule. Adaptive exponential smoothing for improved filtering close to shifts is proposed in [88], constructing a convex combination of the current observation and the previous level estimate with weights depending on the last time point of a shift. However, such schemes are sensitive to outliers. As pointed out by the author himself, a robustification would be desirable.

6.4.4 Impulse Detection

Rules for the detection of spikes can be applied as well. Spikes, also called impulses in the literature, are sometimes interesting for their own sake, or simply because we can replace detected impulses by a cleaned value to increase the robustness of the basic procedure.

In the location context, a couple of approaches for impulse detection have been suggested. Distribution-free rules can be based on the rank of the inspected observation within the time window since outliers are expected to be among the most extreme observations [27]. Difficulties are a high false-detection rate if not only the smallest and the largest observations are regarded as outlying, as well as a lack of detection power in the case of outlier sequences. Another possibility is to use the distance to the median for measuring outlyingness, but such rules cannot distinguish between outliers and shifts. To overcome these problems, rank-based and distance-based rules should be combined [89].

The robustness of the repeated median can be further increased by adding automatic rules for outlier detection and replacement based on robust scale estimators like the MAD [86]. We can check whether the incoming observation $y_{t+\tilde{m}+1}$ is outlying by comparing its residual $r_{\tilde{m}+1} = y_{t+\tilde{m}+1} - \hat{\mu}_t - \hat{\beta}_t(\tilde{m}+1)$ wrt the current regression line to the estimate $\hat{\sigma}_t$ of the standard deviation about the regression line. A promising alternative to the classical MAD for robust scale estimation is [90]

$$\hat{\sigma}_t^{QN} = d_n \cdot \{|r_i - r_j|: \ -m \leqslant i < j \leqslant \tilde{m}\}_{(h)}, \quad h = \binom{m+1}{2}, \tag{6.24}$$

where $r_i = y_{t+i} - \tilde{\mu}_t - i\tilde{\beta}_t$, $i = -m, \ldots, \tilde{m}$. This estimate shows excellent performance at the occurrence of level shifts and performs better then the MAD in the presence of identical measurements (inliers) due to, e.g., rounding. Here, d_n is another finite-sample correction factor depending on the window width $n = m + \tilde{m} + 1$. Replacing detected outliers by their prediction $\hat{\mu}_t + \hat{\beta}_t(\tilde{m}+1)$ gives almost the same robustness as LMS regression even in extreme situations, but additional rules need to be added since, e.g., level shifts remain undetected otherwise because all shifted observations are replaced. Such combined procedures seem preferable to the LMS because of the much better performance in moderate outlier situations and the smaller computational costs.

Many outlier detection rules like the previous one are based on a single difference between the inspected observation and a level estimate. For location-based filters multiple comparison to several weighted medians has been proposed in [91].

6.5 Conclusions

Starting with running medians, many filters have been suggested for detail-preserving robust signal extraction from noisy time series. Many contributions in the literature focus on the attenuation of different types of noise, just imposing that desired signal details like trends are preserved under idealized conditions like the complete absence of observational noise, or that certain signals are roots of the filter. These restrictions are rather weak. Thus, a substantial loss of filtering quality, namely both bias and increased variability, may occur. Requiring appropriate equivariances and invariances whenever possible allows us to construct filters which keep their performance at the occurrence of the interesting signal details.

In particular, locally linear trends can be dealt with using robust regression. Such techniques additionally allow us to overcome the inherent delay which hampers the online application of location-based filters to signals which are not piecewise constant. The repeated median has been regarded a promising method for time series filtering in a couple of investigations. Fast update algorithms are available allowing their application even online and to high frequency data. Similar as for the standard median, modifications are possible for the better preservation of shifts and local extremes. Repeated median hybrid filters offer excellent performance in this respect, but they lose robustness and Gaussian efficiency. A reasonable compromise can be achieved by double-window trimmed repeated medians. Weighted repeated medians seem very promising for the online analysis.

Many interesting aspects could not be addressed in this chapter. Like many other studies we have restricted to the case of independent errors. Here it can be said that the positive autocorrelations found in many applications further increase the efficiencies of robust estimators under Gaussian assumptions as compared to least-squares techniques [15, 30, 51, 92]. The filters discussed here are designed to improve the preservation of certain signal details. Specially designed adaptive order statistic filters even allow us to recover certain signal details, which have been lost before, e.g., due to linear filtering [93]. Finally, repeated medians can also be applied for the highly robust frequency domain analysis, fitting robust sine and cosine coefficients [94].

Acknowledgements

Financial support of the Deutsche Forschungsgemeinschaft (SFB 475, "Reduction of complexity in multivariate data structures") is gratefully acknowledged.

References

[1] J. W. Tukey. *Exploratory Data Analysis*. Addison-Wesley, Reading, MA, 1977.

[2] D. L. Donoho and P. J. Huber. *A Festschrift for Erich Lehmann*. Belmont, Wadsworth, 1983.

[3] P. J. Huber. *Robust Statistics*. Wiley, New York, 1981.

[4] P. L. Davies. *Ann. Stat.*, 21:1843, 1993.

[5] T. Cipra. *J. Forecast.*, 11:57, 1992.

[6] J. Astola and P. Kuosmanen. *Fundamentals of Nonlinear Digital Filtering*. CRC Press, Boca Raton, FL, 1997.

[7] M. Gabbouj, E. J. Coyle, and N. C. Gallagher, Jr. *Circ. Syst. Sig. Proc.*, 11:7, 1992.

[8] S. A. Kassam and H. V. Poor. *Proc. IEEE*, 73:433, 1985.

[9] I. Pitas and A. N. Venetsanopoulous. *Nonlinear Digital Filters: Principles and Applications*. Kluwer, Boston, 1990.

[10] I. Pitas and A. N. Venetsanopoulous. *Proc. IEEE*, 80:1893, 1992.

[11] A. C. Bovik, T. S. Huang, and D. C. Munson, Jr. *IEEE Trans. Pattern Anal. Machine Intellig.*, 9:181, 1987.

[12] A. C. Bovik. *IEEE Trans. Acoust. Speech Sig. Proc.*, 35:493, 1987.

[13] G. R. Arce and R. E. Foster. *IEEE Trans. Acoust. Speech Sig. Proc.*, 37:83, 1989.

[14] A. Nieminem, P. Heinonen, and Y. Neuvo. *IEEE Trans. Patt. Anal. Machine Intellig.*, 9:74, 1987.

[15] R. Fried, T. Bernholt, and U. Gather. *Comp. Stat. Data Anal.*, 50:2313, 2006.

[16] C. A. Pomalaza-Raez and C. D. McGillem. *IEEE Trans. Acoust. Speech Sig. Proc.*, 32:571, 1984.

[17] H. M. Lin and A. N. Willson, Jr. *IEEE Trans. Circ. Syst.*, 35:675, 1988.

[18] V. Katkovnik, K. Egiazarin, and J. Astola. *Sig. Proc.*, 83:251, 2003.

[19] H. Hwang and R. A. Haddad. *IEEE Trans. Image Proc.*, 4:499, 1995.

[20] W.-J. Song and W. A. Pearlman. *IEEE Trans. Circ. Sys.*, 35:1048, 1988.

[21] T. Loupas, W. N. McDicken, and P. L. Allan. *IEEE Trans. Circ. Syst.*, 36:129, 1989.

[22] N. C. Gallagher and G. L. Wise. *IEEE Trans. Acoust. Speech Sig. Proc.*, 29:1136, 1981.

[23] L. Yin, R. Yang, M. Gabbouj, and Y. Neuvo. *IEEE Trans. Circ. Syst. II*, 43:157, 1996.

[24] P. Koivisto, O. Yli-Harja, A. Niemisto, and I. Shmulevich. *Sig. Proc.*, 81:227, 2001.

[25] R. A. Stein and T. J. Fowlow. *Proc. ISCAS*, 85:1331, 1985.

[26] A. C. Bovik, T. S. Huang, and D. C. Munson, Jr. *IEEE Trans. Acous. Speech Sig. Proc.*, 31:1342, 1983.

[27] A. B. Hamza and H. Krim. *IEEE Trans. Sig. Proc.*, 49:3045, 2001.

[28] R. Öten and R. J. P. Figueiredo. *IEEE Trans. Sig. Proc.*, 51:193, 2003.

[29] H. G. Longbotham and A. C. Bovik. *IEEE Trans. Acoust. Speech Sig. Proc.*, 37: 275, 1989.

[30] R. Fried. *J. Stat. Comp. Simul.*, to appear (2006).

[31] J. B. Bednar and T. L. Watt. *IEEE Trans. Acoust. Speech Sig. Proc.*, 32:145, 1984.

[32] Y. H. Lee and S. A. Kassam. *IEEE Trans. Acoust. Speech Sig. Proc.*, 33:672, 1985.

[33] S. R. Peterson, Y. H. Lee, and S. A. Kassam. *IEEE Trans. Acoust. Speech Sig. Proc.*, 36:707, 1988.

[34] A. Restrepo and A. C. Bovik. *IEEE Trans. Acoust. Speech Sig. Proc.*, 36:1326, 1988.

[35] N. Himayat and S. A. Kassam. *IEEE Trans. Sig. Proc.*, 41:2764, 1993.

[36] R. Ding and A. N. Venetsanopoulous. *IEEE Trans. Circ. Syste.*, 34:948, 1987.

[37] B. I. Justusson. *Two-Dimensional Digital Signal Processing II*. Springer, New York, 1981.

[38] J. Nieweglowski, M. Gabbouj, and Y. Neuvo. *Sig. Proc.*, 34:149, 1987.

[39] M. K. Prasad and Y. H. Lee. *Proc. IEEE Int. Symp. Circ. Syst. ISCAS*, 89:425, 1989.

[40] R. Yang, L. Yin, M. Gabbouj, J. Astola, and Y. Neuvo. *IEEE Trans. Sig. Proc.*, 43:591, 1995.

[41] C. L. Mallows. *Ann. Stat.*, 8:695, 1980.

[42] S. Hoyos, J. Bacca, and G. R. Arce. *IEEE Trans. Sig. Proc.*, 53:1045, 2005.

[43] G. R. Arce. *IEEE Trans. Sig. Proc.*, 46:3195, 1989.

[44] K.-S. Choi, A. W. Morales, and S.-J. Ko. *IEEE Trans. Sig. Proc.*, 49:1940, 2001.

[45] I. Djurovic, V. Katkovnik, and L. Stankovic. *Sig. Proc.*, 81:1771, 2001.

[46] P. P. Gahndi and S. A. Kassam. *IEEE Trans. Sig. Proc.*, 39:1524, 1991.

[47] A. Flaig, G. R. Arce, and K. E. Barner. *IEEE Trans. Sig. Proc.*, 46:2101, 1998.

[48] F. Palmieri and C. G. Boncelet. *IEEE Trans. Acoust. Speech Sig. Proc.*, 37:691, 1989.

[49] P. P. Gahndi, I. Song, and S. A. Kassam. *IEEE Trans. Acoust. Speech Sig. Proc.*, 37:1359, 1989.

[50] P. L. Davies, R. Fried, and U. Gather. *J. Stat. Plan. Infer.*, 122:65, 2004.

[51] T. Bernholt, R. Fried, U. Gather, and I. Wegener. Modified repeated median filters. Technical report 46/2004, SFB 475, University of Dortmund, Germany, 2004.

[52] J. Fan, T.-C. Hu, and Y. K. Truong. *Scand. J. Stat.*, 21:433, 1994.

[53] U. Gather, K. Schettlinger, and R. Fried. *Comp. Stat.*, 21:33, 2006.

[54] A. F. Siegel. *Biometrika*, 68:242, 1982.

[55] F. R. Hampel. *Bull. Int. Stat. Inst.*, 46:375, 1975.

[56] P. J. Rousseeuw. *J. Am. Stat. Assoc.*, 79:871, 1984.

[57] S. P. Ellis and S. Morgenthaler. *J. Am. Stat. Ass.*, 87:143, 1991.

[58] P. J. Rousseeuw and A. M. Leroy. *Robust Regression and Outlier Detection*. Wiley, New York, 1987.

[59] K.-M. Lee, P. Meer, and R.-H. Park. *IEEE Trans. Pattern Anal. Machine Intellig.*, 20:200, 1998.

[60] H. Edelsbrunner and D. L. Souvaine. *J. Am. Stat. Assoc.*, 85:115, 1990.

[61] T. Bernholt and R. Fried. *Information Proc. Lett.*, 88:111, 2003.

[62] J. Einbeck and G. Kauermann. *J. Stat. Comp. Simul.*, 73:913, 2003.

[63] U. Gather and R. Fried. *Proceedings in Computational Statistics COMPSTAT 2004*. Physica-Verlag, Heidelberg, 2004.

[64] P. J. Rousseeuw. *Proceedings of the 4th Pannonian Symposium on Mathematical Statistics and Probability*, volume B. Reidel, Dordrecht, 1983.

[65] V. Koivunen. *IEEE Trans. Image Proc.*, 4:569, 1995.

[66] P. J. Rousseeuw and M. Hubert. *J. Am. Stat. Assoc.*, 94:388, 1999.

[67] U. Gather and R. Fried. *Tatra Mount. Math. Pub.*, 26:87, 2003.

[68] R. Fried, J. Einbeck, and U. Gather. Weighted repeated median smoothing and filtering. Technical report 33/2005, SFB 475, University of Dortmund, Germany, 2005.

[69] P. Heinonen and Y. Neuvo. *IEEE Trans. Acoust. Speech Sig. Proc.*, 35:832, 1987.

[70] P. Heinonen and Y. Neuvo. *IEEE Trans. Acoust. Speech Sig. Proc.*, 36:892, 1988.

[71] J. Astola, P. Heinonen, and Y. Neuvo. *IEEE Trans. Circ. Syst.*, 36:1430, 1989.

[72] R. Wichman, J. T. Astola, P. J. Heinonen, and Y. A. Neuvo. *IEEE Trans. Acoust. Speech Sig. Proc.*, 38:2108, 1990.

[73] A. Nieminem, Y. Neuvo, and U. Mitra. *Sig. Proc.*, 18:1, 1989.

[74] N. Himayat and S. A. Kassam. *IEEE Trans. Image Proc.*, 3:265, 1994.

[75] Y. Neuvo, P. Heinonen, and I. Defee. *IEEE Trans. Circ. Syst.*, 34:1337, 1987.

[76] K. K. Yuen and K. K. *Biometrika*, 61:165, 1974.

[77] Z. Hou and T. S. Koh. *Patt. Recogn.*, 36:2083, 2003.

[78] H. Hwang and R. A. Haddad. *IEEE Trans. Sig. Proc.*, 42:249, 1994.

[79] A. C. Bovik and D. C. Munson, Jr. *Comp. Vision Graph. Image Proc.*, 33:377, 1986.

[80] A. Kundu and W.-R. Wu. *IEEE Trans. Acoust. Speech Sig. Proc.*, 37:1293, 1989.

[81] Y. H. Lee and S. Tantaratana. *IEEE Trans. Acoust. Speech Sig. Proc.*, 38:406, 1990.

[82] T. Sun, M. Gabbouj, and Y. Neuvo. *Sig. Proc.*, 38:331, 1994.

[83] X. Z. Sun and A. N. Venetsanopoulos. *IEEE Trans. Circ. Syst.*, 35:57, 1988.

[84] A. C. Bovik, T. S. Huang, and D. C. Munson, Jr. *Pattern Recogn.*, 19:209, 1986.

[85] R. Fried and U. Gather. Robust shift detection in time series. Working paper, Department of Statistics, University of Dortmund, Germany, 2006.

[86] R. Fried. *J. Nonparam. Stat.*, 16:313, 2004.

[87] S. A. Kassam. *J. Time Ser. Anal.*, 3:185, 1982.

[88] E. Yashchin. *Nonlin. Anal. Theory Meth. Appl.*, 30:3997, 1997.

[89] I. Aizenberg and C. Butakoff. *IEEE Sig. Proc. Lett.*, 11:363, 2004.

[90] P. J. Rousseeuw and Ch. Croux. *J. Am. Stat. Assoc.*, 88:1273, 1993.

[91] T. Chen and H. R. Wu. *IEEE Sig. Proc. Lett.*, 8:1, 2001.

[92] R. Fried and U. Gather. *Austrian J. Stat.*, 34:139, 2005.

[93] Y. H. Lee and A. T. Fam. *IEEE Trans. Acoust. Speech Sig. Proc.*, 35:680, 1987.

[94] L. G. Tatum and C. M. Hurvich. *J. Roy. Stat. Soc. Ser. B*, 55:881, 1993.

7 Coupled Oscillators Approach in Analysis of Bivariate Data

Michael Rosenblum, Laura Cimponeriu, and Arkady Pikovsky

We discuss the usage of model-based and nonmodel-based techniques in the analysis of bivariate data. In particular, we consider in detail the coupled oscillators approach for the identification of a weak interaction between two oscillators from signals measured at their output. Our framework allows one to detect and quantify the strength and directionality of weak interaction, as well as to estimate the delay(s) in coupling. We present both theoretical description of the technique and its algorithmic implementation. We illustrate the technique by its application to the analysis of the cardiorespiratory interaction.

7.1 Bivariate Data Analysis: Model-Based Versus Nonmodel-Based Approach

Multichannel measurements are ubiquitous in experimental studies in all branches of natural sciences and, hence, processing of bivariate (or, generally, multivariate) experimental records is a typical task of the data analysis. This task can include a separate processing of two channels by all possible univariate techniques, as well as an application of a true *bivariate* technique which performs a joint analysis of two channels. The goals of the bivariate analysis can be different. So, for example, there exist numerous techniques—linear and nonlinear—which provide information on an interrelation between two signals. However, quite often the analysis goes beyond this task and aims at revealing some information about the *system (or systems)*, which generates the data. Certainly, by making such a step one cannot consider the system as a black box, but requires a certain knowledge or assumption about it. Typically, one assumes (explicitly or implicitly) that the system can be described by a certain *class of models*, e.g., by an input–output system, a delay line, a set of coupled active oscillators, etc. (We emphasize that we mean exactly a model of the data source, but not a model of signals, such as ARMA, etc.) The respective analysis technique that exploits such an assumption can be denoted as *model based*. The interpretation of the results then crucially depends on the correctness of the assumption concerning the model of the data source.

Handbook of Time Series Analysis. Björn Schelter, Matthias Winterhalder, Jens Timmer

ISBN: 3-527-40623-9

For illustration let us consider a common and a power tool of the data analysis, the cross-correlation analysis and its frequency domain counterpart—the cross-spectral analysis. It is well known that this technique provides a complete description of a linear input–output system, namely its transfer function; in this case the interpretation of the results is unambiguous. The technique certainly can (and often must) be used also for the analysis of nonlinear input–output systems, or systems of coupled active oscillators, but the interpretation of the results becomes more complicated and ambiguous. So, in latter complex cases the cross-correlation (spectral) analysis still determines reliably whether certain frequency components of given signals are interrelated (coherent) to a certain degree. However, a computation of the transfer function becomes of a limited, or of no use, and the conclusion about coherence cannot be extrapolated for the case when, say, amplitudes of signals will change.

Another example is related to an estimation of the transmission delay τ. If there is an *a priori* knowledge that two signals represent the input and the output of a delay line, then the delay can be estimated from the position of the maximum of the cross-correlation function. (Sometimes in the biomedical literature the delay is obtained from the phase shift at the characteristic frequency, $\tau = \Delta\varphi/\omega$, which implicitly uses an additional assumption that the delay is smaller than the oscillation period.) However, if we are uncertain about the structure of the system under study, then this technique cannot be used, as it does not distinguish between the delay and the phase shift.

Two above-considered examples shed light on the main difference between nonmodel-based and model-based analyses. Note that the same algorithm, e.g., the computation of the cross-correlation function, can be used for both nonmodel-based and model-based analyses. The model-based analyses provides additional information about the systems which generate signals, but this is true if and only if the assumptions about the data source are correct. Otherwise, the results may be misleading. In contrast, the nonmodel-based analysis can always be employed, but the price for this is the reduced information or ambiguity in the interpretation of the results.

In this chapter we discuss several data analysis tools based on the assumption that the bivariate data originate from two coupled self-sustained oscillators (Fig. 7.1). Below we also discuss the extension of the approach to the multivariate case. These tools are designed to provide the solutions for the following tasks:

- to detect and quantify an interaction between the systems,
- to reveal the direction of coupling,
- to estimate delay(s) in coupling,

provided the following assumptions are fulfilled:

- we deal with two self-sustained oscillators which can be *weakly coupled,*
- we know how to ascribe the signals to systems,

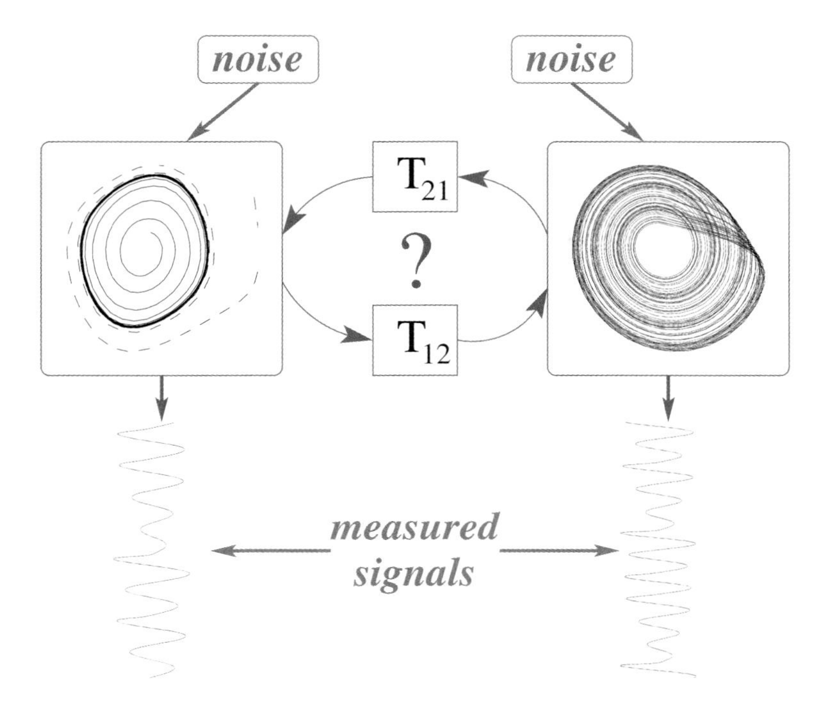

Fig. 7.1: Coupled oscillators' approach to the analysis of bivariate data explicitly assumes that the data are generated by two weakly coupled self-sustained oscillators. The systems can be either periodic or chaotic and are assumed to be perturbed by independent noise sources. The coupling can be uni- or bidirectional, and can occur with delays T_{21} and T_{12}.

- the signals are appropriate for phase estimation.

We emphasize that we prefer to speak about coupling between the systems and interrelation between the signals. Coupling in this context means some physical connection between oscillators which may or may not result in an interrelation between the signals measured at the output of these oscillators.

The chapter is organized as follows. In the rest of this section we briefly discuss the main facts of the coupled oscillators theory. Next, we discuss a particular example of the bivariate data analysis, namely cardiorespiratory interaction in a healthy baby. In the following sections we describe techniques of phase estimation and phase dynamics reconstruction from data. Finally, we present the analysis tools and illustrate them by an application to cardiorespiratory data.

7.1.1 Coupled Oscillators: Main Effects

Active systems, capable of producing long-term sustained rhythmical activity, are known in physics as *self-sustained* oscillators. These are autonomous nonlinear dissipative systems, which compensate the loss of energy at the expense of an internal nonoscillatory energy source. Mathematically, such systems are described by autonomous differential equations. The image of periodic self-sustained oscillations in the phase space is a closed attracting trajectory, called limit cycle; the image of chaotic self-sustained oscillations is a strange attractor. See, e.g., [1–3] for a discussion. The motion of the phase point along the limit cycle or along the flow of a chaotic system is parameterized by a variable, called *phase*. For limit cycle oscillators it is defined as the monotonically growing variable which gain 2π during one oscillation period

$$\dot{\phi} = \omega \,, \tag{7.1}$$

where ω is the natural frequency. The notion of phase and amplitude(s) can be extended, though not rigorously, to some chaotic oscillators [3, 4]. As will be discussed below, the notion of phase is crucial for a description of interaction between self-sustained systems.

Models of coupled self-sustained oscillators appear in various fields of science and engineering, as well as in live nature. An important effect is *synchronization*, when two (or many) weakly interacting systems adjust their phases $\phi_{1,2}$ and average frequencies $\Omega_{1,2} = \langle \dot{\phi}_{1,2} \rangle$, where $\langle\rangle$ denotes averaging over time, so that the following conditions of *phase and frequency locking* are fulfilled

$$|n\phi_1 - m\phi_2| < \text{const} \,, \quad n\Omega_1 = m\Omega_2 \,. \tag{7.2}$$

This nonlinear phenomenon [1–3, 5–7] is often observed in man-made and natural systems. In the latter case it is found on a level of single cells, physiological subsystems, organisms, and even on the level of populations [3, 8, 9]. Sometimes, synchronization is essential for a normal functioning of a system, e.g., for a coordinated motion of several limbs or for the performance of a pacemaker, where many cells fire synchronously, and in this way produce a macroscopic rhythm that governs respiration, heart contraction, etc. Sometimes, the onset of synchrony leads to a severe pathology, e.g., in case of Parkinson's disease, when locking of many neurons results in the tremor activity. Quite often, the functional role of synchrony is yet unknown, e.g., in the case of cardiorespiratory coordination [10–14] or in the case of mutual entrainment of respiration and locomotion; possibly its appearance is just a manifestation of a general property of self-sustained oscillators—to adjust their rhythms due to a weak interaction. However, an onset or a cessation of synchrony reflects variation in the state of the complex system, and therefore may provide important physiological information.

The concept of synchronization can be also applied for the description of interaction of noisy self-sustained oscillators (and natural systems are inevitably noisy). In this case the conditions, Eq. 7.2, are fulfilled only in a statistical sense and the distinction between synchronous and asynchronous regimes is generally ambiguous (see discussion in [3] and references therein). Furthermore, the notions of phase and frequency locking can be extended for a class of chaotic self-sustained systems; in the context of interacting chaotic systems this effect is called phase synchronization [4]. From an experimentalist's viewpoint, the dynamics of weakly coupled noisy and chaotic systems is quite similar, and therefore in the context of the data analysis they can be treated in a similar way.

Note that an interaction can result not only in the adjustment of frequencies of coupled systems, i.e., in the tendency of the systems to become synchronized, but can also lead to the variation of their frequencies (and amplitudes). It means that their frequencies are not constant but oscillate with time. For example, in the context of the cardiorespiratory interaction such a variation of the heart rate with the frequency of respiration is called respiratory sinus arrhythmia. It is, therefore, important to distinguish between two different kinds of interaction. The first is illustrated in Fig. 7.1; in general, such an interaction can affect the frequencies of systems as well as cause their variation. In the second case, which we denote as *modulation*, the modulating source does not act directly on the system, but only on the channel, where its output is being transmitted (see a discussion in [3]). Obviously, this kind of interaction cannot shift the frequency of the driven system but only cause its modulation. In other words, we denote by modulation the action that is called the phase modulation in the engineering literature. However, we do not have to separate what is called in engineering the frequency modulation from a general case of interaction treated in the synchronization theory and illustrated in Fig. 7.1. The distinction of two kinds of interaction from data remains an open question and will be not treated here; below we always assume that the coupling acts directly on the systems, in accordance with Fig. 7.1.

7.1.2 Weakly Coupled Oscillators: Phase Dynamics Description

An important theoretical idea, widely explored below, says that a weak interaction of limit cycle oscillators affects only their phases, whereas the amplitudes can be considered as unchanged [1]. This happens due to the fact that the amplitude is a variable, corresponding to the direction in the phase space which is transversal to the limit cycle, and, therefore, corresponding to the negative Lyapunov exponents of the dynamical system. Hence, the amplitude is a stable variable and cannot be adjusted by weak forcing or interaction. For chaotic systems the amplitudes correspond to negative and positive Lyapunov exponents.

In contrast, the phase corresponds to the direction along the limit cycle (or to the flow of a chaotic system); this direction is characterized by the zero Lyapunov exponent. As a consequence, the phase is a marginally stable variable that can be adjusted by a very weak interaction. The main conclusion is that the description of weakly coupled oscillators can be reduced to the phase dynamics

$$\begin{aligned} \dot{\phi}_1 &= \omega_1 + f_1(\phi_1, \phi_2) + \xi_1(t)\,, \\ \dot{\phi}_2 &= \omega_2 + f_2(\phi_2, \phi_1) + \xi_2(t)\,, \end{aligned} \tag{7.3}$$

where $\omega_{1,2}$ are frequencies of uncoupled systems and functions $f_{1,2}$ describe the coupling; obviously they are 2π-periodic with respect to their arguments. This property will play a very important role in the reconstruction of phase Eqs. (7.3) from data, to be described below, because it naturally restricts the class of test functions for fitting. Noise terms in Eqs. (7.3) are considered as phase independent. Note that Eqs. (7.3) also describe the dynamics of weakly coupled chaotic systems; in this case the irregular terms $\xi_{1,2}$ correspond to perturbations to the phase dynamics due to the chaotic nature of amplitudes.

It is often convenient (in particular, it will be used below for estimations of directionality and delay in coupling) to use instead of continuous time equations (7.3) a corresponding mapping for phase increments $\Delta\phi_{1,2}(t) = \phi_{1,2}(t + \Delta t) - \phi_{1,2}(t)$

$$\begin{aligned} \Delta\phi_1 &= F_1(\phi_1, \phi_2) + \zeta_1(t) \,, \\ \Delta\phi_2 &= F_2(\phi_2, \phi_1) + \zeta_2(t) \,, \end{aligned} \tag{7.4}$$

where the functions $F_{1,2}$ are also 2π-periodic with respect to their arguments.

7.1.3 Estimation of Phases from Data

Prior to the analysis of phase relations we have to estimate phases from data. There exist three main approaches to the problem. One is based on the construction of the complex *analytic signal* $\zeta(t)$ [15] from a scalar experimental time series $s(t)$ via the Hilbert transform (HT)

$$\zeta(t) = s(t) + i s_H(t) = A(t) e^{i\phi(t)} \,, \quad s_H(t) = \pi^{-1} \text{P.V.} \int_{-\infty}^{\infty} \frac{s(\tau)}{t - \tau} d\tau \,, \tag{7.5}$$

where $s_H(t)$ is the HT of $s(t)$. Equation (7.5) unambiguously provides an instantaneous phase $\phi(t)$ and an amplitude $A(t)$. We use the same notation ϕ for the true phase satisfying Eq. (7.1) and its estimate obtained from a scalar time series. Note that HT is parameter free. Practical hints for the computation and usage of the HT, as well as further citations can be found in [3, 16]. Here we briefly mention the crucial points:

- Mathematically, HT is defined for an arbitrary signal. However, $\phi(t)$ and $A(t)$ admit a clear physical interpretation only for narrow band signals. If the signal has no well-expressed peak in its power spectrum, then the computation of the phase and application of the synchronization approach is highly doubtful. We recommend to always perform a simple test, namely to plot $s_H(t)$ versus $s(t)$ and to look whether the trajectory in this presentation always rotates around the origin; only in this case one can meaningfully compute the instantaneous phase. Note that often the origin should be shifted to a point different from zero.

- A complex, broadband, signal that can be considered as a mixture of several narrow band processes should be first decomposed into oscillatory components which can be then considered as signals with slowly varying amplitude and frequency; as the next step, the phases of these components can be obtained via HT. Note that sometimes it is difficult to decide whether a peak in the spectrum represents another process or a harmonic. Decomposition can be done by means of a band-pass filter or by more sophisticated techniques like the independent component analysis.

- Determination of $\phi(t)$ is very sensitive to low-frequency trends, which makes the preprocessing of the data a crucial step in the analysis.

The second approach exploits the wavelet analysis with a complex wavelet function and provides a phase (and an amplitude) as functions of time for a certain spectral frequency band [17, 18]

$$A(t; f)e^{i\phi(t;f)} = \int_{-\infty}^{\infty} s(\tau)\Psi^*(t, \tau; f)\,d\tau\,, \tag{7.6}$$

where $\Psi(t, \tau; f)$ is the Morlet, or Gabor, wavelet

$$\Psi(t, \tau; f) = \sqrt{f}\exp(i \cdot 2\pi f(\tau - t))\exp\left(-\frac{(\tau - t)^2}{2\sigma^2}\right)\,.$$

This procedure is equivalent to a band-pass filtration and subsequent HT of the signal $s(t)$ [19]. The central frequency of the filter is f, and its width is determined by the parameter σ.

Third, the phase can be very easily introduced for processes that can be treated as a series of well-defined events taking place at times t_k (point processes). Examples include signals characterizing heart contraction or neuron firing. If the interval between two events can be considered as a cycle, then it is natural to say that the phase increment between the events is exactly 2π. Hence, we can assign to the times t_k the values of phase $\phi(t_k) = 2\pi k$, and for an arbitrary instant of time $t_k < t < t_{k+1}$ take

$$\phi(t) = 2\pi k + 2\pi\frac{t - t_k}{t_{k+1} - t_k}\,. \tag{7.7}$$

We emphasize that the definition and the practical determination of a phase of a complex signal in the context of the synchronization analysis remains an open problem. One approach, called locking-based frequency measurement, was suggested in [20]. The idea of this approach is to use the signal under study in order to drive a set of uncoupled limit cycle oscillators with different natural frequencies. A subset of these probe oscillators can be entrained by the common forcing, and therefore synchronize in between; the frequency and the phase of these locked oscillators can be taken as an estimate of the frequency and the phase of the original signal.

Note that in theoretical studies of coupled systems the phase can be rigorously defined only for limit cycle oscillators, whereas a rigorous definition of the phase of noisy/chaotic oscillators remains a theoretical challenge. For an autonomous limit cycle oscillator the phase is defined as a uniformly growing function of time, cf. Eq. (7.1) [1, 3]. However, phase estimates according to Eqs. (7.5–7.7) generally do not meet this requirement. As a result the estimated phase obeys (if we neglect numerical errors)

$$\dot{\phi} = \omega + \tilde{f}(\phi), \tag{7.8}$$

where the function $\tilde{f}(\phi)$ reflects the nonuniformity of the motion of the phase along the limit cycle, and the equation for the coupled systems reads (cf. Eq. 7.3)

$$\begin{aligned} \dot{\phi}_1 &= \omega_1 + \tilde{f}_1(\phi_1) + \hat{f}_1(\phi_1, \phi_2) + \xi_1(t), \\ \dot{\phi}_2 &= \omega_2 + \tilde{f}_2(\phi_2) + \hat{f}_2(\phi_2, \phi_1) + \xi_2(t), \end{aligned} \tag{7.9}$$

where the coupling is described by the functions $\hat{f}_{1,2}$. Similarly, if we want to describe the coupled system by a discrete mapping, then the mapping obtained from phase estimates differs from the mapping, Eq. (7.4), for true phases.

Finally, we note that, theoretically, phase is defined on the real line. In the following we call such a phase "unwrapped," while the phase defined on the $(0, 2\pi)$ interval is called "wrapped." The use of wrapped or unwrapped phase depends on the application and is often a crucial point.

7.1.4 Example: Cardiorespiratory Interaction in a Healthy Baby

We choose the study of the cardiorespiratory interaction as a primary example for the illustration of the applicability of our theoretical framework and techniques for the experimental data analysis, for two reasons. The first is based on the *a priori* physiological evidence that the two vital rhythms are self-sustained and interact rather weakly. The second reason is that, despite extensive investigations at both theoretical and experimental levels, the nature of cardiorespiratory interaction remains controversial. In particular, it remains an open question whether the effects of interaction (e.g., the frequency and the phase entrainment between the two rhythms) can be solely attributed to the well-known modulation of the heart rate by the respiratory rhythm (the so-called respiratory sinus arrhythmia), or a reciprocal form of coupling may coexist. To gain insight into this question, appropriate modeling and experimental data analysis tools are needed.

In the following sections, our framework is presented using a case study analysis of the interaction between human cardiac and respiratory systems. The experimental data consist of a single segment of bivariate, artifact-free, cardiorespiratory measurements (the cardiac and respiratory signals) recorded from a 6-month healthy infant during quiet sleep. The data set has been kindly provided by R. Mrowka and A. Patzak, Department of Physiology, Charité, Humboldt University, Berlin. A detailed description of the experimental setup and data preprocessing can be found in [13, 21]. For the computation of the phase of the cardiac signal we assume that the time occurrence of each R-wave in the electrocardiogram (ECG) marks the onset of a new cardio-cycle and that during each cardio-cycle the phase increases in a monotonic uniform way. This translates into the computation of instantaneous phase ϕ_h of the cardiac signal by linear interpolation between successive R-wave peaks (cf. Eq. 7.7). Whereas the cardiac phase has thus a unique determination from R-wave timings, we face several alternatives in the determination of the phase of the respiratory signal. It is known that, in normal physiological conditions, measurements of respiration during the quiet sleep provide a narrow-band signal, characterized by a certain degree of breath-to-breath variability in both amplitude and timing of the onset of the inspiration/expiration. In order to get a signal well behaved with respect to Hilbert transform, and therefore having well-defined instantaneous attributes, a smoothing filter must be employed. The choice of the filter and its parameters results in a compromise between signal distortion due to an excessive filtering (which may provide an almost sinusoidal waveform) and smoothing (measurement noise suppression). With a correct choice of filter parameters, the instantaneous phase computed via HT preserves the information about cycle-to-cycle variability. Alternatively, the instantaneous phase of the respiratory signal can be obtained in a fashion similar to the way the phase of the cardiac signal has been determined. For the respiratory oscillator, the onset time of inspiration/expiration may serve as a physiologically relevant marker event. Figure 7.2 shows the instantaneous phase of the cardiac signal and the phase of the respiration derived via both HT and marker events for the data set considered for the analysis. Note that although the estimates obtained in two different ways differ on the time scale of one cycle, they provide same average frequencies.

7.2 Reconstruction of Phase Dynamics from Data

The first step in the reconstruction of the phase dynamics is a computation of the bivariate series of phases $\phi_{1,2}(j) = \phi_{1,2}(t_j)$, where index $j = 1 \cdots M$ denotes a discrete set of time points $t_j = j \cdots \delta t$, with the help of one of the algorithms described in Section 7.1.3. The next step depends on whether we want to reconstruct the continuous or discrete phase model (see Eqs. (7.3) and (7.4)). In the first case we have to estimate the time derivatives $\dot{\phi}_{1,2}$. For this goal we first compute the phase increments $\Delta\phi_{1,2}$ over the sampling interval. Because the data are noisy, one has to use a smoothening/interpolation technique, based, e.g., on a Savitzky–Golay filter, see the appendix in [3]. In the second case we just compute $\Delta\phi_{1,2}$ over a fixed time interval which can be much larger than the sampling interval (e.g., it can be of the order of the oscillation period; certainly, it is a multiple of the sampling interval).

The main and final step is to approximate the dependences $\Delta\phi_1(j) = \Delta\phi_1(\phi_1(j), \phi_2(j))$, $\Delta\phi_2(j) = \Delta\phi_2(\phi_1(j), \phi_2(j))$ with a model Eq. (7.3) or Eq. (7.4).

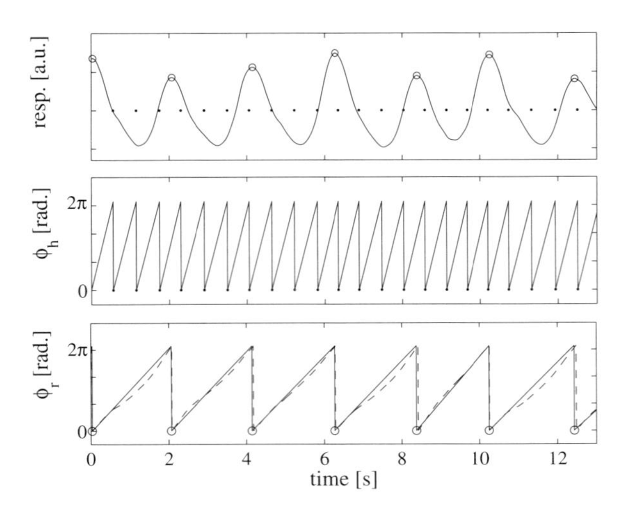

Fig. 7.2: Top panel: respiratory signal. Middle panel: phase of the cardiac signal ϕ_h linearly increases and grows by 2π between two heart beats. Bottom panel: phase of respiration ϕ_r obtained via the Hilbert transform (dashed line) and via the linear interpolation of the time between two onsets of expiration.

Because continuous functions $f_{1,2}$ and $F_{1,2}$ are 2π periodic in arguments, they admit a natural Fourier series representation, and we can in both cases seek for the dependences in the form

$$\Delta\phi_1(\phi_1,\phi_2) = \sum_{m=0,l=-N}^{N} a_{m,l}\cos(m\phi_1 + l\phi_2) + b_{m,l}\sin(m\phi_1 + l\phi_2)\,, \quad (7.10)$$

and similarly for $\Delta\phi_2(\phi_1,\phi_2)$.

Practically, we can use the standard linear least-square regression [22] to fit the data with a truncated Fourier series model. A minimization of

$$\sum_{j=1}^{M}(\Delta\phi_1(j) - \sum_{m=0,l=-N}^{N} a_{m,l}\cos(m\phi_1(j) + l\phi_2(j)) \\ + b_{m,l}\sin(m\phi_1(j) + l\phi_2(j)))^2$$

leads to a linear system

$$\sum_{s=0,n=-N}^{N} a_{s,n}A_{snml} + b_{s,n}B_{snml} = D_{ml}\,,$$

$$\sum_{s=0,n-N}^{N} a_{s,n}B_{mlsn} + b_{s,n}C_{snml} = E_{ml}\,,$$

where

$$A_{snml} = \sum_{j=1}^{M} \cos(s\phi_1(j) + n\phi_2(j)) \cos(m\phi_1(j) + l\phi_2(j)) ,$$

$$C_{snml} = \sum_{j=1}^{M} \sin(s\phi_1(j) + n\phi_2(j)) \sin(m\phi_1(j) + l\phi_2(j)) ,$$

$$B_{snml} = \sum_{j=1}^{M} \sin(s\phi_1(j) + n\phi_2(j)) \cos(m\phi_1(j) + l\phi_2(j)) ,$$

$$D_{ml} = \sum_{j=1}^{M} \Delta\phi_1(j) \cos(m\phi_1(j) + l\phi_2(j)) ,$$

$$E_{ml} = \sum_{j=1}^{M} \Delta\phi_1(j) \sin(m\phi_1(j) + l\phi_2(j)) .$$

Generally, a solution of this problem is rather sensitive to a choice of parameter N. Therefore, we apply a preliminary estimation of the Fourier coefficients based on the assumption that the matrices A and C are diagonal and B vanishes. This assumption is reasonable only when noise in the otherwise synchronous oscillators or quasiperiodic dynamics ensures a quite uniform scattering of phase points over the $[0, 2\pi) \times [0, 2\pi)$ square. In this case $a_{m,l}$ and $b_{m,l}$ are just the real and imaginary parts of the Fourier transform

$$Q(m, l) = \frac{1}{M} \sum_{j=1}^{M} \Delta\phi_1(j) e^{i(m\phi_1(j) + l\phi_2(j))} . \tag{7.11}$$

In order to make use of the FFT algorithm, the irregularly sampled $\Delta\phi_{1,2}$ should be resampled onto a regular grid, by employing some form of interpolation or, in the presence of noise, estimation. After the transform (Eq. (7.11)) has been performed, one can select the dominant modes as the modes with the largest values of $|Q(m, l)|$. Then one can restrict the summation in Eq. (7.10) to these modes only, which significantly improves the reliability of found Fourier coefficients $a_{m,l}$ and $b_{m,l}$.

To exemplify this approach to phase dynamics model reconstruction, we take for the analysis the bivariate cardiorespiratory data set mentioned in the previous section. A segment of ≈ 350 average cardiac cycles length is selected (see Fig. 7.4), and the phases of cardiac ϕ_h and respiratory ϕ_r oscillations along with their finite difference approximations $\Delta\phi_{h,r}$ over time interval 0.05 s are computed. In order to make use of the FFT algorithm for the 2D Fourier transform, we perform a Delaunay-triangulation-based cubic interpolation of $\Delta\phi_{h,r}$ on a uniform grid on the square $[0, 2\pi) \times [0, 2\pi)$ with the grid step $2\pi/128$. In this

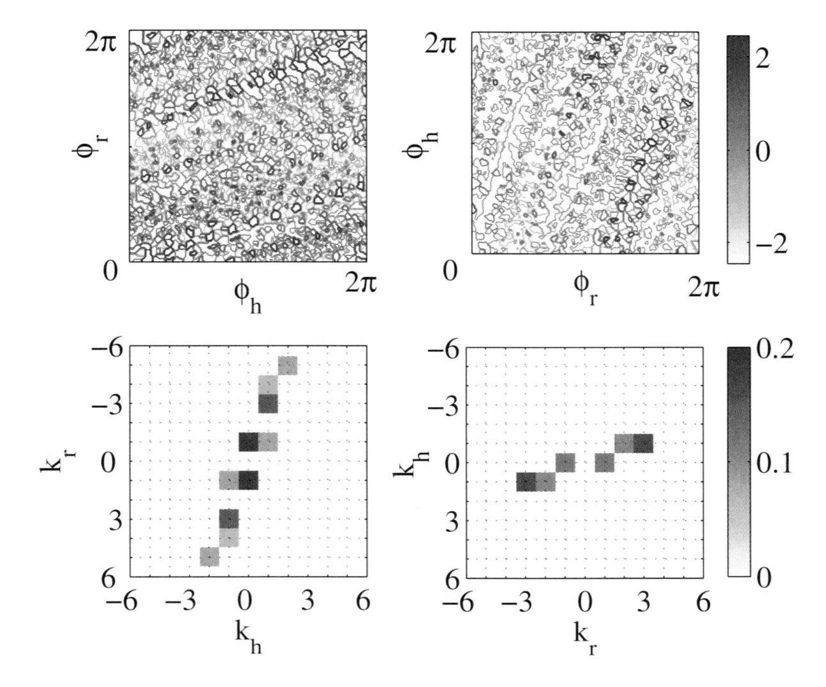

Fig. 7.3: Top panel: two-dimensional contour plots of the resampled $\Delta\phi_h(\phi_h, \phi_r)$ (left) and $\Delta\phi_r(\phi_r, \phi_h)$ (right). Bottom panel: their corresponding two-dimensional Fourier transforms; gray scales code the absolute value of the corresponding Fourier coefficients.

way the Nyquist theorem provides the upper limit of the frequencies resolved by these data as $M = 64$, which, under the assumption that the underlying coupling functions are smooth, can be considered as sufficiently large to prevent aliasing. The next step is the identification of the dominant spatial modes, which will allow us to fit a more parsimonious Fourier series model. For this purpose we employ the surrogate hypothesis testing. Namely, for testing the null hypothesis of no coupling from the respiration to the heart, we compute the Fourier coefficients $|\tilde{Q}(k_h, k_r)|$ for 100 realizations of the randomly shuffled $\Delta\phi_h$ and take $\langle\max(|\tilde{Q}(k_h, k_r)|)\rangle$ as the threshold value, where $\langle\rangle$ means averaging over the realizations of surrogates. It means that for the model fitting we use only the terms which satisfy $|Q(k_1, k_2)| \geqslant \max(|\tilde{Q}(k_h, k_r)|)$. In the same way, we identify the dominant modes of interaction between cardiac and respiratory oscillators. The results of this analysis are given in Fig. 7.3.

The reconstructed model for the specified segment of the cardiorespiratory data reads (we use here the notations $\phi_1 = \phi_h$ and $\phi_2 = \phi_r$)

$$\begin{aligned}
\Delta\phi_h \approx\ & 0.078 + 0.039\sin(\phi_r) + 0.481\cos(\phi_r) \\
& - 0.017\sin(\phi_r - \phi_h) + 0.133\cos(\phi_r - \phi_h) \\
& - 0.248\sin(3\phi_r - \phi_h) + 0.031\cos(3\phi_r - \phi_h) \\
& - 0.064\sin(5\phi_r - 2\phi_h) + 0.036\cos(5\phi_r - 2\phi_h)\,, \\
\Delta\phi_r \approx\ & 0.11 - 0.572\sin(\phi_r) - 0.114\cos(\phi_r) \\
& + 0.004\sin(\phi_h - 2\phi_r) + 0.172\cos(\phi_h - 2\phi_r) \\
& + 0.073\sin(\phi_h - 3\phi_r) + 0.339\cos(\phi_h - 3\phi_r)\,.
\end{aligned} \tag{7.12}$$

We recall that for the reconstruction we used the phase estimates, which, contrary to true phases, do not fulfill (for uncoupled systems) the condition $\dot{\phi} = \omega$. The oscillation of the estimated phase around a uniform growth is especially pronounced if the Hilbert transformation is used. This reflects in the appearance of the terms $\sim \sin(\phi_r)$, $\sim \cos(\phi_r)$ in the equation for $\Delta\phi_r$, cf. Eq. (7.8). The appearance of the same terms in the equation for $\Delta\phi_h$ may, however, have an important physical meaning. Indeed, these terms in addition to the terms $\sim \sin(n\phi_r \pm m\phi_h)$, $\sim \cos(\phi_r \pm \phi_h)$ possibly indicate the presence of two mechanisms of interaction—a modulating one and a synchronizing one.

7.3 Characterization of Coupling from Data

Having estimated the phases of interacting objects from bivariate data we can proceed with the characterization of the intensity and the directionality of interaction as well as of the delay in coupling. Generally speaking, there are two ways to do it. On the one hand, we can directly analyze relations between the phases. On the other hand, we can reconstruct phase Eqs. (7.3) and use their parameters in order to quantify the coupling. The latter, truly model-based approach, is more dependent on the correctness of the assumptions made, but can be more informative, e.g., providing a more precise estimate of the delay, as shown below.

7.3.1 Interaction Strength

We have assumed that the interaction between the systems tends to synchronize them, i.e., to lock their phases and frequencies (cf. Eq. (7.2)). The degree of $n : m$ locking and therefore (indirectly) the degree of interaction can be characterized by a *synchronization index*. A convenient choice is to use the parameter-free index computed as [16, 17]

$$\rho^2_{n,m} = \langle\cos(n\phi_1 - m\phi_2)\rangle^2 + \langle\sin(n\phi_1 - m\phi_2)\rangle^2\,, \tag{7.13}$$

where $\langle\rangle$ denotes time average. The index varies from zero (independent phases) to 1 (see Fig. 7.4). The latter case corresponds to a constant phase difference, which is a more strict condition than that in Eq. (7.2). Generally, the phase difference in a synchronous state oscillates around a constant (especially if the phases

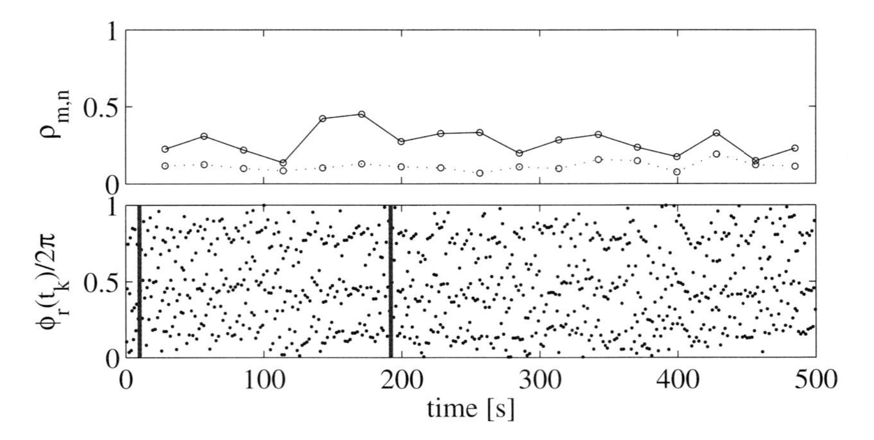

Fig. 7.4: Top panel: synchronization indices $\rho_{1,3}$ (solid line) and $\rho_{2,5}$ (dotted line) for the cardiorespiratory data of a baby. The indices are computed in a running window and therefore are plotted as the functions of time (the window length is equal to 100 average cardiocycles, windows overlap by 50 %). Bottom panel: cardiorespiratory synchrogram. In this representation the phase of the respiration (wrapped to $(0, 2\pi)$ interval) is shown at the instances of appearance of R-peaks in the electrocardiogram, i.e., when the phase of the cardiac systems attains 2π. Note three stripes in the time interval $130\,\text{s} \lesssim t \lesssim 180\,\text{s}$: This is an indication of an interaction that tends to induce the $1:3$ locking. Note also the increase of the $\rho_{1,3}$ index in this time interval. Similar stroboscopic observation of the respiratory phase wrapped to the interval $(0, n \cdot 2\pi)$ can help us reveal an $n:m$ interaction. Vertical lines mark the segment of data used for modeling and identification of coupling.

are estimated from data). Hence, the index is less than 1 even in the synchronous state. Therefore, if the goal of the analysis is to detect a very weak interaction, then it is advisable to use the *stroboscopic approach*. The stroboscopic approach is an application of the well known in the nonlinear dynamics method of Poincaré section to the data analysis. It implies that one fixes some value of the phase $\tilde{\phi}$ and observes the phase of, say, the second system at the times when ϕ_1 attains $\tilde{\phi}$. Next, for these stroboscopically observed values of ϕ_2 one computes

$$\lambda_{\tilde{\phi}} = \langle\cos(\phi_2)\rangle^2 + \langle\sin(\phi_2)\rangle^2 \,. \qquad (7.14)$$

Averaging $\lambda_{\tilde{\phi}}$ over $\tilde{\phi}$ one obtains a stroboscopic synchronization index λ which attains the unit value in the synchronous state even if the phase difference in this state strongly oscillates. The discussion of the stroboscopic index of order $n:m$ and the related graphical tool called "synchrogram" (Fig. 7.4) can be found in [3, 12, 13, 16].

The synchronization index quantifies the end effect of interaction, but not exactly the strength of coupling. The latter is directly related to the amplitudes (norms) of functions $f_{1,2}$ in Eq. (7.3), while the degree of phase locking is determined by both these amplitudes and the frequency mismatch, i.e., the relation between the frequencies of the system. A more detailed information about the strength of interaction may be obtained from the analysis of the coefficients of the reconstructed phase model. This approach is tightly related to the quantification of directionality of coupling, described below. It is also important to emphasize that a synchronization index can be high not only in the case of an interaction that may lead to synchronization, but also in the case of a modulating interaction. Hence, a computation of the index alone does not allow one to draw the conclusion about the synchrony in the coupled system but rather demonstrates the presence of an interaction. The distinction between two types of interactions may be probably done from the analysis of the reconstructed model.

7.3.2 Directionality of Coupling

An estimation of directionality and causality in coupling is an important issue of data analysis. Many techniques used for this goal go back to the Granger's causality concept [23], which can be briefly formulated as follows: If, say, signal 1 depends on signal 2, i.e., there is a directional relation $2 \rightarrow 1$, then the future of 1 can be better predicted if the information on 2 is taken into account; if 2 does not depend on 1, there will be no predictability improvement. Different algorithms, related to this approach, can be found in [24–26]. An extension of this idea in terms of entropy measures has been performed by T. Schreiber [27]; in particular, he applied this approach to the analysis of cardiorespiratory from the bivariate series of the breath rate and the instantaneous heart rate of a sleeping human suffering from sleep apnea.

Another approach, arising from the studies of generalized synchronization, exploited the idea of mutual predictability in the phase space: It quantified the ability to predict the state of the first system from the knowledge of the second one [28, 29]. While both approaches are rather complicated to implement and interpret, neither requires any assumptions on the systems under investigation.

Before presenting the algorithms of the coupled oscillators approach, let us make several notes on the concepts of directionality and causality. First, we note that the assumption of weakly coupled oscillators implies that coupling, say from 2 to 1, is not a cause of oscillation of 1, but a weak perturbation to this oscillation. A second important issue is that the *quantification of directionality is generally ambiguous*. While everything is clear in the case of unidirectional driving, the definition of symmetric interaction in bidirectionally coupled systems

$$\mathbf{X} = f_1(\mathbf{X}) + p_1(\mathbf{Y}) , \quad \mathbf{Y} = f_2(\mathbf{Y}) + p_2(\mathbf{X}) , \qquad (7.15)$$

cannot be unique. Indeed, is the coupling symmetric if $p_1 = p_2$ but $f_1(\cdot) \neq f_2(\cdot)$? Obviously, this question cannot be answered in a unique way, and, hence, different measures of directionality can be proposed and used in different experimental situations.

Directionality from Phases: Mutual Predictability Approach

In the case of weakly coupled oscillators the concept of mutual predictability can be very easily implemented, because we have to deal with two scalar signals only, namely with the time series $\phi_{1,2}$. Let us take one series, say, $\phi_1(t_k)$, and use some scheme to predict a future of its points. For the kth point we compute the *univariate prediction error* $E_1(t_k) = |\phi_1'(t_k) - \phi_1(t_k + \tau)|$, where $\phi_1'(t_k)$ is the τ-step ahead prediction of the point $\phi_1(t_k)$; note that phases are unwrapped. Next, we repeat the prediction for $\phi_1(t_k)$, but this time we use both signals ϕ_1, ϕ_2 for the construction of the predictor. In this way we obtain the *bivariate prediction error* $E_{12}(t_k)$. If system 2 influences the dynamics of system 1, then we expect $E_{12}(t_k) < E_1(t_k)$, otherwise (for sufficient statistics) $E_{12}(t_k) = E_1(t_k)$. The root mean squared $E_1(t_k) - E_{12}(t_k)$, computed over all possible k and denoted by I_{12}, quantifies the *predictability improvement* for the first signal. This measure characterizes the degree of influence of the second system on the first one. Computing in the same way I_{21}, we end with the directionality index

$$p_{(1,2)} = \frac{I_{21} - I_{12}}{I_{12} + I_{21}} . \tag{7.16}$$

This approach has been suggested and applied to cardiorespiratory interaction in [30]. The same algorithm formulated in terms of conditional mutual information has been later used in [31].

Directionality from Phases: Model-Based Approach

In quantification of the directionality from the reconstructed equations of the phase dynamics we follow our previously developed approach [30, 32]. We recall that there is no unique way to quantify the directionality of coupling, even if Eq. (7.3) are known. One way to quantify the directionality is as follows. We quantify the influence of system 2 on system 1 by the coefficient

$$c_1^2 = \|\partial\dot{\phi}_1/\partial\phi_2\| , \tag{7.17}$$

where the norm

$$\|(\cdot)\| = \iint_0^{2\pi} (\cdot)^2 \, d\phi_1 \, d\phi_2 . \tag{7.18}$$

Note that $c_{1,2}$ can be easily obtained from the model coefficients, e.g.,

$$c_1^2 = \sum n^2 (a_{m,n}^2 + b_{m,n}^2) \tag{7.19}$$

[33]. c_1 is an integrative measure of how strongly oscillator 1 is driven and how sensitive it is to the driving. Computing in the same way c_2, we quantify asymmetry in interaction by one number

$$d_{(1,2)} = \frac{c_2 - c_1}{c_1 + c_2} , \tag{7.20}$$

that we call *directionality index*. It varies from 1 in the case of unidirectional coupling ($1 \to 2$) to -1 in the opposite case ($2 \to 1$), while intermediate values correspond to a bidirectional coupling configuration. If two oscillators are structurally identical and differ only by natural frequencies then $f_1(\cdot) = f_2(\cdot)$ and $d^{(1,2)} = (\varepsilon_2 - \varepsilon_1)/(\varepsilon_1 + \varepsilon_2)$. Alternative solutions of the directionality estimates have been discussed and experimentally verified in [30, 34].

We emphasize that the presented algorithm fails if the oscillators are phase locked, which mathematically corresponds to the appearance of a functional dependence between the two phase variables. On the other hand, if the coupling is too weak, so that the systems cannot be distinguished from uncoupled ones, the directionality cannot be estimated as well. Note also that coefficients $c_{1,2}$ are always overestimated; indeed, if the coefficient is zero, its estimate

$$\sqrt{\langle(\partial\dot{\phi}_1/\partial\phi_2)^2\rangle} \tag{7.21}$$

is positive. A way to correct the estimate was suggested in [33].

We remark that in the quantification of the directionality we are not interested in the mostly exact reconstruction of the model equations, but only in the recovery of interdependences in the phase dynamics. In this context it is more appropriate to work with discrete mappings (cf. Eqs. (7.4)). Computation of a phase increment over a relatively large time interval (it can be of the order of oscillation period) helps us to reduce the effect of noise, see discussions in [32, 35] for more details.

Application of the directionality algorithms to cardiorespiratory data can be found in [21, 30, 36]. Here we present the results for the sample data set. The mutual prediction algorithm provides the directionality index

$$p_{h\to r} \approx -0.84\,.$$

The directionality index obtained from coefficients of the model Eq. (7.12) is

$$d_{h\to r} \approx -0.42\,.$$

This means that the coupling is bidirectional, though not symmetrical: The action from respiration to the cardiac system dominates over the reverse action. However, in the interpretation of the results it is important to have in mind that in the case of $n:m$ coupling with *equal* strength, the coefficients c_1 and c_2 are generally different. For an illustration, let us consider a simple model $\dot{\phi}_1 = \omega_1 + \varepsilon\sin(3\phi_2 - \phi_1)$, $\dot{\phi}_2 = \omega_2 + \varepsilon\sin(\phi_1 - 3\phi_2)$. It is easy to see that $c_1 = 3c_2$, which gives $d_{(1,2)} = 0.5$.

7.3.3 Delay in Coupling from Data

We now consider the last problem, namely an estimation of the delay in coupling. There are two ways to treat this problem. First, one can compute from the time series of phases the synchronization index according to Eq. (7.13), and then shift the first series with respect to the second one and compute the index for different, positive and negative, shifts τ. It is natural to expect that this *shift-dependent synchronization index* [37]

$$\rho^2(\tau) = \langle \cos(\phi_1(t) - \phi_2(t-\tau)) \rangle^2 + \langle \sin(\phi_1(t) - \phi_2(t-\tau)) \rangle^2 \tag{7.22}$$

maximizes if the shift corresponds to the (unknown) delay in coupling.

The second model-based approach exploits a generalization of the Models, Eq. (7.3) and Eq. (7.4)

$$\begin{aligned} \dot{\phi}_1 &= \omega_1 + \varepsilon f_1(\phi_1(t), \phi_2(t - T_{21})) + \xi_1(t)\,, \\ \dot{\phi}_2 &= \omega_2 + \varepsilon f_2(\phi_1(t - T_{12}), \phi_2(t)) + \xi_2(t)\,, \end{aligned} \tag{7.23}$$

and

$$\begin{aligned} \Delta\phi_1(t) &= F_1(\phi_1(t), \phi_2(t - T_{21})) + \zeta_1\,, \\ \Delta\phi_2(t) &= F_2(\phi_2(t), \phi_1(t - T_{12})) + \zeta_2\,, \end{aligned} \tag{7.24}$$

where the coupling function in the first equation (map) contains a retarded value of the phase of the second oscillator, and vice versa. The idea is to reconstruct the model, as discussed in the previous sections, fit it to the bivariate data where one series is shifted with respect to the other, and to quantify the fit quality by the root mean square errors $E_{1,2}$ for different shifts τ (errors $E_{1,2}$ describe the quality of modeling of $\dot{\phi}_1$ and $\dot{\phi}_2$, respectively). The dependences $E_{1,2}(\tau)$ should take a minimum at $\tau = T_{12}$, $\tau = T_{21}$. Note that for our goal it is not required to reconstruct the phase dynamics very precisely, because we are not interested in the absolute value of $E_{1,2}(\tau)$ but only in its variation with τ.

Analytical and numerical treatment of these two approaches performed in [35] shows that the position of the maximum of the dependence of the synchronization index ρ on the time shift τ systematically overestimates the delay. Moreover, in the case when the oscillators are far from synchrony, the synchronization index is small for all shifts τ and therefore does not yield the estimate of the delay. Thus, the advantages of this approach, namely its simplicity and the absence of parameters, are accompanied by several drawbacks which can be overcome by the technique based on the model reconstruction. However, if the systems are very close to synchrony, then the model reconstruction fails due to a functional relation between the phases and only the method based on the synchronization index can be used.

The results of the analysis for the cardiorespiratory data set are shown in Fig. 7.5. The values of delay, estimated from the positions of the minima of the dependence $E_{1,2}(\tau)$, are $\mathcal{T}_1 \approx 0.4$ s (delay in coupling from respiration to heart) and $\mathcal{T}_2 \approx 1.4$ s (delay in coupling from heart to respiration). As the system is far from synchrony, the dependence of the synchronization on shift is not efficient

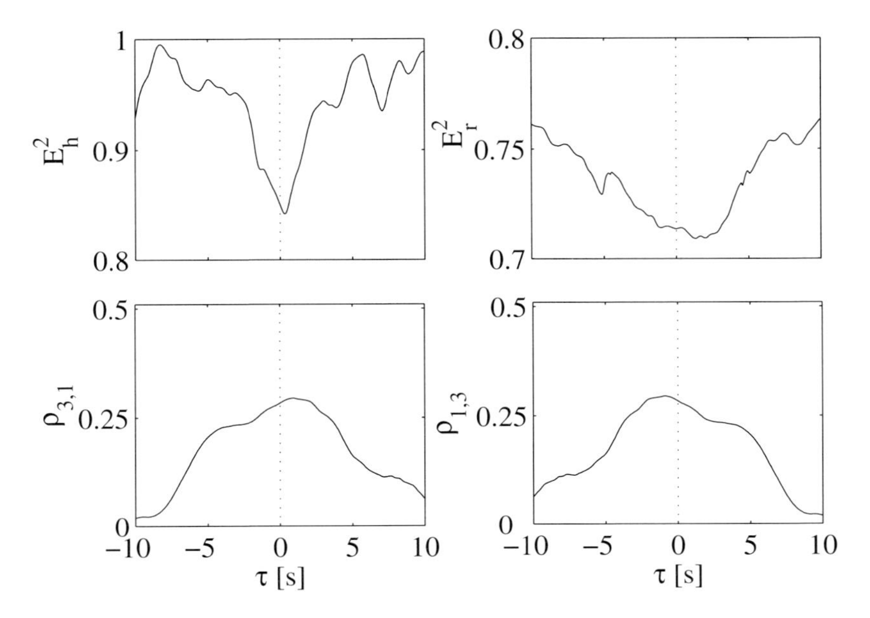

Fig. 7.5: Estimation of the delays in bidirectional cardiorespiratory coupling. Top panels show the (normalized) errors of fit versus time shift between the series. Minima of the dependence indicate the values of delays. Bottom panels show the $\rho_{1,3}$ synchronization index. Dependence of the synchronization on shift (bottom panels) is not efficient in delay estimation.

in delay estimation. Our estimate of the time delay in coupling between the respiratory and cardiac oscillators falls well within the range of documented in [38] latencies in the human cardiac baroreflex response.

We note that the most common tool that can be tested for the detection of the delay is the cross-correlation function. If the fluctuations of the amplitudes of signals are small, then the cross-correlation function $C(\tau)$ has a very simple relation to the synchronization index, namely $\rho(\tau)$ is the envelope of $C(\tau)$ [35]. Hence, the analysis of $C(\tau)$ provides the biased estimate of the delay as well.

7.4 Conclusions and Discussion

In this chapter we have presented a model-based approach for the identification and quantification of an interaction between two coupled systems from experimental data. The approach relies on the assumption that we deal with weakly interacting self-sustained oscillators, and that the measured signals represent the dynamics of different oscillatory systems. We discussed in detail how to estimate the phase data, the object of the analysis, and how to quantify the main characteristics of the interaction, namely, the strength, the directionality, and the delays in coupling. The methods have been exemplified by examining the nature of the interaction between the cardiac and respiratory oscillators of a healthy infant.

Several remarks on the applicability of the presented modeling framework and potential pitfalls of the presented methods for the data analysis are in order. First, one should verify whether the assumptions about the data generating processes are valid. Then, an attention must be paid to the preprocessing of signals (e.g., filtering) required for the computation of the phases. In particular, a careful search for optimal filter parameters must be undertaken, especially for signals with nonsinusoidal shape and/or cycle-to-cycle variability (e.g., in the case of the respiratory signal). Nonoptimal filtering can greatly affect the accuracy of the derivative approximation and of the model reconstruction process. Finally, we recall that the model reconstruction and the subsequent application of the algorithms for directionality and time delay estimation requires that the interaction between two observed oscillator is not strong enough in order to bring them to synchrony. At the other extreme, when the interaction is too weak (on the level of noise) the estimation of the directionality and the delay becomes impossible, too.

Although in this chapter we focused on the study of interaction between two oscillators, a natural question arises about the possibility of exploiting the presented approach for the study of several interacting systems. A preliminary analysis performed in [30, 36] demonstrates that a study of multivariate data can be partially accomplished by a pairwise analysis of bivariate data and the main effects of interaction can be identified. However, a clear distinction between direct and indirect interactions cannot be made in a straightforward manner, and a further work on the extension of our approach is required.

We conclude by expressing our belief that the presented theoretical and methodological framework of interacting self-sustained oscillators provides a useful basis for further development of techniques of the multivariate data analysis.

Acknowledgements

We acknowledge financial support from EU (NEST Project BRACCIA) and DFG (SFB 555).

References

[1] Y. Kuramoto. *Chemical Oscillations, Waves and Turbulence.* Springer, Berlin, 1984.

[2] P. S. Landa. *Nonlinear Oscillations and Waves in Dynamical Systems.* Kluwer, Dordrecht, Boston, London, 1996.

[3] A. Pikovsky, M. Rosenblum, and J. Kurths. *Synchronization. A Universal Concept in Nonlinear Sciences.* Cambridge University Press, Cambridge, 2001.

[4] M. Rosenblum, A. Pikovsky, and J. Kurths. *Phys. Rev. Lett.*, 76:1804, 1996.

[5] I. I. Blekhman. *Synchronization in Science and Technology.* Nauka, Moscow, 1981.

[6] E. Mosekilde, Yu. Maistrenko, and D. Postnov. *Chaotic Synchronization. Applications To Living Systems.* World Scientific, Singapore, 2002.

[7] V. Anishchenko, A. Neiman, V. Astakhov, T. Vadiavasova, and L. Schimansky-Geier. *Chaotic and Stochastic Processes in Dynamic Systems.* Springer, Berlin, 2002.

[8] L. Glass and M. C. Mackey. *From Clocks to Chaos: The Rhythms of Life.* Princeton University Press, Princeton, NJ, 1988.

[9] L. Glass. *Nature*, 410:277, 2001.

[10] H. Pessenhofer and T. Kenner. *Pflügers Arh.*, 355:77, 1975.

[11] T. Kenner, H. Pessenhofer, and G. Schwaberger. *Pflügers Arh.*, 363:263, 1976.

[12] C. Schäfer, M. Rosenblum, J. Kurths, and H.-H. Abel. *Nature*, 392:239, 1998.

[13] R. Mrowka, A. Patzak, and M. Rosenblum. *Int. J. Bif. Chaos*, 10:2479, 2000.

[14] M. Bračič Lotric and A. Stefanovska. *Physica A*, 283:451, 2000.

[15] D. Gabor. *J. IEE London*, 93:429, 1946.

[16] M. Rosenblum, A. Pikovsky, J. Kurths, C. Schäfer, and P. A. Tass. *Neuroinformatics and Neural Modeling*, volume 4 of *Handbook of Biological Physics.* Elsevier, Amsterdam, 2001.

[17] E. Rodriguez, N. George, J.-P. Lachaux, J. Martinerie, B. Renault, and F. J. Varela. *Nature*, 397:430, 1999.

[18] J. P. Lachaux, E. Rodriguez, M. Le van Quyen, A. Lutz, J. Martinerie, and F. J. Varela. *Int. J. Bif. Chaos*, 10:2429, 2000.

[19] R. Quian Quiroga, A. Kraskov, T. Kreuz, and P. Grassberger. *Phys. Rev. E*, 65:041903, 2002.

[20] M. Rosenblum, A. Pikovsky, J. Kurths, G. Osipov, I. Kiss, and J. Hudson. *Phys. Rev. Lett.*, 89:264102, 2002.

[21] R. Mrowka, L. Cimponeriu, A. Patzak, and M. Rosenblum. *Am. J. Physiol. Regul. Integr. Comp. Physiol.*, 285:R1395, 2003.

[22] A. L. Edwards. *An Introduction to Linear Regression and Correlation.* Freeman, San Francisco, CA, 1976.

[23] C. W. J. Granger. *Econometrica*, 37:424, 1969.

[24] M. Wiesenfeldt, U. Parlitz, and W. Lauterborn. *Int. J. Bif. Chaos*, 11:2217, 2001.

[25] U. Feldmann and J. Bhattacharya. *Int. J. Bif. Chaos*, 14:505, 2004.

[26] W. Hesse, E. Moller, M. Arnold, and B. Schack. *J. Neurosci. Methods*, 124:27, 2003.

[27] T. Schreiber. *Phys. Rev. Lett.*, 85:461, 2000.

[28] S. J. Schiff, P. So, T. Chang, R.E. Burke, and T. Sauer. *Phys. Rev. E*, 54:6708, 1996.

[29] R. Quian Quiroga, J. Arnhold, and P. Grassberger. *Phys. Rev. E*, 61:5142, 2000.

[30] M. Rosenblum, L. Cimponeriu, A. Bezerianos, A. Patzak, and R. Mrowka. *Phys. Rev. E*, 65:041909, 2002.

[31] M. Paluŝ and A. Stefanovska. *Phys. Rev. E*, 67:055201(R), 2003.

[32] M. Rosenblum and A. Pikovsky. *Phys. Rev. E*, 64:045202(R), 2001.

[33] D. A. Smirnov and B. P. Bezruchko. *Phys. Rev. E*, 68:046209, 2003.

[34] B. P. Bezruchko, V. Ponomarenko, A. Pikovsky, and M. Rosenblum. *Chaos*, 13:179, 2003.

[35] L. Cimponeriu, M. Rosenblum, and A. Pikovsky. *Phys. Rev. E*, 70:046212, 2004.

[36] L. Cimponeriu, M. Rosenblum, T. Fieseler, J. Dammers, M. Schiek, M. Majtanik, P. Morosan, A. Bezerianos, and P. A. Tass. *Progr. of Theoretical Physics Suppl.*, 150:22, 2003.

[37] D. Rybski, S. Havlin, and A. Bunde. *Physica A*, 320:601, 2003.

[38] H. Seidel, H. Herzel, and D. L. Eckberg. *Am. J. Physiol.*, 272:H2040, 1997.

8 Nonlinear Dynamical Models from Chaotic Time Series: Methods and Applications

Dmitry A. Smirnov and Boris P. Bezruchko

The construction of mathematical models from experimental data is a topical field in mathematical statistics and nonlinear dynamics. It has a long history and still attracts increasing attention. We briefly discuss key problems in nonlinear modeling for typical problem settings ("white," "gray," and "black boxes") and illustrate several contemporary approaches to their solution with simple examples. Finally, we describe a technique for the determination of weak directional coupling between oscillatory systems from short time series based on empirical modeling of their phase dynamics and present its applications to climatic and neurophysiological data.

8.1 Introduction

Ubiquitous use of analog-to-digital converters and fast development of computing power have stimulated considerable interest in methods for modeling discrete sequences of experimental data. The construction of mathematical models from "the first principles" is not always possible. In practice, available information about an object dynamics is often represented in the form of experimental measurements of a scalar or vector quantity η, which is called "observable," at discrete time instants. Such a data set is called "a time series" and denoted by $\{\eta_i\}_{i=1}^N \equiv \{\eta_1, \eta_2, \ldots, \eta_N\}$ where $\eta_i = \eta(t_i)$, $t_i = i\Delta t$, Δt is a sampling interval, N is a time series length. Modeling from experimental time series is known as "system identification" in mathematical statistics and automatic control theory [1] or "reconstruction of dynamical systems" in nonlinear dynamics [2].

Dynamical systems' reconstruction has its roots in the problems of *approximation* and *statistical investigation* of dependences. Initially, observed processes were modeled as explicit functions of time which approximated experimental dependences on the plane (t, η). The purpose of modeling was either predicting the future evolution (via extrapolation) or smoothing the data. A significant advance in empirical modeling of complex processes was achieved in the beginning of the twentieth century when *linear stochastic* autoregressive models were introduced [3]. It gave an origin to ARIMA models technology which became

Handbook of Time Series Analysis. Björn Schelter, Matthias Winterhalder, Jens Timmer

ISBN: 3-527-40623-9

a predominant approach for half a century (1920s–1970s) and found numerous applications, especially in automatic control [1, 4]. Subsequently, birth of the concept of "deterministic chaos" and fast progress of computational power led to the appearance of a different framework. Currently, empirical modeling is often performed with the use of *nonlinear* difference and differential equations, see pioneering works [5–10]. Such empirical models are demanded in many fields of science and practice such as physics, meteorology, seismology, economy, biomedicine, etc. [11].

In this chapter a brief overview of the problems and techniques for the construction of dynamical models from noisy chaotic time series is given. It supplements existing surveys [12–18] due to the use of a special systematization of the variety of problem settings and methods. Also, we try to provide a clear explanation of the key points with simple examples and illustrate some specific problems with our own results. For the most part, we examine finite-dimensional models in the form of difference equations (maps)

$$\mathbf{x}_{n+1} = \mathbf{f}(\mathbf{x}_n, \mathbf{c}) \tag{8.1}$$

or ordinary differential equations (ODEs)

$$d\mathbf{x}/dt = \mathbf{f}(\mathbf{x}, \mathbf{c}), \tag{8.2}$$

where $\mathbf{x}$ is a D-dimensional state vector, $\mathbf{f}$ is a vector-valued function, $\mathbf{c}$ is a P-dimensional parameter vector, n is the discrete time, and t is the continuous time.

We expose the problems "from simple to complex," as the amount of *a priori* information about an object decreases. We start from a situation where only concrete values of model parameters are to be found ("transparent box" or "white box," Section 8.3). Then, we go via the case where a model structure is partly known ("gray box," Section 8.4) to the case of no *a priori* information ("black box," Section 8.5). Throughout the chapter, we refer to a unified scheme of the empirical modeling process outlined in Section 8.2. Some applications of empirical modeling, in particular, to climatic and neurophysiological data are described in Section 8.6.

8.2 Scheme of the Modeling Process

Despite an infinite number of specific situations, objects, and purposes of modeling, one can single out basic stages of the modeling process and present them using a scheme shown in Fig. 8.1 which generalizes similar schemes given in [1, 4]. It starts with the consideration of available *a priori* information about an object under investigation, formulation of the goals of modeling, acquisition and preliminary analysis of experimental data (stage 1). It ends with a desired application of a constructed model. However, the modeling process typically involves multiple reiterations and a step-by-step approach to a "good" model.

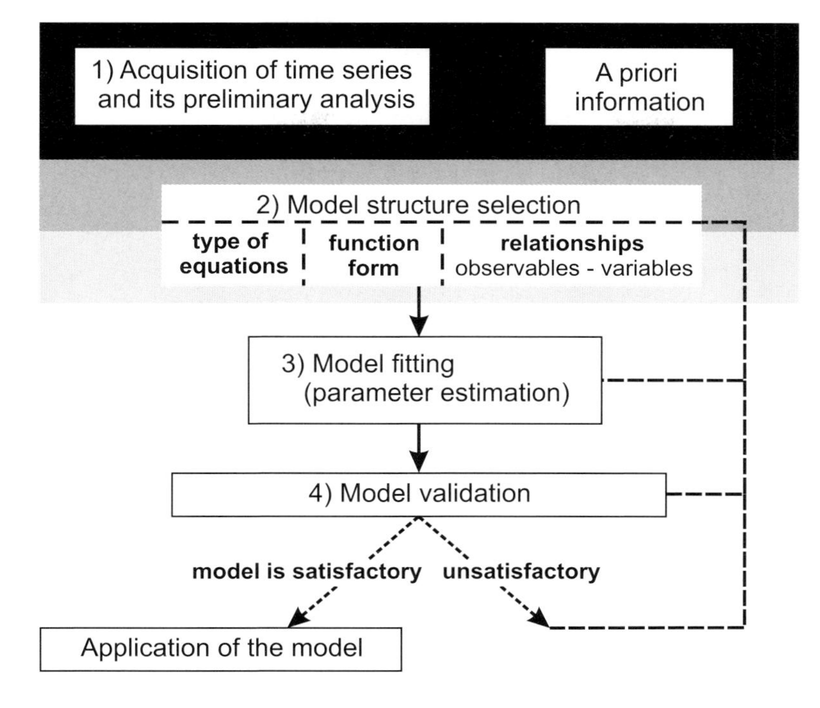

Fig. 8.1: A general scheme of the process of modeling from time series.

At the second stage, a model structure is specified. One chooses the type and number of model equations, the form of functions entering their right-hand sides (components of the function $\mathbf{f}$), and dynamical variables (components of the vector $\mathbf{x}$). As for the latter, one can use just the observable quantities as model variables, but in general the relationship among the observables and dynamical variables may be specified separately. Usually, it takes the form $\eta = h(\mathbf{x})$, where h is called "measurement function." Moreover, the observable values may be corrupted with noise. Stage 2 is often called "structural identification."

At the third stage, the values of the model parameters $\mathbf{c}$ are to be determined. One often speaks of *parameter estimation* or *model fitting*. In the theory of system identification this is a stage of "parametric or nonparametric identification." To perform the estimation, one usually looks for a global extremum of an appropriate *cost function*. For example, the sum of squared deviations of a model time realization from the observed data is often minimized.

Finally, the quality of a model is checked, as a rule, based on a specially reserved test part of a time series. In respect of the final goal of modeling, one can distinguish between two settings: "cognitive identification" (the goal is to obtain an adequate model and to understand better the object behavior) and "practical identification" (a practical goal is to be achieved with the aid of the model, e.g., a forecast). Depending on the setting, one checks either model *adequacy* in respect of some properties (this step is also called model validation or verification) or model *efficiency* in respect of the practical goal. If a model is found satisfactory (adequate or efficient) then it may be exploited. Otherwise, one must return to one of the previous stages of the scheme.

The background colors in (Fig. 8.1) change from black to white reflecting the degree of *a priori* uncertainty. The worst situation is called "black box" problem: information about an appropriate model structure is completely lacking and one must start the modeling process from the very top of the scheme. The more information about a possible model structure is available, the more probable is the success of modeling: the "box" becomes "gray" and even "transparent" ("white"). In any case, one cannot avoid the stage of parameter estimation. Therefore, we start our consideration with the simplest situation when one knows everything about an object, except for the concrete values of the model parameters. It corresponds to white background color in Fig. 8.1.

8.3 "White Box" Problems

If a model structure is completely known, the problem reduces to the estimation of model parameters $\mathbf{c}$ from the observed data. Such a setting is encountered in different applications and, therefore, attracts considerable attention. There are two basic tasks:

1. to obtain parameter estimates with a desired accuracy; this is especially important if the parameters cannot be measured directly under the conditions of experiment, i.e., the modeling procedure acts as "a measurement device" [19–24];
2. to obtain reasonable parameter estimates when time courses of some model state variables x_k can neither be measured directly nor calculated from the available time series of the observable η, i.e., some model variables are "hidden" [25, 26].

Let us discuss both points in turn.

8.3.1 Parameter Estimates and Their Accuracy

As a basic test example, we consider parameter estimation in a nonlinear map from its time series. The object is a quadratic map in a chaotic regime

$$x_{n+1} = f(x_n, c) + \xi_n, \quad \eta_n = x_n + \zeta_n, \tag{8.3}$$

where $f(x_n, c) = 1 - cx_n^2$, the only parameter c is considered unknown, ξ_n, ζ_n are random processes. The process ξ_n is called "dynamical noise" since it affects the evolution of the system, while ζ_n is referred to as "measurement noise" since it corrupts only the observations. In the absence of any noise, one has $\eta_n = x_n$ so that all experimental data points on the plane (η_n, η_{n+1}) lie exactly on

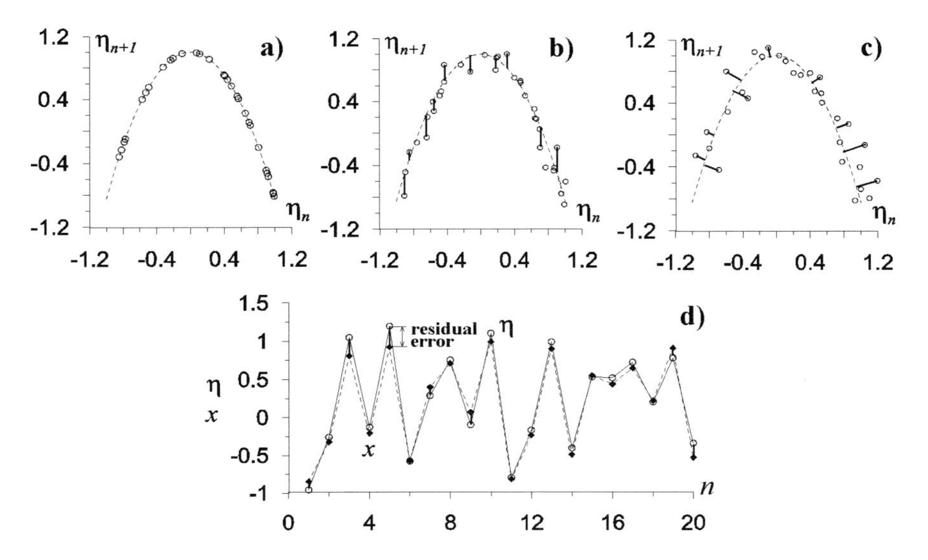

Fig. 8.2: Parameter estimation in the quadratic map (8.3); the true value is $c = 1.85$. Open circles denote observed data. (a) Noise-free case, the dashed line is an original parabola. (b) Uniformly distributed dynamical noise. The dashed line is a model parabola obtained via minimization of the vertical distances. (c) Gaussian measurement noise. The dashed line is a model parabola obtained via minimization of the orthogonal distances. (d) Gaussian measurement noise. Rhombs indicate a model time realization which is the closest one to the observed data in the least-squares sense.

the quadratic parabola (Fig. 8.2(a)). The value of c can be determined from an algebraic equation whose solution takes the form $\hat{c} = (1 - \eta_{n+1})/\eta_n^2$ (throughout the paper, a "hat" denotes quantities calculated from a time series). It is sufficient to use any pair of successive observed values with $\eta_n \neq 0$. As a result, the model is practically ideal.

In the presence of any noise, one must speak of statistical estimates instead of precise calculation of the parameter value. There are various estimation techniques [27]. Below, we describe several of them, which are most widespread.

Maximum Likelihood Approach

The maximum likelihood (ML) approach is the most efficient under

quite general conditions [27]. It is most often announced as a method of choice. However, additional assumptions about the properties of an object and noise are typically accepted in practice reducing the ML approach to a version of the least-squares (LS) technique.

Let us start with the simplest situation when only dynamical noise is present in the system, Eq. (8.3). Let ξ_n be a sequence of independent identically distributed random values whose one-dimensional probability density function is $p_\xi(z)$. Then, an ML estimate is such a value of c which maximizes logarithmic likelihood function

$$\ln L(c) \equiv \ln p(\eta_1, \ldots, \eta_N | c) \approx \sum_{n=1}^{N-1} \ln p_\xi(\eta_{n+1} - f(\eta_n, c)) , \quad (8.4)$$

which is, roughly speaking, a logarithm of a conditional probability to observe the available time series $\{\eta_1, \ldots, \eta_N\}$ at a given c. To apply the ML method, one needs to know the distribution law $p_\xi(z)$ *a priori*. This is rarely the case, therefore, Gaussian distribution is often assumed. It is not always the best idea but it is reasonable both from theoretical (central limit theorem) and practical (successful results) points of view.

Dynamical Noise: Ordinary Least-Squares Technique

For Gaussian noise, the ML estimation, Equation (8.4), reduces to the "ordinary" LS (OLS) technique. The LS method is the most popular estimation technique due to the relative simplicity of implementation, bulk of available theoretical knowledge about the properties of the LS estimates, and many satisfactory practical results. The OLS technique consists in the minimization of the sum of squared deviations

$$S(c) = \sum_{n=1}^{N-1} (\eta_{n+1} - f(\eta_n, c))^2 \to \min . \quad (8.5)$$

Geometrically, it means that a curve of a specified functional form is drawn on the plane (η_n, η_{n+1}) in such a way that the sum of squared *vertical distances* from experimental data points to this curve is minimized (Fig. 8.2(b)). The OLS technique often gives acceptable accuracy of the estimates even if noise is not Gaussian, which is justified by the robust estimation theory, see e.g., [28]. Therefore, it is valuable on its own, apart from being a particular case of the ML approach.

A technical problem in the application of the ML and the OLS estimation arises if a "relief" of the cost function to be optimized exhibits multiple local extrema. It may be the case for the problem, Eq. (8.5), if f is nonlinear in parameter c. Then, the optimization problem is solved with the aid of iterative techniques which require a starting guess for the estimated parameter. Whether a global extremum will be found depends typically on the closeness of the starting guess to the true value of the parameter. The function f is linear in c for the example, Eq. (8.3), therefore the cost function S is quadratic in c and has the only minimum which is easily found via the solution of a linear algebraic equation. Such a simplicity of the LS problem solution is a reason for the widespread use of the models which are linear in parameters, the so-called *pseudo-linear* models, see also Section 8.5.

The error in the estimate $\hat{c}$ decreases with the time series length. Namely, for the dynamical noise case, both ML and OLS techniques give asymptotically unbiased and consistent estimates, i.e., error in the estimate vanishes as $N \to \infty$. Moreover, it can be shown that the variance of the estimates decreases as N^{-1} [27, 28].

Measurement Noise: Monotonically the Total Least-Squares Technique and Others

If only measurement noise is present, the estimation problem becomes more difficult. The OLS technique, Eq. (8.5), provides biased estimates for arbitrary long time series, since it is developed under the assumption of the dynamical noise. However, it is simple in implementation and still may be used sometimes to get a crude approximation. Roughly speaking, if the measurement noise level is not high, namely up to 1%, then the OLS estimates are reasonably good [20]. Throughout the chapter, we define the noise level as the ratio of the noise root-mean-squared value to the signal root-mean-squared value.

At a higher noise level, to enhance the accuracy of the estimates is partly possible with the aid of the total LS (TLS) method [19] where the sum of squared *orthogonal distances* is minimized, see Fig. 8.2(c). But this is only a partial solution since the bias in the estimates is not completely eliminated. A more radical approach is to write the "honest" likelihood function taking into account the effect of measurement noise. To accomplish it, one must include an initial condition of a model map into the set of estimated quantities. Thus, for Gaussian measurement noise the problem reduces to a version of the LS technique where a model *time realization* is made as close to *the observed time series* as possible (Fig. 8.2(d))

$$S(c, x_1) = \sum_{n=0}^{N-1} \left(\eta_{n+1} - f^{(n)}(x_1, c)\right)^2 \to \min\,, \tag{8.6}$$

where $f^{(n)}$ stands for the nth iteration of the map $x_{n+1} = f(x_n, c)$, $f^{(0)}(x, c) \equiv x$.

As an orbit of a chaotic system is highly sensitive to initial conditions and parameters, the variance of such an estimate decreases very quickly with time series length N, even exponentially for specific examples [22, 23]. But it holds true only if a global minimum of the cost function, Eq. (8.6), is guaranteed to be found. However, the graph of the cost function S becomes so "jagged" for a large N that it appears practically impossible to find its global minimum (see Fig. 8.3(a)) because it would require unrealistically lucky starting guesses for c and x_1. It is also difficult to speak of the asymptotic properties of such estimates since the cost function, Eq. (8.6), is no longer smooth in the limit $N \to \infty$. Therefore, modifications of the direct ML approach have been developed for this problem setting [20, 21, 23, 24].

In particular, it was suggested to divide an original time series into segments of moderate length L, minimize Eq. (8.6) for each segment separately, and aver-

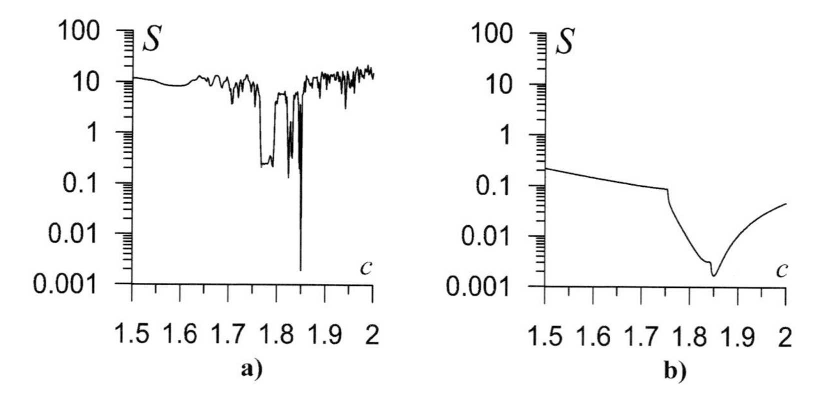

Fig. 8.3: Cost functions for the example of the quadratic map (8.3) at $N = 20$ and true values $c = 1.85$, $x_1 = 0.3$. (a) for the forward iteration approach, Eq. (8.6), (b) for the backward iterations, Eq. (8.7). Trial values of x_1 and x_N are kept equal to their true values for illustration purposes.

age the segment estimates (a piecewise approach). This is a practically reasonable technique but the resulting estimate may remain asymptotically biased. Its variance decreases again only as N^{-1}. Several tricks to enhance the accuracy of the estimates are described below (Section 8.3.2). Here, we would like to note a specific version of the LS technique suggested in [24] for one-dimensional maps. It relies upon the property that the only Lyapunov exponent of a one-dimensional map becomes negative under the time reversal so that a "reverse-time" orbit is no longer highly sensitive to parameters and an "initial" condition. Therefore, one minimizes

$$S(c, x_N) = \sum_{n=0}^{N-1} \left(\eta_{N-n} - f^{(-n)}(x_N, c)\right)^2 \to \min\,, \tag{8.7}$$

where $f^{(-n)}$ is the nth backward iteration of the map. The graph of this cost function looks rather smooth and gradually changing (as in Fig. 8.3(b)) even for arbitrary long time series so that its global minimum can be readily found. At low and moderate noise levels (up to 5–15 %), the error in the estimates obtained via Eq. (8.7) turns out less than for the piecewise approach. Moreover, for sufficiently low noise levels the backward iteration technique gives asymptotically unbiased estimates whose variance decreases generically as N^{-2}. The latter property is determined by close returns of the map orbit to an arbitrary small vicinity of the extrema of the function f [24].

8.3.2 Hidden Variables

If the measurement noise level is considerable, the state variable x can be treated as "hidden" since its true values are not known. But even "more hidden" are those variables whose values can neither be measured directly nor calculated from the observed time series. The latter case is encountered in practice very often. To estimate model parameters is much more problematic in such a situation than for the settings considered in Section 8.3.1. However, if one succeeds, there appears a possibility of getting time courses of the hidden variables as a by-product of the estimation procedure. Hence, a modeling procedure acts as a measurement device in respect of dynamical variables.

Let us briefly mention available techniques. To a significant extent, all of them rely on the idea, Eq. (8.6), i.e., one looks for initial conditions and parameters of a model which provide the least deviation of a model time realization from the observed data. The naive solution of the problem, Eq. (8.6), directly is called "initial value approach" [18]. As we already mentioned, such a method is inapplicable already for moderately long chaotic time series, while simple division of the time series into segments decreases the accuracy of the estimates and the backward iterations are not appropriate for multidimensional dissipative systems.

To overcome the difficulties and exploit longer time series (than allowed by the initial value approach) is partly possible with the aid of Bock's algorithm [18, 25]. It is often called "multiple shooting approach" since it replaces the Cauchy problem with a set of boundary-value problems to get a model orbit. Namely, the idea is to divide the time series into shorter segments of the length L and consider "initial conditions" of the model on each of them as additional quantities to be estimated. Optimization problems, Eq. (8.6), are solved for each segment *while keeping* model parameter values $\mathbf{c}$ *the same* for all segments *and* imposing constraints of "sewing the segments together" to finally obtain a model orbit which is continuous over the entire observation period. Thus, the number of free parameters ("independent" estimated quantities) remain the same as in the initial value approach but intermediate trial values for all estimated quantities may pass through a domain which corresponds to a discontinuous model orbit and is, therefore, forbidden for the initial value approach. The latter property provides higher flexibility of Bock's algorithm [25].

The multiple shooting approach softens the demands to the choice of starting guesses for the estimated quantities. However, for a longer time series it can also become inefficient since the requirement of closeness of a chaotic model orbit to the observed time series over the entire observation interval can appear very strict. One can overcome some difficulties if final discontinuity of a model orbit at some fixed time instants within the observation period is allowed. It increases the number of free parameters and, hence, leads to the growth of the variance of their estimates, but simultaneously the probability of finding a global minimum of the cost function increases. Such a modification allows the use of arbitrary long chaotic time series. The undesirable "side effect" is that a model with inadequate structure can sometimes be regarded "good" due to its ability to reproduce only short segments of a time series. Therefore, one must avoid the use of too short continuity segments [18].

We note that there exist and are currently developed several methods for parameters and hidden variables' estimation which are suitable even for the case of simultaneous presence of dynamical and measurement noise. They are based on the Bayesian approach [29] and Kalman filtering [18, 30]. But that broad field of research is beyond the scope of this chapter.

Model validation for the "white box" problems can be performed via one of the two basic lines: (1) analysis of residual model errors, i.e., checking the agreement among their statistical properties and expected theoretical properties of the noise (typically, Gaussianity and temporal uncorrelatedness) [4]; (2) comparison of dynamical, geometrical, and topological characteristics of a model attractor with the corresponding properties of an object [2].

8.3.3 What Do We Get from Successful and Unsuccessful Modeling Attempts?

Success of the methods described above provides both estimates of model parameters and time courses of hidden variables. It promises exciting applications such as validation of the "physical" ideas underlying a specified model structure, "indirect measurement" of quantities inaccessible for a device of an experimentalist, and restoration of the lost or distorted segments of an observed time realization. However, unsuccessful modeling attempts also give useful information. Let us elaborate.

In practice, one never encounters a purely "white box" problem. A researcher may only have faith that a trial model structure is adequate to an object. Therefore, the result of modeling may well appear negative, i.e., reveal an impossibility to get an adequate model with the specified structure. In such a case, a researcher has to claim falseness of his/her ideas about underlying mechanisms of the investigated process and return to the stage of structural identification.

If there are several alternative model structures, then the results of time series modeling may reveal the most adequate among them. In other words, a modeling procedure provides opportunity to falsify or verify (or, possibly, make more accurate) substantial notions about the dynamics of an object. An impressive example of such a modeling process and substantial conclusions about the mechanism underlying a biochemical signaling process in cells is given in [31]. In a similar way, Horbelt and co-authors validated concepts about a gas laser behavior and reconstructed interdependences among transition rates and pumping current which are difficult to measure directly [32]. However, despite these and some other successful practical attempts, an estimation problem can often appear technically unsolvable: the more hidden variables and unknown parameters involved, the weaker are the chances for the success and the lower is the accuracy of the obtained estimates.

8.4 "Gray Box" Problems

From our point of view, the most promising line of research in the field of dynamical systems' reconstruction is related to the "gray box" problems when one knows a lot about an appropriate model structure except for some components of the function $\mathbf{f}$ in Eqs. (8.1) or (8.2). These components are, in general, nonlinear functions which can often be meaningfully interpreted as *equivalent characteristics* of certain elements of an object under investigation.

One has to choose some *approximating* functions for the characteristics. In this section we focus on the approximation of univariate dependences. Such a case is much simpler than multivariate approximation addressed in Section (8.5). Despite models deduced from physical considerations most often take the form of differential equations, let us consider a model map as the first illustration for the sake of clarity.

8.4.1 Approximation and "Overlearning" Problem

Let the object be a one-dimensional map $x_{n+1} = F(x_n)$. We pretend that the form of the function F is unknown. Let the observable coincide with the dynamical variable x: $\eta_n = x_n$. One has to build a one-dimensional model map $x_{n+1} = f(x_n, \mathbf{c})$. The problem reduces to the selection of a model function $f(x, \mathbf{c})$ and its parameters $\mathbf{c}$ so that it could approximate F to the best possible accuracy. It is the matter of agreement to attribute this problem setting to the "gray box" class. We do so since the knowledge that *one-dimensional* model is appropriate can be considered as an important *a priori* information.

Usually, the OLS technique, Eq. (8.5), is used to calculate parameter values. However, the interpretation of the results differs. Now, one speaks of approximation and its mean-squared error rather than of the estimates and noise. Typically, an individual model parameter is not physically meaningful, only the entire model function $f(x, \hat{\mathbf{c}})$ can make sense as a nonlinear characteristic. A key question is how to choose the form of the model function f.

One may choose it intuitively via looking at the experimental data points on the plane (η_n, η_{n+1}). However, this way is not always possible. Thus, it is practically excluded if an unknown univariate function is only a component of a multidimensional model. A more general and widespread approach is to use a functional basis for approximation. For example, the celebrated Weierstrass theorems state that any continuous function over a finite interval can be uniformly approximated to arbitrary high accuracy with an algebraic polynomial (or a trigonometric polynomial under an additional condition). An algebraic polynomial $f(x, \mathbf{c}) = c_1 + c_2 x + \cdots + c_{K+1} x^K$ is one of the most efficient constructions for approximation of smooth univariate dependences. Therefore, we use it below for illustration.

Theoretically, any smooth function can be accurately approximated with a polynomial of *sufficiently high* order K. What value of the order must be chosen

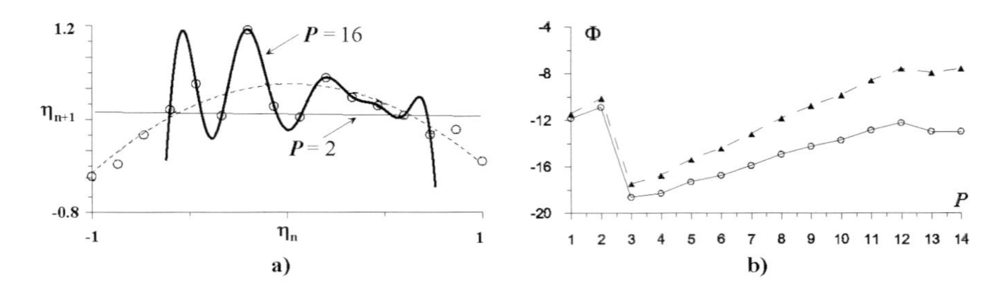

Fig. 8.4: Approximation based on the noisy quadratic map data. (a) Observed data points are shown with circles. Graphs of model polynomials of different orders K are presented. The dashed line for $K = 2$, the thin line for $K = 1$, and the thick line corresponds to $K = 15$. (b) Different cost functions (Eq. (8.8)) versus model size: Circles for the Akaike criterion and triangles for the Schwartz criterion. Both cost functions indicate an optimal model size $P = 3$ corresponding to the true polynomial order $K = 2$.

in practice given a time series of the finite length N, i.e., $N - 1$ data points on the plane (η_n, η_{n+1})? It is a bad idea to specify a very small polynomial order since a model function could not reasonably reproduce an observed nonlinearity (Fig. 8.4(a), the thin line). It is a bad idea to choose very big order as well: e.g., at $K = N - 2$ the graph of the model polynomial on the plane (η_n, η_{n+1}) can pass through all the experimental data points *exactly*, but typically it would extremely badly predict additional (test) observations. In the latter case, the model is said to be *overlearned* or *overtrained* [28]. It does not generalize, rather it just reproduces the observed $N - 1$ data points (Fig. 8.4(a), the thick line).

In practice, one often tries different polynomial orders, starting from a very small one and successively increasing it. One stops when a model gives more or less satisfactory description of an object dynamics and/or the results of approximation saturate. This is a subjective criterion, but it is the only one which is generally applicable, since any "automatic" approach to the order selection is based on a specific well-formalized practical requirement and may not recognize the most adequate model. Such automatic criteria were developed, e.g., in the framework of the information theory. They are obtained from different considerations, but formally reduce to the minimization of a cost function

$$\Phi(P) = (\text{model error}) + (\text{model size}) \to \min \,. \tag{8.8}$$

Here, the model error rises monotonically with the mean-squared approximation error $\varepsilon^2 = S/(N-1)$. The model size is an increasing function of the number of model parameters P. Thus, the first term in the sum, Eq. (8.8), may be very large for small polynomial orders, while the second term dominates for big orders. One often observes a minimum of the cost function, Eq. (8.8), for an intermediate K. The minimum corresponds to an optimal model size. The cost function $\Phi(P) = (N/2)\ln\varepsilon^2(\hat{\mathbf{c}}) + P$ is called the Akaike criterion, $\Phi(P) = (N/2)\ln\varepsilon^2(\hat{\mathbf{c}}) + P\ln N/2$

is the Schwartz criterion, and $\Phi(P) = \ln \varepsilon^2(\hat{\mathbf{c}}) + P$ is a model entropy [5]. More "cumbersome" is a formula for a cost function named *description length* [33]. Description length minimization is currently the most widely used approach to the model size selection, e.g., [34]. It is based on the ideas of optimal information compression, the Schwartz criterion is an asymptotic expression for the description length. In Fig. 8.4(b) we present an example of a polynomial order selection for approximation of quadratic function from a short time series of the quadratic map, Eq. (8.3), with dynamical noise.

If an approximating function is defined in a closed form for the entire range of the argument (e.g., an algebraic polynomial) then the approximation and the model are called *global* [9]. An alternative approach is a *local* (piecewise) approximation where a model function is defined through a simple formula whose parameters' values differ for different small domains within the range of the argument [7, 9]. The most popular examples of the latter approach are piecewise-constant functions, piecewise-linear functions, and cubic splines. Local models are superior for the description of "complicated" nonlinear dependences (strongly fluctuating dependences, dependences with knees and discontinuities, etc.), but they are less robust to noise influence and require larger amount of data than global models of moderate size.

8.4.2 Model Structure Selection

As a rule, one needs to supplement a procedure for model size selection with a technique to search for an optimal model of a specified size. Thus, according to the technique described above the polynomial order is increased starting from zero and the procedure is stopped at a certain value of K, i.e., the terms are added to a model structure in a predefined order. Therefore, a final model inevitably comprises all power of x up to K, inclusively. However, some of the low-order terms might be "superfluous." Hence, it would be much better to exclude them from the model. Different approaches have been suggested to realize a more flexible way of the model structure selection. They are based either on successive selective complication of a model [34] or its selective simplification starting from the biggest size [16, 35–37], see also [38]. Let us describe briefly a version of the latter strategy [36].

One of the efficient principles to recognize "superfluous" model terms is to look at the behavior of the corresponding coefficient estimates when reconstruction is performed from different segments of a time series, i.e., from the sets of data points occupying different domains in the model state space. Typically, it is realized in the most efficient way of a time series corresponding to a transient process is used. The idea is that the parameter values of an adequate global model of a *dynamically stationary* system must not depend on the reconstruction segment. However, the estimates of parameters corresponding to superfluous terms may exhibit significant changes when a reconstruction segment is moved along a time series. A procedure for model structure selection can be based on successive removal of the terms whose coefficients are the least stable being estimated from different segments. In [36] the degree of instability of a coefficient is defined as the ratio of its standard deviation to its empirical mean. Removal is stopped, e.g., when model ability to reproduce an object behavior in a wide domain of state space starts to worsen.

8.4.3 Reconstruction of Regularly Driven Systems

In many cases uncertainty in a model structure can be reduced if *a priori* knowledge about object properties is taken into account. We illustrate it with an example of systems under regular (periodic or quasiperiodic) driving. Indication to the presence of external driving can be often seen in the power spectrum which typically exhibits pronounced discrete peaks for regularly driven systems, even though it is neither a necessary nor a sufficient sign. Having the hypothesis about the presence of external regular driving, one can incorporate functions explicitly depending on time into the model structure to describe the assumed driving. For the first time, it was done for nonlinear two-dimensional oscillators under sinusoidal driving in [39]. In the same work, the successful reconstruction of nonlinear dynamical characteristics of a capacitor with ferroelectric was demonstrated.

In a more general setting, the reconstruction of regularly driven systems was considered in [40, 41]. For harmonical additive driving, it is reasonable to construct a model in the form

$$d^D x/dt^D = f(x, dx/dt, \ldots, d^{D-1}x/dt^{D-1}, \mathbf{c}) + a\cos\omega t + b\sin\omega t\,, \quad (8.9)$$

where f is an algebraic polynomial and the number of variables D is less than for a corresponding *standard* model by 2 (see Section 8.5 about the standard structure).

In the case of arbitrary additive regular driving (either complex periodic or quasiperiodic one), it is convenient to use the model form

$$d^D x/dt^D = f(x, dx/dt, \ldots, d^{D-1}x/dt^{D-1}, \mathbf{c}) + g(t, \mathbf{c})\,, \quad (8.10)$$

where the function g describes driving and also depends on unknown parameters. It may take the form of a sum of trigonometric polynomials [41]

$$g(t, \mathbf{c}) = \sum_{i=1}^{k}\sum_{j=1}^{K_i} c_{i,j}\cos(j\omega_i t + \varphi_{i,j})\,. \quad (8.11)$$

We note that adequate models with trigonometric polynomials can be obtained even for a very large number of involved harmonics (K_i of the order of hundreds), while the use of a high-order algebraic polynomial K leads typically to model orbits diverging to infinity.

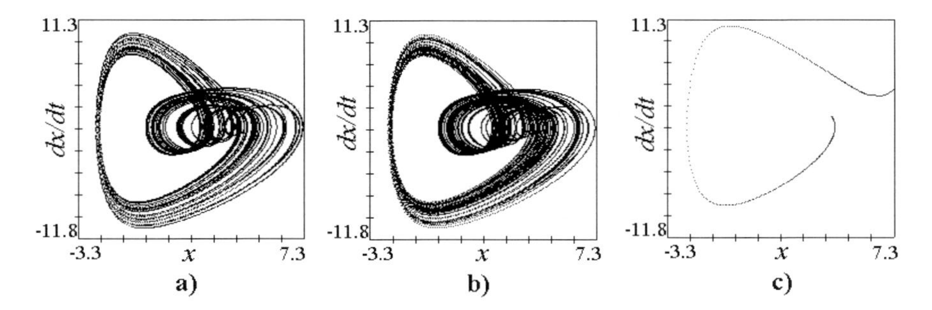

Fig. 8.5: The reconstruction of the driven Toda oscillator $d^2x/dt^2 = -0.45\,dx/dt + (5+4\cos t)(e^{-x}-1) + 7\sin t$. (a) an attractor of the original system; (b) an attractor of a model of the type, Eq. (8.9), with $D = 2, K = 9$, and sinusoidal dependence of time introduced into all polynomial coefficients, (c) a diverging phase orbit of a standard model, Eqs. (8.13) and (8.14) with $D = 4, K = 6$.

Besides, the explicit time dependence can be introduced into all the coefficients of the algebraic polynomial f to allow the description of not just additive driving [40], Fig. 8.5. Efficiency of all these approaches was shown in numerical experiments with the reconstruction of equations of exemplary oscillators from their noise-corrupted chaotic time series for pulse periodic, periodic with subharmonics, and quasiperiodic driving.

8.5 "Black Box" Problems

If nothing is known about an appropriate model structure, one must appeal to universal constructions. They usually involve huge number of parameters that do not allow the use of majority of the estimation techniques described in Section 8.3. In particular, the hidden variables problem is unsolvable in such a case. Therefore, time series of all dynamical variables must be either measured directly or calculated from the observed data. The latter is called "reconstruction of state vectors." Then, one constructs a multidimensional model of the form, Eq. (8.1) or Eq. (8.2), where the multivariate function **f** takes one of the universal forms comprising many parameters. In practice, to estimate these parameters is reasonable with the aid of the OLS technique. To simplify the problem further, it is desirable to choose functions **f** which are linear in parameters **c** (pseudo-linear models). Considerable efforts of many researchers were devoted to the development of such techniques.

8.5.1 Universal Structures of Model Equations

A theoretical background for different approaches to the reconstruction of model state variables from a scalar observable time realization is the celebrated Takens theorem [42]. One of them states that for almost any deterministic dynamical system of the form, Eq. (8.1) or (8.2), with a sufficiently smooth function on the right-hand side, its dynamics on an m-dimensional smooth manifold can be topologically equivalently described in terms of vectors constructed as D-plets of successive values of almost any observable $\eta = h(\mathbf{x})$ separated with an almost arbitrary fixed time interval τ. The equivalent description is (almost) guaranteed if dimensionality of these vectors is high enough, namely, $D > 2m$. One says that the original manifold is *embedded* into the new state space which is often called "embedding space." Rigorous formulations, detailed discussions, and generalizations of the theorems can be found in [43–45].

Thus, the vectors $(\eta_n, \eta_{n+\tau}, \ldots, \eta_{n+(D-1)\tau})$, where τ is a time delay, can serve as state vectors. This approach is very popular since it does not involve any transformation of the observed time series. It is usually employed for the construction of model maps in the form

$$\eta_n = f(\eta_{n-\tau}, \ldots, \eta_{n-D\tau}, \mathbf{c}) . \tag{8.12}$$

Theoretically, the value of τ may be almost arbitrary. However, in practice it is undesirable to use both very small delays (to avoid strong correlations among the state vector components) and very big ones (to avoid complication of the structure of the reconstructed attractor). Therefore, an optimal choice of τ is possible. There are several recipes such as to take the first zero of the autocorrelation function of the time series [46], the first minimum of the mutual information function [47], etc. [48]. It was also suggested to use a nonuniform embedding where time intervals separating successive components of a state vector are not the same [49, 50]. Finally, a variable embedding is possible where the set of time delays and even dimensionality of a state vector depends on the location in state space [50].

Since the value of m is not known *a priori*, it is not clear what value of model dimension to specify. There are several approaches which can give a hint: false nearest neighbors technique [51], correlation dimension estimation [52], or principal component analysis [53]. However, in practice one usually tries different model dimensions, starting from a very small value and successively increasing it until a satisfactory model is obtained or the results saturate. Therefore, the choice of the model dimension and even of the time delays may become an integral part of a monolithic modeling process, rather than a separate first stage.

Different approaches have been suggested to choose the form of the function f in Eq. (8.12). Algebraic polynomials perform extremely badly already for the approximation of bivariate functions [16, 40], while for the "black box" problem one must often exploit the value of D in the range 5–10. Therefore, algebraic polynomials are rarely used in practice. They represent an example of *weak approximation* technique [34] since their number of parameters and errors rise very quickly with model dimension D. Weak approximation techniques also involve trigonometric polynomials and wavelets.

Much attention has been paid to the search for *strong approximation* techniques which behave almost equally well for small and rather big model dimensions. They involve, in particular, local methods [7, 9, 10, 54]. Strong global approximation can be achieved using radial, cylindrical, and elliptic basis functions [34, 50, 55], and artificial neural networks [8]. See also [56] for examples of different approaches. We do not discuss them in details but note that these constructions involve many parameters and the problem of model structure selection (Section 8.4.2) is especially important here.

Another Takens theorem considers continuous-time dynamical systems, Equation (8.2), with much smoother functions on their right-hand side. It states that one can perform embedding into the space of successive derivatives of the observable, i.e., state vectors can be constructed as $\eta, d\eta/dt, \ldots, d^{D-1}\eta/dt^{D-1}$. This approach does not involve a parameter τ which is an advantage. However, it is more difficult to realize in practice since even weak measurement noise is a serious obstacle in the calculation of high-order derivatives. Sometimes, this problem can be solved with the aid of filtering, e.g., Savitsky–Golay filter, but for a sufficiently strong noise it becomes unsolvable. In practice, it is realistic to use the values of $D = 2$–3; rare successes are reported for $D = 5$ [16]. In combination with these state vectors, one constructs usually a model ODE in the form

$$d^D\eta/dt^D = f(\eta, d\eta/dt, \ldots, d^{D-1}\eta/dt^{D-1}, \mathbf{c})\,. \tag{8.13}$$

The situation with the choice of approximating function is the same as discussed above for the model, Eq. (8.12). However, when using the successive derivatives, there are more chances to observe a gradually varying experimental dependence, Eq. (8.13). Therefore, additional reasons to use algebraic polynomials appear. So, in Eq. (8.13) f often takes the form

$$f(x_1, x_2, \ldots, x_D, \mathbf{c}) = \sum_{l_1, l_2, \ldots, l_D = 0}^{K} c_{l_1, l_2, \ldots, l_D} \prod_{j=1}^{D} x_j^{l_j}\,, \quad \sum_{j=1}^{D} l_j \leqslant K\,. \tag{8.14}$$

The structure, Eq. (8.13), with algebraic polynomial, Eq. (8.14), or rational function on the right-hand side is even called *standard* [57] since, theoretically, any smooth dynamical system can be transformed into such a form for a sufficiently large D and K. The values of coefficients in both Eq. (8.12) and Eq. (8.13) are estimated with the aid of the OLS technique. This is valid for a sufficiently low measurement noise level.

Successful results of constructing a model in the form (8.12) can be found, e.g., in [50, 54, 56]. Examples of successful modeling with the aid of Eq. (8.13), we are aware of, are even more rare [16]. As a rule, the structure, Eqs. (8.13) and (8.14), leads to very cumbersome equations tending to exhibit orbits diverging to infinity. It is especially inefficient in the case of multidimensional models. We stress that all the approaches described in this section are rigorously justified only in the case of absence of both measurement and dynamical noise. Their generalizations to the noisy cases are quite problematic [58].

8.5.2 Choice of Dynamical Variables

Let us pay more attention to the important problem of the choice of dynamical (state) variables, i.e., components of the state vectors $\mathbf{x}$. There are very many techniques to obtain time series of state variables from an observable η. Having only a scalar observable, one can use either successive differentiation or time delay embedding (Section 8.5.1). Besides, there are techniques of weighted summation [59] and integration [60] appropriate for strongly nonuniform signals. Further, one can restore a phase of the signal as an additional variable using the analytic signal approach implemented either via the Hilbert transform or the complex wavelet transform [61]. It is also possible to use combinations of all the techniques, e.g., to obtain several variables with the time-delay embedding, several others with integration, and the rest with differentiation [59]. If one observes more than one quantity characterizing a process under investigation, then it is possible to obtain dynamical variables from a time realization of each observable using any combination of the mentioned techniques so that the number of variants rises extremely quickly, see also [62]. It may appear possible that some of the observables should better be ignored in modeling. For example, it may well happen that a better model can be constructed with successive derivatives of the only observable if it turns out easy to find an appropriate approximating function f in Eq. (8.13) for such a choice.

After the reconstruction of state vectors $\{\mathbf{x}(t_i)\}$, an experimental time series of "left-hand sides" of model equations $\{\mathbf{y}(t_i)\}$ is obtained from the time series $\{\mathbf{x}(t_i)\}$ via the numerical differentiation of $\{\mathbf{x}(t_i)\}$ for model ODEs, Eq. (8.2), or the time shift of $\{\mathbf{x}(t_i)\}$ for model maps, Eq. (8.1). "Unlucky" choice of dynamical variables can make the approximation of the model dependence $\mathbf{y}(\mathbf{x})$ with a smooth function more difficult, or even impossible if the relationship among $\mathbf{y}$ and $\mathbf{x}$ appears nonunique.

Taking into account the importance of the stage of the state variables selection [63, 64] and multiple alternatives available, an actual problem is to look for the best (or, at least, for a reasonable) set of state variables. It is, of course, possible just to try different variants and look for the best model in each case. However, this procedure would be too time consuming. Moreover, it may remain unclear why a good model is not achieved for a given set of dynamical variables: Whether it is due to inappropriate model function or due to inappropriate state variables.

A procedure suggested in [65] allows us to test different sets of dynamical variables and select variants which are more promising for the global modeling purposes. It is based on the ideas of [66, 67] and consists in a nonparametric test of an approximated dependence $y(\mathbf{x})$ for uniqueness and continuity. A domain V comprising the set of vectors $\{\mathbf{x}(t_i)\}$ is divided into "hypercubic" boxes of the size δ (Fig. 8.6(a)). Then, all the boxes $s_1, s_2, \ldots, s_M$ comprising at least two vectors are selected. The difference between maximal and minimal values of the "left-hand side" variable y within a box s_k is called *a local variation* ε_k. Maximal local

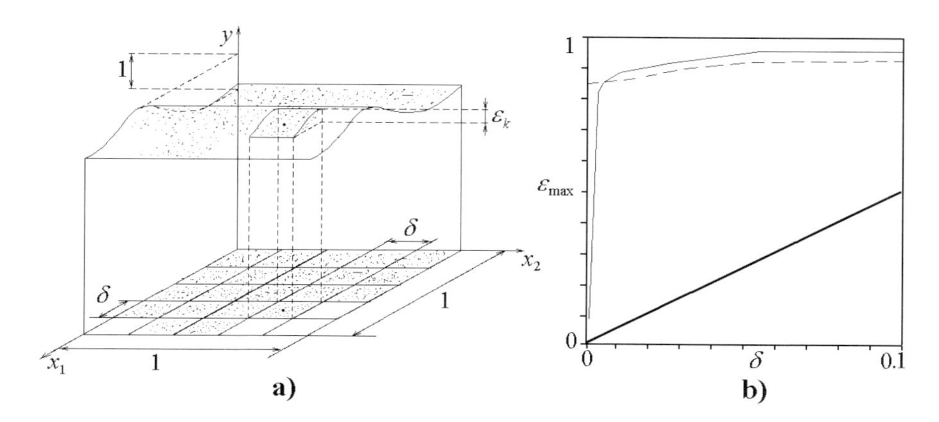

Fig. 8.6: (a) An illustration for the technique of testing a dependence $y(\mathbf{x})$ for uniqueness and continuity, $D = 2$. (b) The plots $\varepsilon_{max}(\delta)$ for different sets of dynamical variables. The thick line corresponds to the best variant, the dashed line to the worst one (nonuniqueness).

variation $\varepsilon_{max} = \max\{\varepsilon_1, \varepsilon_2, \ldots, \varepsilon_M\}$ and the plot $\varepsilon_{max}(\delta)$ are used as the main characteristics of the investigated dependence $y(\mathbf{x})$. Suitability of the considered quantities $\mathbf{x}$ and y for global modeling is estimated as follows. One must choose the variables so that the plot $\varepsilon_{max}(\delta)$ tend to the origin gradually, without "knees" (Fig. 8.6(b), the lowest curve) for each of the approximated dependences $y(\mathbf{x})$.

8.6 Applications of Empirical Models

Probably, the most famous application is a forecast of the future evolution based on the available time series. This intriguing task is considered, e.g., in [4, 7, 9–11, 54–56]. Weather and climate forecasts, prediction of earthquakes, currency exchange rates and stock prices are often in the center of attention. Up to now, empirical models of the type described here are rarely useful to predict such complex processes due to "the curse of dimensionality" (difficulties in modeling quickly grow with dimensionality of the investigated dynamics), deficit of experimental data, and noise. But chances for a successful forecast are higher in simpler situations.

An adequate empirical model may provide a deeper insight into mechanisms underlying the process under investigation [5, 16]. A positive result of model construction (high model quality) may validate physical ideas underlying the model structure. Such a conclusion is of an all-sufficient basic value and may inspire later practical applications.

Below, we consider other applications of empirical models. Namely, we focus on the problem of determination of a directional coupling between oscillators from short time series (Section 8.6.1) and present its applications to climatic signals (Section 8.6.2) and electroencephalograms (Section 8.6.3). Finally, we mention different practical applications and give references for further reading (Section 8.6.4).

8.6.1 Method to Reveal Weak Directional Coupling Between Oscillatory Systems from Short Time Series

One can extract different useful information from the estimates of model parameters. Thus, a sensitive approach to the determination of directionality of coupling between two oscillatory systems solely from their bivariate time series, a problem which is important in many practical and scientific fields, was suggested recently in [68]. It is based on the construction of model equations for *the phase dynamics* of the systems. Its main idea is to estimate how strong future evolution of the first system's phase depends on the second system's phase and vice versa. A detailed discussion can be found in the chapter written by M. Rosenblum (Chap. 7 in this volume). We describe only several points necessary to explain our modification of the method for the case of short time series and its applications.

First, one restores time series of the oscillations phases $\{\phi_1(t_1), \phi_1(t_2), \ldots, \phi_1(t_N)\}$ and $\{\phi_2(t_1), \phi_2(t_2), \ldots, \phi_2(t_N)\}$ from the original signals $\{x_1(t_1), x_1(t_2), \ldots, x_1(t_N)\}$ and $\{x_2(t_1), x_2(t_2), \ldots, x_2(t_N)\}$. We do it below with the analytic signal approach implemented via complex wavelet transform [61]. Given a signal $X(t)$, one defines signal $W(t)$ as

$$W(t) = \frac{1}{\sqrt{s}} \int_{-\infty}^{\infty} X(t')\psi^*\left((t-t')/s\right) dt' , \tag{8.15}$$

where $\psi(\eta) = \pi^{-1/4} \exp(-j\omega_0\eta)\exp(-\eta^2/2)$ is Morlet wavelet, s is a fixed time scale. For $\omega_0 = 6$ used below, $\mathrm{Re}W(t)$ can be regarded as $X(t)$ band-pass filtered around the frequency $f \approx 1/s$ with the relative bandwidth of 1/8. The phase is defined as $\phi(t) = \arg W(t)$. It is the angle of rotation of the radius vector on the plane $(\mathrm{Re}W, \mathrm{Im}W)$ which increases by 2π after each complete evolution. To avoid edge effects while estimating Eq. (8.15) from a time series, we ignore segments of the length 1.4 s at each edge after the phase calculation.

Second, one constructs a global model relating phase increments over a time interval τ to the phases. Similarly to [37, 68], we use the form

$$\begin{aligned} \phi_1(t+\tau) - \phi_1(t) &= F_1\left(\phi_1(t), \phi_2(t+\Delta_1)\right) + \xi_1(t) , \\ \phi_2(t+\tau) - \phi_2(t) &= F_2\left(\phi_2(t), \phi_1(t+\Delta_2)\right) + \xi_2(t) , \end{aligned} \tag{8.16}$$

where $\xi_{1,2}$ are zero-mean random processes, $\Delta_{1,2}$ stand for possible time delays in coupling, F_1 is a trigonometric polynomial

$$F_1 = \sum_{m,n} \left[a_{m,n} \cos(m\phi_1 + n\phi_2) + b_{m,n} \sin(m\phi_1 + n\phi_2)\right] , \tag{8.17}$$

F_2 is defined analogously. The strength of the influence of system 2 on system 1 ($2 \to 1$) is quantified as

$$c_1^2 = \frac{1}{2\pi^2} \int_0^{2\pi} \int_0^{2\pi} (\partial F_1 / \partial \phi_2)^2 \, d\phi_1 \, d\phi_2 \\ = \sum_{m,n} n^2 \left(a_{m,n}^2 + b_{m,n}^2 \right) . \tag{8.18}$$

The influence $1 \to 2$ is quantified "symmetrically" (c_2^2). We use the third-order polynomials for $F_{1,2}$ and set τ equal to a basic oscillation period.

Given a time series, one estimates the coefficients $a_{m,n}, b_{m,n}$ via the OLS technique. Then, one can get the estimate of $\hat{c}_1^2$ by replacing the true values of $a_{m,n}$, $b_{m,n}$ in Eq. (8.18) with their estimates. A reliable detection of the weak directional coupling can only be achieved in nonsynchronous regimes. The latter can be diagnosed if the mean phase coherence

$$\rho(\Delta) = \sqrt{\langle \cos(\phi_1(t) - \phi_2(t+\Delta)) \rangle_t^2 + \langle \sin(\phi_1(t) - \phi_2(t+\Delta)) \rangle_t^2} \tag{8.19}$$

[69] is much less than 1.

The estimators $\hat{c}_1$ and $\hat{c}_2$ are quite precise only for long signals (about 1000 basic periods for moderate noise levels). However, in practice one must often deal with much shorter signals of about several dozens of basic periods. Thus, to analyze a nonstationary time series (e.g., in physiology) one must divide it into relatively short segments and estimate coupling characteristics from each segment separately. An attempt to apply the technique without modifications to such short series leads to biased estimates. Unbiased estimators γ_1 and γ_2 have been proposed in [70] instead of $\hat{c}_1^2$ and $\hat{c}_2^2$, respectively, and an index $\delta = \gamma_2 - \gamma_1$ is used to characterize coupling directionality. Expressions for their 95% confidence bands have also been derived. The latter allows us to trace significance of the estimates obtained from each short segment. (We do not show the formulas here since they are rather cumbersome.) For moderate coupling strength and phase nonlinearity, γ_1 and γ_2 guarantee the probability of erroneous conclusions about the presence of coupling less than 0.025 [71]. Additional tests with exemplary oscillators show that $\gamma_1(\Delta_1)$ and $\gamma_2(\Delta_2)$ are applicable for a time series as short as 20 basic periods if $\rho(\Delta) < 0.4$. The latter condition excludes synchronous-like signals. Other available techniques for coupling direction identification and conditions for superiority of the described technique are reported in [72].

8.6.2 Application to Climatic Data

Using the above technique, we investigated the dynamics of the North Atlantic oscillation (NAO) and El Niño/Southern oscillation (ENSO) processes for the second half of the twentieth century. ENSO and NAO represent the leading modes of interannual climate variability for the globe and Northern Hemisphere (NH),

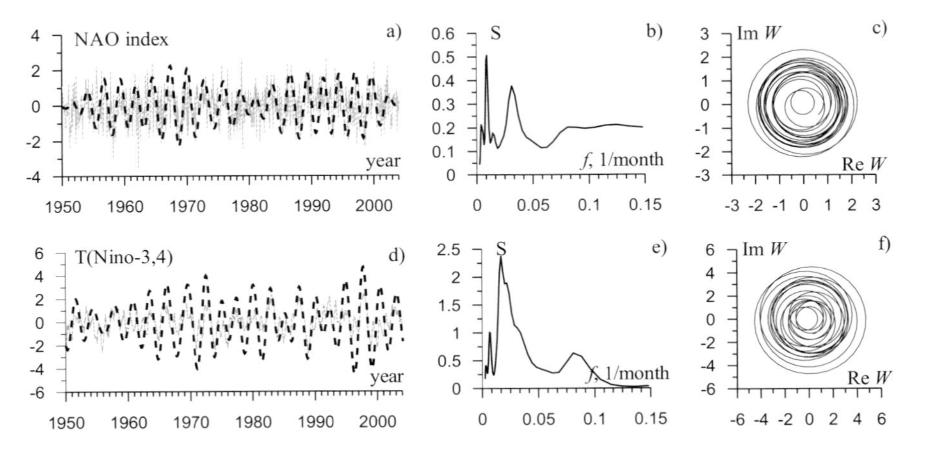

Fig. 8.7: Individual characteristics of the NAO index and T(Niño-3,4). (a) NAO index (the gray line) and ReW for $s = 32$ months (the dashed line). (b) Global wavelet spectrum of the NAO index. (c) An orbit $W(t)$ for the NAO index, $s = 32$ months. (d)–(f) The same as (a)–(c), but for T(Niño-3,4).

respectively [73, 74]. Different tools have been used for the analysis of their interaction, in particular, cross-correlation function and Fourier and wavelet coherence, e.g., [75]. However, all the climatic signals are rather short that has made confident inference about the character of interaction between those processes difficult.

Here, we present the results only for a specific pair of climatic indices. The first one is NAO index `http://www.ncep.noaa.gov` defined as the leading decomposition mode of the field of 500 hPa geopotential height in NH based on the "rotated principal component analysis" [76]. The second one is T(Niño-3,4) which characterizes sea surface temperature in an equatorial region of the Pacific Ocean (5°N–5°S, 170°W–120°W) [77]. These time series cover the period 1950–2004 (660 monthly values).

Figure 8.7 demonstrates individual characteristics of the NAO index (Figure 8.7(a)) and T(Niño-3,4) (Fig. 8.7(d)). Global wavelet spectra of the NAO index and T(Niño-3,4) exhibit several peaks (Figs. 8.7(b)and (e)). One can assume that the peaks correspond to some oscillatory processes for which the phase can be adequately introduced. To extract phases of "different rhythms" in NAO and ENSO, we tried several values of s in Eq. (8.15) corresponding to the different spectral peaks. We estimated coupling between all the rhythms pairwise. The only case when substantial conclusions about the presence of coupling are inferred is the "rhythm" with $s = 32$ months for both signals, see the dashed lines in Figs. 8.7(a) and 8.7(d). The phases of 32-month rhythms in both signals are well defined since clear rotation of the orbits around the origin on the complex plane takes place (Figs. 8.7(c) and 8.7(f)).

The results of the phase dynamics modeling are shown in Fig. 8.8 for $s = 32$

and model, Eq. (8.16), with $\tau = 32$. Figure 8.8(a) shows that the technique is applicable only for $\Delta_1 > -30$ where $\rho < 0.4$. The influence ENSO $\rightarrow$ NAO is pointwise significant for $-30 \leqslant \Delta_1 \leqslant 0$ and maximal for $\Delta_1 = -24$ months (Fig. 8.8(b)). Apart from the pointwise p-level, one can infer the presence of the influence ENSO $\rightarrow$ NAO as follows. Probability of a random erroneous conclusion about coupling presence based only on a pointwise significant γ_1 for a specific Δ_1 is 0.025. Taking into account that the values of $\gamma_1(\Delta_1)$ separated with Δ_1 less than τ are strongly correlated, one can consider as "statistically independent" the values of γ_1 from the two groups: $-30 \leqslant \Delta_1 \leqslant 0$ and $0 < \Delta_1 \leqslant 32$. Then, the probability of erroneous conclusion based on pointwise significant γ_1 at least in one of the two groups as observed in Fig. 8.8(b) is approximately twice as large and, hence, equal to 0.05. Thus, we conclude with confidence probability of 0.95 that the influence ENSO $\rightarrow$ NAO is present. Most probably, it is delayed by 24 months. However, the latter conclusion is not so reliable. No signs of the influence NAO $\rightarrow$ ENSO are detected (Fig. 8.8(c)).

We note that large ρ for $\Delta < -30$ do not indicate strong coupling. For such short time series and close basic frequencies of oscillators, the probability to get $\rho > 0.4$ for uncoupled processes is greater than 0.5 as observed in numerical experiments with exemplary oscillators. More details can be found in [78].

We stress that the conclusion about the presence of the influence ENSO $\rightarrow$ NAO is quite reliable here. Confidence probability 0.95 was not accessible for traditional techniques. It can be attributed to high sensitivity of the phases to weak coupling.

8.6.3 Application to Electroencephalogram Data

Here, we present an application of the estimators to analyze a two-channel human intracranial epileptic electroencephalogram (EEG) recording with the purpose of epileptic focus localization.

The data were recorded from intracranial depth electrodes implanted in a patient with medically refractory temporal lobe epilepsy as part of routine clinical investigations to determine candidacy for epilepsy surgery (provided by Dr. Richard Wennberg, Toronto Western Hospital). The recordings included several left temporal neocortical $\rightarrow$ hippocampal seizures that occurred over the course of a long partial status epilepticus, see an example in Figs. 8.9(a) and (b). Two channels were analyzed: the first channel situated in the left hippocampus, and the second channel in the left temporal neocortex, where the "interictal" activity between seizures at the time comprised pseudoperiodic epileptiform discharges. The visual analysis of the interictal–ictal transitions (shown with vertical dashed lines) determined that the seizures all started first in the neocortex, with an independent seizure subsequently beginning at the ipsilateral hippocampus. We analyzed four recordings, but here we present the results for only one of

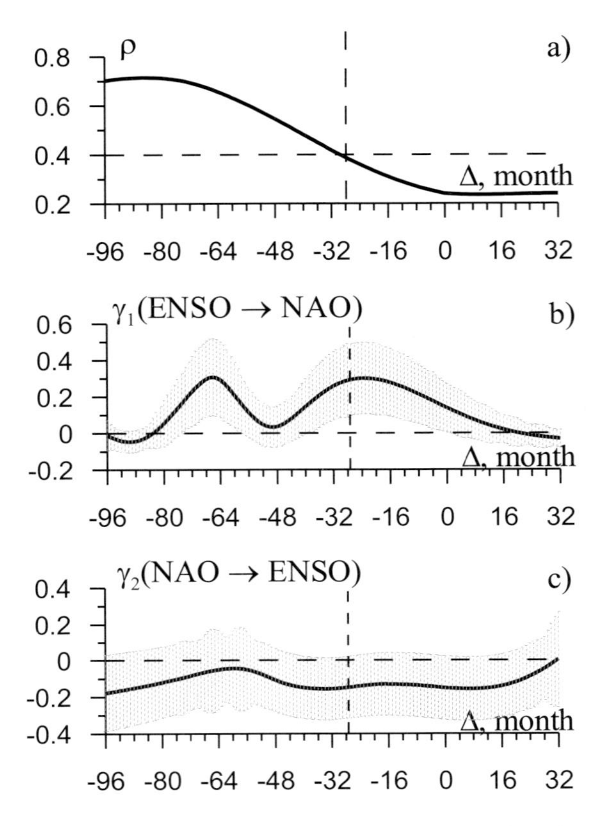

Fig. 8.8: Analysis of coupling from the NAO index and T(Niño-3,4). (a) Mean phase coherence. (b, c) The estimators of the strength of the influence ENSO → NAO (Δ means Δ_1) and NAO → ENSO (Δ means Δ_2), respectively, with their 95% confidence bands.

them for the sake of brevity, as an illustration of application of the method to a nonstationary real-world system.

The time series of Figs. 8.9(a) and (b) contains 4.5 min of depth electrode EEG (referential recording to scalp vertex electrode) recorded at a sampling frequency of 250 Hz. There are more or less significant peaks in power spectra for both channels (not shown). For the hippocampal channel: at frequency 3.2 Hz before the seizure (starting approximately at the 100th second and finishing approximately at the 220th second), 2.3 Hz after the seizure, and 7.1 Hz during the seizure. For the neocortex channel: at frequency 1.4 Hz before the seizure, 1.6 Hz after the seizure, and 7.1 Hz during the seizure. We have computed coupling characteristics in a running window. The length of running window was changed from $N = 10^3$ data points to $N = 10^4$ data points. Time delays $\Delta_{1,2}$ were set equal to zero. The phases were determined using Eq. (8.15) with $\omega_0 = 2$ and different

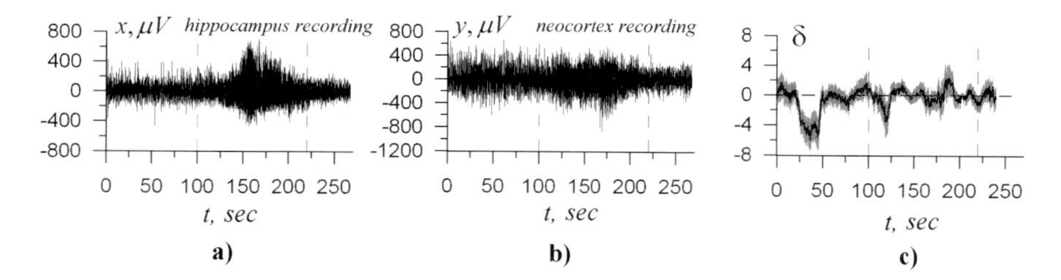

Fig. 8.9: Intracranial EEG recordings: (a) from the hippocampus, (b) from the temporal lobe of the neocortex. (c) Coupling directionality index δ with its 95% confidence band (the gray train), t is the starting time instant of a running window of the length of 8 s. Negative values of δ indicate influence of the neocortex on the hippocampus. The vertical dashed lines indicate a seizure onset and offset. Index δ is significantly less than zero during a period of 25–55 s before the seizure.

time scales s. In particular, we tried the time scales corresponding to the main peak of the scalogram for each signal which is $s = 0.14$ s for the hippocampal signal, and $s = 0.19$ s for the neocortex signal, see Fig. 8.9(c) (where $\tau = 33\Delta t$).

We present only one set of results in Fig. 8.9(c) (gray tail denotes 95% confidence bands) obtained for $N = 6000$. Coupling is regarded as significant if the confidence band does not include zero, e.g., gray tail does not intersect the abscissa axis. The preliminary results seem promising for the localization of the epileptic focus, because a long interval (30 s length for the example shown) of significant predominant coupling direction neocortex → hippocampus is observed before the seizure. It can be considered as an indication that epileptic focus is located near the neocortex channel that agrees with *a priori* clinical information. Despite we presented only one example, we note that the results are sufficiently robust and are observed for a significant range of values of the above-mentioned window lengths and parameters.

Similar results are observed for the three of the four analyzed recordings and not observed for one of them. Right now, we do not draw any definite conclusions about the applicability of the method to localize epileptic focus. This is only the first attempt and, of course, more EEG recordings should be processed to quantify the method's sensitivity and specificity. This is a subject of ongoing research. Therefore, the results presented here should not be overestimated, being rather an illustration of the way how to apply the method in practice and what kind of information one can expect from it.

8.6.4 Other Applications

We should mention several other useful applications of the reconstruction methods. They include detection of quasistationary segments in nonstationary signals [79–82], prediction of bifurcations in weakly nonautonomous systems [83], multichannel confidential transmission of information [84, 85], signal classification [86], testing for nonlinearity and determinism [87], and adaptive nonlinear noise reduction [88–90]. Among the very interesting applications, we stress again on the reconstruction of characteristics of nonlinear elements in electric circuits and other systems with the aid of a modeling procedure in the "gray box" setting when such characteristics may not be accessible to direct measurements. This approach is successfully brought about during the investigation of dynamical properties of a ferroelectric [39], semiconductor diodes [91], and optical fiber ring [92].

8.7 Conclusions

Seemingly, mathematical modeling will always remain an art to a significant extent. However, there may be developed some general principles and particular recipes increasing our chances to obtain a "good" model. Some results of this type related to the time series modeling are discussed in this chapter. Besides, we systematized many available techniques based on the scheme of Fig. 8.1 whose different items were illustrated with different problem settings: from "white box" via "gray box" to "black box" problems. We outlined different techniques which were tested in numerical experiments with the reconstruction of exemplary equations from their noise-corrupted solutions. Many of the techniques were already successfully applied to the investigation of laboratory and real-world systems such as nonlinear electric circuits, climatic processes, functional systems of living organisms, etc. In particular, we reported the results of the analysis of the interaction between complex processes in climatology and neurophysiology based on their empirical modeling.

We have not discussed modeling of spatially distributed systems, even though it attracts considerable attention [93–97]. As well, we have omitted discussion of time-delay systems [92, 98, 99] and only briefly touched on stochastic nonlinear models [29, 100]. Many methods for the construction of finite-dimensional deterministic models are also just mentioned. Instead, we have tried to give simple illustrations of some key points and provide multiple references to the works comprising more detailed discussion for the further reading. Therefore, this survey is only an "excursus into ...," rather than an irrefragable treatment of the empirical modeling problems.

Acknowledgements

We acknowledge fruitful collaboration with our colleagues Ye. P. Seleznev, V. I. Ponomarenko, M. D. Prokhorov, T. V. Dikanev, M. B. Bodrov, I. V. Sysoev, A. S. Karavaev, V. S. Vlaskin, R. A. Wennberg, J.-L. Perez Velazquez, and I. I. Mokhov. Our research in the field of time series modeling was supported by the Russian Foundation for Basic Research (grant 05-02-16305), the President of Russia (MK-

1067.2004.2), Program "Basic Sciences for Medicine" of the Presidium of Russian Academy of Sciences, Program BRHE of the American Civilian Research and Development Foundation and Russian Ministry of Education (REC-006), and Russian Science Support Foundation.

References

[1] L. Ljung. *System Identification. Theory for the User*. Prentice-Hall, New Jersey, 1987.

[2] G. Gouesbet, S. Meunier-Guttin-Cluzel, and O. Menard, editors. *Chaos and Its Reconstructions*. Nova, New York, 2003.

[3] G. U. Yule. *Phil. Trans. R. Soc. London A*, 226:267, 1927.

[4] G. Box and G. Jenkins. *Time Series Analysis. Forecasting and Control*. Holden-Day, San-Francisco, 1970.

[5] J. P. Crutchfield and B. S. McNamara. *Complex Syst.*, 1:417, 1987.

[6] J. Cremers and A. Hubler. *Z. Naturforschung A*, 42:797, 1987.

[7] J. D. Farmer and J. J. Sidorowich. *Phys. Rev. Lett.*, 59:845, 1987.

[8] D. S. Broomhead and D. Lowe. *Complex Syst.*, 2:321, 1988.

[9] M. Casdagli. *Physica D*, 35:335, 1989.

[10] H. D. I. Abarbanel, R. Brown, and J. B. Kadtke. *Phys. Lett. A*, 138:401, 1989.

[11] J. B. Kadtke and Yu. A. Kravtsov, editors. *Predictability of Complex Dynamical Systems*. Springer, Berlin, 1996.

[12] H. D. I. Abarbanel, R. Brown, J. J. Sidorowich, and L. S. Tsimring. *Rev. Mod. Phys.*, 65:1331, 1993.

[13] H. D. I. Abarbanel. *Analysis of Observed Chaotic Data*. Springer, New York, 1996.

[14] H. Kantz and T. Schreiber. *Nonlinear Time Series Analysis*. Cambridge University Press, Cambridge, 1997.

[15] P. E. Rapp, T. I. Schmah, and A. I. Mees. *Physica D*, 132:133, 1999.

[16] G. Gouesbet, S. Meunier-Guttin-Cluzel, and O. Menard, editors. *Chaos and Its Reconstructions*, pages 1–160. Nova, New York, 2003.

[17] C. R. Shalizi. *arXiv:nlin.AO/*, 0307015:3, 2003. URL `http://www.arxiv.org/abs/nlin.AO/0307015`.

[18] H. U. Voss, J. Timmer, and J. Kurths. *Int. J. Bif. Chaos*, 14:1905, 2004.

[19] L. Jaeger and H. Kantz. *Chaos*, 6:440, 1996.

[20] P. E. McSharry and L. A. Smith. *Phys. Rev. Lett.*, 83:4285, 1999.

[21] K. Judd. *Phys. Rev. E*, 67:026212, 2003.

[22] W. Horbelt and J. Timmer. *Phys. Lett. A*, 310:269, 2003.

[23] V. F. Pisarenko and D. Sornette. *Phys. Rev. E*, 69:036122, 2004.

[24] D. A. Smirnov, V. S. Vlaskin, and V. I. Ponomarenko. *Phys. Lett. A*, 336:448, 2005.

[25] E. Baake, M. Baake, H. J. Bock, and K. M. Briggs. *Phys. Rev. A*, 45:5524, 1992.

[26] U. Parlitz. *Phys. Rev. Lett.*, 76:1232, 1996.

[27] I. A. Ibragimov and R. Z. Has'minskii. *Asymptotic Theory of Estimation*. Nauka, Moscow, 1979. In Russian.

[28] V. N. Vapnik. *Estimation of Dependencies Based on Empirical Data*. Springer, Berlin, Heidelberg, 1982.

[29] C. L. Bremer and D. T. Kaplan. *Physica D*, 160:116, 2001.

[30] A. Sitz, U. Schwartz, J. Kurths, and H. U. Voss. *Phys. Rev. E*, 66:016210, 2002.

[31] I. Swameye, T. G. Muller, J. Timmer, O. Sandra, and U. Klingmuller. *Proc. Natl. Acad. Sci. USA*, 100:1028, 2003.

[32] W. Horbelt, J. Timmer, M. J. Bunner, R. Meucci, and M. Ciofini. *Phys. Rev. E*, 64:016222, 2001.

[33] J. Rissanen. *Stochastic Complexity in Statistical Inquiry*. World Scientific, Singapore, 1989.

[34] K. Judd and A. I. Mees. *Physica D*, 82:426, 1995.

[35] L. A. Aguirre, U. S. Freitas, C. Letellier, and J. Maquet. *Physica D*, 158:1, 2001.

[36] B. P. Bezruchko, T. V. Dikanev, and D. A. Smirnov. *Phys. Rev. E*, 64:036210, 2001.

[37] L. Cimponeriu, M. Rosenblum, and A. Pikovsky. *Phys. Rev. E*, 70:046213, 2004.

[38] T. Nakamura, D. Kilminster, and K. Judd. *Int. J. Bif. Chaos*, 14:1129, 2004.

[39] R. Hegger, H. Kantz, F. Schmuser, M. Diestelhorst, R.-P. Kapsch, and H. Beige. *Chaos*, 8:727, 1998.

[40] B. P. Bezruchko and D. A. Smirnov. *Phys. Rev. E*, 63:016207, 2001.

[41] B. P. Bezruchko, Ye. P. Seleznev, D. A. Smirnov, and I. V. Sysoev. *Sov. Tech. Phys. Lett.*, 29:69, 2003.

[42] F. Takens. *Lec. Notes Math.*, 898:366, 1981.

[43] T. Sauer, J. A. Yorke, and M. Casdagli. *J. Stat. Phys.*, 65:579, 1991.

[44] M. Casdagli, S. Eubank, J. D. Farmer, and J. Gibson. *Physica D*, 51:52, 1991.

[45] C. J. Cellucci, A. M. Albano, and P. E. Rapp. *Phys. Rev. E*, 67:066210, 2003.

[46] J. F. Gibson, J. D. Farmer, M. Casdagli, and S. Eubank. *Physica D*, 57:1, 1992.

[47] A. M. Fraser and H. L. Swinney. *Phys. Rev. A*, 33:1131, 1986.

[48] W. Liebert and H. G. Schuster. *Phys. Lett. A*, 142:107, 1989.

[49] J. P. Eckmann and D. Ruelle. *Rev. Mod. Phys.*, 57:617, 1985.

[50] K. Judd and A. I. Mees. *Physica D*, 120:273, 1998.

[51] M. B. Kennel, R. Brown, and H. D. I. Abarbanel. *Phys. Rev. A*, 45:3403, 1992.

[52] P. Grassberger and I. Procaccia. *Physica D*, 9:189, 1983.

[53] D. S. Broomhead and G. P. King. *Physica D*, 20:217, 1986.

[54] D. Kugiumtzis, O. C. Lingjaerde, and N. Christophersen. *Physica D*, 112: 344, 1998.

[55] K. Judd and M. Small. *Physica D*, 136:31, 2000.

[56] N. A. Gerschenfeld and A. S. Weigend, editors. *Time Series Prediction: Forecasting the Future and Understanding the Past*, volume XV of *SFI Studies in the Science of Complexity*. Addison-Wesley, Reading, MA, 1993.

[57] G. Gouesbet and C. Letellier. *Phys. Rev. E*, 49:4955, 1994.

[58] J. Stark, D. S. Broomhead, M. Davies, and J. Huke. *Nonlinear Analysis. Theory, Methods and Applications*. Elsevier, The Netherlands, 1997.

[59] R. Brown, N. F. Rulkov, and E. R. Tracy. *Phys. Rev. E*, 49:3784, 1994.

[60] N. B. Janson, A. N. Pavlov, and V. S. Anishchenko. *Int. J. Bif. Chaos*, 8:825, 1998.

[61] J. P. Lachaux, E. Rodriguez, M. Le Van Quyen, A. Lutz, J. Martienerie, and F. J. Varela. *Int. J. Bif. Chaos*, 10:2429, 2000.

[62] L. Cao, A. I. Mees, and K. Judd. *Physica D*, 121:75, 1998.

[63] C. Letellier, J. Macquet, L. Le Sceller, G. Gouesbet, and L. A. Aguirre. *J. Phys. A: Math. Gen.*, 31:7913, 1998.

[64] C. Letellier and L. A. Aguirre. *Chaos*, 12:549, 2002.

[65] D. A. Smirnov, B. P. Bezruchko, and Ye. P. Seleznev. *Phys. Rev. E*, 65:026205, 2002.

[66] D. T. Kaplan. *Physica D*, 73:738, 1994.

[67] N. F. Rulkov, M. M. Sushchik, L. S. Tsimring, and H. D. I. Abarbanel. *Phys. Rev. E*, 51:980, 1995.

[68] M. G. Rosenblum and A. S. Pikovsky. *Phys. Rev. E*, 64:R045202, 2001.

[69] F. Mormann, K. Lehnertz, P. David, and C. E. Elger. *Physica D*, 144:358, 2000.

[70] D. Smirnov and B. Bezruchko. *Phys. Rev. E*, 68:046209, 2003.

[71] D. A. Smirnov, M. B. Bodrov, J. L. Perez Velazquez, R. A. Wennberg, and B. P. Bezruchko. *Chaos*, 15:024102, 2005.

[72] D. A. Smirnov and R. G. Andrzejak. *Phys. Rev. E*, 71:036207, 2005.

[73] Clivar Initial Implementation Plan. Technical report, WCRP No. 103. WMO/TD No.869. ICPO No.14, 1998. URL `http://www.clivar.dkrz.de/hp.html`.

[74] J. T. Houghton, Y. Ding, D. J. Griggs, and M. Noguer et al., editors. *Climate Change 2001: The Scientific Basis. Intergovernmental Panel on Climate Change*. Cambridge University Press, Cambridge, 2001.

[75] S. Jevrejeva, J. Moore, and A. Grinsted. *J. Geophys. Res.*, 108:4677, 2003.

[76] A. G. Barnston and R. E. Livezey. *Mon. Wea. Rev.*, 115:1083, 1987.

[77] K. Arpe, L. Bengtsson, G. S. Golitsyn, I. I. Mokhov, V. A. Semenov, and P. V. Sporyshev. *Geophys. Res. Lett.*, 27:2693, 2000.

[78] D. A. Smirnov and I. I. Mokhov. *Geophys. Res. Lett.*, 33:L03708, 2006. doi: 10.1029/2005GL024557.

[79] T. Schreiber. *Phys. Rev. Lett.*, 78:843, 1997.

[80] T. Schreiber. *Phys. Rep.*, 308:3082, 1999.

[81] D. Gribkov and V. Gribkova. *Phys. Rev. E*, 61:6538, 2000.

[82] T. Dikanev, D. Smirnov, R. Wennberg, J. L. Perez Velazquez, and B. Bezruchko. *Clin. Neurophysiol.*, 116:1796, 2005.

[83] A. M. Feigin, Y. I. Molkov, D. N. Mukhin, and E. M. Loskutov. *Faraday Discussions*, 120:105, 2002.

[84] V. S. Anishchenko and A. N. Pavlov. *Phys. Rev. E*, 57:2455, 1998.

[85] V. I. Ponomarenko and M. D. Prokhorov. *Phys. Rev. E*, 66:026215, 2002.

[86] M. Kremliovsky, J. Kadtke, M. Inchiosa, and P. Moore. *Int. J. Bif. Chaos*, 8: 813, 1998.

[87] M. Small, K. Judd, and A. I. Mees. *Stat. Comp.*, 11:257, 2001.

[88] J. D. Farmer and J. J. Sidorowich. *Physica D*, 47:373, 1991.

[89] E. J. Kostelich and T. Schreiber. *Phys. Rev. E*, 48:1752, 1993.

[90] M. E. Davies. *Physica D*, 79:174, 1994.

[91] I. V. Sysoev, D. A. Smirnov, Ye. P. Seleznev, and B. P. Bezruchko. *Proc. 2nd IEEE International Conference on Circuits and Systems for Communications*, volume 140. Moscow, Russia, 2004.

[92] H. U. Voss, A. Schwache, J. Kurths, and F. Mitschke. *Phys. Lett. A*, 256:47, 1999.

[93] J. Timmer, H. Rust, W. Horbelt, and H. U. Voss. *Phys. Lett. A*, 274:123, 2000.

[94] H. U. Voss, M. Bunner, and M. Abel. *Phys. Rev. E*, 57:2820, 1998.

[95] M. Baer, R. Hegger, and H. Kantz. *Phys. Rev. E*, 59:337, 1999.

[96] U. Parlitz and C. Merkwirth. *Phys. Rev. Lett.*, 84:1890, 2000.

[97] A. Sitz, J. Kurths, and H. U. Voss. *Phys. Rev. E*, 68:016202, 2003.

[98] M. J. Bunner, M. Ciofini, A. Giaquinta, R. Hegger, H. Kantz, R. Meucci, and A. Politi. *Eur. Phys. J. D*, 10:165, 2000.

[99] M. D. Prokhorov, V. I. Ponomarenko, A. S. Karavaev, and B. P. Bezruchko. *Physica D*, 203:209, 2005.

[100] J. Timmer. *Chaos, Solit. Fract.*, 11:2571, 2000.

9 Data-Driven Analysis of Nonstationary Brain Signals

Mario Chavez, Claude Adam, Stefano Boccaletti, and Jacques Martinerie

Many neurobiological processes generally result from the interaction of many oscillators with different time scales, and it often arises that the frequency content of the observed oscillations changes rapidly across time. In such a case, traditional methods of the spectral analysis may be insufficient to provide a meaningful characterization of the dynamics. Empirical mode decomposition (EMD) has been recently introduced as an adaptive and fully data-driven method for the analysis of nonlinear and nonstationary time series. Instead of using an *a priori* choice of filters or basis functions to separate a frequency component, the EMD technique expands the time series into a set of functions defined by the signal itself. The signal is represented as the sum of amplitude- and frequency-modulated components called intrinsic oscillation modes. As the major feature of these modes is their local time–frequency discrimination, they may detect embedded nonstationary oscillations and their possible interactions. When applied to the general case of coupled oscillators with multiple time scales, we found that the motions are captured in a finite number of phase-locked time scales. This feature may be used to detect the time scales involved in the synchronization of complex oscillators with several spectral components. This approach is illustrated on electric intracranial signals recorded from an epileptic patient. Despite the time-varying spectrum displayed by the recorded signals, epileptic dynamics was characterized by a finite number of modes. Further, seizure onset was characterized by transient periods of synchronization at different time scales. Numerical and experimental results suggest that this data-driven approach can be a useful technique for the analysis of nonstationary and noisy time series.

9.1 Introduction

The Fourier transform is probably the most used technique for the spectral analysis. By means of this linear technique, a reliable estimation of the spectral components of a signal can be obtained, provided the observed process is stationary. Nevertheless, many neurobiological processes are nonstationary, and the frequency content of the recorded signals changes often rapidly across time. Fol-

Handbook of Time Series Analysis. Björn Schelter, Matthias Winterhalder, Jens Timmer

ISBN: 3-527-40623-9

lowing this rationale, a decomposition based on local characteristic time scales of the data is necessary to correctly characterize nonstationary oscillations and their possible interactions.

The study of the synchronization mechanisms between neural populations is one of the most active topics in neurosciences [1, 2]. Nonlinear dynamics theory has provided a number of useful tools for the analysis of interdependences [3, 4]. Based on theoretical studies of coupled dynamical systems, the concept of phase synchronization has offered a new framework for the analysis of interactions between neurobiological signals [5–8]. An important question is whether these synchronization mechanisms can be properly characterized from nonstationary and noisy brain signals.

To characterize a phase locking, a continuous phase variable is currently estimated from a time series by means of its representation as an analytical signal [9]. However if the signal possesses a multicomponent or a nonstationary spectrum, this representation may fail and a phase cannot be straightforwardly defined [10, 11]. The usual approach consists in a band-pass filtering in order to properly isolate a time-scale oscillation, to which the analytic signal representation can be applied to extract the phase variable. However, potential problems associated with filtering bandwidth in the estimation of phase interactions between nonstationary time series have been pointed out [12, 13].

In this work, we address this problem by using the recently introduced empirical mode decomposition [14]. The empirical mode decomposition (EMD) is an adaptive and fully data-driven method for the analysis of nonlinear and nonstationary time series. Instead of using an *a priori* choice of filters or basis functions to separate a frequency component, the EMD technique expands the time series into a set of functions defined by the signal itself. The signal is represented as the sum of amplitude- and frequency-modulated components. The local time–frequency discrimination of these modes is a suitable property to estimate an instantaneous phase and thus, to detect possible time-scale synchronization of nonstationary signals.

9.1.1 EMD-Related Work

Within the framework of the time-series analysis, the EMD procedure has provided a powerful framework of the time-series analysis in different fields ranging from engineering and physics to biology [15–45].

When applied to purely stochastic processes, the EMD has been found to act like a dyadic filter bank [46]. In contrast, the EMD analysis of autonomous deterministic oscillators has revealed that a chaotic flow is composed of a small number of intrinsic oscillation modes for each of which the phase fluctuates as a fractional Brownian motion around a uniform rotation [47–49]. Furthermore, the chaotic regime was found to be characterized by well-localized distributions of the instantaneous frequencies estimated from each rotation modes.

In the case of forced or interacting oscillators, the EMD has also been used to characterize time-scale correlations between nonlinear and nonstationary systems [22, 28, 41]. Indeed, we have recently shown that the synchronization of multitime scales' oscillators can be characterized by a phase-locking condition of the oscillators' intrinsic time scales [50].

9.2 Intrinsic Time-Scale Decomposition

The key procedure of the EMD algorithm is a sifting process that expands the signal into a set of zero-mean amplitude- and frequency-modulated components called intrinsic oscillation modes or functions (IMFs). The sifting process for extracting these modes from a given time series $x(t)$ can be summarized as follows [14]:

1. identify all extrema of $x(t)$;
2. interpolate between minima (resp. maxima) to get two envelopes $x_{\min}(t)$ (resp. $x_{\max}(t)$);
3. compute the mean envelope $m(t) = \big(x_{\max}(t)+x_{\min}(t)\big)/2$ and extract the residual $d(t) = x(t) - m(t)$;
4. iterate on $d(t)$ until this latter can be considered as zero mean according to a stopping criterion.

Once this process is achieved the resulting signal is considered as an IMF. The obtained intrinsic mode C_1 is extracted from $x(t)$ and steps (1)–(4) are repeated to obtain the second mode C_2. This sifting process continues until the last mode shows no apparent variation.[1] At the end of the sifting process, the original signal is decomposed in a finite number of modes as $x(t) = r(t) + \sum_i C_i(t)$, where $r(t)$ stands for a residual trend, and the intrinsic modes $C_i(t)$s are nearly orthogonal to each other [14].

An oscillation must verify two criteria to be considered as an IMF:

1. the mean envelope defined by the local maxima and the envelope of the local minima is zero at any time; and
2. the number of extrema and thus the number of zero crossings are equal or they differ at most by 1. This latter property is similar to a local (in time) narrow band requirement [51]. By construction, the spectral supports are decreased when going from one mode to the next. Nevertheless their frequency discrimination applies only locally (in time) and they cannot correspond to a sub-band filtering [46].

[1] MATLAB codes for the EMD algorithm and some examples shown in this work are fully online available in `http://www.ens-lyon.fr/~flandrin/software.html` and `http://perso.wanadoo.fr/e.delechelle/codes.html`.

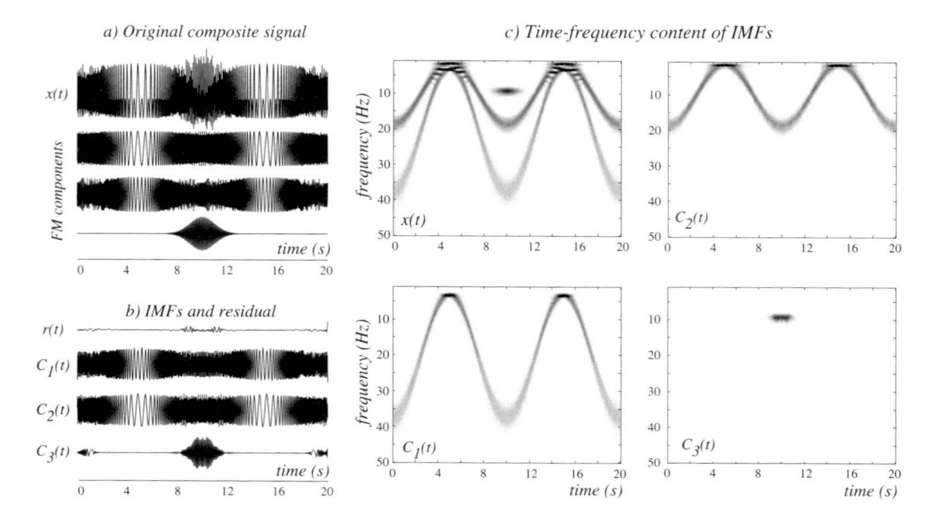

Fig. 9.1. EMD of a composite nonstationary signal: (a) the analyzed signal $x(t)$ formed by frequency-modulated components. (b) Intrinsic oscillation modes $C_i(t)$ and residual $r(t)$ obtained by the EMD algorithm. (c) Time–frequency structure of the signal $x(t)$ and the intrinsic oscillation modes $C_i(t)$. For all the examples, time–frequency distributions were obtained by means of the wavelet transform as in [6].

For illustration, let us consider a composite signal obtained by the superposition of two sinusoidal frequency-modulated (FM) signals and one FM tone modulated by a Gaussian [52]. As illustrated in Fig. 9.1, the components of the signal $x(t)$ overlap in the time–frequency plane which renders difficult their decomposition by traditional spectral techniques as the Fourier transform. As the EMD is based on the local characteristic time scales of the data, the modes obtained are well localized in time and frequency which enables a successful separation of the different nonstationary components.

9.2.1 EMD and Instantaneous Phase Estimation

A time series $x(t)$ with a time-varying spectrum can be characterized by a complex representation of the form $z(t) = A(t)\exp(i\phi(t))$, where the pair of functions $\{A(t), \phi(t)\}$ are related to the instantaneous amplitude and phase the signal [10]. In practice, this representation on the complex plane is usually obtained by means of the analytical signal defined as [9]

$$\psi_x(t) = x(t) + i\mathcal{H}(x(t)) = A(t)\exp(i\phi(t)) . \quad (9.1)$$

The imaginary part of $\psi_x(t)$ is the Hilbert transform of $x(t)$ defined as

$$\mathcal{H}(x(t)) = \frac{1}{\pi}\,\text{p.v.}\int_{-\infty}^{+\infty} \frac{x(t)}{t-\tau}\,d\tau \quad (9.2)$$

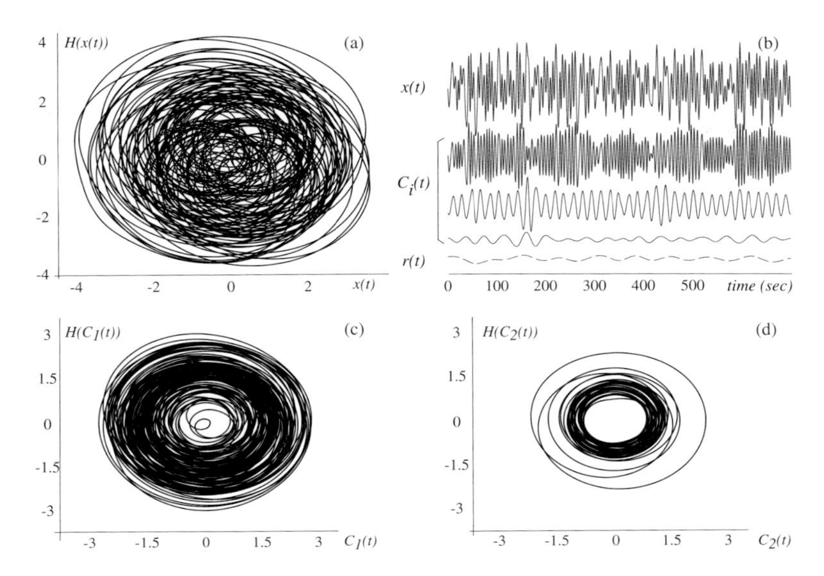

Fig. 9.2. Instantaneous phase of a multicomponent and nonstationary signal: (a) trajectory of $x(t)$ on the complex plane (see the multiple centers of rotation). (b) IMFs obtained by the EMD algorithm. (c)–(d) Analytic signals of $C_i(t)$ yield a unique rotation center necessary for a properly definition of a phase.

where p.v. indicates that the integral is taken in the sense of Cauchy principal value. This complex representation yields a trajectory in the complex plane $\{x(t), i\mathcal{H}(x(t))\}$ whose phase $\phi(t)$ can be defined through the representation $\psi_x(t) = A(t)\exp(i\phi(t))$. An associated instantaneous frequency can be thus obtained at each time by $f(t) = \frac{1}{2\pi}\frac{d\phi(t)}{dt}$.

Although the Hilbert transform can be applied to any arbitrary signal, instantaneous phase has a clear physical meaning only for *monocomponent* signals with a unique center of rotation on the complex plane. In fact, if the time series possesses a time–frequency structure with multiple overlapping components this representation may fail and a phase cannot be straightforwardly defined [10, 11]. In the EMD decomposition, the resulting intrinsic modes $C_i(t)$ are zero mean and the number of extrema and the number of zero crossings of each IMF are equal. This ensures that trajectories in the complex plane of modes $C_i(t)$ rotate around a unique rotation center (not necessarily at a constant frequency) and a phase can thus be defined [14, 46–48].

For illustration, we consider a simple case encountered in real biological systems: a nonstationary process under the influence of periodic forces with different time scales, and under the influence of noise. Let us firstly consider the following Van der Pol oscillator with a randomly varying parameter w [53, 54]

$$\begin{aligned}\dot{x} &= y\,,\\ \dot{y} &= \epsilon(1-x^2)y - w^2 x + C\sin(\Omega t)\,,\\ w &= w_0 + \eta(t)\,,\end{aligned} \tag{9.3}$$

with $\epsilon = 0.1$, $w_0 = 1$, $C = 0.2$ and $\Omega = 1/3$. Random perturbation is given by $\eta(t)$ which is an exponentially correlated colored noise, $\langle\eta(t)\rangle = 0$ and $\langle\eta(t)\eta(t')\rangle = \frac{D}{\tau}\exp(-\frac{|t-t'|}{\tau})$, with $D = 0.1$. The eigenfrequency of the oscillator thus exhibits a slow random variation given by $\eta(t)$ with a correlation time $\tau = 200$. Numerical examples were simulated by Euler's technique with the time step $\delta t = 0.005$. For all simulations, a transient of 10^4 points was discarded.

The interaction of the nonstationary autonomous oscillator and the driving force results in nonstationary oscillations at different time scales. This multicomponent spectrum induces a major difficulty in the estimation of an instantaneous phase: The trajectory of the analytic signal $\psi_x(t)$ thus exhibits multiple centers of rotation in the complex plane and a phase cannot be straightforwardly defined (Fig. 9.2(a)). In contrast, the IMFs display a clear and unique rotation center in the complex plane what allows a proper estimation of an instantaneous phase. This is illustrated in Figs. 9.2(b), (c), (d).

9.2.2 Drawbacks of the EMD

Though a signal can be fully decomposed in a finite number of modes, a careful interpretation of IMFs is necessary. Let us consider the composite signal whose time–frequency structure is depicted in Fig. 9.3. In the EMD procedure, the signal is considered as a fast oscillation locally (in time) superimposed to slow oscillations. The time–frequency distributions show that, at each step of the sifting process, the low-frequency content of the time series is basically what it remains after the iterative extraction of the fast components. For this reason, the EMD may provide, in some extreme cases, oscillations without a clear physical meaning.

The choice of the interpolation technique also plays an important role in the decomposition. Although the original algorithm uses a spline interpolation, this technique often produces overshoots in order to achieve the second derivative, and new extrema points, not present in the original signal, may therefore be introduced. Nevertheless, other interpolation techniques tend to spread spurious components over adjacent modes, increasing the number of sifting iterations. Recently, the effects on the EMD of different interpolation methods have been studied in detail and compared in [23, 55, 56].

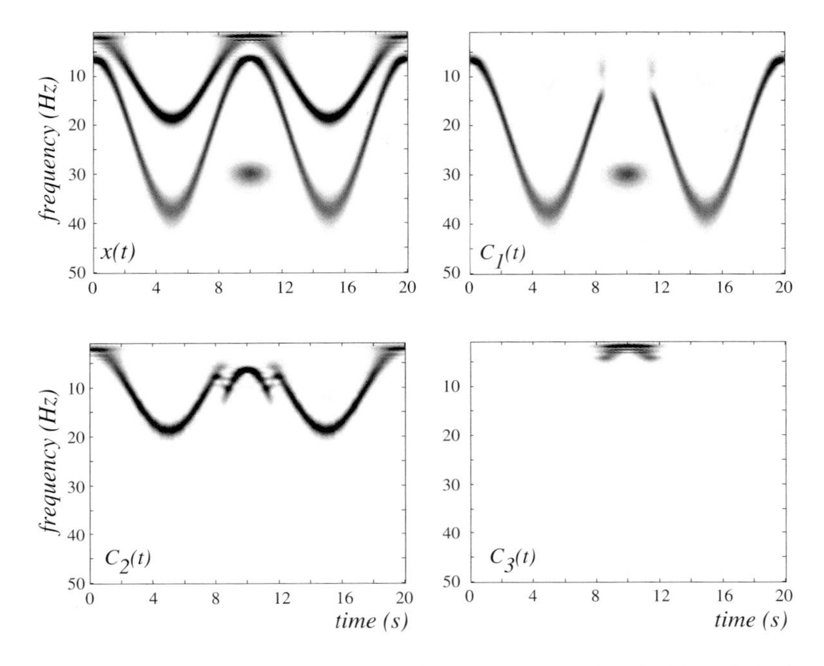

Fig. 9.3. Time–frequency structure of the composite signal $x(t)$ and the intrinsic oscillation modes $C_i(t)$.

9.3 Intrinsic Time Scales of Forced Systems

The entrainment of a system may be detected by a phase-locking index between the driving signal and an appropriate scalar observation of the forced oscillator [3, 4]. To quantify

the phase entrainment of nonlinear oscillators, a data-driven method was proposed in [53, 54] for the analysis of univariate data. In this work, an analytical model was described to relate the zero crossings of the time series to the phase of a single external periodic forcing. For the case of multiple driving forces, the authors used a coarse and simplified version of the EMD procedure. However, in the case of several driving forces with different time scales, this approach may fail.

When applied to periodically forced systems, the oscillations captured by the IMF were found to be phase locked with the driving forces [50]. This is illustrated in Fig. 9.4(a). Modes $C_i(t)$ were computed from the scalar variable $x(t)$ of the system, Eq. (9.3), for a noise level of $D = 0.15$. Despite the nonstationary behavior of the signal, the EMD yields two oscillations at different time scales: The first mode C_1 corresponds to the eigenfrequency of the autonomous oscillator while the second mode C_2 is in phase locked with the driving force.

We consider now a case where different external driving forces interact with

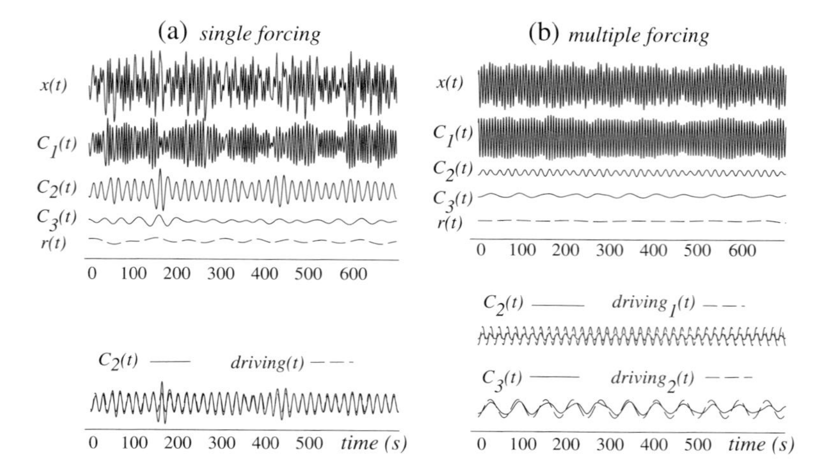

Fig. 9.4. EMD analysis of forced oscillators: (a) with a random variation of the parameter w and (b) with multiple forces and noise. $x(t)$ indicates the scalar observation and $C_i(t)$ the IMFs. The driving force(s) and the corresponding phase-locked modes are depicted at the bottom plots by the dotted and solid lines, respectively.

the internal time scale of the autonomous oscillator. We apply the EMD to the following forced and noisy Van der Pol system [53, 54]:

$$\begin{aligned}\dot{x} &= y\,,\\ \dot{y} &= \epsilon(1-x^2)y - w_0^2 x + C_1\sin(\Omega_1 t) + C_2\sin(\Omega_2 t) + \xi(t)\,,\end{aligned} \tag{9.4}$$

where $\epsilon = 0.1$, $w_0 = 1$, $C_1 = C_2 = 0.1$, $\Omega_1 = 0.5$ and $\Omega_2 = 0.1$. A Gaussian noise is used here as random perturbation such that $\langle\xi(t)\rangle = 0$ and $\langle\xi(t)\xi(t')\rangle = 2D\delta(t-t')$ with $D = 0.01$. As depicted in Fig. 9.4(b) the time scales of the external driving forces are perfectly captured by different modes (C_2 and C_3), while the oscillation at the eigenfrequency of the oscillator is captured by other mode (C_1 for this example).

9.4 Intrinsic Time Scales of Coupled Systems

Recent works have suggested that different synchronization phenomena (phase synchronization, lag synchronization, and generalized synchronization [3, 4]) are particular cases of the so-called time-scale synchronization [57, 58]. Within this framework, we have recently found that the IMFs obtained from two synchronized multitime scales' oscillators may display distinct phase-locking behavior [50].

To illustrate these time-scale correlations, let us consider a coupled system formed by two coupled chaotic oscillators with different time scales. Equations of motion read [59]

$$\begin{aligned}
\dot{x} &= y\,,\\
\dot{y} &= A_{xy}y(1-x^2) - Bx^3 + C\sin(\omega_{xy}t)\,,\\
\dot{u} &= v\,,\\
\dot{v} &= A_{uv}v(1-u^2) - Bu^3 + C\sin(\omega_{uv}t) + \epsilon(x-u)\,,
\end{aligned} \tag{9.5}$$

with $A_{xy} = 0.6$, $A_{uv} = 0.2$, $B = 1$, $C = 2$, $\omega_{xy} = 0.6$ and $\omega_{uv} = 0.65$. Subscripts xy and uv refer to the oscillators described by the variables (x, y) and (u, v), respectively. The system describes a pair of forced Van der Pol oscillators unidirectionally coupled. Parameters were set such that both oscillators exhibit a chaotic motion for the uncoupled case. The domain of coupling values where different synchronization phenomena arise has been studied in [50, 59].

To evaluate the mutual entrainment, we have computed the phase-locking index [3, 4]

$$\Gamma = \frac{1}{N}\left|\sum_{t=1}^{N} e^{i\Delta\varphi_{x,u}(t)}\right| \tag{9.6}$$

where $\Delta\varphi_{x,u}(t) = \varphi_x(t) - \varphi_u(t)$ stands for the difference between the instantaneous phase of modes $C_k(t)$ obtained from signals (over a time window of length N) $x(t)$ and $u(t)$. Weak synchronization yields a nearly uniform distribution of the phase differences on the unit circle and a small value of Γ. In contrast, a phase-locked condition results in a distribution of $\Delta\varphi_{x,u}(t)$ concentrated around a preferred value, so that $\Gamma \sim 1$.

Figure 9.5 illustrates the behavior of the unwrapped variable $\Delta\Psi(t)$ at different intrinsic time scales for different coupling strengths. In the absence of coupling, the IMFs are not phase locked because of the mismatch of the external frequencies. Thus, instantaneous variable $\Delta\Psi(t)$ diffuses at all the intrinsic time scales, which yields the phase-locking values (mean $\pm$ s.d. computed over 20 realizations) of $\Gamma = 0.021 \pm 0.001$, $\Gamma = 0.03 \pm 0.002$, and $\Gamma = 0.0016 \pm 0.002$, respectively. At a coupling value $\epsilon = 3$, the fastest IMF are unsynchronized ($\Gamma = 0.2061 \pm 0.02$), whereas the phase locking of slower IMFs increases to $\Gamma = 0.6472 \pm 0.03$ and $\Gamma = 0.5433 \pm 0.03$, respectively. When coupling is further increased, the synchronization is established. For a coupling strength $\epsilon = 5$, the phase-locking value between slow oscillation modes is also increased, as reflected by the indices of $\Gamma = 0.915 \pm 0.02$ and $\Gamma = 0.580 \pm 0.03$, respectively. One must note that the fastest time scales display only intermittent short periods of synchronization. During rather small intervals of time, the phase difference changes by 2π which yields phase slips and a $\Gamma = 0.6703 \pm 0.03$.

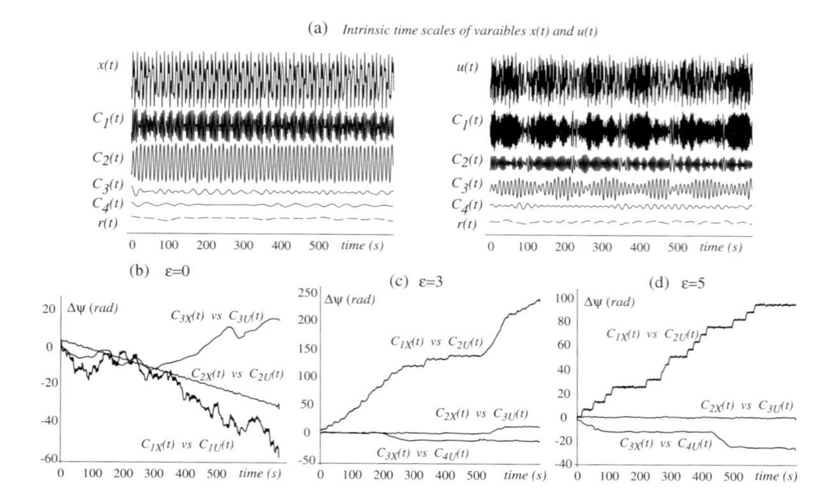

Fig. 9.5. Time-scale synchronization of the coupled system given by Eq.(9.5) as revealed by the EMD: (a) example of the intrinsic time scales obtained from the variable $u(t)$ for a coupling value $\epsilon = 5$; (b)–(d) phase differences $\Delta\Psi(t)$ of the IMFs for the coupling values of $\epsilon = 0$, $\epsilon = 3$, and $\epsilon = 5$. $C_{ix}(t)$ and $C_{iu}(t)$ stand for the ith mode obtained from the variables $x(t)$ and $u(t)$, respectively.

9.5 Intrinsic Time Scales of Epileptic Signals

9.5.1 Intracerebral Activities

To illustrate the method on experimental data, we have applied the EMD to electromagnetic signals recorded from epileptic patients candidate for a surgical treatment. Intracerebral electrical activities (or SEEG) were recorded directly from brain areas suspected to be involved in seizure generation. The number and the position of the depth electrodes were determined by electrophysiologists and were not chosen for the purpose of this study. SEEG signals were recorded by means of depth electrodes using an external reference, sampled at 400 Hz and bandpass filtered between 0.1 Hz and 90 Hz.

The multicomponent spectrum observed in the epileptic SEEG signals is depicted in Fig. 9.6(a). One can note that the frequency content of some oscillations may change rapidly across time over a wide range of frequencies. In this case the analytic signal approach yields a trajectory in the complex plane with multiple centers of rotation and a phase cannot be properly defined (Fig. 9.6(b)).

Examples of the intrinsic time scales obtained from SEEG signals are illustrated in Fig. 9.7(a). Despite the nonstationary behavior of data, the epileptic dynamics was characterized by a small number of intrinsic oscillations. The different IMFs capture the different oscillations (often with a time-varying spectrum) embedded in the original signal (Fig. 9.7(b)). The mode $C_1(t)$, for instance, corresponds to the low-voltage fast discharges observed at the seizure onset, whereas

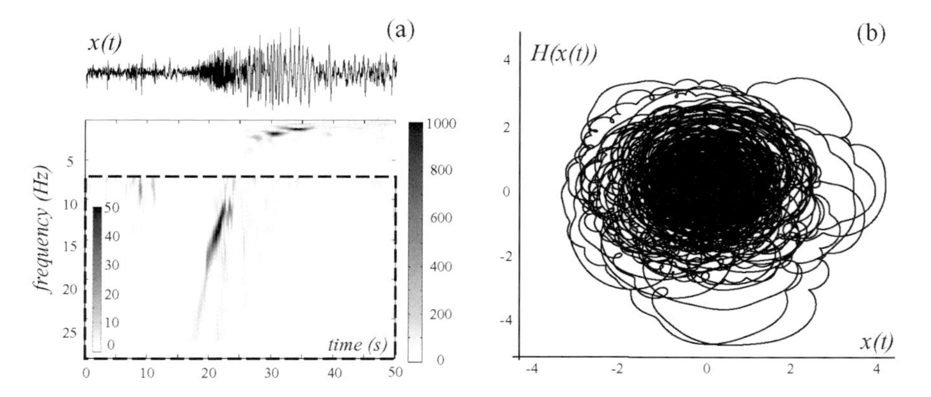

Fig. 9.6. Time scales embedded in the epileptic SEEG signals: (a) time–frequency content of a recorded SEEG signal (the fast oscillation component within the dotted box is zoomed). (b) Trajectory of the original signal on the complex plane (see the multiple ripples of the rotation).

the spike and wave discharge is mainly captured by the mode $C_4(t)$. One can note that the local time–frequency localization of the IMFs ensures that trajectories in the complex plane of analytic signals obtained from IMFs rotate around a unique rotation center and thus a phase can be properly defined.[2]

9.5.2 Magnetoencephalographic Data

Magnetoencephalography (MEG) data were recorded from an epileptic patient suffering from absence epileptic seizures. MEG signals were sampled at 1250 Hz and bandpass filtered between 2 Hz and 80 Hz. This modality of acquisition has the major feature that collective neural behavior, as synchronization of large and sparsely distributed cortical assemblies, are reflected as interactions between MEG signals [60].

The dynamics of absence seizures is characterized by two possible states: a steady state of ongoing activity, apparently random, and another one characterized by a sudden discharge of paroxysmal spike-wave components occurring over the entire cortex [61]. When applied to the MEG data, the EMD yields a finite number of proper rotation modes $C_i(t)$ and a trend $r(t)$ as those illustrated in Fig. 9.8(a). All the modes display an oscillatory type-like burst activity mainly during the seizure, whereas a few modes show some activity before the onset. The instantaneous frequency content of IMFs was found to be similar over all channels and for the three seizures analyzed here. The transition to

[2] Phases obtained from real data may look ill-defined without a clear "hole" in the center of the trajectory. This problem of visualization is due to the low amplitudes of some oscillations present in the modes. However, the number of extrema and the number of zero crossings of these small oscillations verify IMF's criteria.

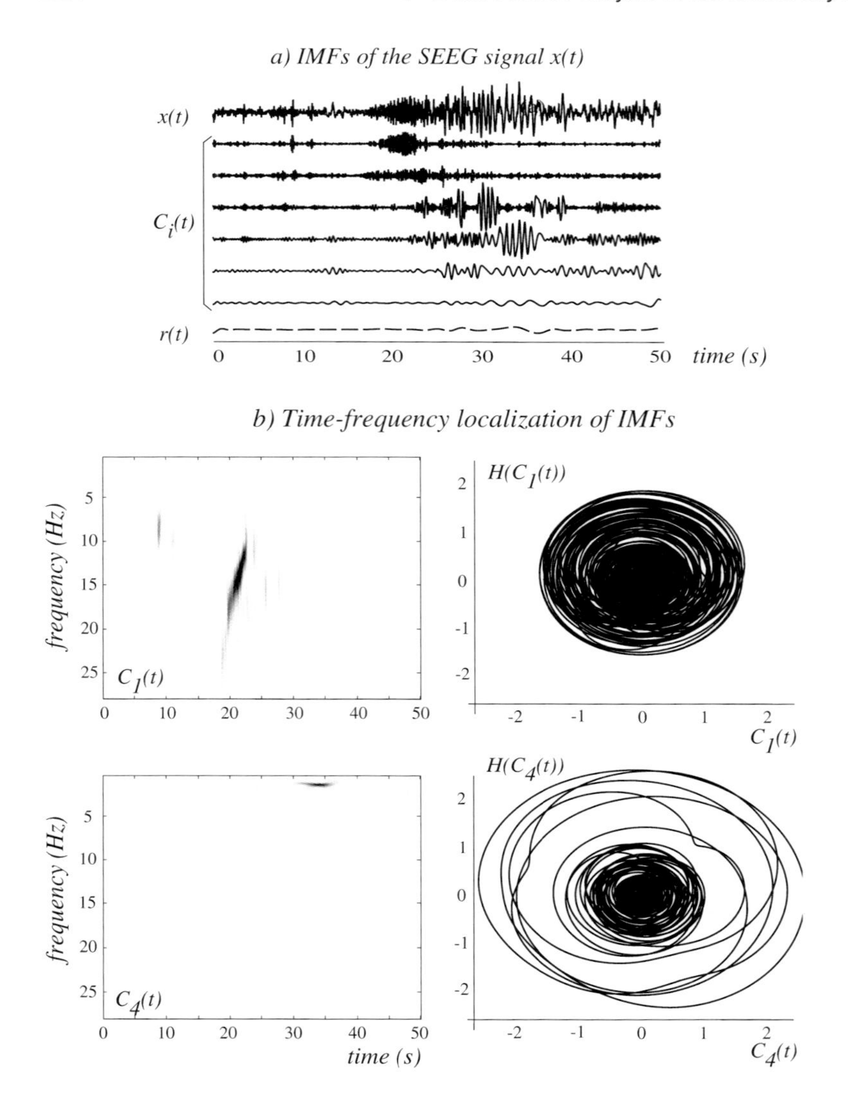

Fig. 9.7. Time-scale analysis of a SEEG signal: (a) IMFs obtained by the EMD algorithm; (b) example of time–frequency localizations of intrinsic oscillation modes $C_i(t)$, and their corresponding representations as analytic signals.

seizures was found to be mainly characterized by changes in the average instantaneous frequencies of some of the intrinsic modes. The distributions plotted in Fig. 9.8(b) suggest that the emergence of seizures has a greater influence on rotations corresponding to modes $C_4(t)$ and $C_1(t)$ (indicated by the dashed circles). Changes in those modes were statistically significant ($p < 0.001$; two-tailed t-test; $t = 16.5$ and $t = 5.8$ for $C_4(t)$ and $C_1(t)$ respectively).

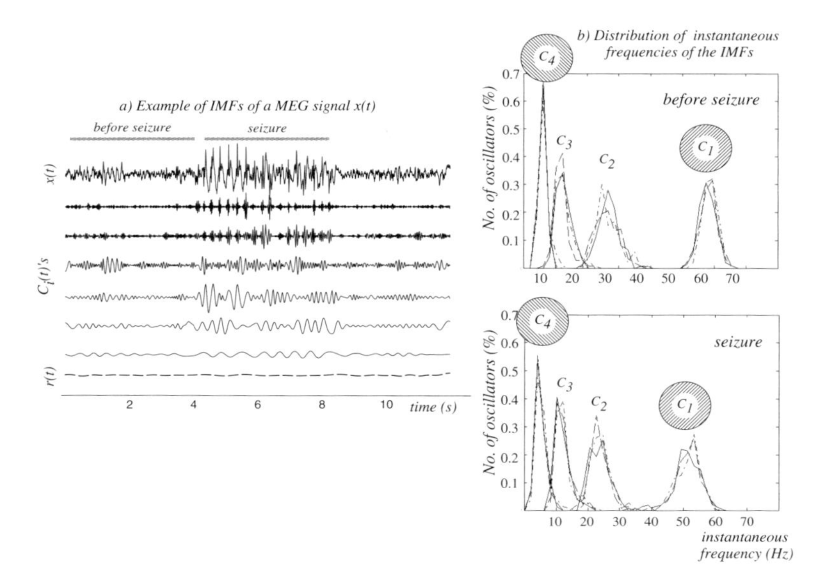

Fig. 9.8. (a) Example of an MEG signal $x(t)$ and the corresponding intrinsic rotation modes $C_i(t)$. (b) Distributions, over all the MEG sensors, of the average instantaneous frequencies calculated from different modes before and during the seizure. Distributions of three different seizures are depicted by the solid, dotted, and slashed curves respectively.

9.6 Time-Scale Synchronization of SEEG Data

In Fig. 9.9 we illustrate the synchronization of the time scales detected by the EMD in the SEEG signals. Time-scale interactions, quantified by the Γ index, were computed during three periods of the original epileptic dynamics (Fig. 9.9(b)): a seizure-free period (I), during the low-voltage fast discharges observed at the seizure onset (II), and the sustained spikes and waves (III).

The example in Fig. 9.9(a) shows that fast activities were mainly localized at the seizure onset (at $t > 15$ s). The matrices of Γ values suggest that the synchronization of fast oscillations (modes $C_1(t)$) at the seizure onset (period II) mainly involves brain areas corresponding to signals A–E. These fast time scales were not synchronized during the beginning of the recording, or during and after the spike and wave discharge. The time windows used for these modes (5 s) approximately correspond to the fast discharge at seizure onset.

The behavior of slow time scales is reported in Fig. 9.9(c): Slow modes $C_4(t)$ are clearly localized during the spike and wave discharges observed at $25 < t < 36$ s. The strong synchronization at period III is widely extended and it involves all the signals, whereas the rest of the recording is characterized by a weak synchronized state. Time intervals used (10 s) basically correspond to the spikes and waves discharge.

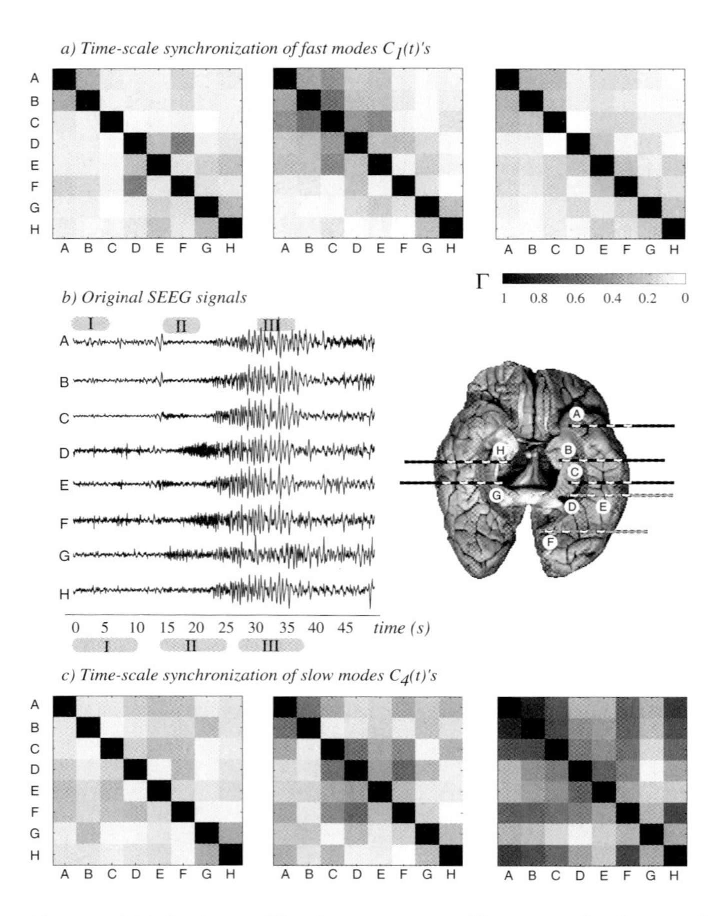

Fig. 9.9. (a) Matrices of Γ values computed between the fast modes $C_1(t)$ at each of the intervals indicated by the top gray boxes in (b). (b) Schematic example of the brain structures explored by intracerebral electrodes and the corresponding original SEEG recordings. (c) Values of the Γ index between the slow modes $C_4(t)$ computed inside each of the intervals indicated by the bottom gray boxes in (b).

9.7 Conclusions

In this work, we have presented the empirical mode decomposition as a method for the analysis of neurobiological signals. The EMD is an adaptive and fully data-driven method for the analysis of nonlinear and nonstationary time series. Instead of using an *a priori* choice of filters or basis functions to separate a frequency component from the broadband activity, the EMD extracts the basis functions directly from the data. As the major feature of the intrinsic modes is their local time–frequency localization, an instantaneous phase can be properly estimated from each of the intrinsic modes, even if they exhibit a time-varying spectrum.

When applied to the general case of coupled oscillators with multiple time scales, we found that the motions are captured in a finite number of phase-locked modes. Despite the nonstationary behavior of the data, the epileptic dynamics was characterized by a small number of IMFs with a well time–frequency localization. This feature allowed us to detect transient periods of synchronization at different time scales which may display a time-varying spectrum. The analysis of the mode oscillations agrees with the hypothesis of multitime-scale interactions underlying the seizure dynamics: The spike discharges were found to be accompanied by a widespread strong synchronization at slow time scales, whereas the entrainment of fast time scales onset was found to involve a reduced number of electrodes at the seizure onset.

To summarize, the EMD method should be widely applicable in different situations of biological interest. A nonstationary decomposition as the EMD, based on the local characteristic time scales of the data, may be a useful tool for the analysis of nonstationary interactions as those resulting from a frequency modulation. Though the main drawback of the EMD is the lack of a theoretical framework and that it is limited to numerical simulations [46, 62–65], some theoretical aspects begin to be explored [66–68]. This formalism remains therefore an exciting challenge for the signal processing community.

Acknowledgements

The authors thank D. Schwartz, A. Ducorps, and J. C. Bourzeix for technical support.

References

[1] F. Varela, J. P. Lachaux, E. Rodriguez, and J. Martinerie. *Nature Rev. Neurosci.*, 2:229, 2001.

[2] A. K. Engel, P. Fries, and W. Singer. *Nature Rev. Neurosci.*, 2:704, 2001.

[3] A. Pikovsky, M. Rosenblum, and J. Kurths. *Synchronization. A Universal Concept in Nonlinear Systems.* Cambridge Nonlinear Science Series 12. Cambridge University Press, Cambridge, UK, 2001.

[4] S. Boccaletti, J. Kurths, G. Osipovd, D. L. Valladares, and C. S. Zhou. *Phys. Rep.*, 366:1, 2002.

[5] P. Tass, M. G. Rosenblum, J. Weule, J. Kurths, A. S. Pikovsky, J. Volkmann, A. Schnitzler, and H. J. Freund. *Phys. Rev. Lett.*, 81:3291, 1998.

[6] E. Rodriguez, N. George, J. P. Lachaux, J. Martinerie, B. Renault, and F. J. Varela. *Nature*, 397:430, 1999.

[7] J. Fell, P. Klaver, K. Lehnertz, T. Grunwald, C. Schaller, C. E. Elger, and G. Fernández. *Nat. Neurosci.*, 4:1259, 2001.

[8] C. Tallon-Baudry, S. Mandon, W. A. Freiwald, and A. K. Kreiter. *Cerebral Cortex*, 14:713, 2004.

[9] D. Gabor. *IEE J. Comm. Eng.*, 93:429, 1946.

[10] B. Boashash. *Proc. IEEE.*, 80:520, 1992.

[11] P. M. Oliveira and V. Barroso. *IEEE Signal Proc. Lett.*, 6:81, 1999.

[12] D. J. DeShazer, R. Breban, E. Ott, and R. Roy. *Phys. Rev. Lett.*, 87:044101, 2001.

[13] M. Chavez, M. Le Van Quyen, V. Navarro, M. Baulac, and J. Martinerie. *IEEE Trans. Biomed. Eng.*, 50:571, 2003.

[14] N. E. Huang, Z. Shen, S. R. Long, M. C. Wu, H. H. Shih, Q. Zheng, N.-C. Yen, and C. C. Tung. *Proc. R. Soc. Lond. A*, 454:903, 1998.

[15] N. E. Huang, M.-L. Wu, W. Qu, S. R. Long, and S. S. P. Shen. *Appl. Stochastic Models Bus. Ind.*, 19:245, 2003.

[16] H. Huang and J. Pan. Speech pitch determination based on Hilbert–Huang transform. *Sig. Proc.*, 86:792, 2006.

[17] X. Zhu, Z. Shen, S. D. Eckermann, M. Bittner, I. Hirota, and J.-H. Yee. *J. Geophys. Res.*, 102:16545, 1997.

[18] N. E. Huang, Z. Shen, and S. R. Long. *Annu. Rev. Fluid Mech.*, 31:417, 1999.

[19] J. I. Salisbury and M. Wimbush. *Nonlinear Proc. Geoph.*, 9:341, 2002.

[20] M. E. Montesinos, J. L. Muñoz-Cobo, and C. Pérez. *Ann. Nucl. Energy*, 30: 715, 2003.

[21] K. Coughlin and K. K. Tung. *Adv. Space Res.*, 34:323, 2004.

[22] H. El-Askary, S. Sarkar, L. Chiu, M. Kafatos, and T. El-Ghazawi. *Adv. Space Res.*, 33:338, 2004.

[23] M. Dätig and T. Schlurmann. *Ocean Eng.*, 31:1783, 2004.

[24] I. M. Jánosi and R. Müller. *Phys. Rev. E*, 71:056126, 2005.

[25] Z. K. Peng, P. W. Tse, and F. L. Chu. *Mech. Syst. Signal Pr.*, 19:974, 2005.

[26] Z. K. Peng, P. W. Tse, and F. L. Chu. *J. Sound Vib.*, 286:187, 2005.

[27] N. Huang and S. Shen, editors. *Hilbert–Huang Transform: Introduction and Applications*. World Scientific, Singapore, 2005.

[28] I. Z. Kiss and J. L. Hudson. *Phys. Rev. E*, 64:046215, 2001.

[29] S. C. Phillips, R. J. Gledhill, J. W. Essex, and C. M. Edge. *J. Phys. Chem.*, 107: 4869, 2003.

[30] W.-S. Lam, W. Ray, P. N. Guzdar, and R. Roy. *Phys. Rev. Lett.*, 94:010602, 2005.

[31] W. Huang, Z. Shen, N. E. Huang, and Y. C. Fung. *Proc. Natl. Acad. Sci.*, 95: 4816, 1998.

[32] W. Huang, Z. Shen, N. E. Huang, and Y. C. Fung. *Proc. Natl. Acad. Sci.*, 95: 12766, 1998.

[33] W. Huang, Z. Shen, N. E. Huang, and Y. C. Fung. *Proc. Natl. Acad. Sci.*, 96: 1834, 1999.

[34] H. Liang, Z. Lin, and R. W. McCallum. *Med. Biol. Eng. Comput.*, 38:35, 2000.

[35] J. C. Echeverria, J. A. Crowe, M. S. Woolfson, and B. R. Hayes-Gill. *Med. Biol. Eng. Comput.*, 39:471, 2001.

[36] M. A. Bray and J. P. Wikswo. *Phys. Rev. E*, 65:051902, 2002.

[37] D. A. T. Cummings, R. A. Irizarry, N. E. Huang, T. P. Endy, A. Nisalak, K. Ungchusak, and D. S. Burke. *Nature*, 427:344, 2004.

[38] R. Balocchi et al. *Chaos Solitons Fract.*, 20:171, 2004.

[39] J. I. Salisbury and Y. Sun. *Ann. Biomed. Eng.*, 32:1348, 2004.

[40] E. P. Souza-Neto, M. A. Custaud, J. C. Cejka, P. Abry, J. Frutoso, C. Gharib, and P. Flandrin. *Methods Inf. Med.*, 43:60, 2004.

[41] V. I. Ponomarenko et al. *Chaos Solitons Fract.*, 23:1429, 2005.

[42] H. Liang, S. L. Bressler, R. Desimone, and P. Fries. *Neurocomp.*, 65:801, 2005.

[43] H. Liang, S. L. Bressler, E. A. Buffalo, R. Desimone, and P. Fries. *Biol. Cybern.*, 92:380, 2005.

[44] H. Liang, Q. Lin, and J. D. Z. Chen. *IEEE Trans. Biomed. Eng.*, 52:1692, 2005.

[45] R. Roulier, A. Humeau, T. P. Flatley, and P. Abraham. *Phys. Med. Biol.*, 50: 5189, 2005.

[46] P. Flandrin, G. Rilling, and P. Gonçalvès. *IEEE Signal Proc. Lett.*, 11:112, 2004.

[47] T. Yalçinkaya and Y. C. Lai. *Phys. Rev. Lett.*, 79:3885, 1997.

[48] Y. C. Lai. *Phys. Rev. E*, 58, 1998.

[49] Y. C. Lai and N. Ye. *Int. J. Bif. Chaos*, 13:1383, 2003.

[50] M. Chavez, C. Adam, V. Navarro, S. Boccaletti, and J. Martinerie. *Chaos*, 15: 023904, 2005.

[51] N. M. Blachman. *Trans. Inf. Theory*, 45:2115, 1999.

[52] G. Rilling, P. Flandrin, and P. Gonçalvès. On empirical mode decomposition and its algorithms. In *Proc. IEEE EURASIP Workshop Nonlinear Signal Image Processing, Italy*, 2003.

[53] N. B. Janson, A. G. Balanov, V. S. Anishchenko, and P. V. E. McClintock. *Phys. Rev. Lett.*, 86:1749, 2001.

[54] N. B. Janson, A. G. Balanov, V. S. Anishchenko, and P. V. E. McClintock. *Phys. Rev. E*, 65:036211, 2002.

[55] A. Linderhed. *Adaptive Image Compression with Wavelet Packets and Empirical Mode Decomposition*. PhD thesis, Linköping University, Sweden, 2004.

[56] C. D. Blakely. A fast empirical mode decomposition technique for nonstationary nonlinear time series. Reprint CSCAMM-05-10. Center of Scientific Computation and Mathematical Modeling, University of Maryland, 2005.

[57] A. E. Hramov and A. A. Koronovskii. *Chaos*, 14:603, 2004.

[58] A. A. Koronovskii and A. E. Hramov. *Tech. Phys. Lett.*, 30:587, 2004.

[59] I. Bove, S. Boccaletti, J. Bragard, J. Kurths, and H. Mancini. *Phys. Rev. E*, 69: 016208, 2004.

[60] M. Hämäläinen, R. Hari, R. Ilmoniemi, J. Knuutila, and O. V. Lounasmaa. *Rev. Mod. Phys.*, 65:413, 1993.

[61] V. Crunelli and N. Leresche. *Nat. Rev. Neurosci.*, 3:371, 2002.

[62] N. E. Huang, M.-L. C. Wu, S. R. Long, S. S. P. Shen, W. Qu, P. Gloersen, and K. L. Fan. *Proc. R. Soc. Lond. A*, 459:2317, 2003.

[63] P. Flandrin and P. Gonçalvès. *Int. J. Wavelets, Multires. Info. Proc.*, 2:477, 2004.

[64] Z. Wu and N. E. Huang. *Proc. R. Soc. Lond. A*, 460:1597, 2004.

[65] S. Olhede and A. T. Walden. *Proc. R. Soc. Lond. A*, 460:955, 2004.

[66] R. C. Sharpley and V. Vatchev. Analysis of the intrinsic mode functions. Technical Report No. 2004:12, Industrial Mathematics Institute, University of South Carolina, 2004.

[67] T. Qian, Q. Chen, and L. Li. *Physica D*, 203:80, 2005.

[68] E. Delechelle, J. Lemoine, and O. Niang. *IEEE Signal Proc. Lett.*, 12:764, 2005.

10 Synchronization Analysis and Recurrence in Complex Systems

Maria Carmen Romano, Marco Thiel, Jürgen Kurths, Martin Rolfs, Ralf Engbert, and Reinhold Kliegl

We discuss an approach to detect and quantify phase synchronization in the case of coupled non-phase-coherent oscillators, which is based on the recurrence properties of the underlying system. First, we present an index which detects phase synchronization without computing the phase directly. We show that this index is also appropriate for non-phase-coherent systems, i.e., systems with a rather broad power spectrum. Furthermore, we illustrate the applicability of this index for time series strongly contaminated by noise.

Second, we present an algorithm, which is also based on recurrence to generate surrogates to test for phase synchronization. The generated surrogates correspond to independent copies of the underlying system. Hence, computing a phase synchronization index between one observed oscillator and the surrogate of the second oscillator, we can test for phase synchronization.

Finally, we apply the recurrence-based index, as well as the recurrence-based surrogates to fixational eye movements and find strong indications that both the left and right fixational eye movements are synchronized.

10.1 Introduction

The study of synchronization goes back to the seventeenth century and begins with the analysis of synchronization of nonlinear periodic systems. The synchronization phenomenon was probably discovered first by Huygens in 1673, who noticed that two pendulum clocks that hang on the same beam can synchronize. This discovery can be considered as the beginning of Nonlinear Science. The synchronization of the flashing of fireflies, the peculiarities of adjacent organ pipes which can almost annihilate each other or speak in unison, or the synchronization of diodes are other well known examples.

However, the research of synchronization in complex systems did not begin until the end of the eighties. It has been studied extensively during the last years [1–4], as this phenomenon has found numerous applications in natural (cardiorespiration, Parkinson patients, ecology, El Niño-Monsoon, etc.) [5–10]

Handbook of Time Series Analysis. Björn Schelter, Matthias Winterhalder, Jens Timmer

ISBN: 3-527-40623-9

and engineering (lasers, plasma, tubes, etc.) systems [11–13]. Two systems are said to be phase synchronized when their respective frequencies and phases are locked. Note that synchronization is a process (of adapting rhythms) and not a state. Till now phase synchronization (PS) of chaotic systems has been mainly observed for attractors with rather coherent phase dynamics. These attractors have a relatively simple topology of oscillations and a well-pronounced peak in the power spectrum, which allows to introduce the phase and the characteristic frequency of motions, Eq. (10.2). However, some difficulties appear when dealing with non-coherent attractors characterized by a rather broad band power spectra. Then it might not be straightforward to define a phase of the oscillations, and in general no single characteristic time scale exists. In contrast to phase coherent attractors, it is quite unclear whether some phase synchronized state can be achieved (Fig. 10.1).

To treat this problem, we propose a method based on another basic property of complex chaotic systems: recurrences in phase space. The concept of recurrence in dynamical systems goes back to Poincaré [14], when he proved that after a sufficiently long time interval, the trajectory of an isolated mechanical system will return arbitrarily close to each former point of its route. We will show that the concept of recurrence allows to detect indirectly synchronization and works even in the case of noisy non-phase-coherent oscillators. Instead of defining directly the phase, we consider the coincidence of certain recurrence structures of both coupled subsystems. By means of this comparison we are able to detect synchronization in complex systems.

Another important problem in the synchronization analysis is that even though the synchronization measures may be normalized, experimental time series often yield values which are not at the borders of the interval and hence are difficult to interpret. This problem can be overcome if the coupling strength between the two systems can be varied systematically and a rather large change in the measure can be observed, i.e., we have a so called active experiment [1–4]. However, there are other kind of experiments (passive ones), in which it is not possible to change the coupling strength systematically, e.g., the synchronization of the heart beats of a mother with her fetus [15]. In some cases, this problem has been tackled by interchanging the pairs of oscillators [15], for example the EEGs of other pregnant women were used as "natural surrogates." These surrogates are independent and hence not in PS with the original system. Hence, if the synchronization index obtained with the original data is not significantly higher than the index obtained with the natural surrogates, there is no sufficient evidence to claim synchronization. But even this rather innovative approach has some drawbacks. The natural variability and also the frequency of the heart beats of the surrogate mothers are usually lightly different from the ones of the real mother. Furthermore, the data acquisition can be expensive and at least in some cases problematic or even impossible (e.g., geophysical time series). In these cases it would be convenient to perform a hypothesis test based on surrogates generated by a mathematical algorithm.

Therefore, we present a technique for the generation of surrogates, which is based on the recurrences of a system. These surrogates mimic the dynamical behavior of the system. Then, computing the synchronization index between one subsystem of the original system and the other subsystem of the surrogate, and comparing it with the synchronization index obtained for the original system, we can test for PS.

In Section 10.2, we introduce the concept of recurrence, as well as the synchronization index based on the recurrence properties of the system. In Section 10.3 we show how to detect another kind of synchronization, namely generalized synchronization (GS) by means of recurrences and in Section 10.4 we show that the recurrence-based indices indicate the transition to PS and GS in accordance with other known theoretical methods. In Section 10.5 we present the twin surrogates technique and apply it to test for synchronization in the paradigmatic two coupled Rössler systems. In Section 10.6 we show an application of the recurrence-based index and surrogates to measured physiological data, namely fixational eye movements.

10.2 Phase Synchronization by Means of Recurrences

First, we exemplify the problem of defining the phase in systems with rather broad power spectrum by the paradigmatic system of two coupled nonidentical Rössler oscillators

$$\begin{aligned}
\dot{x}_{1,2} &= -\omega_{1,2} y_{1,2} - z_{1,2} \,, \\
\dot{y}_{1,2} &= \omega_{1,2} x_{1,2} + a y_{1,2} + \mu(y_{2,1} - y_{1,2}) \,, \\
\dot{z}_{1,2} &= 0.1 + z_{1,2}(x_{1,2} - 8.5) \,,
\end{aligned} \tag{10.1}$$

where μ is the coupling strength and $\omega_{1,2}$ determine the mean intrinsic frequency of the (uncoupled) oscillators in the case of phase coherent attractors. In our simulations we take $\omega_1 = 0.98$ and $\omega_2 = 1.02$. The parameter $a \in [0.15 : 0.3]$ governs the topology of the chaotic attractor. When a is below a critical value a_c ($a_c \approx 0.186$ for $\omega_1 = 0.98$ and $a_c \approx 0.195$ for $\omega_2 = 1.02$), the chaotic trajectories always cycle around the unstable fixed point $(x_0, y_0) \approx (0, 0)$ in the (x, y) subspace, i.e., $\max(y) > y_0$ (Fig. 10.1(a)). In this case, simply the rotation angle

$$\phi = \arctan \frac{y}{x} \tag{10.2}$$

can be defined as the phase, which increases almost uniformly. The oscillator has a coherent phase dynamics, i.e., the diffusion of the phase dynamics is very low (10^{-5} to 10^{-4}). In this case, other phase definitions, e.g., based on the Hilbert transform or on the Poincaré section, yield equivalent results [1–4]. However, beyond the critical value a_c, the trajectories no longer always completely cycle

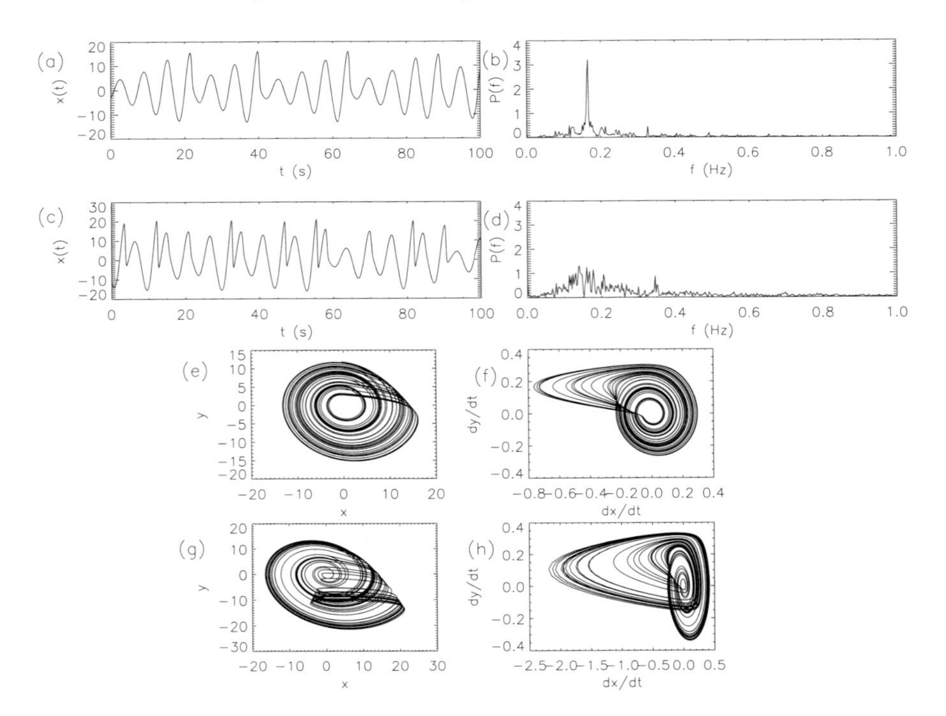

Fig. 10.1: (a,c) Segment of the x_1-component of the trajectory of the Rössler systems, Eq. (10.1). (b,d) periodogram of the x-component of the trajectory. (e,g) projection of the attractor onto the (x, y) plane. (g,h) projection onto the $(\dot{x}, \dot{y})$ plane. (a,b,e,f) computed for $a = 0.16$ and (c,d,g,h) computed for $a = 0.2925$.

around (x_0, y_0), and some $\max(y) < y_0$ occur, which are associated with faster returns of the orbits; the attractor becomes a funnel one. Such earlier returns in the funnel attractor happen more frequently with increasing a (Fig. 10.1(b)). It is clear that for the funnel attractors, usual (and rather simple) definitions of phase, such as Eq. (10.2), are no longer applicable [1–4].

Rosenblum et al. have proposed in [16] to use an ensemble of phase coherent oscillators which is driven by the non-phase-coherent oscillator in order to estimate the frequency of the last, and hence detect PS in such kind of systems. However, depending on the component one uses to couple the non-phase-coherent oscillator to the coherent ones, the result of the obtained frequency can be different.

Furthermore, Osipov et al. [17] have proposed another approach which is based on the general idea of the curvature of an arbitrary curve [18]. For any two-dimensional curve $\mathbf{r}_1 = (u, v)$ the angle velocity at each point is $\nu = \frac{ds}{dt}/R$, where $ds/dt = \sqrt{\dot{u}^2 + \dot{v}^2}$ is the velocity along the curve and $R = (\dot{u}^2 + \dot{v}^2)^{3/2}/[\dot{v}\ddot{u} - \ddot{v}\dot{u}]$ is the curvature. If $R > 0$ at each point, then $\nu = \frac{d\phi}{dt} = \frac{\dot{v}\ddot{u} - \ddot{v}\dot{u}}{\dot{u}^2 + \dot{v}^2}$ is always positive and therefore the variable ϕ defined as $\phi = \int \nu \, dt = \arctan \frac{\dot{v}}{\dot{u}}$, is a monotonically growing angle function of time and can be considered as a phase of the oscillations. Geometrically it means that the projection $\mathbf{r}_2 = (\dot{u}, \dot{v})$ is a curve cycling monotonically around a certain point.

These definitions of ϕ and ν hold in general for any dynamical system if the projection of the phase trajectory onto some plane is a curve with a positive curvature. This approach is applicable to a large variety of chaotic oscillators, such as the Lorenz system [19], the Chua circuit [20] or the model of an ideal four-level laser with periodic pump modulation [21].

This is clear for phase-coherent as well as funnel attractors in the Rössler oscillator. Here projections of chaotic trajectories on the plane $(\dot{x}, \dot{y})$ always rotate around the origin (Fig. 10.1(c) and (d)) and the phase can be defined as

$$\phi = \arctan \frac{\dot{y}}{\dot{x}} . \tag{10.3}$$

We have to note that for funnel-like chaotic attractors the coupling may change their topology. As a consequence the strong cyclic structure of orbits projection in the $(\dot{x}, \dot{y})$-plane may be destroyed and the phase measurement by Eq. (10.3) fails occasionally for intermediate values of coupling. But for small coupling and for coupling near the transition to PS, the phase is well-defined by Eq. (10.3) [22].

We consider two criteria to detect the existence of PS: Locking of the mean frequencies $\Omega_1 = \langle \nu_1 \rangle = \Omega_2 = \langle \nu_2 \rangle$, and locking of the phase $|\phi_2(t) - \phi_1(t)| \leqslant$ const (we restrict here to 1 : 1 synchronization). Applying the new definition of the phase Eq. (10.3) to the system defined by Eq. (10.1) for $a = 0.2925$ (strongly noncoherent) and $\mu = 0.179$, one obtains the phase difference represented in Fig. 10.2.

We find two large plateaus in the evolution of the difference of the phases with time, i.e., we detect PS, but we also find a phase slip associated to a different number of oscillations in the two oscillators in the represented period of time. This means, we observe the rare occurrence of phase slip. It is interesting to note that in this system PS occurs after one of the positive Lyapunov exponents passes to negative values, i.e., it is also a transition to generalized chaotic synchronization (GS).

Although this approach works well in non-phase-coherent model systems, we have to consider that one is often confronted with the computation of the phase in experimental time series, which are usually corrupted by noise. In this case, some difficulties may appear when computing the phase by Eq. (10.3), because derivatives are involved in its definition.

Hence, we propose a different approach based on recurrences in phase space to detect PS indirectly. We define a recurrence of the trajectory of a dynamical system $\{\mathbf{x}(i)\}_{i=1}^{N}$ in the following way: We say that the trajectory has returned at time $t = j\delta t$ to the former point in phase space visited at $t = i\delta t$ if

$$R_{i,j}^{(\varepsilon)} = \Theta(\varepsilon - \|\mathbf{x}(i) - \mathbf{x}(j)\|) = 1 , \tag{10.4}$$

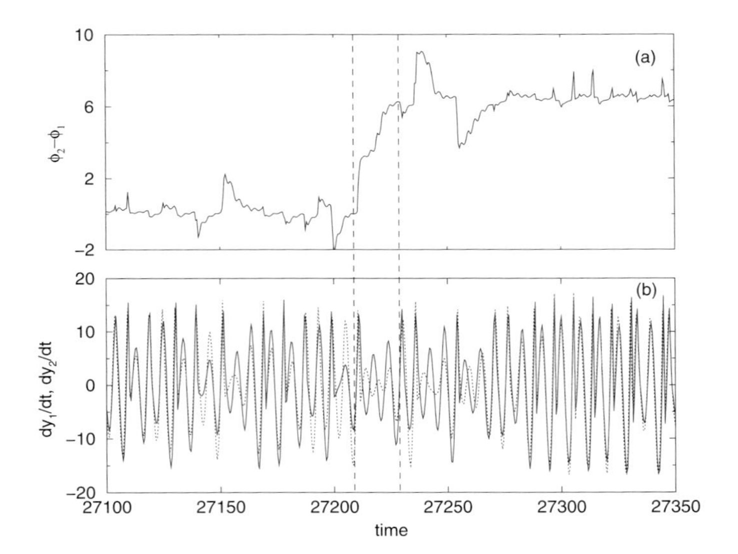

Fig. 10.2: (a) Time evolution of phase difference of the system of Eq. (10.1). (b) Variables $\dot{y}_{1,2}$ in system (10.1) for $a = 0.2925$ and $\mu = 0.179$. Solid and dotted lines correspond to the first and the second oscillator, respectively. In the time interval between dashed lines the first oscillator produces four rotations in the $(\dot{x}_1, \dot{y}_1)$-plane around the origin, but the second one generates only three rotations, which leads to a phase slip in (a).

where ε is a pre-defined threshold, $\Theta(.)$ is the Heaviside function and δt is the sampling rate. A "1" in the matrix at i, j means that $\mathbf{x}(i)$ and $\mathbf{x}(i)$ are neighboring, a "0" that they are not. The black and white representation of this binary matrix is called recurrence plot (RP). This method has been intensively studied in the last years: Different measures of complexity have been proposed based on the structures obtained in the RP and have found numerous applications in, e.g., physiology and earth science [23–27]. Furthermore, it has been even shown that some dynamical invariants can be estimated by means of the recurrence structures [28].

Based on this definition of recurrence, we want to tackle the problem of performing a synchronization analysis in the case of non-phase-coherent systems. We avoid the direct definition of the phase and instead use the recurrence properties of the systems in the following way: The probability $P^{(\varepsilon)}(\tau)$ that the system returns to the neighborhood of a former point $\mathbf{x}(i)$ of the trajectory[1] after τ time steps can be estimated as follows:

$$\hat{P}^{(\varepsilon)}(\tau) = \frac{1}{N-\tau} \sum_{i=1}^{N-\tau} \Theta(\varepsilon - \|\mathbf{x}(i) - \mathbf{x}(i+\tau)\|) = \frac{1}{N-\tau} \sum_{i=1}^{N-\tau} R^{(\varepsilon)}_{i,i+\tau} . \quad (10.5)$$

[1] The neighborhood is defined as a box of size ε centered at $\mathbf{x}(i)$, as we use the maximum norm.

This function can be regarded as a generalized autocorrelation function, as it also describes higher order correlations between the points of the trajectory in dependence on the time delay τ. A further advantage with respect to the linear autocorrelation function is that $\hat{P}^{(\varepsilon)}(\tau)$ is defined for a trajectory in phase space and not only for a single observable of the system's trajectory. Further, we have recently shown that it is possible to reconstruct the attractor by only considering the recurrences of single components of the system [29] and it is also possible to estimate dynamical invariants of the system (e.g., entropies and dimensions) by means of recurrences in phase space [28]. Hence, the recurrences of the system in phase space contain information about higher order dependencies within the components of the system.

For a periodic system with period length T in a two-dimensional phase space, it can be easily shown that

$$P(\tau) = \lim_{\varepsilon \to 0} \hat{P}^{(\varepsilon)}(\tau) = \begin{cases} 1 & \tau = T \\ 0 & \text{otherwise}\,. \end{cases}$$

For coherent chaotic oscillators, such as Eq. (10.1) for $a = 0.16$, $\hat{P}^{(\varepsilon)}(\tau)$ has well-expressed local maxima at multiples of the mean period, but the probability of recurrence after one or more rotations around the fixed point is less than one (Fig. 10.5).

Analyzing the probability of recurrence, it is possible to detect PS for non-phase-coherent oscillators, too. This approach is based on the following idea: Originally, a phase ϕ is assigned to a periodic trajectory $\mathbf{x}$ in phase space, by projecting the trajectory onto a plane and choosing an origin, around which the trajectory oscillates all the time. Then an increment of 2π is assigned to ϕ, when the point of the trajectory has returned to its starting position, i.e., when $\|\mathbf{x}(t + T) - \mathbf{x}(t)\| = 0$. Analogously to the case of a periodic system, we can refer an increment of 2π to ϕ to a complex nonperiodic trajectory $\mathbf{x}(t)$, when $\|\mathbf{x}(t + T) - \mathbf{x}(t)\| \sim 0$, or equivalently when $\|\mathbf{x}(t + T) - \mathbf{x}(t)\| < \varepsilon$, where ε is a predefined threshold. That means, a recurrence $R^{(\varepsilon)}_{t,t+\tau} = 1$ can be interpreted as an increment of 2π of the phase in the time interval τ[2].

$\hat{P}^{(\varepsilon)}(\tau)$ can be viewed as a statistical measure on how often ϕ in the original phase space has increased by 2π or multiples of 2π within the time interval τ. If two systems are in PS, in the mean, the phases of both systems increase by $k \cdot 2\pi$, with k a natural number, within the same time interval τ. Hence, looking at the coincidence of the positions of the maxima of $\hat{P}^{(\varepsilon)}(\tau)$ for both systems, we can quantitatively identify PS (from now on, we omit (ε) and $\hat{}$ in $\hat{P}^{(\varepsilon)}(\tau)$ to simplify the notation). The proposed algorithm then consists of two steps:

- Compute $P_{1,2}(\tau)$ of both systems based on Eq. (10.5).

[2] This can be considered as an alternative definition of the phase to Eqs. (10.2) and (10.3).

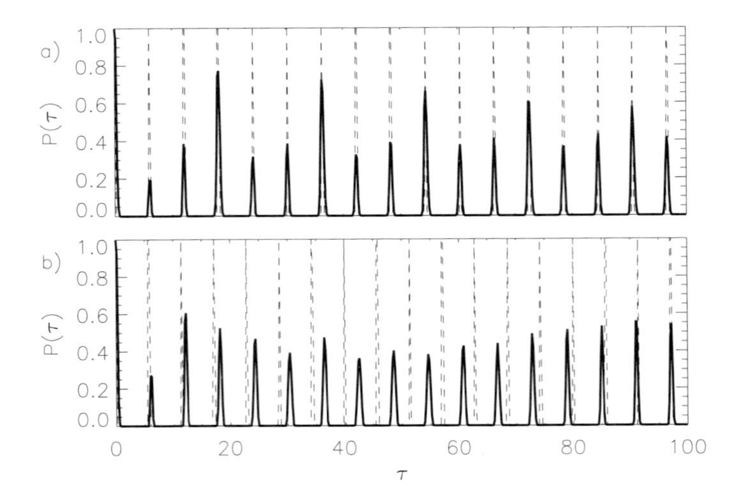

Fig. 10.3: $P(\tau)$ for a periodically driven Rössler (Eqs. (10.7)) in PS (a) and in non-PS (b). Solid line: $P(\tau)$ of the driven Rössler, dashed line: $P(\tau)$ of the periodic forcing.

- Compute the cross-correlation coefficient between $P_1(\tau)$ and $P_2(\tau)$ (correlation between probabilities of recurrence, CPR)

$$\mathrm{CPR}^{1,2} = \frac{\langle \bar{P}_1(\tau)\bar{P}_2(\tau)\rangle_\tau}{\sigma_1\sigma_2}, \tag{10.6}$$

where $\bar{P}_{1,2}$ means that the mean value has been subtracted and σ_1 and σ_2 are the standard deviations of $P_1(\tau)$ and $P_2(\tau)$, respectively.

If both systems are in PS, the probability of recurrence is maximal simultaneously and $\mathrm{CPR}^{1,2} \sim 1$. In contrast, if the systems are not in PS, the maxima of the probability of recurrence do not occur jointly and expect low values of $\mathrm{CPR}^{1,2}$.

10.2.1 Examples of Application

In this section we exemplify the application of CPR to detect PS for four prototypical chaotic systems. The number of data points used for the analysis presented here is 5000.

1. We start with the periodically driven Rössler system [1–4]

$$\begin{aligned}
\dot{x} &= -y - z + \mu\cos(\omega t)\\
\dot{y} &= x + 0.15y\\
\dot{z} &= 0.4 + z(x - 8.5)\,.
\end{aligned} \tag{10.7}$$

For the frequency $\omega = 1.04$ and the coupling strength $\mu = 0.16$, the periodic forcing locks the frequency of the Rössler system. This can be clearly seen in

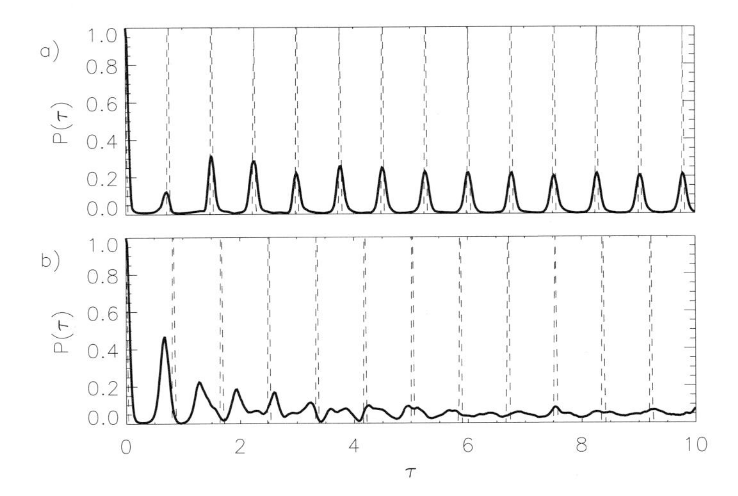

Fig. 10.4: $P(\tau)$ for a periodically driven Lorenz in PS (a) and in non-PS (b). Solid line: $P(\tau)$ of the driven Lorenz, dashed line: $P(\tau)$ of the periodic forcing.

Fig. 10.3(a). The position of the maxima coincide. The value of the recurrence-based PS index (Eq. (10.6)) is CPR $= 0.862$.

For the parameters $\omega = 1.1$ and $\mu = 0.16$, the periodic forcing does not synchronize the Rössler system. Hence, the joint probability of recurrence is very low, which is reflected in the drift of the peaks of the corresponding $P(\tau)$ (Fig. 10.3(b)). In this case, CPR $= -0.00241$.

2. We continue our considerations with the periodically driven Lorenz system for the standard parameters

$$\begin{aligned} \dot{x} &= 10(y - x) \\ \dot{y} &= 28x - y - xz \\ \dot{z} &= -8/3z + xy + \mu\cos(\omega t)\,. \end{aligned} \tag{10.8}$$

In Fig. 10.4(a) the probabilities of recurrence $P(\tau)$ in the PS case ($\mu = 10$, $\omega = 8.35$) are represented. We see that the position of the local maxima of the Lorenz oscillator coincide with the ones of the periodic forcing. However, the local maxima are not as high as in the case of the Rössler system, and they are broader. This reflects the effective noise which is intrinsic in the Lorenz system [1–4]. Therefore, the phase synchronization is not perfect: An exact frequency locking between the periodic forcing and the driven Lorenz cannot be observed [30]. In this case, we obtain CPR $= 0.667$. In the non-PS case ($\mu = 10$, $\omega = 7.5$), we obtain CPR $= 0.147$ (Fig. 10.4(b)).

3. Next, we consider the case of two mutually coupled Rössler systems in the phase coherent regime, more precisely we analyze Eqs. (10.1) with $a = 0.16$.

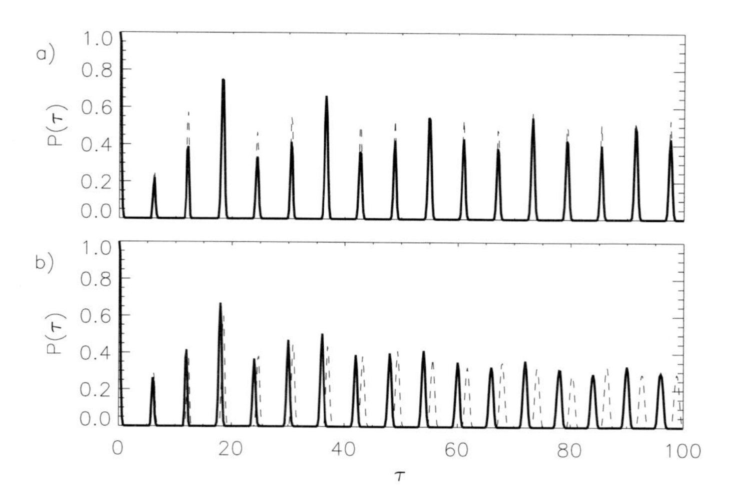

Fig. 10.5: $P(\tau)$ for two mutually coupled Rössler systems (Eqs. (10.1)) in phase coherent regime ($a = 0.16$) for $\mu = 0.05$ (a) and for $\mu = 0.02$ (b).

According to [17], for $\omega_1 = 0.98, \omega_2 = 1.02$ and $\mu = 0.05$ both systems are in PS. We observe that the local maxima of P_1 and P_2 occur at $\tau = n \cdot T$, where T is the mean period of both Rössler systems (Fig. 10.5(a)). The heights of the local maxima are in general different for both systems if they are only in PS and not in, e.g., complete synchronization or generalized synchronization. But the positions of the local maxima of $P(\tau)$ coincide. In this case, we obtain $CPR = 0.998$.

At a coupling strength of $\mu = 0.02$ the systems are not in PS and the positions of the maxima of $P(\tau)$ do not coincide anymore (Fig. 10.5(b)), clearly indicating that the frequencies are not locked. In this case, we obtain $CPR = 0.115$.

4. As a last example with simulated data, we analyze the challenging case of two mutually coupled Rössler systems in the funnel regime. Therefore, we study Eqs. (10.1) with $a = 0.2925$, $\omega_1 = 0.98$, and $\omega_2 = 1.02$. We analyze two different coupling strengths: $\mu = 0.2$ and $\mu = 0.05$. We observe that the structure of $P(\tau)$ in the funnel regime (Fig. 10.6) is rather different from the one in the phase coherent Rössler system (Fig. 10.5). The peaks in $P(\tau)$ are not as well pronounced as in the coherent regime, reflecting the different time scales that play a crucial role and the broad-band power spectrum of this system. However, we notice that for $\mu = 0.2$ the locations of the local maxima coincide for both oscillators (Fig. 10.6(a)), indicating PS, whereas for $\mu = 0.05$ the positions of the local maxima do not coincide anymore (Fig. 10.6(b)), indicating non-PS. These results are in accordance with [17].

   In the PS case, we obtain $CPR = 0.988$, and in the non-PS case, $CPR = 0.145$. Note that the position of the first peak in Fig. 10.6(b) coincides, although the

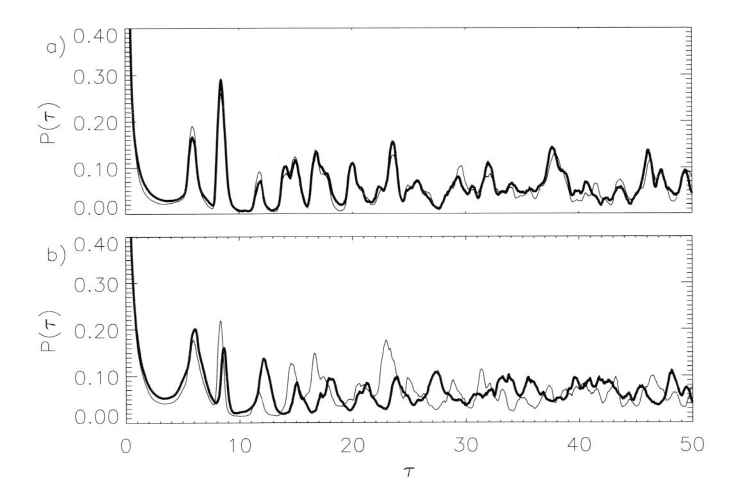

Fig. 10.6: $P(\tau)$ for two mutually coupled Rössler systems (Eqs. (10.1)) in funnel regime ($a = 0.2925$) for $\mu = 0.2$ (a) and for $\mu = 0.05$ (b). Bold line: $P_1(\tau)$, solid line: $P_2(\tau)$.

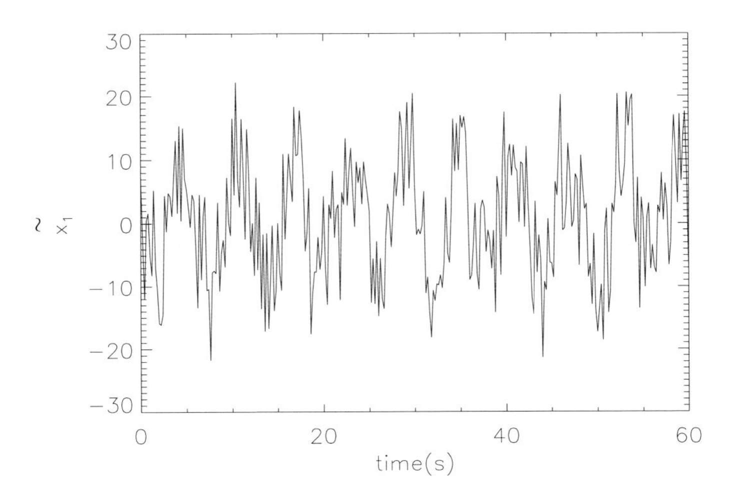

Fig. 10.7: First component x_1 of Eqs. (10.1) with 80% independent Gaussian noise (for $\mu = 0.05$). From the figure it is clearly recognizable that it is difficult to compute the phase by means of, e.g., the Hilbert transformation.

oscillators are not in PS. This is due to the small frequency mismatch ($|\omega_1 - \omega_2| = 0.04$). However, by means of the index CPR we can distinguish rather well between both regimes.

10.2.2 Influence of Noise

Measurement errors are omnipresent in experimental time series. Hence, it is necessary to analyze the influence of noise on CPR (correlation of probability of recurrence).

First, we treat additive or observational white noise. We use Eqs. (10.1) with two different coupling strengths, so that we can compute the deviation which is caused by noise in the nonsynchronized and in the synchronized case.

We add independent Gaussian noise with standard deviation $\sigma_{noise} = \alpha\sigma_j$ to each coordinate j of the system, where σ_j is the standard deviation of the component j and α is the noise level. In Fig. 10.7 the "corrupted" x-component of the first Rössler subsystem $\tilde{x}_1(t) = x_1(t) + \alpha\sigma_1\eta(t)$ is represented. Herein $\eta(t)$ is a realization of Gaussian noise and $\alpha = 0.8$. From Fig. 10.7 it is obvious that it is difficult to compute the phase by means of, e.g., the Hilbert transformation for such a high noise level without filtering.

The choice of ε for the computation of $P_1(\tau)$ and $P_2(\tau)$ in the presence of noise is performed automatically by fixing the recurrence rate RR, i.e., the percentage of recurrence points in the recurrence matrix, Eq. (10.4). The results presented below were computed for RR $= 0.1$, but the results are rather independent of the choice of RR. However, RR should not be chosen too small if the level of noise is very high [23–27].

In order to compute CPR for the noisy oscillators, we calculate first the probabilities of recurrence $P_1(\tau)$ and $P_2(\tau)$ for coupling strengths $\mu = 0.05$ (PS, Fig. 10.8) and $\mu = 0.02$ (non-PS, Fig. 10.9).

We note that the peaks in $P_1(\tau)$ and $P_2(\tau)$ become lower and broader (Figs. 10.8(b) and 10.9(b)) compared with the noise free case (Figs. 10.8(a) and 10.9(a)), which is expected. However, despite of the large level of noise, the positions of the local maxima coincide in the PS case, and they drift away in the non-PS case. This a convenient result, because we can still decide whether the oscillators are synchronized in a statistical sense or not. This is reflected in the obtained values for the CPR index: at a noise level of 80% noise, in the PS case the obtained value for CPR is exactly the same with and without noise, and in the non-PS case it is nearly the same (see Table 10.1). This shows that the index CPR for PS is very robust against observational noise.

Now, we analyze the influence of colored noise on the index CPR. We add a realization of colored noise with a very high noise amplitude to each component of the first system and another realization of colored noise with a smaller noise amplitude to each component of the second system (see Fig. 10.10(a) and (b) and the corresponding caption). Other methods fail determining the phase in this case, as for example the one presented in [17], because it requires the computation of the derivative of the time series, and due to the large level of noise, this is not possible. But by means of $P(\tau)$ we can distinguish PS from non-PS even in this case (Fig. 10.10(c) and (d)): We obtain CPR $= 0.0276$ for the non-PS case and CPR $= 0.530$ for the PS case.

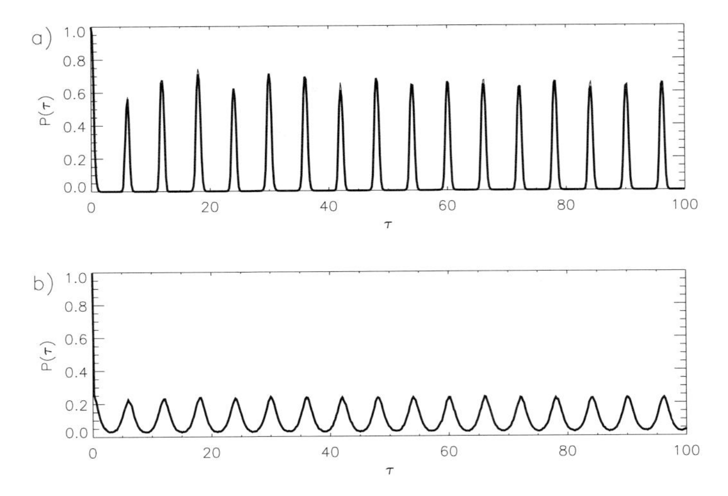

Fig. 10.8: Probabilities of recurrence for two coupled Rössler systems (Eqs. (10.1)) in PS ($\mu = 0.05$) without noise (a) and with 80% Gaussian observational noise (b). Bold line: subsystem 1, solid line: subsystem 2. Note that the position of the peaks of $P_1(\tau)$ and $P_2(\tau)$ coincide in both cases, and hence the solid line is hidden by the bold one.

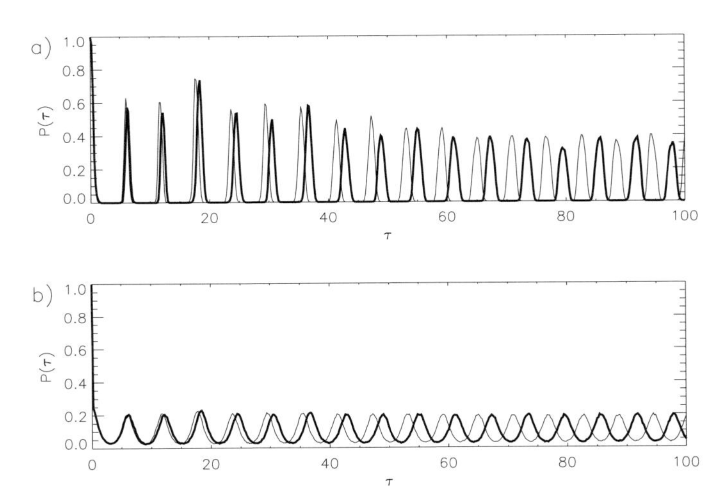

Fig. 10.9: Probabilities of recurrence for two coupled Rössler systems (Eqs. (10.1)) in non-PS ($\mu = 0.02$) without noise (a) and with 80% Gaussian observational noise (b). Bold line: subsystem 1, solid line: subsystem 2.

10.3 Generalized Synchronization and Recurrence

In this section we treat the issue of synchronization of coupled systems which are essentially different. This problem has been addressed first in [31, 32]. In this

Tab. 10.1: Index CPR for PS calculated for two coupled Rössler systems (10.1) with observational noise and without noise, for comparison.

μ	CPR (80% noise)	CPR (0% noise)
0.02 (non-PS)	0.149	0.115
0.05 (PS)	0.998	0.998

case, there is in general no trivial manifold in the phase space which attracts the systems' trajectories. It has been shown that these systems can synchronize in a more general way, namely $y = \psi(x)$, where ψ is a transformation which maps asymptotically the trajectories of x into the ones of the attractor y. This kind of synchronization is called generalized synchronization (GS). The properties of the function ψ depend on the features of the systems x and y, as well as on the attraction properties of the synchronization manifold $y = \psi(x)$ [33]. GS has been demonstrated in laboratory experiments for electronic circuits and laser systems [34–38] and has found applications for the the design of communication devices [39–43] and model verification and parameter estimations from time series [44, 45].

Some statistical measures have been introduced for the detection of GS, such as the method of mutual false nearest neighbors [31, 32] or variations of the method proposed and analyzed in [46–48], which are based on the squared mean distance and conditional distance between mutual nearest neighbors. Some other methods are based on the mutual predictability to detect dynamical interdependence [49, 50]. There, the nearest neighbors of each subsystem are computed separately in the respective (sub)state space.

In this section we present a criterion for the detection of GS, which exploits the relationship between the geometric connection of both systems and their recurrences. The connection between recurrences and GS is even more straightforward than the one between recurrences and PS. One can see that the concept of GS is linked to the one of recurrence, considering the fact that when $x(t)$ and $y(t)$ are in GS, two close states in the phase space of x correspond to two close states in the space of y [31, 32]. Hence, the "neighborhood identity" in phase space is preserved, i.e., they are topologically equivalent. Since the recurrence matrix (Eq. (10.4)) is nothing else but a record of the neighborhood of each point of the trajectory, one can conclude that two systems are in GS if their respective RPs are almost identical. Note that it is possible, under some conditions, to reconstruct the rank order of the time series considering only the information contained in the RP [29]. Therefore, we can use the recurrence properties to detect and quantify GS.

However, in practice we note that the recurrence matrices of two systems in GS are very similar, but not identical. Several reasons can be given to explain this observation: The finite ε-threshold, computational roundoff errors, measurements inaccuracies, etc. Hence, we construct an index that quantifies the degree

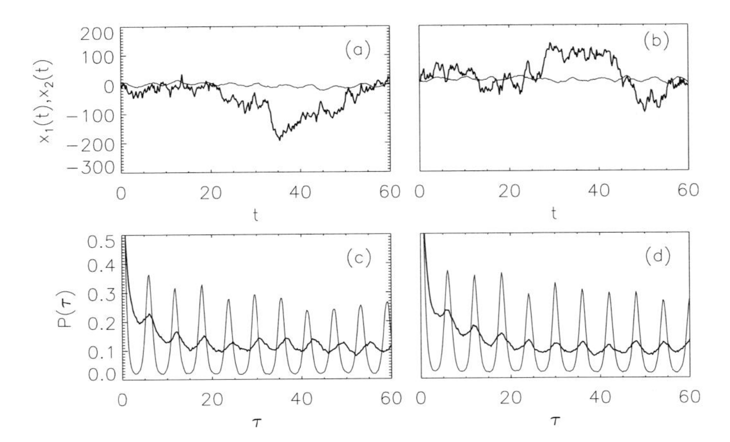

Fig. 10.10: (a,b) Segments of the x-components of the trajectories of two mutually coupled Rössler systems in phase coherent regime ($a = 0.16$) strongly contaminated by colored noise. A realization of $r_{t+1} = 0.99r_t + 10\eta_t$ and $s_{t+1} = 0.982s_t + \xi_t$ were respectively added to each component of the Rössler systems. (a) non-PS ($\mu = 0.02$). (b) PS ($\mu = 0.05$). (c) $P(\tau)$ for the two noisy Rössler for $\mu = 0.02$ (non-PS), (d) $P(\tau)$ for the two noisy Rössler for $\mu = 0.05$ (PS). Solid line: system 1, dashed line: system 2.

of similarity between the respective recurrences of both systems. It compares the recurrences of each point of the first system with the local recurrences of the second system. This index has the advantage that it distinguishes rather well between non-PS, PS, and GS.

This index is based on the average probability of joint recurrence over time, given by

$$RR^{x,y} = \frac{1}{N^2} \sum_{i,j=1}^{N} \Theta(\varepsilon^x - \|x_i - x_j\|)\Theta(\varepsilon^y - \|y_i - y_j\|) . \tag{10.9}$$

If both systems x and y are independent from each other, then the average probability of a joint recurrence[3] is given by $RR^{x,y} = RR^x RR^y$. If the oscillators are on the other hand in GS, we expect an approximate identity of their respective recurrences, and hence $RR^{x,y} = RR^x = RR^y$ [31, 32].

For the computation of the recurrence matrix in the case of essentially different systems that undergo GS, it is more appropriate to use a fixed number of nearest neighbors for each column in the matrix, following the idea presented in [46–48], than using a fixed threshold. This means that the threshold is different for each column in the RP, but subjected to the following condition $\sum_{j=1}^{N} \Theta(\varepsilon^i - \|x_i - x_j\|) = A \; \forall i$, where A is the fixed number of nearest neighbors.

[3] Note that the average probability of a joint recurrence is the recurrence rate of the joint recurrence plot (JRP) [51].

We can automatically fix the RR by means of $RR = AN/N^2 = A/N$, and using the same A for each subsystem $\mathbf{x}$ and $\mathbf{y}$, $RR^x = RR^y = RR$.

Hence, the coefficient $S = \frac{RR^{x,y}}{RR}$ is an index for GS that varies from RR to 1: It is approximately RR for independent systems, and it is close to 1 for systems in GS. However, with the index S we would not detect lag synchronization (LS) ($\mathbf{y}(t+\tau) = \mathbf{x}(t)$). Since LS can be considered as a special case of GS [52], it would be desirable to have an index that also detects LS. For this reason, we include a time lag τ in the similarity and introduce the following quotient:

$$S(\tau) = \frac{1/N^2 \sum_{i,j=1}^{N} \Theta(\varepsilon_x^i - \|\mathbf{x}_i - \mathbf{x}_j\|)\Theta(\varepsilon_y^i - \|\mathbf{y}_{i+\tau} - \mathbf{y}_{j+\tau}\|)}{RR}, \tag{10.10}$$

where the thresholds ε_x^i and ε_y^i fullfil the following conditions: $\sum_{j=1}^{N} \Theta(\varepsilon_x^i - \|\mathbf{x}_i - \mathbf{x}_j\|) = A$ and $\sum_{j=1}^{N} \Theta(\varepsilon_y^i - \|\mathbf{y}_i - \mathbf{y}_j\|) = A \ \forall i$. Then, we choose the maximum value of $S(\tau)$ and normalize

$$JPR = \max_{\tau} \frac{S(\tau) - RR}{1 - RR}. \tag{10.11}$$

We denote this index by JPR because it is based on the average joint probability of recurrence. Since $S(\tau)$ varies between RR and 1, JPR ranges from 0 to 1. The value of RR is a free parameter and its choice depends on the case under study. We consider rather low values of RR, e.g., 1% or 2% as appropriate.

10.3.1 Examples of Application

In this section we show two examples of chaotic systems that undergo GS and compute for them the recurrence-based index JPR (Eq. (10.11)).

1. First we consider a Lorenz system driven by a Rössler system. The equations of the driving system are

$$\begin{aligned} \dot{x}_1 &= 2 + x_1(x_2 - 4) \\ \dot{x}_2 &= -x_1 + x_3 \\ \dot{x}_3 &= x_2 + 0.45x_3, \end{aligned} \tag{10.12}$$

and following are the equations of the driven system:

$$\begin{aligned} \dot{y}_1 &= -\sigma(y_1 - y_2) \\ \dot{y}_2 &= ru(t) - y_2 - u(t)y_3 \\ \dot{y}_3 &= u(t)y_2 - by_3, \end{aligned} \tag{10.13}$$

where $u(t) = x_1(t) + x_2(t) + x_3(t)$ and the parameters were chosen as follows: $\sigma = 10$, $r = 28$, and $b = 2.666$. In [53] it was shown that the systems given by Eqs. (10.12) and (10.13) are in GS, since the driven Lorenz system is asymptotically stable.

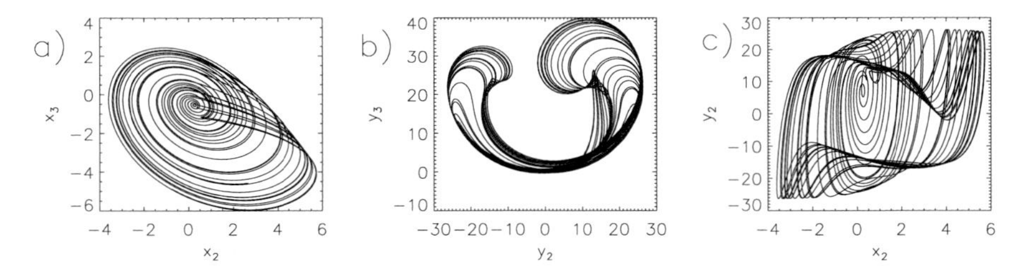

Fig. 10.11: Projection of the Rössler driving system (a), the driven Lorenz system (b) and the diagram x_2 versus y_2 of Eqs. (10.12) and (10.13) (c).

To illustrate that they are completely different systems and that they are not in LS or even complete synchronized, Fig. 10.11 shows the projections of the system (Eqs. (10.12)) (a), of the system (Eqs. (10.13)) (b) and the x_2 versus y_2 diagram (c).

When dealing with experimental time series, usually only one observable of the system is available. Hence, we perform the analysis with just one component of each system to illustrate the applicability of the proposed method (we use 10 000 data points with a sampling time interval of 0.02 s). In this example, we take x_3 and y_3 as observables, respectively. Then, we reconstruct the phase space vectors using delay coordinates [54]. For the subsystem $\mathbf{x}$ we obtain the following embedding parameters [55]: delay time $\tau = 5$ and embedding dimension $m = 3$. For the subsystem $\mathbf{y}$ we find: $\tau = 5$ and $m = 7$. The corresponding RPs and JRP are represented in Fig. 10.12.

We see that despite of the essential difference between both subsystems, their RPs are very similar (Fig. 10.12(a) and (b)). Therefore, the structures are reflected also in the JRP and consequently, its recurrence rate is rather high. In this case, with the choice RR = 0.02 we obtain JPR = 0.605 (the value of JPR is similar for other choices of RR).

In order to illustrate the second case, where both subsystems are independent (Fig. 10.13), we compute the RP of the Rössler system (Eqs. (10.12)) and of the independent Lorenz system,[4] as well as their JRP (Fig. 10.14). Note that the mean probability over time for a joint recurrence is very small, as the JRP has almost no recurrence points. In this case, one obtains JPR = 0.047 using embedding parameters $\tau = 5$ and $m = 3$ for both systems, and RR = 0.02.

For $\sigma = 10$ and $b = 8/3$ they display chaotic behavior.

2. Two mutually coupled Rössler systems (Eqs. (10.1)): for the coupling strength $\mu = 0.11$ both oscillators are in LS, as can be seen from Fig. 10.15.

[4] The Lorenz equations are given by $\dot{x} = -\sigma x + \sigma y$, $\dot{y} = -xz + rx - y$, $\dot{z} = xy - bz$. For $\sigma = 10$ and $b = 8/3$ they display chaotic behavior.

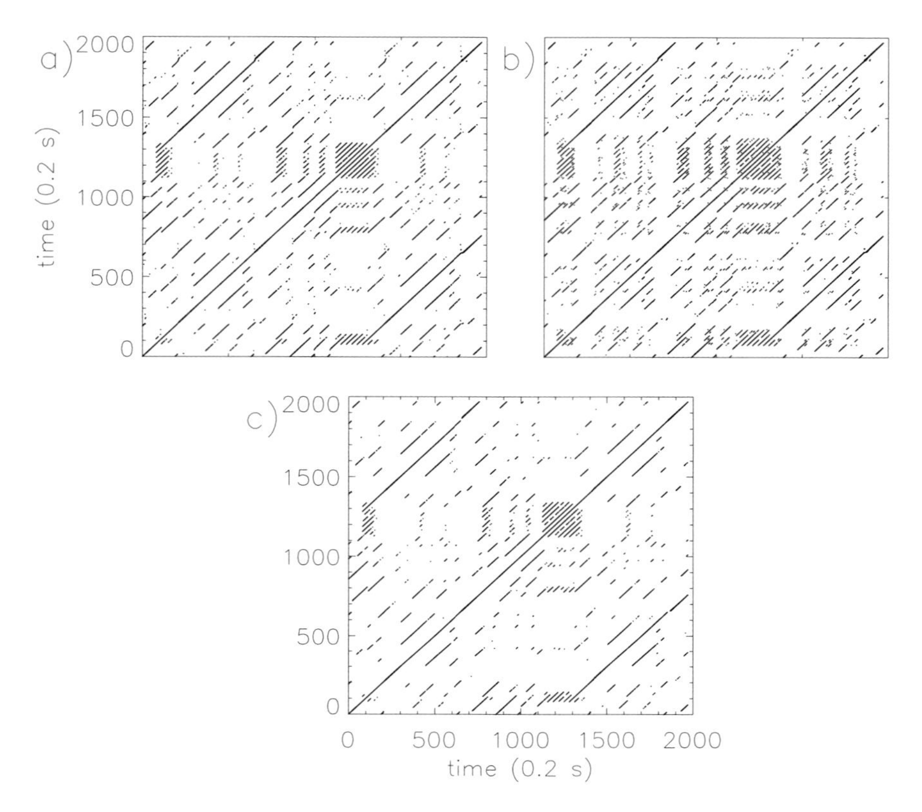

Fig. 10.12: (a) RP of the Rössler subsystem (Eqs. (10.12)). (b) RP of the driven Lorenz subsystem (Eqs. (10.13)). (c) JRP of whole system (Eqs. (10.12) and (10.13)).

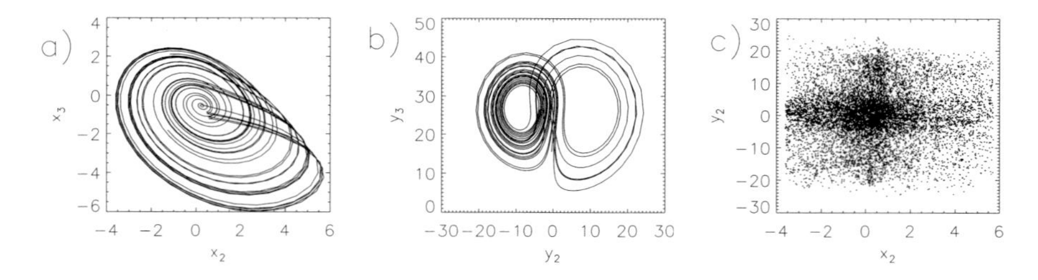

Fig. 10.13: Projection of the Rössler system (Eqs. (10.12)) (a), the independent Lorenz system (see footnote 4) (b) and the diagram x_2 versus y_2, where x_2 is the second component of the Rössler system and y_2 is the second component of the independent Lorenz system (c).

In this case, the RPs of both subsystems are obviously almost identical, except for a displacement on τ in the diagonal direction. Computing the index fol-

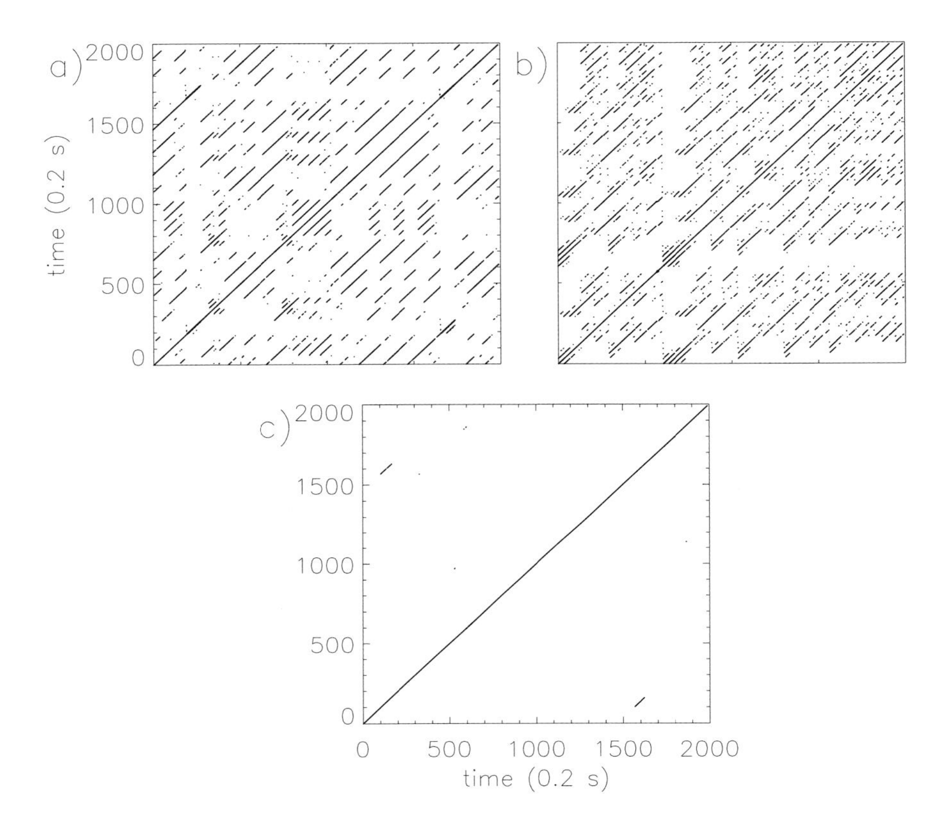

Fig. 10.14: (a) RP of the Rössler subsystem (Eqs. (10.12)). (b) RP of the independent Lorenz system (see footnote 4) (c) JRP of whole system.

lowing Eq. (10.11), we obtain the value JPR $= 0.988$ (JPR in this case is not exactly 1), because we do not have perfect LS, i.e., $\mathbf{x}(t+\tau) \simeq \mathbf{y}(t)$ [52]). For a smaller coupling strength $\mu = 0.02$ the oscillators are not in LS anymore. The obtained value in this is case JPR $= 0.014$.

10.4 Transitions to Synchronization

We have seen in the previous sections that the indices CPR and JPR clearly distinguish between oscillators in PS and oscillators which are not in PS, respectively of GS. On the other hand, the synchronization indices should not only distinguish between synchronized and nonsynchronized regimes, but also clearly indicate the onset of PS, respectively of GS.

In order to demonstrate that the recurrence-based indices fulfill this condition, we exemplify their application in the two cases: Two mutually coupled Rössler systems in a phase coherent regime and in a non-phase-coherent funnel regime (Eqs. (10.1)) with $a = 0.16$, respectively $a = 0.2925$). In both the cases we increase

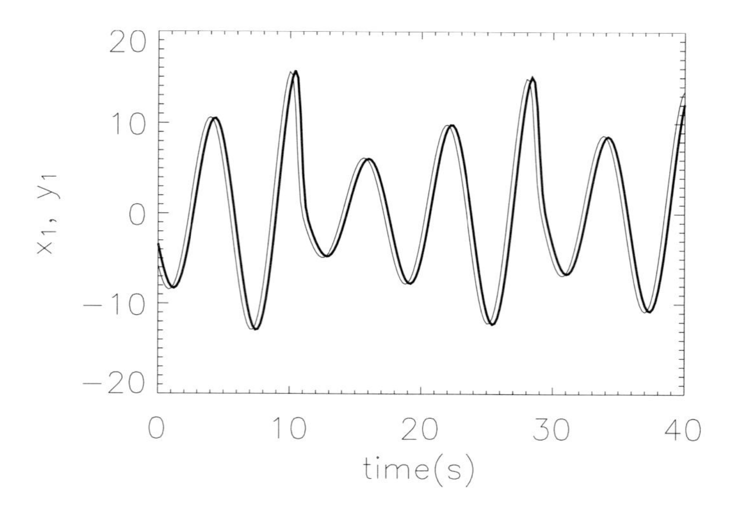

Fig. 10.15: Example of lag synchronization: It is clearly seen that x_1 (bold line) goes behind y_1 (solid line). It holds: $x_1(t+\tau) = y_1(t)$, with $\tau = 4$.

the coupling strength μ continuously and compute for each value of μ the indices CPR and JPR.

On the other hand, in the phase coherent case for a not too large but fixed frequency mismatch between both oscillators and increasing coupling strength, the transitions to PS and LS are reflected by the Lyapunov spectrum [1–4].[5] If both oscillators are not in PS, there are two zero Lyapunov exponents (λ_3 and λ_4), which correspond to the (almost) independent phases. Increasing the coupling strength, the fourth Lyapunov exponent λ_4 becomes negative (Fig. 10.16(c)), indicating the onset of PS. For higher coupling strengths, the second Lyapunov exponent λ_2 crosses zero, which indicates the establishment of a strong correlation between the amplitudes (Fig. 10.16(c)). This last transition occurs almost simultaneously with the onset of LS [52]. Therefore, we compute for our two examples also λ_2 and λ_4 in order to validate the results obtained with CPR and JPR.

In Fig. 10.16 the indices CPR (a) and JPR (b) are represented for increasing coupling strength μ for the phase coherent case. In (c) λ_2 and λ_4 are shown in dependence on μ.

By means of CPR, the transition to PS is detected when CPR becomes of the order of 1. We see from Fig. 10.16(a) that the transition to PS occurs at approximately $\mu = 0.037$, in accordance with the transition of the fourth Lyapunov exponent λ_4 to negative values. The index JPR shows three plateaus in dependence on the coupling strength (Fig. 10.16(b)), indicating the onset of PS at the beginning of the second one. On the other hand, JPR clearly indicates the onset

[5] For other cases, e.g., for a fixed coupling strength and decreasing frequency mismatch, or for a large frequency mismatch and increasing coupling strength, the transition to PS is not always simply reflected in the Lyapunov spectrum [17, 51].

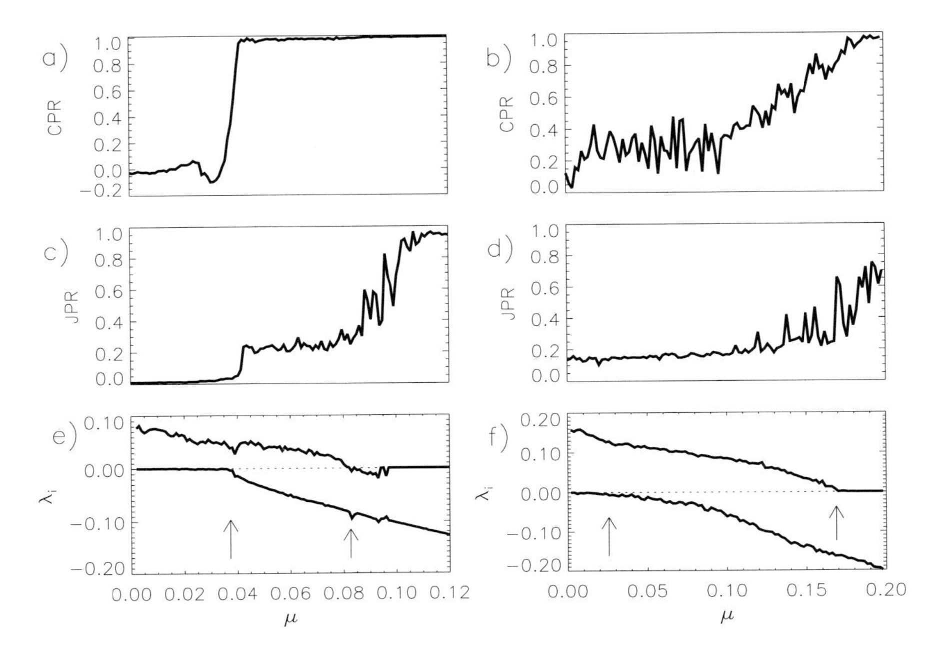

Fig. 10.16: CPR index, JPR index and λ_2 and λ_4 as functions of the coupling strength μ for two mutually coupled Rössler systems in phase coherent regime (a,c,e) and in funnel one (b,d,f). The dotted zero line in (e) and (f) is plotted to guide the eye. Here, we choose ε corresponding to 10% recurrence points in each RP.

of LS because it becomes nearly one (third plateau) at approximately $\mu = 0.1$ (Fig. 10.16(b)), after the transition from hyperchaoticity to chaoticity, which takes place at approximately $\mu = 0.08$ (Fig. 10.16(c)). Between $\mu = 0.08$ and $\mu = 0.1$, the values of JPR have large fluctuations. This reflects the intermittent LS [1–4], where LS is interrupted by intermittent bursts of no synchronization.

Now we regard the more complex case of two coupled Rössler systems in the non-phase-coherent funnel regime, where the direct application of the Hilbert transformation is not possible [17]. In Fig. 10.16 the coefficients CPR and JPR are represented for this case in dependence on the coupling strength μ. Again, λ_2 and λ_4 are also shown (Fig. 10.16(f)).

First, note that for $\mu > 0.02$, λ_4 has already passed to negative values (Fig. 10.16(f)). However, CPR is still rather low, indicating that both oscillators are not in PS yet. CPR does not indicate PS until $\mu = 0.18$ (Fig. 10.16(d)), as found with other techniques [17]. Furthermore, we see from Fig. 10.16(f) that λ_2 vanishes at $\mu \sim 0.17$. This transition indicates that the amplitudes of both oscillators become highly correlated. At approximately the same coupling strength, JPR reaches rather high values, indicating the transition to GS (Fig. 10.16(e)). Then, according to the index CPR the transition to PS occurs after the onset of

GS. This is a general result that holds for systems with a strong phase diffusion, as reported in [17]. For highly non-phase-coherent systems, there is more than one characteristic time scale. Hence, a rather high coupling strength is necessary in order to obtain phase locking of both oscillators. Hence, PS is not possible without a strong correlation in the amplitudes. PS for such non-phase-coherent systems has been recently found and analyzed in electrochemical oscillators [56] and in El Niño-Monsoon system [57].

Note that the synchronization indices presented in these sections based on recurrences are applicable to multivariate time series.

10.5 Twin Surrogates to Test for PS

As we have mentioned in Section 10.1, another essential problem in the synchronization analysis of observed time series is the construction of an appropriate hypothesis test to test for PS. Several approaches in this direction have been published [58, 59]. Usually, these are linear surrogates based on randomization of the Fourier phases [60, 61]. They mimic the individual spectra of the two components of the original bivariate series as well as their cross-spectrum, i.e., their linear properties, but not the higher order moments. In this case, the corresponding null hypothesis is that the putative synchronization in the underlying system can be explained by a bivariate linear stochastic process. The specificity of this test is not always satisfactory, because the concept of PS assumes the mutual adaption of self-sustained oscillators, i.e., nonlinear deterministic systems. On the other hand, pseudo-periodic surrogates (PPS) have been proposed to test the null hypothesis that an observed time series is consistent with an uncorrelated noise-driven periodic orbit [62]. The PPS are in a certain sense closer to the surrogates needed to test for PS as they correspond to trajectories of a deterministic system with noise, but they are still not appropriate to test for PS, as they are not able to model chaotic oscillators. Therefore, we present a technique for the generation of surrogates which are consistent with the null hypothesis of a trajectory of the same underlying system, but starting at different initial conditions [63]. Hence, they can also be used to test for PS in the case of chaotic oscillators.

The main idea consists in exchanging one original subsystem with one surrogate. Then, if the synchronization index obtained for the original system is not significantly different from the one computed for the exchanged subsystems, we have no sufficient evidence to claim synchronization (see Fig. 10.17). One could argue that the same can be achieved using different realizations of the same process and exchanging the subsystems. However, there are cases where it is not possible to measure several realizations, like, e.g., in geophysical systems.

The construction of the surrogates we present in this section is also based on the recurrence matrix (10.4). It is important to note that if the recurrence matrix is computed from a univariate time series, it contains all topological information about the underlying attractor, which therefore can be reconstructed from it [29].

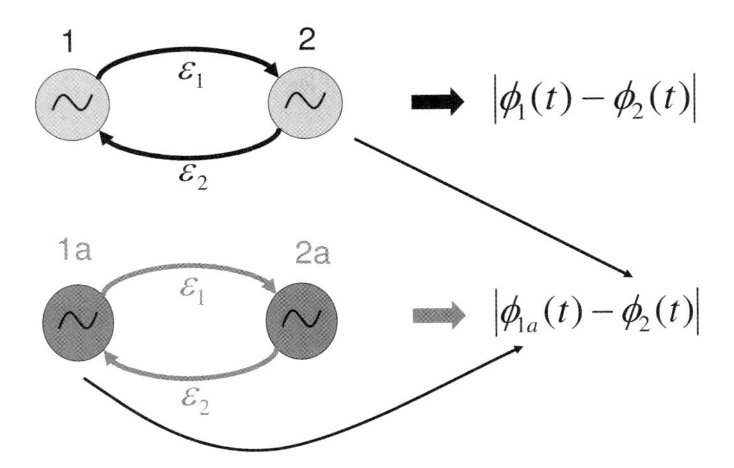

Fig. 10.17: This diagram represents the main idea using twin surrogates to test for PS.

Hence, a first idea for the generation of surrogates is to change the structures in a RP consistently with the ones produced by the underlying dynamical system. In this way one could reconstruct a new realization of the trajectory from the modified $R_{i,j}$. However, one cannot arbitrarily interchange columns in an RP, because such a modification changes the distribution of diagonal lines and hence the entropy and predictability of the system [28].

Therefore, we propose a modified approach. In general, in an RP there are identical columns, i.e., $R_{k,i} = R_{k,j}\ \forall k$ [28]. Thus, there are points which are not only neighbors (i.e., $\|\mathbf{x}_i - \mathbf{x}_j\|_\infty < \varepsilon$), but which also share the same neighborhood. Reconstructing the attractor from an RP, the respective neighborhoods of these points cannot help to distinguish them, i.e., from this point of view they are identical. This is why we will call them *twins*. Twins are special points of the time series as they are indistinguishable considering their neighborhoods but in general different and hence, have different pasts and—more important—different futures. The key idea of how to introduce the randomness needed for the generation of surrogates of a deterministic system is that one can jump randomly to one of the possible futures of the existing twins.

A surrogate trajectory $\mathbf{x}^s(i)$ of $\mathbf{x}(i)$ with $i = 1, \ldots, N$ is then generated in the following way:

1. Identify all pairs of twins.
2. Choose an arbitrary starting point, say $\mathbf{x}^s(1) = \mathbf{x}(k)$.
3. If $\mathbf{x}(k)$ has no twin, the next point of the surrogate trajectory is $\mathbf{x}^s(2) = \mathbf{x}(k+1)$.
4. If $\mathbf{x}(k)$ has a twin, say $\mathbf{x}(m)$, then one can go to either $\mathbf{x}(k+1)$ or to $\mathbf{x}(m+1)$, i.e., $\mathbf{x}^s(2) = \mathbf{x}(k+1)$ or $\mathbf{x}^s(2) = \mathbf{x}(m+1)$ with equal probability[6].

Steps three and four are then iterated until the surrogate time series has the same length as the original one.

This algorithm creates twin surrogates (TS) which are shadows of a (typical) trajectory of the system [64]. In the limit of an infinitely long original trajectory, its surrogates are characterized by the same dynamical invariants and the same attractor. However, if the measure of the attractor can be estimated from the observed finite trajectory reasonably well, its surrogates share the same statistics. Also their power spectra and correlation functions are consistent with the ones of the original system. TS do not only seem to give reasonable results for deterministic systems; the TS of for example an ARMA process also show the typical behavior of a linear Gaussian process.

Next, we use the TS to test for PS. The idea behind this approach is similar to the one by means of "natural surrogates" in the mother–fetus heartbeat synchronization [15]. Suppose that we have two coupled self-sustained oscillators $x_1(t)$ and $x_2(t)$. Then, we generate M TS of the joint system, i.e., $x_1^{s_i}(t)$ and $x_2^{s_i}(t)$, with $i = 1, \dots, M$. These surrogates are independent copies of the joint system, i.e., trajectories of the whole system beginning at different initial conditions. Note that the coupling between $x_1(t)$ and $x_2(t)$ is also mimicked by the surrogates. Next, we compute the differences between the phases of the original system $\Delta\Phi(t) = \Phi_1(t) - \Phi_2(t)$ applying, e.g., the analytical signal approach [1–4] and compare them with $\Delta\Phi^{s_i}(t) = \Phi_1(t) - \Phi_2^{s_i}(t)$ (one can also consider $\Phi_1^{s_i}(t) - \Phi_2(t)$). Then, if $\Delta\Phi(t)$ does not differ significantly from $\Delta\Phi^{s_i}(t)$ with respect to some index for PS, the null hypothesis cannot be rejected and hence, we do not have enough evidence to state PS.

As a test case, we consider two nonidentical, mutually coupled Rössler oscillators

$$\begin{aligned} \dot{x}_{1,2} &= -(1 \pm \nu) y_{1,2} - z_{1,2} + \varepsilon(x_{2,1} - x_{1,2}), \\ \dot{y}_{1,2} &= (1 \pm \nu) x_1 + 0.15 y_{1,2}, \\ \dot{z}_{1,2} &= 0.2 + z_{1,2} + z_{1,2}(x_{1,2} - 10), \end{aligned} \tag{10.14}$$

where $\nu = 0.015$ denotes the frequency mismatch. In this "active experiment", we vary the coupling strength ε from 0 to 0.08 and compute a PS index for the original trajectory for each value of ε. Next we generate 200 TS and compute the PS index between the measured first oscillator and the surrogates of the second one. As PS index we use the mean resultant length R of complex phase vectors [65, 66], which is motivated by Kuramoto's order parameter [67]

$$R = \left| \frac{1}{N} \sum_{t=1}^{N} \exp\left(i\Delta\Phi(t)\right) \right| . \tag{10.15}$$

It takes on values in the interval from 0 (non PS) to 1 (perfect PS) [65, 66]. Let R^{s_i} denote the PS index between the first oscillator and the surrogate i of the second

[6] If triplets occur one proceeds analogously.

one. To reject the null hypothesis at a significance value α, R must be larger than $(1-\alpha)\cdot 100$ percent of all R^{s_i}. Note that this corresponds to computing the significance level from the cumulative histogram at the level $(1-\alpha)$.

Figure 10.18(a) shows the results for R of the original system (bold line) and the 1% (solid) significance level. Figure 10.18(b) displays the difference between R of the original system and the 1%, 2% and 5% significance level. For $\varepsilon < 0.025$, R of the original system is, as expected, below the significance levels and hence the difference is negative, and for higher values of ε the curves cross (the difference becomes positive). This is in agreement with the criterion for PS via Lyapunov exponents λ_i [1–4]: λ_4 becomes negative at $\varepsilon \sim 0.028$ (Fig. 10.18(b)), which approximately coincides with the intersection of the curve of R for the original system and the significance level (zero-crossing of the curves in Fig. 10.18(b)). Therefore, we recognize successfully the PS region by means of the TS.

Note that also the significance limit increases when the transition to PS occurs (Fig. 10.18(a)). As the TS mimic both the linear and nonlinear characteristics of the system, the surrogates of the second oscillator have in the PS region the same mean frequency as the first original oscillator. Hence R^{s_i} is rather high. However, $\Phi_1(t)$ and $\Phi_2^{s_i}(t)$ do not adapt to each other, as they are independent. Hence, the value of R for the original system is significantly higher than the R^{s_i}. We state in conclusion that even though the obtained value for a normalized PS index is higher than 0.97 (right side of Fig. 10.18(a)), this does not offer conclusive evidence for PS. Hence, *the knowledge of the PS index alone does not provide sufficient evidence for PS*. Note that the more phase coherent the oscillators are, the more difficult it is to decide whether they are in PS or not. A certain phase diffusion, which allows to measure the adaptation of the phases of the interacting oscillators is necessary to detect PS. However, the test based on the TS reveals whether there is enough evidence for PS.

Next, we perform an analysis of the specificity and sensitivity of the test for $\varepsilon = 0$ and $\nu = 0$. For 100 random initial conditions of the Rössler system and a significance level of $\alpha = 1\%$, the null hypothesis was erroneously rejected only in 1 out of the 100 cases. This is a rather auspicious result, as due to the identical frequencies, it is extremely difficult to recognize that there is no PS in this case [68]. In the case of $\varepsilon = 0.02$ (e.g., no PS) and $\nu = 0.015$, there were no erroneous rejections of the null hypothesis. Finally, for PS ($\varepsilon = 0.045$ and $\nu = 0.015$), in all 100 test runs the null hypothesis was correctly rejected. These results indicate that the specificity and sensitivity of the test are good.

10.6 Application to Fixational Eye Movements

Next we apply the recurrence approach to check fixational movements of left and right eyes for PS. During fixation of a stationary target our eyes perform small involuntary and allegedly erratic movements to counteract retinal adaptation. If these eye movements are experimentally suppressed, retinal adaptation to the

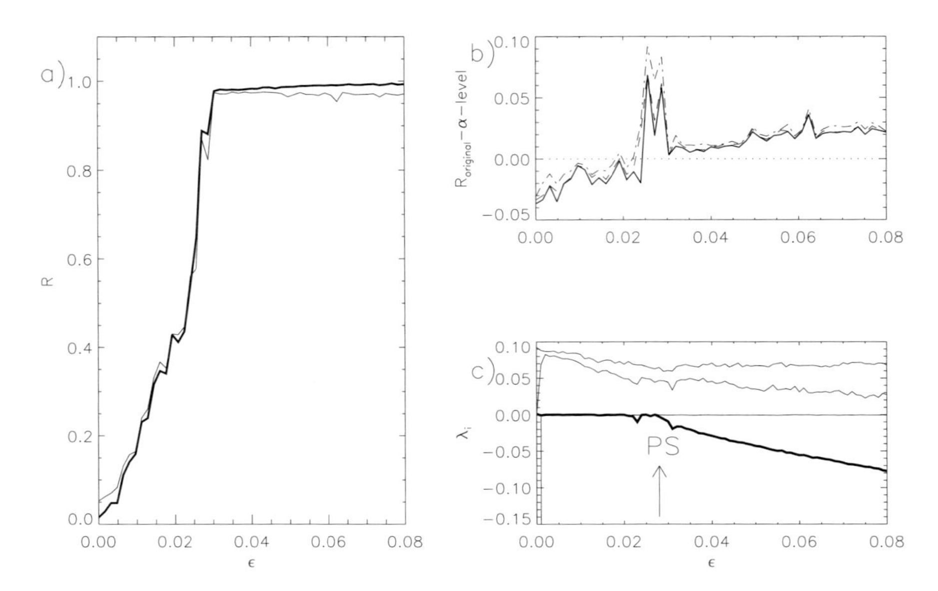

Fig. 10.18: (a) R of the original two mutually coupled Rössler systems with a frequency mismatch of $\nu = 0.015$ (bold) and significance level of 1% (solid). (b) Difference between R of the original data and significance level of 1% (solid), 2% (dashed) and 5% (dashed-dotted). The zero line is plotted (dotted) to guide the eye. (c) Four largest Lyapunov exponents for the six-dimensional system considered. λ_4 is highlighted and the arrow indicates the transition to PS.

constant input induces very rapid perceptual fading [69, 70]. Moreover, statistical correlations show a timescale separation from persistence to antipersistence [71]. Persistence on the short timescale counteracts retinal fading, whereas antipersistence on the long timescale contributes to stability of ocular disparity. According to current textbook knowledge, the fixational movements of the left and right eyes are correlated very poorly at best [72]. Therefore, it is highly desirable to examine these processes from a perspective of PS. We analyze the data of two subjects. Each performed three trials, in which they fixated a small stimulus (black square on a white background, 3× pixels on a computer display) with a spatial extent of 0.12°, or 7.2 arc · min. Eye movements were recorded using an EyeLink-II system (SR Research, Osgoode, Ontario, Canada) with a sampling rate of 500 Hz and an instrument spatial resolution less than 0.005°. Figure 10.19 shows a segment of the horizontal (a) and vertical (b) component of the eye movements for one person.

The data were first high-pass filtered applying a difference filter $\tilde{x}(t) = x(t) - x(t - \tau)$ with $\tau = 40\,\text{ms}$ in order to eliminate the slow drift of the data. After this filtering, we find an oscillatory trajectory, which has maximum spectral power in the frequency range between 3 and 8 Hz (Fig. 10.20(a) and (b)). How-

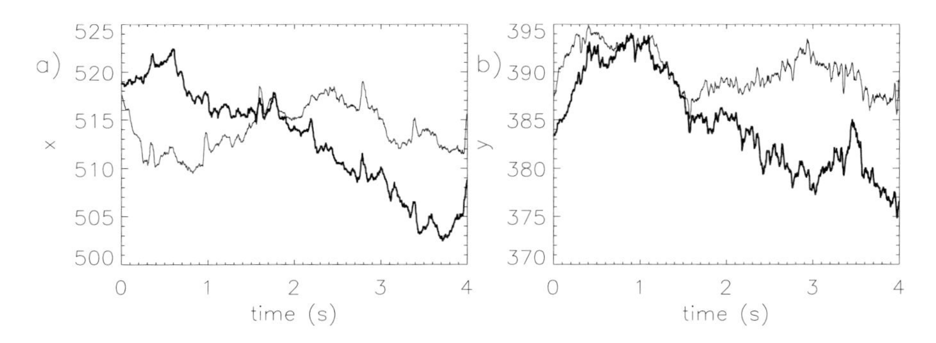

Fig. 10.19: Simultaneous recording of left (bold) and right (solid) fixational eye movements (a) horizontal component (b) vertical component.

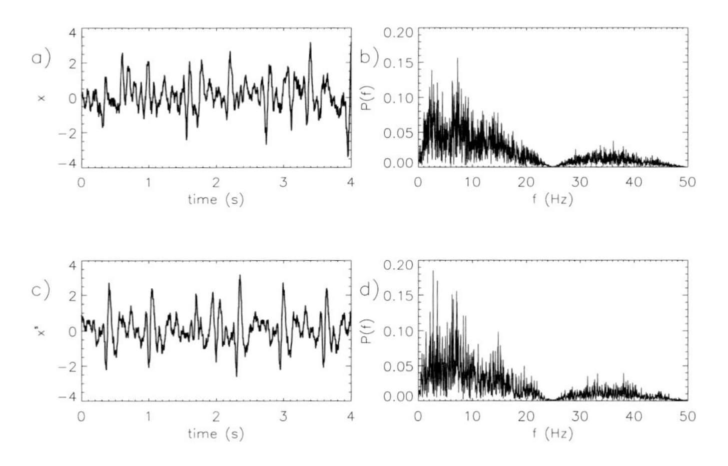

Fig. 10.20: Filtered horizontal component of the left eye of one participant (a) and its corresponding periodogram (b). In (c) the horizontal component of one surrogate of the left eye is represented and in (d) its corresponding periodogram.

ever, the trajectories of the eyes are rather noisy and non-phase-coherent. Therefore, it is cumbersome to estimate the phase of these data. Hence we apply the recurrence-based measure CPR introduced in Section 10.2 and we obtain the values displayed in the first column of Table 10.2. First, we observe that the variability between the different trials is smaller for the first participant as for the second one. Furthermore, the values of CPR are rather high for the first participant but not so high for the second one. Hence, a hypothesis test should be performed in order to get statistically significant results.

Therefore, we compute 200 twin surrogates of the left eye's trajectory. In Fig. 10.20(c) the horizontal component of one surrogate is represented. At a first glance, the characteristics of the original time series are well reproduced by the

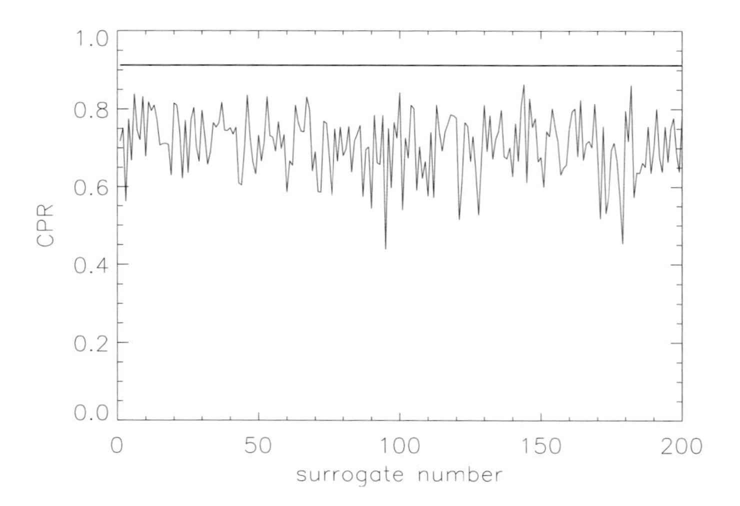

Fig. 10.21: Result of the test performed for one trial of one participant. The PS index for the original data (bold line) is significantly different from the one of the surrogates (solid).

twin surrogate. In Fig. 10.20(d) the corresponding periodogram is displayed. It is also noteworthy that the structure of the original curve (Fig. 10.20(b)) is also qualitatively reproduced. The periodogram of the twin surrogate is of course not identical with the one of the original time series. This is consistent with the null hypothesis of another realization of the same underlying process, respectively another trajectory starting at different initial conditions of the same underlying dynamical system.

Now, we compute the recurrence-based synchronization index CPR^{s_i} between the twin surrogates of the left eye and the measured right eye's trajectory. In Fig. 10.21 the results of the test of one trial are visualized.

The second column of Table 10.2 summarizes the results for both subjects and all trials.

In all cases, the PS index of the original data is significantly different from the ones of the surrogates, which strongly indicates that the concept of PS can be successfully applied to study the interaction between the trajectories of the left and right eyes during fixation. This result also suggests that the physiological mechanism in the brain that produces the fixational eye movements controls both eyes simultaneously, i.e., there might be only one center in the brain that produces the fixational movements in both eyes or a close link between two centers. Our finding of PS between left and right eyes is in good agreement with current knowledge of the physiology of the oculomotor circuitry. In a single-cell study, 66 % of abducens motor neurons fired in relation to the movements of either eye, while premotor neurons in the brainstem encode monocular movements [73]. Thus, motor neurons—as the final common pathway of neural control of eye

Tab. 10.2: Results for the test for PS between the trajectories of the left and right fixational eye movements performed for three trials for the two participants. Two hundred surrogates were used for the test. The null hypothesis was rejected in all cases at a 2% level.

Participant	CPR of the original data	Null hypothesis
M.R.	0.9112	Rejected
	0.9432	Rejected
	0.9264	Rejected
M.T.	0.6080	Rejected
	0.4844	Rejected
	0.3520	Rejected

movements—are candidates for the synchronization of left and right fixational movements. Furthermore, we are interested in whether the fixational movements in the horizontal and vertical direction of one eye are synchronized. Horizontal and vertical saccadic eye movements are controlled in two spatially distinct brainstem nuclei [74]. Therefore, we can expect that, on the level of fixational eye movements, horizontal and vertical components are independent. Applying the synchronization index CPR between the x- and y-component of the left eye of each participant for each trial and generate 200 surrogates of the two-dimensional trajectory of the left eye. Then we compare the synchronization index CPR^{s_i} between the original x-component and the y-component of the surrogates. We find in all but one cases that CPR is not significantly different from CPR^{s_i} (see Table 10.3). Hence, we do not have evidence to claim synchronization between the horizontal and vertical components of the eye movements, as expected.

Tab. 10.3: Results for the test for PS between the horizontal and vertical components of fixational movements of one eye performed for three trials for the two participants. 200 surrogates were used for the test. In all cases but one, we failed to reject the null hypothesis at a 2 % level.

Participant	CPR of the original data	Null hypothesis
M.R.	0.3746	Not rejected
	0.6103	Not rejected
	0.4812	Rejected
M.T.	0.4681	Not rejected
	0.3194	Not rejected
	0.4172	Not rejected

10.7 Conclusions

In conclusion, we have presented solutions to two main problems of the synchronization analysis of measured time series: The detection of PS in non-phase-coherent systems and the hypothesis test for PS, which is interesting especially for passive experiments, where the coupling strength between the two subsystems cannot be varied systematically.

We have given solutions to these two problems based on the concept of recurrence in phase space. First, we have shown that by means of the recurrence properties it is possible to detect indirectly PS even in the case of non-phase-coherent and strong noisy time series. Furthermore, it is also possible to detect GS by means of recurrences. Second, the method of twin surrogates has been presented, which is also based on recurrence, and we have shown that it can be used to test for PS.

We have used the well studied system of two mutually coupled Rössler oscillators in order to validate the techniques proposed. Furthermore, we have tested for PS in experiments of binocular fixational movements and found that the left and right eyes are in PS, in agreement with physiological results about the functional role of motor neurons in the final common pathway for the control of eye movements. Hence, we have shown that the techniques proposed are also applicable to rather noisy observed time series.

Acknowledgements

This work has been suported by the DFG Priority Program 1114 and the "Internationales Promotionskolleg—Helmholtz Center for the Study of Mind and Brain Dynamics" at the University of Potsdam.

References

[1] M. Rosenblum, A. Pikovsky, and J. Kurths. *Phys. Rev. Lett.*, 76:1804, 1996.

[2] A. Pikovsky, M. Rosenblum, G. Osipov, and J. Kurths. *Physica D*, 104:219, 1997.

[3] A. Pikovsky, M. Rosenblum, and J. Kurths. *Synchronization*, volume 12 of *Cambridge Nonlinear Science Series*. Press Syndicate of the University of Cambridge, Cambridge, UK, 2001.

[4] S. Boccaletti, J. Kurths, G. V. Osipov, D. Valladares, and C. Zhou. *Phys. Rep.*, 366:1, 2002.

[5] R. C. Elson et al. *Phys. Rev. Lett.*, 81:5692, 1998.

[6] P. Tass et al. *Phys. Rev. Lett.*, 81:3291, 1998.

[7] C. M. Ticos et al. *Phys. Rev. Lett.*, 85:2929, 2000.

[8] V. Makarenko and R. Llinas. *Proc. Natl. Acad. Sci. USA*, 95:15474, 1998.

[9] B. Blasius, A. Huppert, and L. Stone. *Nature*, 399:354, 1999.

[10] C. Schäfer et al. *Nature*, 392:239, 1998.

[11] D. J. DeShazer et al. *Phys. Rev. Lett.*, 87:044101, 2001.

[12] S. Boccaletti et al. *Phys. Rev. Lett.*, 89:194101, 2002.

[13] I. Kiss and J. Hudson. *Phys. Rev. E*, 64:046215, 2001.

[14] H. Poincaré. *Acta Math.*, 13:1, 1890.

[15] P. Van Leeuwen et al. *BMC Physiol.*, 3:2, 2003.

[16] M. G. Rosenblum, A. S. Pikovsky, J. Kurths, G. V. Osipov, I. Z. Kiss, and J. L. Hudson. *Phys. Rev. Lett.*, 89:264102, 2002.

[17] G. V. Osipov, B. Hu, C. Zhou, M. V. Ivanchenko, and J. Kurths. *Phys. Rev. Lett.*, 91:024101, 2003.

[18] G. Fisher. *Plane Algebraic Curves*. American Mathematical Society, Providence, Rhode Island, 2001.

[19] C. Sparrow. *The Lorenz Equations: Bifurcations, Chaos, and Strange Attractors*. Springer, Berlin, Heidelberg, 1982.

[20] R. N. Madan. *Chua Circuit: A Paradigm for Chaos*. World Scientific, Singapore, 1993.

[21] W. Lauterborn, T. Kurz, and M. Wiesenfeldt. *Coherent Optics. Fundamentals and Applications*. Springer, Berlin, Heidelberg, New York, 1993.

[22] J. Y. Chen et al. *Phys. Rev. E*, 64:016212, 2001.

[23] J.-P. Eckmann, S. O. Kamphorst, and D. Ruelle. *Europhys. Lett.*, 4:973, 1987.

[24] C. L. Weber Jr. and J. P. Zbilut. *J. Appl. Physiol.*, 76:965, 1994.

[25] N. Marwan, N. Wessel, U. Meyerfeldt, A. Schirdewan, and J. Kurths. *Phys. Rev. E*, 66:026702, 2002.

[26] N. Marwan and J. Kurths. *Phys. Lett. A*, 302:299, 2002.

[27] M. Thiel et al. *Physica D*, 171:138, 2002.

[28] M. Thiel, M. C. Romano, P. Read, and J. Kurths. *Chaos*, 14:234, 2004.

[29] M. Thiel, M. C. Romano, and J. Kurths. *Phys. Lett. A*, 330:343, 2004.

[30] E.-H. Park, M. Zaks, and J. Kurths. *Phys. Rev. E*, 60:6627, 1999.

[31] V. S. Afraimovich, N. N. Verichev, and M. I. Rabinovich. *Izvestiya Vysshikh Uchebnykh Zavedenii Radiofizika*, 29:1050, 1986.

[32] N. F. Rulkov, M. M. Sushchik, L. S. Tsimring, and H. D. I. Abarbanel. *Phys. Rev. E*, 51 2:980, 1995.

[33] U. Parlitz and L. Kocarev. Synchronization of chaotic systems. In H. G. Schuster, editor, *Handbook of Chaos Control*. Wiley-VCH, Weinheim, 1999.

[34] L. M. Pecora and T. L. Carroll. *Phys. Rev. Lett.*, 64:821, 1990.

[35] K. M. Cuomo and A. V. Oppenheim. *Phy. Rev. Lett.*, 71:65, 1993.

[36] A. Kittel, A. Parisi, and K. Pyragas. *Physica D*, 112:459, 1998.

[37] G. D. Van Wiggeren and R. Roy. *Science*, 279:1198, 1998.

[38] K. Otsuka, R. Kawai, S.-L. Hwong, J.-Y. Ko, and J.-L. Chern. *Phys. Rev. Lett.*, 84:3049, 2000.

[39] C. W. Wu and L. O. Chua. *Int. J. Bif. Chaos*, 4:1979, 1994.

[40] T. L. Carroll and L. M. Pecora. *Physica D*, 67:126, 1993.

[41] L. Kocarev and U. Parlitz. *Phys. Rev. Lett.*, 74:5028, 1995.

[42] U. Parlitz, L. Kocarev, T. Stojanovski, and H. Preckel. *Phys. Rev. E*, 53:4351, 1996.

[43] L. Kocarev, K. S. Halle, K. Eckert, L. O. Chua, and U. Parlitz. *Int. J. Bif. Chaos*, 2:709, 1992.

[44] T. L. Carroll and L. M. Pecora. *Physica D*, 67:126, 1993.

[45] U. Parlitz, L. Junge, and L. Kocarev. *Phys. Rev. E*, 54:6253, 1996.

[46] J. Arnhold, P. Grassberger, K. Lehnertz, and C. E. Elger. *Physica D*, 134:419, 1999.

[47] R. Q. Quiroga, J. Arnhold, and P. Grassberger. *Phys. Rev. E*, 61:5142, 2000.

[48] A. Schmitz. *Phys. Rev. E*, 62:7508, 2000.

[49] S. J. Schiff, P. So, T. Chang, R. E. Burke, and T. Sauer. *Phys. Rev. E*, 54:6708, 1996.

[50] M. Wiesenfeldt, U. Parlitz, and W. Lauterborn. *Int. J. Bif. Chaos*, 11:2217, 2001.

[51] M. C. Romano, M. Thiel, J. Kurths, and W. von Bloh. *Phys. Lett. A*, 330:214, 2004.

[52] M. G. Rosenblum, A. S. Pikovsky, and J. Kurths. *Phys. Rev. Lett.*, 78:4193, 1997.

[53] L. Kocarev and U. Parlitz. *Phys. Rev. Lett.*, 76:1816, 1996.

[54] F. Takens. Detecting strange attractors in turbulence. In D. A. Rand and L.-S. Young, editors, *Dynamical Systems and Turbulence*, volume 898 of *Lecture Notes in Mathematics*. Springer, Berlin, 1980.

[55] H. Kantz and T. Schreiber. *Nonlinear Time Series Analysis*. Cambridge University Press, Cambridge, 1997.

[56] M. C. Romano, M. Thiel, J. Kurths, I. Z. Kiss, and J. Hudson. *Europhys. Lett.*, 71:466, 2005.

[57] D. Maraun and J. Kurths. *Geophys. Res. Lett.*, 32:15709, 2005.

[58] M. Palus. *Phys. Lett. A*, 235:341, 1997.

[59] M. Palus and A. Stefanovska. *Phys. Rev. E*, 67:055201(R), 2003.

[60] J. Theiler, S. Eubank, A. Longtin, B. Galdrikian, and J. D. Farmer. *Physica D*, 58:77, 1992.

[61] T. Schreiber and A. Schmitz. *Phys. Rev. Lett.*, 77:635, 1996.

[62] M. Small, D. Yu, and R. G. Harrison. *Phys. Rev. Lett.*, 87:188101, 2001.

[63] M. Thiel, M. C. Romano, J. Kurths, M. Rolfs, and R. Kliegl. *Europhys. Lett.* In press.

[64] E. Ott. *Chaos in Dynamical Systems*. Cambridge University Press, Cambridge, 1993.

[65] E. Rodriguez et al. *Nature*, 397:430, 1999.

[66] C. Allefeld and J. Kurths. *Int. J. Bif. Chaos*, 14:405, 2004.

[67] Y. Kuramoto. *Chemical Oscillations, Waves and Turbulence*. Springer, New York, 1984.

[68] M. Peifer, B. Schelter, M. Winterhalder, and J. Timmer. *Phys. Rev. E*, 72: 026213, 2005.

[69] L. A. Riggs, F. Ratliff, J. C. Cornsweet, and T. N. Cornsweet. *J. Opt. Soc. Am.*, 43:495, 1953.

[70] D. Coppola and D. Purves. *Proc. Natl. Acad. Sci. USA*, 93:8001, 1996.

[71] R. Engbert and R. Kliegl. *Psychol. Science*, 15:431, 2004.

[72] K. J. Ciuffreda and B. Tannen. *Eye Movement Basics for the Clinician*. Mosby, St. Louis, 1995.

[73] W. Zhou and W. M. King. *Nature*, 393:692, 1998.

[74] D. L. Sparks. *Nature Rev. Neurosci.*, 3:952, 2002.

11 Detecting Coupling in the Presence of Noise and Nonlinearity

Theoden I. Netoff, Thomas L. Carroll, Louis M. Pecora, and Steven J. Schiff

Establishing the presence of coupling and interaction in weakly coupled systems, especially in the presence of noise and nonlinearity, is a difficult problem. In this chapter, we explore different measures to detect a relationship between two systems. We compare the sensitivity of the different measures to stochastic coupled systems, discontinuous chaotic systems and continuous chaotic systems. We then test the robustness of the detection of coupling in the presence of additive noise. In conclusion, we find that nonlinear methods are more sensitive to detecting coupling under ideal conditions. However, in the presence of noise, linear techniques are more robust.

11.1 Introduction

When are two or more dynamical systems coupled? Although this issue has been extensively studied for linear systems [1–3], the interest in nonlinear dynamics and nonlinear (generalized) synchronization has renewed interest in this issue in recent years [4–6]. Detecting coupling when the underlying equations are unknown, and when an arbitrary amount of measurement or dynamical noise is present is especially unclear [7]. Such is the problem when analyzing data taken from neuronal systems, especially when coupling is weak and noise is high. When detecting coupling between cells or cortical areas in nervous systems, the dynamics of spiking neurons and their synaptic connections are highly nonlinear functions, in biological networks which seem built upon and appear to require a certain level of noise to function properly. In this case, one never knows the underlying equations or the complete network topology, and verification of detected coupling is impossible.

Recent results have highlighted this issue for nonlinear and neuronal systems. It has been shown that the application of certain nonlinear synchronization detection methods may give spurious results when applied to experimental neuronal networks [8]. Furthermore, it has been shown that linear methods may clearly outperform a sensitive nonlinear measure when faced with additive noise for coupled nonlinear systems [7].

Handbook of Time Series Analysis. Björn Schelter, Matthias Winterhalder, Jens Timmer

ISBN: 3-527-40623-9

We will compare how linear and nonlinear methods succeed at detecting coupling under various conditions in known numerical and experimental nonlinear systems with known levels of coupling. We will first test whether various linear and nonlinear methods, described in Section 11.2, can detect the absence of coupling for known uncoupled systems, described in Section 11.3. This is done by quantifying the false positive detection of coupling in such uncoupled systems in Section 11.4. We then compare these methods on linear and nonlinear systems with known levels of coupling and additive noise in Section 11.5, lastly examining these methods on a known set of coupled nonlinear circuits in Section 11.5.4. We offer our conclusions that faced with unknown levels of noise and nonlinearity, in systems where the coupling may be manifest through a variety of dynamical expressions, no solitary linear or nonlinear approach can be relied upon to adequately detect subtle coupling, and that linear methods should always be included in such analysis.

11.2 Methods of Detecting Coupling

Cross-correlation (CC), mutual information (MI), mutual information in two dimensions (MI2D), phase correlation (PC), and continuity measure (CM) will be employed on linear and nonlinear data sets.

11.2.1 Cross-Correlation

Cross-correlation is a linear test that measures the significance of the linear correlation between two data sets. It has several advantages: it is a global measure (using all the points in the time series), its statistics are well understood and it is computationally efficient.

Cross-correlation between two channels was calculated as

$$CC_{1,2}(\tau) = \frac{1}{\sigma_1 \sigma_2 (N - 2\tau)} \sum_{t=\tau}^{N} (X_1(t) - \mu_1)(X_2(t-\tau) - \mu_2), \tag{11.1}$$

where $X_1(t)$ and $X_2(t)$ are the two time series of length N, with sample means μ_1 and μ_2, sample standard deviations σ_1 and σ_2, and time lag τ. It is well known that any finite length set of uncorrelated time series, whose spectra are not both white noise, will have a finite value of cross-correlation which is of course spurious [2, 9]. To compensate for this, we employ an estimator of the expected cross-correlation for uncoupled linear stochastic time series with finite auto-correlation as developed by Bartlett [2, 9]. The expected variance of the CC at a given lag l is

$$\mathrm{var}(l) = \frac{1}{(n - l)} \sum_{\tau=-n}^{n} CC_{1,1}(\tau) CC_{2,2}(\tau), \tag{11.2}$$

where $CC_{i,i}$ is the auto-correlation value of channel i at lag τ. For a given lag, τ, $CC_{1,2}(\tau)$ values were considered significant if they were greater than a significance limit set so that false positives will occur only 5 % of the time. Because multiple time lags are employed, one must compensate for the multiple comparisons tested for significance. We assume that the distribution of $CC_{1,2}(\tau)$ values are normal with a standard deviation estimated using the Bartlett estimator. We then set the significance level at 1.96 times the standard deviation at $\tau = 0$ and expect 5 % false positives for each time lag. But, the correlation is measured at k lags and it is necessary to correct for multiple measures. Therefore, we use the Bonferroni correction to set the probability of the false detection for each lag, p_i, so that the total probability, $p_t = \prod_{i=1}^{k} p_i$, and therefore $p_i \simeq 1 - (1 - p_t)/k$ for small $(1 - p_t)$. If we set $p_t = 0.95$ over ± 20 lags, for a total of 41 lags, including the zero lag, then $p_i = 0.9988$. For a normal distribution this results in a significance cutoff at 3.0 times the expected standard deviation for a two sided t-test (and 2.8 for a one sided t-test). However, there is an expected auto-correlation of the auto-correlation functions (see [2]), this allows for a more sophisticated compensation for multiple comparisons. This significance cutoff works well for weak correaltions. For strongly coupled data sets, the significance cutoff can become larger than one because the distribution of correlation values are limited to the range of -1 to 1 and the distribution is related to a normal distribution through a tanh function [10]. In this case we use a significance cutoff of $\tanh(3.0 \operatorname{atanh}(\sqrt{\operatorname{var}(l)}))$.

11.2.2 Mutual Information

Mutual information (MI) is a nonlinear measure. It is a measure of how much information can be known about time series Y by knowing the distribution of time series X. The information capacity, I, of a single trace, $X(t)$, is

$$I_X = -\sum_{i=1}^{N} P_X(i) \log_2 P_X(i), \tag{11.3}$$

where N bins were used to partition the data, and $P_X(i)$ is the probability that the voltage values of time series X will fall within bin i [11]. The MI from two channels can be calculated as

$$MI_{X,Y} = -\sum_i \sum_j P_{X,Y}(i,j) \log_2 \frac{P_{X,Y}(i,j)}{P_X(i) P_Y(j)}. \tag{11.4}$$

This measure of MI is an estimate that must be less than the true amount of information in the system. This systematic bias can be compensated for by estimating the errors introduced by the partitioning into bins. The corrected MI is as follows

$$MI^{\infty}_{X,Y} = MI_{X,Y} + \frac{B_X + B_Y - B_{XY} - 1}{2N}, \tag{11.5}$$

where B_X, B_Y, $B_{X,Y}$ are the number of bins that have points in them for the X data set, Y data set and the combined data set respectively, and N is the number of points in each time series [12]. Because we lack an analytic method of establishing confidence limits for false positive MI values in uncoupled systems, we will employ a bootstrap statistic. Mutual information at short lags (< 100 ms) were compared to mutual information calculated between the channels with randomly selected lags. These shift surrogate data sets were generated by time shifting one data set relative to the other, and wrapping the extra values to the beginning of each data set. Shift surrogates have an advantage in that they preserve the statistical structure of the original time series, but destroy the short-term correlations between them. MI was tested at 20 positive and negative time shift lags for a total of 41 lags. Each lag was chosen randomly with the restriction that time shifts be longer than four seconds. We considered the MI detected between the two channels significant if the value was greater than 2.8 standard deviations (one-sided t-test) from the mean calculated using 20 shift surrogates.

11.2.3 Mutual Information in Two Dimensions

In multivariate time series from unknown experimental systems, the systems may be higher dimensional and the interactions between them may occur in higher dimensions. Mutual information can be calculated in more than one dimension. If two data sets are each multivariate in two dimensions, or have been embedded in two dimensions by time delay embedding [13, 14], the MI of the combined system must be calculated in higher dimensions. If the systems and their coupling are nonlinear, then MI in higher dimension may reveal the coupling with more sensitivity than the standard univariate approach. While embedding in higher dimensions allows for more complex interactions, it requires more data to fill out the state space to achieve the same level of accuracy. Mutual Information in two dimensions (MI2D) is calculated as

$$\mathrm{MI}_{X_{i,j},Y_{k,l}} = -\sum_{i,j,k,l} P_{X,Y}(i,j,k,l)\log_2 \frac{P_{X,Y}(i,j,k,l)}{P_X(i,j)P_Y(k,l)}\,. \tag{11.6}$$

For time delay embedding, delays were chosen based on the decay of mutual information between a signal and a time shifted version of itself [15]. As with MI, highest significance of the 41 time lags, compared to mean and standard deviation determined using shift surrogates, was used. Significance cutoff was set at 2.8 standard deviations from the mean of the surrogates for a one sided t-test.

11.2.4 Phase Correlation

Similar to CC where correlation between amplitudes are measured, phase correlation (PC) measures the correlation between phases. A growing body of work suggests that PC can detect weak correlations that occur in simultaneous phase shifts of two data sets. This method may be sensitive to detecting coupling in nonlinear systems including neuronal systems where methods that depend on the amplitude may fail [16, 17]. One way to assign a phase to a univariate signal $X(t)$ is to employ the Hilbert transform

$$\mathcal{H}(t) = \frac{1}{\pi} \int_{-\infty}^{\infty} \frac{X(\tau)}{\tau - t} \, d\tau \,, \tag{11.7}$$

where the Cauchy principal value of the integral is used. In practice, for discrete signals, one uses $\mathcal{H}(t) = \text{Im}[2 \int_0^\infty \mathcal{X}(f) e^{i2\pi\omega t} \, df]$, where $\mathcal{X}(f)$ is the Fourier transform of $X(t)$, and ω is frequency [3].

The data can be expressed as a Gabor analytic signal of vectors $X(t) + i\mathcal{H}(t)$. Amplitude $A(t)$ and phase $\phi(t)$ can be measured at each time step as $A(t) = |X(t) + i\mathcal{H}(t)|$, where $|\cdot|$ indicates absolute value, and phase $\phi(t) = \arctan\left(\frac{\mathcal{H}(t)}{X(t)}\right)$ [3]. To quantify phase correlation, mutual information was calculated between the phase angles of the two data sets

$$\text{MI}_{X,Y} = -\sum_{i,j} P(\phi_i^X, \phi_j^Y) \log_2 \frac{P(\phi_i^X, \phi_j^Y)}{P(\phi_i^X) P(\phi_j^Y)}$$

where $P(\phi_i^X, \phi_j^Y)$ is the joint probability that signal X has the phase angle ϕ_i^X while signal Y has the phase angle ϕ_j^Y, at times i, j. Similarly, one time series can be time shifted, and a surrogate phase correlation calculated. The phase correlation between the data sets will be considered significant if the MI for the unshifted phase angles is greater than 2.8 standard deviations from the mean, calculated from 20 surrogates.

To visually display phase differences between channels X and Y, histograms of phase difference

$$p_{\phi_{i,j}}^{X,Y} = p(\text{mod}(\phi_i^X - \phi_j^Y, 2\pi)), \tag{11.8}$$

were calculated modulus 2π . If the signals are uncoupled, such histograms will be flat from uniformly random associations of phase, and if coupled, such histograms will be peaked.

11.2.5 Continuity Measure

The continuity measure (CM) is a method for detecting a functional relationship between two systems. This is done by testing for continuity of mapping between neighboring points in one data set to their corresponding points in the other data set. One advantage of CM is that it makes no distributional assumptions of the data, and another is that we have developed an analytical derivation of significance [7]. CM can also be used to infer directionality of a connection. A drive system has no information about a unidirectionally driven response system, but the responding system, having input from the drive, can predict the activity of the drive system. This method is outlined in detail elsewhere [7].

Briefly, data set X is time delay embedded in N dimensions. A fiduciary point is selected at random at time τ, $X(\tau)$, and a number of neighboring points n_δ

within a small region δ are found. Around the fiduciary point's time corresponding point, $Y(\tau)$, in the second data set, we select n_ϵ neighbors within a small region ϵ. Of the n_δ neighbors about $X(\tau)$, we find how many were amongst the ϵ neighbors about $Y(\tau)$. This is illustrated using coupled Hénon maps in Fig. 11.1. By using the hypergeometric function (for selection without replacement), it can be calculated how many n_δ points need to map into ϵ to be significantly greater than random. Repeating this calculation for all fiduciary points, the number of points that reach significance is counted. The binomial theorem is then employed to assess whether more points around each fiduciary mapped than expected by chance, and a global estimate of continuity significance is obtained. Because the quality of the mapping depends on the magnitude of the noise relative to the magnitude of the dynamics, the mapping is dependent on the size of δ and ϵ, and we test the global mapping for a range of δ and ϵ values, excluding $\delta > \epsilon$. We will use 36 tests of continuity using different δ, ϵ pairs in each direction for a total of 72 tests. Because there are multiple tests, we use the Bonferroni correction to adjust the significance level. Therefore, we set the limit at $1 - (1 - 0.95)/72$, so that only 5 % of the time will any of the 72 tests reach significance.

11.3 Linear and Nonlinear Systems

11.3.1 Gaussian Distributed White Noise

The simplest model of the data is to assume they are independent and completely stochastic processes. Two Gaussian distributed, white noise (with a uniformly flat spectrum and no correlations in time), random data sets were generated using the random number generator from Matlab (Mathworks, Natick, MA).

11.3.2 Autoregressive Model

In Gaussian white noise, there is no functional relationship between the neighboring points in time. To introduce some correlation in the data, and a way to couple them together, we use a second-order autoregressive model [2] to generate random time series with finite autocorrelation

$$X_1(t) = A_1 X_1(t-1) + B_1 X_1(t-2) + \alpha\xi_1 \tag{11.9}$$

$$X_2(t) = A_2 X_2(t-1) + B_2 X_1(t-2) + \alpha\xi_2 + C\left(X_2(t-1) - X_1(t-1)\right) \tag{11.10}$$

where the ξ_i are uniformly distributed random numbers. The coefficients B_1 and B_2 were set equal to -0.99. Because it is very difficult on short time scales to distinguish coupling from the null hypothesis for processes that have the exact same frequency, we frequency shift one of the time series. This allows the two time series to decorrelate over time if they are uncoupled. To give the two time

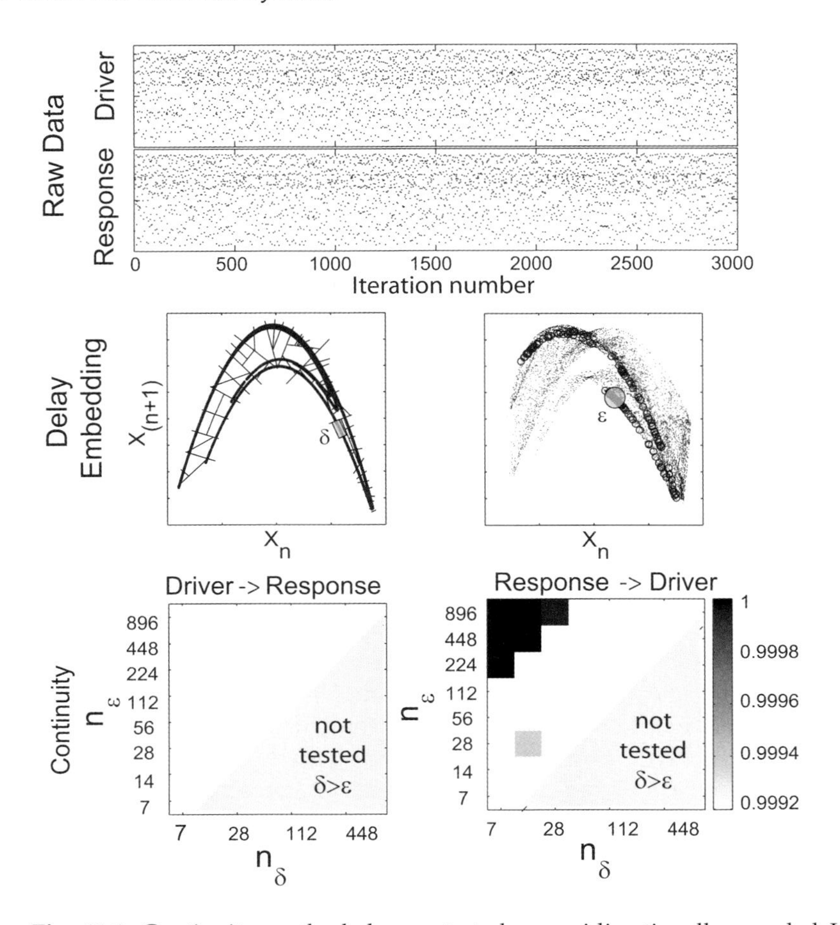

Fig. 11.1: Continuity method demonstrated on unidirectionally coupled Hénon maps. The top two rows show time series from each Hénon map. Equations are given in Section 11.3.3. In the middle panel is plotted the delay embedding of the driver and response system. A fiduciary point is selected in one time series and the n_δ nearest neighbors selected. Their corresponding time points in the other system are indicated with darker circles. The region ϵ around the fiduciary point's corresponding time point is indicated. The number of points that map from δ into ϵ are counted. The calculated probability of mapping about the fiduciary points geometrically averaged and plotted for varying δ and ϵ sizes is shown in the bottom panel. Note that, as expected, significance was found mapping from response to driver but not vice versa in this unidirectionally coupled system.

series slightly different frequencies we set $A_1 = 1.95$ and $A_2 = 1.96$. Unidirectional coupling from the first processes to the second is set through the term with coefficient C.

11.3.3 Hénon Map

The autoregressive model is a linear stochastic system. In contrast, we will next use coupled nonlinear deterministic systems that are not smooth in time – coupled Hénon maps [18]. Hénon maps are chaotic in suitable parameter regimes, giving them very complex time series. Plotting the time sequence of a variable from coupled systems like these can look very similar to nonuniformly distributed noise. By plotting one time series against itself delayed by a time step reveals very precise and deterministic behavior. The Hénon maps were iterated from one time step to the next using the following equation [19]:

$$X(t+1) = 1.4 - X(t)^2 + A_1 X(t-1) \tag{11.11}$$

$$Y(t+1) = 1.4 - (CX(t) + (1-C)Y(t))Y(t) + A_2 Y(t-1)\,. \tag{11.12}$$

The value 1.4 sets the dynamics of the isolated equations within the chaotic regime. The variable C adjusts the unidirectional coupling strength between the first and second map. The coefficients A_1 and A_2 were set to 0.3.

The nervous system is highly nonlinear and complex, but it also contains a stochastic element due to fluctuations of ions, the probabilistic release of neurotransmitter, branch point conductance failure, and the intrinsically stochastic nature of voltage and chemically activated membrane channels. To determine the effect of noise on the detection of synchrony, noise was added to the measured values. Normally distributed noise was added to the variables X and Y with standard deviations of 0, 0.125, 0.25, and 0.5 times the standard deviation of X.

11.3.4 Rössler Attractor

Coupled Rössler systems [5] were used as an example of a system with continuous but chaotic variables. This system is integrated in time rather than iterated like the map. The differential equations are described as follows, distinguishing the two systems by their subscript, 1 or 2

$$dX_1/dt = (-Y_1 - Z_1)S(r_1, s) + \rho\xi_1(t)\sqrt{dt} \tag{11.13}$$

$$dX_2/dt = (-Y_2 - Z_2 - C(X_2 - Y_1))S(r_2, s) + \rho\xi_1(t)\sqrt{dt} \tag{11.14}$$

$$dY_{1,2}/dt = (X_{1,2} + 0.2Y_{1,2})S(r_{1,2}, s) + \rho\xi_{1,2}(t)\sqrt{dt} \tag{11.15}$$

$$dZ_{1,2}/dt = (0.2 + Z_{1,2}(X_{1,2} - \mu_{1,2}))S(r_{1,2}, s) + \rho\xi_{1,2}(t)\sqrt{dt}\,. \tag{11.16}$$

The coefficients $\mu_1 = 5.7$ and $\mu_2 = 6.5$ were set so that both the driving and response attractors were chaotic. C is the coupling strength between the Y variable of the drive attractor to the X variable of the slave attractor, so that they only synchronize out of phase with each other. Because these Rössler systems can be close to periodic, we increased the diffusion rate between the two attractors through a

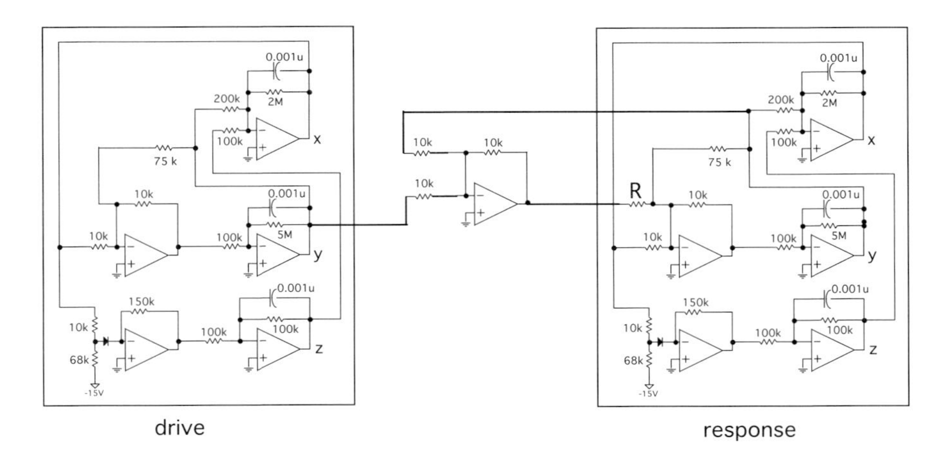

Fig. 11.2: Circuit diagram of coupled scroll attractor circuits.

function $S(r,s) = s(r^2 - \bar{r}^2)$, where $r = \sqrt{X_i^2 + Y_i^2}$ is the distance from the center of the attractor and $\bar{r}$ is the average radius of the attractor and s scales the amplitude of the oscillations. $S(r,s)$ therefore changes the rotation rate as a function of radius, which increases the diffusion rate between two uncoupled attractors.

Dynamical noise was added by adding independent noise $\xi_{1,2}(t)$ to the each term of the equations. We used uniformly distributed noise with range from $-\rho$ to ρ so that the system would not become unstable with a large perturbation. Each noise step was scaled by the square root of the integration time step, so that noise amplitude would be independent of time step size. The attractors were integrated with a time step of 0.01 using a fourth-order Runge–Kutta integrator, and sampled at every 10th time step [20].

11.3.5 Circuit Data

To experimentally reflect measurement noise in the setting of coupled nonlinear systems, two electronic circuits that produce activity similar to Rössler attractors were coupled using resistors of different magnitude (R, Fig. 11.2). These circuits are described in further detail in [21]. Four voltage measurements were recorded, two from each circuit. A selection of 65 536 points from the first channel of each data set were chosen for analysis, corresponding to roughly 3 500 rotations. Circuit data were digitized and stored on computer.

11.4 Uncoupled Systems

Detecting coupling implies that we reject the null hypothesis that the systems are uncoupled. So we begin with a simple question—can each method detect the uncoupled state when confidence limits are applied? In Fig. 11.3, we display time series from five uncoupled data sets: Gaussian distributed random data, AR models, Hénon attractors, Rössler attractors, and uncoupled circuit data.

For each data set CC, MI, and MI2D were measured at ± 20 time lags, for a total of 41 lags including the zero lag. The largest value from all the lags for each trial was plotted with significance limit in each panel. CC values were normalized by three times the Bartlett estimate so that the significance level of the line shown is one. MI values were divided by the mean of 20 surrogates and normalized by the standard deviation of the surrogates to provide (with significance level set at 2.8 standard deviations) the value at which only one of the 41 lags should reach significance 5 % of the time. For phase correlation, information was used as a measure of the nonuniformity of the phase difference histogram. Results for phase correlation were compared to mean and standard deviation of 20 surrogates. To achieve a 95 % significance limit using a one-sided t-test, the limit was set to 1.65 SD above the surrogate mean for the univariate comparisons. For continuity, we measure continuity between the two data sets for eight sizes of δ and ϵ in each domain and range for a total of 72 tests. Therefore, using the Bonferroni correction for multiple samples, we set the significance cutoff to be $1 - (1 - 0.95)/72 = 0.9993$.

For delay embedded measures (MI2D and CM) each point is expressed as a vector $\overline{X}(t) = (X(t), X(t - \tau))$. We used the time lag τ at which the mutual information is $1/e$ (i.e., 0.37) the maximum MI. Values of τ are indicated in the caption of Fig. 11.3.

11.4.1 Correlation Between Gaussian Distributed Random Data Sets

The simplest model of data is that each trace is completely uncorrelated in time. In the first column of Fig. 11.3, we show excerpts from two random time series with Gaussian distribution. We then plot the first trace against itself, in a delay embedding, which reveals no structure between the current point and a previous point in time. CC at varying lags shows points that cross the significance lines approximately 5% of the time, as expected. The Bartlett estimate, used to establish the significance limits, uses the frequency content shared by both traces to calculate the expected amount of cross-correlation under the null hypothesis that they are unconnected. MI, MI2D, and PC also only rarely touch the significance lines with respect to the surrogate data distribution. CM does not show any significant continuity at any group size, and grids of blank CM measures are omitted from this figure.

11.4.2 Correlation Between Uncoupled AR Models

The uncoupled AR models are two stochastic processes filtered with slightly different frequency filters. Unlike the Gaussian distributed white noise, such finite time series appear correlated in time due to their intrinsic auto-correlation. By plotting one system against the other, in the third row, we see ellipses. However, the history of the system cannot yield predictive value beyond the correlation time, determined by the spectral content of the time series. When systems like these, in the uncoupled state, are closely matched in frequency, the amount of spurious cross-correlation increases as the length of the sampled data decreases. The results for MI show crossings of the significance line, but they are outside of the range of lags used to identify correlation (indicated by the dark bar on the significance line). Were more lags used to include these crossings, it would be necessary to raise the significance limit to account for the increased number of lags. Increasing the range of lags used will also decrease the power of the test (the ability to detect correlation when the systems are actually coupled) by increasing the rate of false negatives. All the other tests, MI2D, PC, and CM (data not shown) confirm that these data sets are uncoupled.

11.4.3 Correlation Between Uncoupled Hénon Maps

In the third column of Fig. 11.3 results are shown from two time series from the uncoupled Hénon map. Because the Hénon map is iterated through a map, it produces points that have very little correlation in time. However, because this system is deterministic, nearby points in the same state space can be used to predict the future behavior of the system. The chaotic nature of the Hénon map causes nearby trajectories to diverge, which confines prediction to only local behavior. The chaotic and highly structured nature of this system is demonstrated in the delay embedding plotted on the third row. Testing for coupling between the two uncoupled systems shows no significance with any method.

11.4.4 Correlation Between Uncoupled Rössler Attractors

The Rössler system has both the smooth trajectory, as seen in the AR models, and the deterministic behavior of the Hénon map. Delay embedding of one of the data sets demonstrates their complex yet highly structured nature. When weakly coupled, plotting one system against the other produces similarly complex relationships (not shown). For the uncoupled Rössler data, no significant correlation was detected with any method.

11.4.5 Uncoupled Electrical Systems

The uncoupled electrical circuit demonstrates several real-world problems. The data sets are short and have measurement noise. Nevertheless, the uncoupled circuit data reveal that no significant correlation is detected with any method.

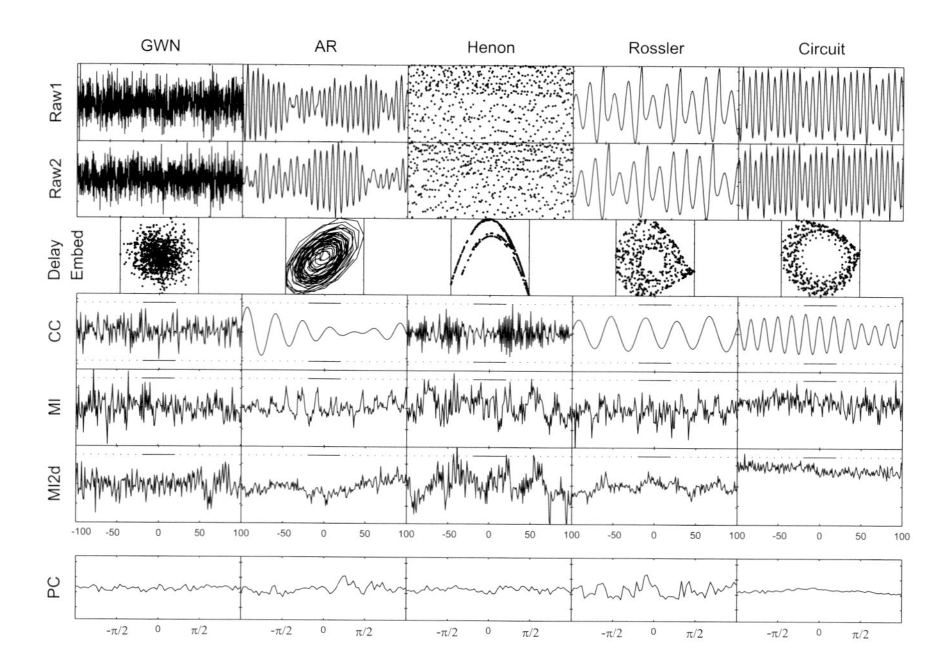

Fig. 11.3: Five different pairs of independent data sets from (in columns from left to right): Gaussian distributed white noise (GWN), autoregressive (AR) model, Hénon map, Rössler attractor, and electrical circuit. Samples of raw data are provided in the top two rows, and in the third row is a time delay embedding plot for one of the data sets, where measurement at time t is plotted against the measurement at time $t - \tau$ ($\tau = 1$ for noise and the Hénon map, $\tau = 10$ for the AR model, $\tau = 7$ for the Rössler attractor, and $\tau = 6$ for the circuit data). Below is plotted results from 100 lags using cross-correlation (CC), mutual information (MI), and mutual information in two dimensions (MI2D). Significance is only tested in the range of ± 20 lags, indicated by the solid portion of the significance line. Significance limits were set for CC using Bartlett's estimator (three standard deviations shown, calculated at 5 % limit for 41 lags and a two sided t-test), for MI 2.8 standard deviations (for 5 % limit of 41 lags and a one-sided t-test) from the mean of shift surrogate data, and MI2D at 2.8 standard deviations from the mean of shift surrogate data (same as MI significance limits). In the bottom row, the histogram of phase differences used to calculate the phase correlation (PC) is plotted. Significance for PC was calculated by measuring the information content of the histogram and comparing it to the information distribution calculated from 20 shift surrogates (significance is not shown in this graph). Continuity (CM) was also measured, but data were not shown because no plots indicated significant results for any of these uncoupled data sets.

11.5 Weakly Coupled Systems

In this section, the sensitivity thresholds of different methods are tested on known levels of weak coupling in different models. The robustness of the methods to noise is also tested. Results are plotted in Fig. 11.4.

For each data set CC, MI, and MI2D were measured at ± 20 time lags, for a total of 41 lags, including the zero lag. The largest value across the lags for each trial was used to test for significance. For each test, ten different data sets were generated and results averaged. Because results for the CM are log normally distributed, the geometric mean across trials is plotted. Otherwise, statistics are the same as in the uncoupled conditions described earlier.

11.5.1 Coupled AR Models

The AR equations are linear stochastic systems. They were coupled for a range of coupling strengths. For the AR model $\tau = 10$ was used. CC was the most sensitive test for detecting coupling in this system. MI, MI2D, and PC did poorly by comparison. Continuity performs poorly for coupled AR systems. This is because the dynamics of the system are stochastic and not deterministic, therefore the continuity between the systems is expected to be poor.

11.5.2 Coupled Hénon Maps

For coupled identical Hénon maps, CM was by far the most sensitive test for the noise free condition. Unexpectedly, even though there was very little autocorrelation in time within each signal, CC was quite effective in detecting weak coupling between the two maps. Additive measurement noise, with Gaussian distribution and standard deviation 0.125, 0.25, and 0.5, was added to the Hénon data. With the introduction of noise into this system, CC and MI2D appeared to be the most robust tools for detecting weak coupling in the presence of noise, while the performance of MI and CM rapidly degraded. Similar qualitative effects from introducing dynamical noise (< 0.125 SD to maintain system stability) were noted but not shown in the figure. In the presence of noise, linear CC was the most sensitive detector of coupling in this nonlinear map system.

11.5.3 Weakly Coupled Rössler Attractors

For coupled Rössler systems, MI, both in 1D and 2D, appeared to be the most robust methods at detecting coupling. Although CC cannot take advantage of the highly complex nature of the interaction, it was very effective at detecting such coupling, although not as robust as MI and MI2D. Dynamical noise, with Gaussian distribution and standard deviation 0.0125, 0.025, and 0.05, was added to the equations. Unlike in the Hénon map, the addition of large amounts of noise to the dynamical system made this system unstable. The small amounts of noise that did not create instability did not result in a substantial loss of sensitivity for CM measure. Similar qualitative effects from introducing dynamical noise (< 0.125 SD to maintain system stability) was noted but not shown in the figure. Surprisingly, we found that additive noise (data not shown) and dynamical noise, up to the level that it created instability, did not effect the ability of the methods to detect coupling (as seen by the similarity of the curves). This may be due to the smoothness of the data caused by oversampling in time.

11.5.4 Experimental Electrical Nonlinear Coupled Circuit

When real systems are encountered, it is almost always the case that the equations specifying the dynamics of the system are unknown, and the coupling strength is determined through experimental measures. Here, we provide an examination of an experimental nonlinear system where the full specification of the dynamics and coupling is available. We chose seven levels of coupling strengths by changing resistors connecting the circuits. The data sets were collected and coded so that they were analyzed blindly with respect to knowledge of the coupling strengths, and only afterward identified and compared.

CC was an inconsistent detector of coupling in this circuit for short data sets with similar frequencies. Significant results are shown for $10\,\mathrm{k}\Omega$, $400\,\mathrm{k}\Omega$, and $1.0\,\mathrm{M}\Omega$ resistances. One reason for this poor performance of CC is caused by the similarities of the frequencies from these circuits, which was resistor dependent. For this low noise system, MI and MI2D found significant coupling at all levels of coupling. PC suffered from the same problems that cross-correlation suffered, we suspect because the systems had such similar frequencies that even the surrogates showed strong correlation. It appears that as the systems became more uncoupled and the phases could shift more, PC method became more sensitive. In contrast, CM showed strong coupling strength dependency for stronger coupling, yet, as resistance increased, the value of CM became lost in the measurement noise.

11.6 Conclusions

In conclusion, it was found that nonlinear methods are indeed very sensitive for detecting complex correlations between nonlinear systems, in noiseless systems. However, because methods such as continuity and mutual information section state spaces into discrete sizes, noise on the order of the sectioning results in a great loss of sensitivity. Global methods, such as cross-correlation, are much more robust to noise. In coupled stochastic systems, the continuity method is insensitive to detecting correlations because historical repeats of a particular activity do not improve the ability to predict the future trajectory of a trace.

Although all methods are subject to false positives in uncoupled data, the use of appropriate confidence limits and corrections for multiple comparisons reduces false positive rates to a minimum.

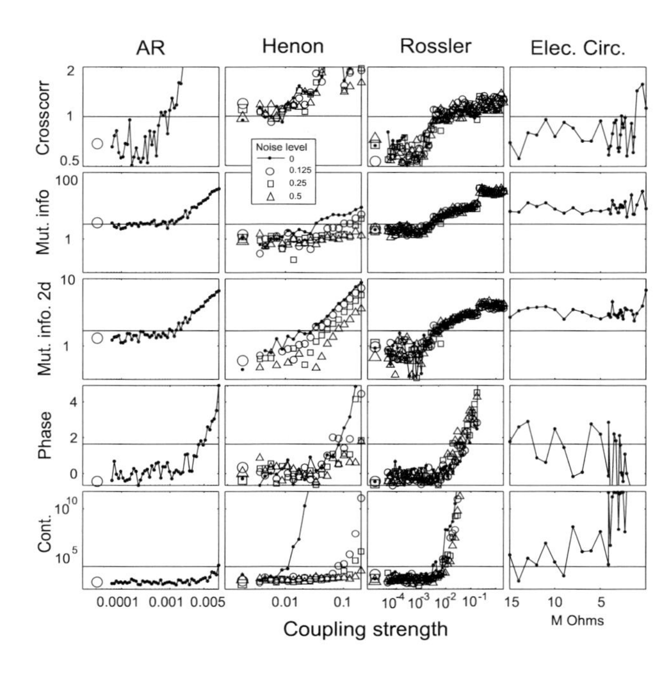

Fig. 11.4: Weak correlations within weakly coupled AR, Hénon, and Rössler equations and Electrical Circuit. In all panels, increased coupling is plotted from left to right and all points above the solid lines are significant. Additive noise is in the Hénon map and dynamical noise (1/10 the amplitude) are indicated by indicated markers. CC results in the top row show maximum cross-correlation for all lags normalized by three standard deviations (for 41 lags, and two sided t-test significance cutoff at 5%) calculated by the Bartlett estimator at different coupling strengths. All points above the heavy line at are considered significant. MI results in the second row showing mutual information at maximum of 41 lags normalized by standard deviation of 20 shift surrogates, plotting values as $t = \frac{MI_{data} - <MI_{surrogates}>}{STD(MI_{surrogates})}$. All points above the heavy line at 2.8 are considered significant at 5% for a one-sided t-test. Third row, mutual information calculated in two dimensions at different coupling strengths, all points above the heavy line at 2.8 are considered significant for 41 lags at 5 % for a one sided t-test. Fourth row, phase correlation results, maximum information of the phase difference histogram measured in standard deviations calculated using 20 shift surrogates, significance level for one lag is at 5 % for single sided t-test is 1.64. Bottom row, results from continuity reported in 1/p, where p is the smallest probability from all measured mappings at various sizes of δ and ϵ. Significance limit set at results from continuity measured at different coupling strengths. The maximum continuity measured is for eight δ and ϵ sizes ranging from eight to 256 points. The significance line is set at 1/0.9993, the value of which the maximum value for all δ and ϵ sizes is only expected to reach 5 % of the time for uncoupled systems. Values above the line indicates significant continuity.

Detecting correlations between systems that are broadband with short auto-correlations is straightforward. When systems have similar frequencies and long auto-correlations it becomes difficult to distinguish coupling from systems that take a very long time to diverge in time. This can result in false negatives.

For smooth systems, like the Rössler attractor, we found that dynamical noise, up to the point of causing instability, (results shown) and additive noise (results not shown), had little effect on sensitivity of all the methods used in detecting coupling. Qualitatively, oversampling renders the detection of correlation fairly robust to noise.

11.7 Discussion

We compared the performance of linear and nonlinear methods of coupling detection on known numerical and experimental data. Although it seems natural to assume that nonlinear methods would generically be more suitable for coupling detection in nonlinear systems, especially for map data, this may not be true in practice. Indeed, the primary conclusion that we would stress is that when the level of coupling needs to be measured from a system whose degree of nonlinearity, noise, and coupling nature is unspecified, the most plausible approach would be to probe for coupling employing a variety of independent methods as presented here.

Defining significance cutoffs is a difficult part of determining coupling. Although there are some analytical methods of estimating the degree of apparent coupling for uncoupled linear [9] or nonlinear [7] systems, for many methods there is much value in being able to test the measure on the uncoupled state. Ideally data should be taken to test this null hypothesis, since this will detect fundamental problems with the data acquisition or analysis. In many cases, uncoupling two data sets may not be possible; a good alternative is to test correlations taken from separate recordings under identical recording conditions. If this is not possible, then a time shifted surrogate of the two data sets can be used. We recommend not only testing that the null hypothesis is correctly identified, but also that the false positive rate occurs at the rate expected.

A single positive or negative result should never be considered conclusive in isolation – there seems much value in demonstrating the reproducibility of such measurements, and further in the the use of independent strategies to confirm the validity of coupling. In contrast, it should also be cautioned that the failure of a single method in detecting coupling is not evidence that the systems are not connected, but rather that the method was unable to detect connectivity. Ideally, failure of one method and success of another should help characterize the nature of the coupling. Unfortunately, our ability to handle conflicting results from these different coupling tests remains incomplete.

To analytically determine significance limits, two approaches were used: Bartlett's estimator [2, 9] and the CM confidence limit, introduced elsewhere [7]. Unfortunately, the simple form of Bartlett's estimator is problematic. It is a long-lag estimator, whose inaccuracy increases as the length of the data set decreases, the similarity of frequencies increases, the ratio of fundamental frequency to data set length decreases, and degree of irrationality in the relationship between fundamental frequencies from different data sets decreases. A more complex form of this cross-correlation expectation can be found in the discussion in [22]. Although we have refined an analytical approach to CM detection, this method relies on an accurate state space reconstruction, and is extremely sensitive to noise and nonstationarity [7].

When analytical statistical methods for defining significance limits are difficult to construct, the use of shift surrogates as a boot strap method can be very effective for determining coupling. Although there is some reduction in low frequency cross-spectrum from such shift surrogate data, the statistical properties of the data sets are largely retained.

Although our data were statistically stationary, real systems are generally not. When faced with such data, any method that relies on an accurate state space reconstruction (such as MI2D or CM) will be inherently at a disadvantage over measures whose computation admits some tolerance for variation in the data sets as a function of time, such as CC and PC.

We have been as guilty as any of our colleagues in being fascinated by the theory and methods of nonlinear dynamics. Hence we have continually been surprised by the robust capabilities of linear CC to detect weak coupling in nonlinear systems, especially in the presence of noise [7]. CC was even effective in detecting weak coupling in map data, where it was the most effective in the presence of noise. Our findings here further strengthen our view that robust linear methods should always be included in an analysis of coupling in arbitrary systems.

Acknowledgements

Supported by NIH grants F31MH12421 (TIN), K02MH14093 (SJS), R01MH50006 (SJS).

References

[1] M. R. Jarvis and P. P. Mitra. *Neural Comput.*, 13:717, 2001.

[2] G. E. P. Box and G. M. Jenkins. *Time Series Analysis: Forecasting and Control.* Holden-Day, San Francisco, revised edition, 1976.

[3] J. S. Bendat and A. G. Piersol. *Random Data: Analysis and Measurement Procedures.* Wiley, New York, 1986.

[4] L. M. C. Pecora and T. L. Carroll. *Phys. Rev. Lett.*, 64:821, 1990.

[5] N. F. Rulkov, M. M. Sushchik, L. S. Tsimring, et al. *Phys. Rev. E*, 512:980, 1995.

[6] S. J. Schiff, T. Chang, R. E. Burke, et al. *Phys. Rev. E*, 54:6709, 1996.

[7] T. I. Netoff, L. M. Pecora, and S. J. Schiff. *Phys. Rev. E*, 69:017201, 2004.

[8] T. I. Netoff and S. J. Schiff. *J. Neurosci.*, 22:7297, 2002.

[9] M. S. Bartlett. *J. Roy. Stat. Soc. B*, 8:27, 1946.

[10] R. A. Fischer. *Statistical Methods for Research Workers.* Oliver and Boyd, London, 1925.

[11] C. E. Shannon and W. Weaver. *The Mathematical Theory of Communication.* University of Illinois Press, Urbana, 1964.

[12] M. S. Roulston. *Physica D*, 125:285, 1999.

[13] H. D. I. Abarbanel. *Analysis of Observed Chaotic Data.* Springer, New York, 1996.

[14] F. Takens. *Dynamical systems and turbulence, Warwick 1980: proceedings of a symposium held at the University of Warwick 1979/80*, page 366. Springer, Berlin; New York, 1981.

[15] A. M. Frazer and H. L. Swinney. *Phys. Rev. A*, 33:1134, 1986.

[16] F. Varela, J. P. Lachaux, E. Rodriguez, et al. *Nat. Rev. Neurosci.*, 2:229, 2001.

[17] P. Tass, M. G. Rosenblum, J. Weule, et al. *Phys. Rev. Lett.*, 81:3291, 1998.

[18] K. T. Alligood, T. D. Sauer, and J. A. Yorke. *Chaos, an Introduction to Dynamical Systems.* Springer, New York, 1996.

[19] T. Chang, T. Sauer, and S. J. Schiff. *Chaos*, 51:118, 1995.

[20] W. H. Press, S. A. Teukolsky, W. T. Vetterling, et al. *Numerical Recipes in C: the Art of Scientific Computing.* Cambridge University Press, Cambridge (Cambridgeshire), New York, 1992.

[21] T. L. Carroll. *Am. J. Phys.*, 63:337, 1995.

[22] G. M. Jenkins and D. G. Watts. *Spectral Analysis and its Applications.* Holden Day, San Francisco, 1968.

12 Linear Models for Mutivariate Time Series

Manfred Deistler

This contribution is concerned with system identification, i.e., with data driven modeling, for multivariate time series. Linear dynamic models in the framework of stationary processes are considered. After an introduction to stationary processes, two topics are treated: The first is identification of multivariate state space- and ARMA(X) systems, with focus on modern approaches to state space system identification. It is argued that, in this case, opposed to the AR(X) case, a rather deep understanding of problems of realization and parameterization is required for construction of powerful identification procedures and for their evaluation. Subspace procedures and maximum likelihood estimation using data driven local coordinates are described. The second topic is concerned mainly with modeling high dimensional time series, for cases where "full" state space modeling would result into, in relation to sample size, too high dimensional parameter spaces and where therefore lower dimensional parameterizations are needed. Dynamic principal component analysis, linear dynamic factor models with idiosyncratic noise, and generalized linear dynamic factor models are discussed.

12.1 Introduction

In areas such as economics, finance, business, engineering, biology and medicine, often several single time series are collected and information exceeding the univariate information in every single time series is of interest. The main reasons for joint or conditional modeling of multivariate time series are:

1. The analysis of the dynamic relations between time series.

2. Extraction of factors or features common to all time series.

3. The improvement of forecasts.

Here we only consider discrete-time, equally spaced observations $y_t, t = 1, \dots, T$, $y_t = (y_t^{(i)})^{i=1,\dots,s} \in \mathbb{R}^s$. Of course there are many possibilities to model multivariate time series. In this contribution we consider two groups of model classes:

1. "Full" state space- and ARMA(X) models.

Handbook of Time Series Analysis. Björn Schelter, Matthias Winterhalder, Jens Timmer

ISBN: 3-527-40623-9

2. Factor type models.

Both cases are dealt within a stationary context. This contribution consists of three parts. The first part, Section 12.2, is concerned with results from the theory of stationary processes, which are necessary for an understanding of the two main parts. The reader familiar with the basic facts of this theory may skip this part.

The second part, Section 12.3, is concerned with identification (in the sense of data driven modeling) of full state space and ARMA(X) systems, where the state space point of view is emphasized. Despite the fact that state space and ARMA(X) identification is, in a certain sense, a mature subject now, in many applications, in particular in an automatized context, still severe problems arise. This is particularly true for the multivariable case. A good part of these problems does not show up in the usual asymptotic analysis. Because of these problems AR(X) modeling still dominates in many applications.

In this part our aim is not to present an extensive survey on multivariable state space and ARMA(X) identification, but instead we present two novel approaches, namely a special subspace estimation procedure (as an important representative for subspace procedures) and maximum likelihood estimation using data driven local coordinates. We claim that, opposed to the AR(X) case, both for a proper understanding and for the development of powerful identification algorithms, a rather deep understanding of the underlying structure theory for state space and ARMA(X) systems is needed. For this reason the relevant parts of this theory are reported. Given the importance of the subject, it is not surprising that there exist several other novel approaches to multivariable ARMA(X) or state space estimation, see e.g., [1–3], to mention a few.

The third part, Section 12.4, is concerned with factor models for time series. Despite the fact that factor models and the related errors-in-variables have a long history, this subject is much less mature. Recently, there has been a resurging interest in factor models, in particular for modeling and forecasting of high dimensional time series, e.g., in finance and macroeconomics. We present three important classes of factor models, dynamic principal components, linear dynamic factor models with idiosyncratic noise and generalized linear dynamic factor models and discuss identification for these classes.

12.2 Stationary Processes and Linear Systems

Stationary processes are extremely important as models for time series. The theory of stationary processes was developed in the thirties and fourties of the last century; the extensions to the multivariable case have been made a few decades later; [4, 5] are major references which include the multivariate case. Here we only give a very brief account of the main results needed for a better understanding of this contribution. Let $(y_t \mid t \epsilon \mathbb{Z}) = (y_t)$ denote a stochastic process over an underlying probability space $(\Omega, \mathcal{A}, \mathcal{P})$. Here $\mathbb{Z}$ denotes the integers and

$y_t \colon \Omega \to \mathbb{R}^s$ are random variables. A process (y_t) is called *(wide sense) stationary* if $\mathbb{E}y_t' y_t < \infty$, $t \in \mathbb{Z}$; $\mathbb{E}y_t = \text{const.}$ and if $\mathbb{E}y_{t+r}y_t'$ does not depend on t. Let a' denote the transpose of a vector or a matrix a.

For simplicity of presentation, we assume that $\mathbb{E}y_t = 0$ holds. For our purposes, the main information about a stationary process is contained in the covariance function

$$\gamma \colon \mathbb{Z} \to \mathbb{R}^{s \times s} \colon \gamma(r) = \mathbb{E}y_{t+r}y_t' \,. \tag{12.1}$$

A central result in the theory of stationary processes states that every stationary process admits a spectral representation

$$y_t = \int_{[-\pi,\pi]} e^{i\lambda t}\, dz(\lambda), \tag{12.2}$$

where the stochastic process $(z(\lambda) \mid \lambda \epsilon [-\pi,\pi])$, $z(\lambda) \colon \Omega \to \mathbb{C}^s$ ($\mathbb{C}$ denotes the complex numbers) satisfies $\mathbb{E}z(\lambda)^* z(\lambda) < \infty$, $z(-\pi) = 0$, $\lim_{\varepsilon \downarrow 0} z(\lambda + \varepsilon) = z(\lambda)$ and $\mathbb{E}\big(z(\lambda_4) - z(\lambda_3)\big)\big(z(\lambda_2) - z(\lambda_1)\big)^* = 0$ for $\lambda_1 < \lambda_2 \leqslant \lambda_3 < \lambda_4$. Here $*$ denotes the conjugate transpose and if not extra mentioned, limits of random variables are understood in mean squares sense.

The *spectral distribution function* is defined as $F : [-\pi,\pi] \to \mathbb{C}^{s \times s} : F(\lambda) = \mathbb{E}z(\lambda)z(\lambda)^*$. The spectral representation of a stationary process leads to the spectral representation

$$\gamma(t) = \int_{[-\pi,\pi]} e^{i\lambda t}\, dF, \tag{12.3}$$

of the covariance function and this constitutes a one-to-one relation between γ and F. Thus F and γ contain the same information about the underlying process. In many cases F is absolutely continuous w.r.t. the Lebesque measure ν; then the spectral density $f \colon [-\pi,\pi] \to \mathbb{C}^{s \times s}$ exists and satisfies

$$F(\lambda) = \int_{-\pi}^{\lambda} f(\nu)\, d\nu \,. \tag{12.4}$$

A sufficient condition for the existence of a spectral density is that

$$\sum_{t=-\infty}^{\infty} \|\gamma(t)\|^2 < \infty \tag{12.5}$$

holds, where $\| \; \|$ denotes a (matrix) norm. In this case, the relation between the covariance function γ and the spectral density f is given by

$$\gamma(t) = \int_{-\pi}^{\pi} e^{i\lambda t} f(\lambda)\, d\lambda \tag{12.6}$$

and

$$f(\lambda) = (2\pi)^{-1} \sum_{t=-\infty}^{\infty} e^{-i\lambda t}\gamma(t) \tag{12.7}$$

where the infinite sum on the right-hand side of Eq. (12.7) is defined in the sense of mean squares convergence.

Consider a linear transformation of a stationary process (x_t)

$$y_t = \sum_{j=-\infty}^{\infty} k_j x_{t-j} \quad k_j \in \mathbb{R}^{s \times m} . \tag{12.8}$$

Equation (12.8) can be interpreted as a (noise-free) linear system with input (x_t) and output (y_t). As can be easily seen, the stationarity of (x_t) implies (joint) stationarity of $(x_t', y_t')'$ and, using an obvious notation, from the spectral representation we obtain

$$y_t = \int_{[-\pi,\pi]} e^{i\lambda t}\, dz_y(\lambda) = \int_{[-\pi,\pi]} e^{i\lambda t}\Big(\sum_{j=-\infty}^{\infty} k_j e^{-i\lambda j} \Big)\, dz_x(\lambda) . \tag{12.9}$$

The *transfer function* of the linear system (12.8) is defined by

$$k(z) = \sum_{j=\infty}^{\infty} k_j z^j , \quad z \in \mathbb{C} . \tag{12.10}$$

If the spectral density f_x of (x_t) exists, then the spectral density f_y of (y_t) and the cross-spectral density f_{yx} between (y_t) and (x_t) exist and are given by

$$f_y(\lambda) = k(e^{-i\lambda}) f_x(\lambda) k(e^{-i\lambda})^* \tag{12.11}$$

and

$$f_{yx}(\lambda) = k(e^{-i\lambda}) f_x(\lambda) . \tag{12.12}$$

A linear transformation (Eq. (12.8)) is called *causal*, if $k_j = 0$, for $j < 0$ holds. An important special case is causal, linear transformations

$$y_t = \sum_{j=0}^{\infty} k_j \epsilon_{t-j} , \quad \sum_{j=0}^{\infty} \|k_j\|^2 < \infty \tag{12.13}$$

of white noise (ϵ_t) (i.e., $\mathbb{E}\epsilon_t = 0$, $\mathbb{E}\epsilon_s \epsilon_t' = \delta_{st}\Sigma$). Then, by Eq. (12.11)

$$f_y(\lambda) = (2\pi)^{-1} k(e^{-i\lambda}) \Sigma k(e^{-i\lambda})^* \tag{12.14}$$

and thus the information contained in the second moments of (y_t) is contained in the transfer function k and the variance matrix Σ.

Important insight in the structure of stationary processes is provided by the *Wold decomposition:* Let $\hat{y}_{t+h|t}$ denote the best linear least-squares forecast of y_{t+h}

given y_s, $s \leqslant t$. Then a stationary process (y_t) is called (linearly) *singular*, if $\hat{y}_{t+h|t} = y_{t+h}$ (for one and thus for all $t \in \mathbb{Z}$, $h > 0$) holds and (linearly) *regular* if

$$\lim_{h \to \infty} \hat{y}_{t+h|t} = 0$$

(for one and thus for all t) holds. Now the Wold decomposition says that every stationary process (x_t) can be uniquely decomposed as

$$x_t = y_t + z_t$$

where $\mathbb{E} y_s z_t' = 0$ for all $s, t \in \mathbb{Z}$ and both (y_t) and (z_t) are obtained as causal linear transformations (or as limits of such transformations) from (x_t) and where (y_t) is regular and (z_t) is singular. In addition, every regular process (y_t) can be represented *(Wold representation)* as a causal linear transformation of white noise , Eq. (12.13), where in addition also (ϵ_t) is a causal linear transformation of (y_t) (or a limit of such transformations).

By the Wold decomposition the regular and the singular component can be forecasted separately, and for the regular component we have

$$\hat{y}_{t+h|h} = \sum_{j=h}^{\infty} k_j \epsilon_{t+h-j} \,. \tag{12.15}$$

The spectral factorization problem is concerned with finding $k(z)$ corresponding to the Wold representation (and Σ) from the spectral density f_y. If we assume $\Sigma > 0$ and (w.l.o.g.) $k_0 = I$ then $k(z)$ and Σ are uniquely determined from f_y.

In many cases the observed outputs are not exact transformations of the observed inputs. Then we consider linear systems with noise of the form

$$\hat{y}_t = \sum_{j=-\infty}^{\infty} l_j z_{t-j} \,, \quad l_j \in \mathbb{R}^{s \times m} \tag{12.16}$$

and

$$y_t = \hat{y}_t + u_t \tag{12.17}$$

where

$$u_t = \sum_{j=0}^{\infty} k_j \epsilon_{t-j} \quad k_j \in \mathbb{R}^{s \times s} \,. \tag{12.18}$$

Here z_t are observed inputs, $\hat{y}_t$ are unobserved outputs, y_t are observed outputs, (u_t) is a regular (unobserved) noise process and Eq. (12.18) is in Wold representation. By

$$l(z) = \sum_{j=-\infty}^{\infty} l_j z^j \quad \text{and} \quad k(z) = \sum_{j=0}^{\infty} k_j z^j ,$$

we denote the input-to-unobserved output and the noise transfer function, respectively. Throughout we assume that

$$\mathbb{E}z_s u_t' = 0 \quad \text{for all} \quad s, t$$

holds, which is equivalent to assume that $\hat{y}_t$ is the best linear squares approximation of y_t by (z_t). In addition we assume that Eq. (12.16) is causal.

The relation between the transfer functions $l(z)$, $k(z)$, and the innovation variance $\Sigma = \mathbb{E}\epsilon_t\epsilon_t'$ on the one side and the second moments of the observed processes on the other side is given by (compare Eqs. (12.11), (12.12))

$$f_{yz}(\lambda) = l(e^{-i\lambda})f_z(\lambda) \tag{12.19}$$

$$f_y(\lambda) = l(e^{-i\lambda})f_z(\lambda)l(e^{-i\lambda})^* + (2\pi)^{-1}k(e^{-i\lambda})\Sigma k(e^{-i\lambda})^* \tag{12.20}$$

using an obvious notation. In particular, if $f_z(\lambda) > 0$ for all $\lambda \in [-\pi, \pi]$, then

$$l(e^{-i\lambda}) = f_{yz}(\lambda)f_z(\lambda)^{-1} \tag{12.21}$$

holds. Equation (12.21) is the so-called Wiener filter formula.

12.3 Multivariable State Space and ARMA(X) Models

AR(X), ARMA(X), and (linear) state space (SS) systems are the most important models for time series. In this section we consider the case of "full models" where no overidentifying or structural *a priori* restrictions in addition to stability, the miniphase assumption and minimality are imposed.

In many applications AR(X) models still dominate. The main advantages of AR(X) modeling when compared to ARMA(X) and SS modeling are:

- There are no problems with identifiability in the AR(X) case. More generally, the structure theory is so simple that it does not have to be explicitly considered.
- For estimation of parameters least-squares-type procedures can be used, which are explicitly given, asymptotically efficient and numerically fast and reliable.

On the other hand SS and ARMA(X) systems, (both describe the same class of transfer functions) are more flexible compared to AR(X) systems and thus often fewer parameters have to be used in order to obtain a good model.

For the multivariate case, i.e., when the output dimension s is larger than one, additional problems arise:

- The "curse of dimensionality": Even for the AR case, when the specified maximum lag is denoted by p, the parameter space for the system parameters $(a_1, \ldots, a_p)$ in Eq. (12.29) has dimension s^2p and thus depends quadratically on s.

- When compared to the univariate case, for multivariate ARMA(X) and SS systems, problems of parameterization are both more intricate and more important.

We claim that a proper understanding of the structure theory for ARMA(X) and SS systems leads to better identification procedures, which, in turn, will extend the range of applications for such systems. The basic references for Section 12.3 are [6, 7].

12.3.1 State Space and ARMA(X) Systems

We consider linear state space systems in the prediction error form ([6] chapter 1)

$$x_{t+1} = Ax_t + B\epsilon_t \quad (+Lz_t) \tag{12.22}$$

$$y_t = Cx_t + \epsilon_t \quad (+Dz_t) \tag{12.23}$$

where x_t is the n-dimensional state, (ϵ_t) is the s-dimensional white noise, y_t are the s-dimensional observed outputs, z_t the m-dimensional observed inputs and $A \in \mathbb{R}^{n\times n}$, $B \in \mathbb{R}^{n\times s}$, $L \in \mathbb{R}^{n\times m}$, $C \in \mathbb{R}^{s\times n}$, and $D \in \mathbb{R}^{s\times m}$ are the parameter matrices. Throughout we assume that the *stability condition*

$$|\lambda_{\max}(A)| < 1, \tag{12.24}$$

where $\lambda_{\max}$ denotes an eigenvalue of maximum modulus and the *miniphase condition*

$$|\lambda_{\max}(A - BC)| \leqslant 1 \tag{12.25}$$

hold. In addition we assume that

$$Ez_s\epsilon_t' = 0 \quad \text{for all} \quad s, t\,.$$

The steady-state solution of Eqs. (12.22)–(12.23) is given by

$$y_t = C(Iz^{-1} - A)^{-1}(B\epsilon_t + Lz_t) + \epsilon_t + Dz_t, \tag{12.26}$$

where z is used for a complex variable as well as for the backward shift on the integers $\mathbb{Z}$, i.e., $z(y_t|t \in \mathbb{Z}) = (y_{t-1}|t \in \mathbb{Z})$. Thus the solution (12.26) is a system of the form, Eqs. (12.16-12.18). In particular, by Eq. (12.25)

$$k(z)\varepsilon_t = C(Iz^{-1} - A)^{-1}B\varepsilon_t + \varepsilon_t \tag{12.27}$$

is already in Wold representation. Note that the transfer function coefficients k_j are of the form

$$k_j = CA^{j-1}B \quad \text{for} \quad j > 0 \quad \text{and} \quad k_0 = I \tag{12.28}$$

and an analogous result holds for l_j.

In the following, for the sake of brevity, unless the contrary is explicitly stated, we will assume that we have no observed inputs.

An important notion for state space systems is *minimality*; a state space system is called minimal, if there is no other state space system with the same transfer function having smaller state dimension. A state space system is minimal if and only if the matrices

$$C_n = (B, AB, \ldots, A^{n-1}B)$$

and

$$O_n = (C', A'C', \ldots, (A')^{n-1}C')'$$

both have rank n. For the case of observed inputs, B in C_n is replaced by (B, L). Throughout we assume that $\Sigma = \mathbb{E}\epsilon_t\epsilon_t'$ is nonsingular.

ARMA(X) systems are (vector-) difference equations of the form

$$a(z)y_t = b(z)\varepsilon_t(+d(z)z_t) \tag{12.29}$$

where

$$\begin{aligned} a(z) &= \sum_{j=0}^{p} a_j z^j\,, \quad a_j \in \mathbb{R}^{s\times s}\,; \\ b(z) &= \sum_{j=0}^{q} b_j z^j\,, \quad b_j \in \mathbb{R}^{s\times s}\,; \\ d(z) &= \sum_{j=0}^{r} d_j z^j\,, \quad d_j \in \mathbb{R}^{s\times m}\,. \end{aligned} \tag{12.30}$$

We assume that the *stability condition*

$$\det a(z) \neq 0 \quad |z| \leqslant 1 \tag{12.31}$$

and the miniphase condition

$$\det b(z) \neq 0 \quad |z| < 1 \tag{12.32}$$

hold, and again we assume

$$Ez_s\epsilon_t' = 0$$

and that Σ is nonsingular. The steady-state solution then is given by

$$y_t = a^{-1}(z)[b(z)\epsilon_t(+d(z)z_t)]\,. \tag{12.33}$$

Again we see that this gives a system of the form, Eqs. (12.16-12.18). Minimality for ARMA(X) systems is expressed as left coprimeness of $a(z)$ and $b(z)$ (and $d(z)$), see [6] chapter 2. Equations (12.16)–(12.18) are sometimes called the *input–output representation*, Eqs. (12.22)–(12.23) the *state space representation* and Eq. (12.29) the *ARMA*(X) *representation*.

For the case of no observed inputs we have the following relation between these representations ([6], chapter 1):

Under our assumptions,

- Every state space systems (12.22)–(12.23) and every ARMA system (12.29) has a rational transfer function $k(z)$ which is analytic in a disk containing the closed unit disk (and thus is causal and stable) and which satisfies $\det k(z) \neq 0$, $|z| < 1$.
- Conversely, for every rational transfer function $k(z)$ which is analytic in a disk containing the closed unit disk and which satisfies $\det k(z) \neq 0$, $|z| < 1$ and $k(0) = I$ there is a stable and miniphase state space-, and a stable and miniphase ARMA representation.

Thus, in particular, SS- and ARMA representations are two alternative ways to describe the same class of input/output behaviors $k(z)$. Note that the assumption $k(0) = I$ is a normalizing condition defining Σ. We have ([6], chapter 1).

Any rational and a.e. nonsingular spectral density matrix f_y may be uniquely factorized as in Eq. (12.14), where $k(z)$ is rational, analytic within a circle containing the closed unit disk, $\det k(z) \neq 0$, $|z| < 1$ and $k(0) = I$ and where $\Sigma > 0$.

12.3.2 Realization of State Space and ARMA Systems

Realization is concerned with the construction of a state space or an ARMA system from the observed process (y_t), or from its population second moments, or from the transfer function $k(z)$ corresponding to the Wold representation Eq. (12.13). Thus realization is concerned with an idealized identification problem, commencing, e.g., from the observed process (or in the ergodic case, from an infinite data string) rather than from a finite number of observations.

Formula (12.13) can be rewritten as the following infinite-dimensional state space system

$$\tilde{x}_{t+1} = \tilde{A}\tilde{x}_t + \tilde{B}\epsilon_t \tag{12.34}$$

$$y_t = \tilde{C}\tilde{x}_t + \epsilon_t \tag{12.35}$$

where

$$\tilde{x}_t = \underbrace{\begin{pmatrix} \hat{y}_{t|t-1} \\ \hat{y}_{t+1|t-1} \\ \hat{y}_{t+2|t-1} \\ \vdots \end{pmatrix}}_{\hat{Y}_t^+} = \underbrace{\begin{pmatrix} k_1 & k_2 & k_3 & \cdots \\ k_2 & k_3 & k_4 & \cdots \\ \cdots & \cdots & \cdots & \cdots \\ \cdots & \cdots & \cdots & \cdots \end{pmatrix}}_{H_\infty} \underbrace{\begin{pmatrix} \epsilon_{t-1} \\ \epsilon_{t-2} \\ \epsilon_{t-3} \\ \vdots \end{pmatrix}}_{E_t^-} \tag{12.36}$$

and where

$$\tilde{A} = \begin{pmatrix} 0 & I & 0 & \dots \\ 0 & 0 & I & 0 \\ \dots & \dots & \dots & \dots \end{pmatrix}, \quad \tilde{B} = \begin{pmatrix} k_1 \\ k_2 \\ \vdots \end{pmatrix}, \quad \tilde{C} = (I, 0, 0, \dots). \tag{12.37}$$

The matrix H_∞ is called the Hankel matrix of the transfer function. It can be shown that, since $k(z)$ is rational, H_∞ must have finite rank equal to the dimension n (called the order) of the state of a minimal state space system with this transfer function (see, e.g., [6], chapter 2). Such a minimal state space system can be obtained from H_∞ as follows (see, e.g., [8]): Let $S \in \mathbb{R}^{n \times \infty}$ be a matrix such that the rows of SH_∞ form a basis for the row space of H_∞. Then from Eq. (12.36) we obtain

$$x_{t+1} = S\tilde{x}_{t+1} = SH_\infty E^-_{t+1} = S \begin{pmatrix} k_2 & k_3 & \dots \\ k_3 & k_4 & \dots \\ \dots & \dots & \dots \end{pmatrix} E^-_t + S \begin{pmatrix} k_1 \\ k_2 \\ \vdots \end{pmatrix} \epsilon_t\,. \tag{12.38}$$

Now, determine (A, B, C) from

$$S \begin{pmatrix} k_2 & k_3 & \vdots \\ k_3 & k_4 & \vdots \\ \vdots & \vdots & \vdots \end{pmatrix} = ASH_\infty \tag{12.39}$$

$$B = S(k_1', k_2', \dots)' \tag{12.40}$$

$$(k_1, k_2, \dots) = CSH_\infty\,. \tag{12.41}$$

Note that for given S, the system (A, B, C) is uniquely defined. From Eq. (12.38)–Eq. (12.41) we obtain

$$x_{t+1} = Ax_t + B\epsilon_t \tag{12.42}$$

$$y_t = \tilde{C}\tilde{x}_t + \epsilon_t = \tilde{C}H_\infty E^-_t + \epsilon_t = CSH_\infty E^-_t + \epsilon_t = Cx_t + \epsilon_t \tag{12.43}$$

To repeat, the state space representation (12.42–12.43) is minimal.

Two minimal state space systems (A, B, C) and $(\bar{A}, \bar{B}, \bar{C})$ are observationally equivalent (i.e., they have the same transfer function) if and only if there exists a nonsingular matrix T such that

$$\bar{A} = TAT^{-1}; \quad \bar{B} = TB; \quad \bar{C} = CT^{-1} \tag{12.44}$$

hold.

The realization procedure described above has a nice interpretation in the Hilbert space spanned by the one-dimensional components $y_t^{(i)}$, $i = 1 \dots s$, $t \in \mathbb{Z}$ of the process (y_t), see [9]: From Eq. (12.36) we see that the linear dependence structure of the rows of H_∞ and of the one-dimensional components of $\tilde{x}_t$ is identical. Thus a minimal state x_t is obtained as a basis for the space obtained by projecting the space spanned by the future variables $y_r^{(i)}$, $i = 1, \ldots s$, $r \geqslant t$ on the space spanned by the past variables $y_r^{(i)}$, $i = 1, \ldots s$, $r < t$. This space is called the state space. Thus the state makes the future and the past of the process (y_t) conditionally orthogonal. This is the so-called splitting property of the state. The state contains the information from the past of the inputs relevant for the future outputs.

In order to obtain identifiability we have to choose representatives from the equivalence classes described by Eq. (12.44). One example is *echelon forms*, where we select a special basis for the row space of H_∞, namely the first rows of H_∞ which form a basis for its row space ([6], chapter 2). In an analogous way, echelon forms for ARMA systems are defined (see again [6], chapter 2). In this way a nice (homeomorphic and diffeomorphic) bijection between minimal ARMA and minimal state space systems is defined. In particular, once a state space system has been estimated, we can transform it to the state space echelon form and then to the ARMA echelon form.

12.3.3 Parametrization and Semi-Nonparametric Identification

Structure theory in general is concerned with the relation between (properties of) observed processes and internal parameters; here it is concerned with the relation between transfer functions and state space or ARMA parameters; our focus will be on the state space case, where this relation is given by Eq. (12.28). Parametrization is concerned with describing sets of transfer functions by state space or ARMA parameters. Parametrization and realization are part of structure theory. For a more detailed presentation of the ideas described here we refer to [10]. From now on, for simplicity of notation, we will assume that the strict miniphase assumption holds, i.e., that inequality (12.25) is strict.

In semi-nonparametric identification, estimation consists of two steps:

- In the first step, the model selection step, a subclass of the whole model class is determined from the data. In the case considered here this is done by estimating the order n, e.g., by information criteria such as AIC or BIC (see [6], chapter 5). Let $S_m(n)$ denote the set of all $(A, B, C) \in \mathbb{R}^{n^2+2sn}$ satisfying Eq. (12.24) and the strict miniphase assumption and which are in addition minimal. Additionally imposing minimality leaves an open dense set and $S_m(n)$ is open in $\mathbb{R}^{n^2+2sn}$. Then $S_m(n)$ is such a subclass. By $M(n)$ we denote the set of the corresponding transfer functions. $M(n)$ is endowed with the so-called pointwise topology, which corresponds to the relative topology in the product space $(\mathbb{R}^{s\times s})^{\mathbb{N}_0}$ for the coefficients $(k_j | j \in \mathbb{N}_0)$ and $M(n)$ can be shown to be a real analytic manifold of dimension $2sn$, see [6], chapter 2. Let $\pi : S_m(n) \to M(n)$ denote the mapping attaching transfer functions to state space matrices, defined by Eq. (12.28). The mapping π is not injective, and by Eq. (12.44) the classes of observational equivalence are n^2 dimensional manifolds. Thus, in a certain sense, we have n^2 too many coordinates if we use $S_m(n)$ directly as parameter space. For $s > 1$, more than one chart is needed to describe $M(n)$. Identifiable parameter spaces for parts of $M(n)$ are obtained either from an overlapping description of $M(n)$ or from canonical forms, such as echelon form. These parts of $M(n)$, V_α say, are characterized by a vector $\alpha = (n_1, \dots, n_s)$ of integers n_i, which also have to be estimated. For this and for further details we refer to [6] and [10].

- In the second step, the state space matrices (A, B, C) (or the free parameters for (A, B, C)) are estimated. The (Gaussian) likelihood function is given by

$$L_T(A, B, C, \Sigma) = T^{-1} \log \det \Gamma_T(A, B, C, \Sigma) + T^{-1} y(T)' \Gamma_T^{-1}(A, B, C, \Sigma) y(T) \quad (12.45)$$

where $y(T) = (y_1', \dots, y_T')'$ is the stacked sample and

$$\Gamma_T(A, B, C, \Sigma) = \left(\int_{-\pi}^{\pi} e^{-i\lambda(r-t)} f_y(\lambda; A, B, C, \Sigma) \, d\lambda \right)_{r,t=1,\dots,T}$$

where $f_y(\lambda; A, B, C, \Sigma)$ is the spectral density of a process given by Eq. (12.26). To be precise, Eq. (12.45) is $-2T^{-1}$ times the log-likelihood function up to a constant.

Note that $\Gamma_T(A, B, C, \Sigma)$ and thus the likelihood function depends on (A, B, C) only via the transfer function. Thus we can define a coordinate free maximum likelihood estimator (MLE) $(\hat{k}_T, \hat{\Sigma}_T)$, which does not depend on the specific parameterization under consideration.

For the asymptotic properties of the MLE we refer to [6], chapter 4. Under general conditions, the coordinate free MLEs $\hat{k}_T$, $\hat{\Sigma}_T$ over $M(n)$ are consistent. If $S_m(n)$ is used as a parameter space, we have, as has been stated, a nonuniqueness problem for the corresponding parameter estimators. For overlapping descriptions or for, e.g., echelon forms, the mapping from transfer functions to parameters is continuous and thus consistency for the transfer functions implies consistency for the MLEs for the system parameters. In this case also, under general assumptions, these parameter estimators are asymptotically normal and asymptotically efficient. Even, if the true transfer function is not contained in $M(n)$, the MLEs have a nice asymptotic behavior ([6], chapters 4 and 7).

A number of problems in actual estimation of SS and ARMA systems do not show up in asymptotic theory and even not in the usual statistical analysis at all, since this analysis deals with the exact MLE (defined by the exact optimum of the likelihood function). In general, the MLE is not explicitly given, but has to be obtained by a numerical optimization procedure, typically by a gradient search procedure, e.g., a Gauss–Newton procedure. The optimization problem is nonconvex, the choice of a good initial estimator is important and problems of multiple local optima occur. It turns out that numerical properties of optimization algorithms strongly depend on the choice of the parameterization. As has been shown in [11], "traditional" parameterizations such as echelon forms, from a certain order onwards, face severe numerical problems, which can be overcome by the so-called data driven local coordinates discussed in Section 12.3.5. This is quite remarkable, since for the univariate ARMA case, echelon forms correspond to the usual parameterization in terms of the coefficients of numerator and denominator polynomials of the transfer functions. One advantage of SS compared to ARMA systems is that for the SS case, typically, the classes of observational equivalence are larger, so we can select a numerically better representative.

Explicitly given estimators, such as subspace estimators or instrumental variables estimators are used either to obtain initial estimators for the numerical optimization of the likelihood function or as an alternative to MLE. The so-called Hannan–Rissanen procedure and its multivariable generalization (see, e.g., [6]) is an integrated approach consisting of initial estimation commencing from fitting a long autoregression, a Gauss–Newton step and order estimation.

In the following two subsections we describe two modern estimation procedures, the CCA subspace procedure and ML estimation based on data driven local coordinates.

12.3.4 CCA-Subspace Estimators

Subspace estimators for state space systems are based on a realization algorithm combined with a model reduction step, see, e.g., [12–16]. The main advantage of subspace procedures is that they are numerically fast and reliable. Throughout we assume that n has already been estimated.

Here we describe the CCA (canonical correlations analysis) procedure proposed by [13]. This procedure consists of two steps:

- In the first step an estimator $\hat{x}_t$ of the state x_t is obtained as follows: As has been explained in Section 12.3.2, a minimal state is a basis for the space obtained by projecting the space spanned by the future variables on the space spanned by the past variables: Let $Y_t^+ = (y_t', y_{t+1}', \dots)'$ and $Y_t^- = (y_{t-1}', y_{t-2}', \dots)'$. We write the Wold representation (12.13) as

$$Y_t^- = \underbrace{\begin{pmatrix} I & k_1 & k_2 & \dots \\ 0 & I & k_1 & \dots \\ \dots & \dots & \dots & \dots \end{pmatrix}}_{T} E_t^- \tag{12.46}$$

then we obtain from Eq. (12.36)

$$Y_t^+ = \underbrace{H_\infty T^{-1}}_{P} Y_t^- + \nu_t, \tag{12.47}$$

where $\nu_t = Y_t^+ - \hat{Y}_t^+$ is the infinite vector of prediction errors. Since H_∞ has rank n, also P has rank n. Every decomposition $P = OK$ where $O \in \mathbb{R}^{\infty \times n}$,

$K \in \mathbb{R}^{n\times\infty}$, where O and K both have rank n, then fixes a basis for the state space and $x_t = KY_t^-$ defines a minimal state.

The statistical analogon for this procedure is as follows: We estimate the northwest corner of P, say $\beta \in \mathbb{R}^{sf\times sp}$, from the truncated analogon of Eq. (12.47), from a finite future and a finite past

$$Y_{t,f}^+ = \beta Y_{t,p}^- + \tilde{\nu}_t, \tag{12.48}$$

where $Y_{t,f}^+ = (y_t' \ldots, y_{t+f-1}')'$; $Y_{t,p}^- = (y_{t-1}', \ldots, y_{t-p}')'$; $f, p > n$, by ordinary least squares, to obtain an estimate $\hat{\beta}$, say.

Now, typically, $\hat{\beta}$ has rank equal to $\min(fs, ps)$ whereas β has rank n. A model reduction step is now performed as follows: Let $\hat{W}_f^+ = (\hat{\Gamma}_f^+)^{-1/2}$ denote a square root (e.g., a Cholesky factor) of the inverse of the sample covariance matrix $\hat{\Gamma}_f^+$ of $Y_{t,f}^+$ and let $\hat{W}_f^-$ denote a square root of the sample covariance matrix of $Y_{t,p}^-$ (also other choices for weighting matrices are used). Now consider the singular value decomposition of the weighted estimate

$$\hat{W}_f^+ \hat{\beta} \hat{W}_p^- = \hat{U}\hat{\Lambda}\hat{V}' = \hat{U}_n \hat{\Lambda}_n \hat{V}_n' + R, \tag{12.49}$$

where $\hat{\Lambda}_n$ is the diagonal matrix consisting of the n largest singular values of $\hat{W}_f^+ \hat{\beta} \hat{W}_p^-$ (i.e., the n largest elements of the diagonal matrix $\hat{\Lambda}$) and $\hat{U}_n \in \mathbb{R}^{fs\times n}$ and $\hat{V}_n \in \mathbb{R}^{ps\times n}$ are the matrices consisting of the corresponding left and right singular vectors, respectively. The matrix R corresponds to the neglected smaller singular values. In this way we define a rank n approximization to $\hat{\beta}$ by $\hat{O}_f \hat{K}_p$, where $\hat{O}_f = (\hat{W}_f^+)^{-1} \hat{U}_n \hat{\Lambda}_n^{1/2} \in \mathbb{R}^{fs\times n}$ and $\hat{K}_p = \hat{\Lambda}_n^{1/2} \hat{V}_n' (\hat{W}_p^-)^{-1} \in \mathbb{R}^{n\times ps}$ and the estimator for the state is given by $\hat{x}_t = \hat{K}_p Y_{t,p}^-$.

- In the second step, given the state estimator $\hat{x}_t$, the matrix C is estimated by the least-squares formula

$$\hat{C}_T = \left(\frac{1}{T}\sum_{t=1}^{T} y_t \hat{x}_t'\right)\left(\frac{1}{T}\sum_{t=1}^{T} \hat{x}_t \hat{x}_t'\right)^{-1} \tag{12.50}$$

and ε_t is estimated by $\hat{\varepsilon}_t = y_t - \hat{C}_T \hat{x}_t$. In the same way, A and B are estimated by regressing $\hat{x}_{t+1}$ on $\hat{x}_t$ and $\hat{\varepsilon}_t$. The matrix Σ is estimated by

$$\hat{\Sigma}_T = (T-p)^{-1} \sum_{t=p+1}^{T} \hat{\varepsilon}_t \hat{\varepsilon}_t' .$$

If the estimated system is not miniphase, the corresponding estimated spectral density has to be factorized again to obtain a stable miniphase factor.

In the case of observed inputs the effect of the future observed inputs on the forecasts has to be taken into account, in addition.

As has been stated already the advantage of subspace methods lies in the substantial reduction of computational effort compared to MLEs obtained by numerical optimization. Typically subspace methods do not use canonical forms, or, more generally, no *a priori* prescribed representatives from the equivalence classes.

Consistency and asymptotic normality of the CCA method, partially also for the case of observed inputs, have been shown in [17–21]. In [22] it is shown that CCA, for the case of no observed inputs and when the true system is described in $M(n)$, is asymptotically equivalent to MLE in the sense that by transforming the CCA estimates to the echelon form, $\sqrt{T}$ times their difference to the corresponding MLE converges to zero in probability. Unfortunately, this is not the case when observed inputs are present.

Recently, an EM algorithm based on a state estimation step has been proposed in [3] which seems to superior to CCA in a number of cases.

12.3.5 Maximum Likelihood Estimation Using Data Driven Local Coordinates

Data driven local coordinates (DDLs) have been introduced in [11] and [23]. A closely related idea has been developed in [24]. Properties of DDLCs have been derived in [25]. The basic idea of DDLCs is as follows: We commence from the model class $S_m(n)$ and an initial estimate, (A_0, B_0, C_0) say. $(A_0, B_0, C_0) \in S_m(n)$ is obtained, e.g., by a subspace procedure. Consider the class $\mathcal{E}(A_0, B_0, C_0)$, of all (minimal) systems observationally equivalent to (A_0, B_0, C_0), choose a point $(\tilde{A}_0, \tilde{B}_0, \tilde{C}_0)$ in this class (the choice of such a point is a design parameter for the procedure), construct the tangent space (in $\mathbb{R}^{n^2+2sn}$) to $\mathcal{E}(A_0, B_0, C_0)$ at this point and take the orthocomplement (in $\mathbb{R}^{n^2+2sn}$) to this tangent space as a preliminary parameter space. $\mathcal{E}(A_0, B_0, C_0)$ has dimension n^2 and the orthocomplement is of dimension $2sn$. Let $Q^{\perp}$ denote a $(n^2 + 2sn) \times 2sn$ matrix whose columns form an orthonormal basis for this orthocomplement. Then we consider the mapping

$$\gamma_D\colon \mathbb{R}^{2sn} \rightarrow \mathbb{R}^{n^2+2sn}\colon \gamma_D(\tau_D) = \mathrm{vec}\begin{pmatrix} \tilde{A}_0 \\ \tilde{B}_0 \\ \tilde{C}_0 \end{pmatrix} + Q^{\perp}\tau_D\,, \quad \tau_D \in \mathbb{R}^{2sn}\,. \tag{12.51}$$

The corresponding parameter space $T_D \subset \mathbb{R}^{2sn}$ is defined by removing the nonminimal, the unstable and the not strictly miniphase systems and the corresponding space of transfer functions is $V_D = \pi(\gamma_D(T_D))$. Now, e.g., a Gauss–Newton step is performed in T_D for optimizing the likelihood. This gives a new, second initial estimate and the procedure is iterated.

The procedure can be interpreted as optimization of the likelihood over $M(n)$. The asymptotic properties are just the properties of the MLE. The advantage of the procedure compared, e.g., to MLE using echolon forms lies in its numerical properties: The intuitive motivation behind DDLC is that, due to the orthogonality of the parameter space to the tangent space, the numerical properties of optimization procedures are, at least locally, favorable. Comparisons with other parameterizations corroborate this notion, see, e.g., [23, 26]. In particular these comparisons show that echolon forms are clearly outperformed by DDLC. DDLC, together with a subspace initial estimator is now the default option in the "system identification" toolbox of MATLAB 6.x.

As can be shown, the parameter space T_D is not identifiable, however there exist open neighborhoods $T_D^{loc} \subset T_D$ of $0 \in T_D$ and V_D^{loc} of $\pi(A_0, B_0, C_0) \in M(n)$ such that T_D^{loc} is identifiable and the restriction of the mapping $\pi \circ \gamma_D$ to T_D^{loc} is a homeomorphism.

One way to reduce the dimension of the parameter space over which (a suitable version of) the likelihood function has to be optimized numerically, is to concentrate out parameters which appear linearly in the prediction error by a (ordinary or generalized) least-squares step [27]. For the concentrated likelihood again the DDLC approach is used, see [26, 28, 29]. We call this the separable least squares (sls) DDLC approach.

We commence from the inverse state space system

$$x_{t+1} = \bar{A}x_t + \bar{B}y_t \tag{12.52}$$

$$\varepsilon_t = \bar{C}x_t + y_t \tag{12.53}$$

where $\bar{A} = (A - BC)$, $\bar{B} = B$ and $\bar{C} = -C$ and $(\bar{A}, \bar{B}, \bar{C})$ and (A, B, C) are in a one-to-one relation. The conditional (Gaussian) likelihood function, which is asymptotically equivalent to Eq. (12.45), is given as

$$\tilde{L}_T(\bar{A}, \bar{B}, \bar{C}, \Sigma) = \log\det\Sigma + T^{-1}\sum_{t=1}^{T} \mathrm{tr}\{\varepsilon_t(\bar{A}, \bar{B}, \bar{C})\varepsilon_t(\bar{A}, \bar{B}, \bar{C})'\Sigma^{-1}\}. \tag{12.54}$$

Here tr denotes the trace and ε_t is the function of $\bar{A}$, $\bar{B}$, $\bar{C}$ and the observation $y_1, \ldots, y_t$ defined by Eq. (12.52) and Eq. (12.53). Now either $\bar{C}$ or $\bar{B}$ appear linearly in the prediction error and can be concentrated out in a first step. For example, if $\bar{C}$ is concentrated out, then the system parameters are written as $\tau = (\tau_1', \tau_2')'$, where $\tau_1 = \left((\mathrm{vec}\,\bar{A})', (\mathrm{vec}\,\bar{B})'\right)'$ and $\tau_2 = (\mathrm{vec}\,\bar{C})'$. Concentrating out in addition Σ, leads to the doubly concentrated likelihood

$$L_T^{cc}(\tau_1) = \log\det\sum_{t=1}^{T} \varepsilon_t(\tau_1)\varepsilon_t(\tau_1)'. \tag{12.55}$$

Thus the nonlinear optimization problem for Eq. (12.54) has been reduced to a nonlinear optimization problem in τ_1. For $\bar{A}, \bar{B}$ the equivalence classes are given by

$$\{T\bar{A}T^{-1}, T\bar{B} \mid \det T \neq 0\} \tag{12.56}$$

compare Eq. (12.44) and the DDLC idea is applied in the $\bar{A}, \bar{B}$ space, thus reducing the dimension from $n^2 + sn$ to sn. Simulations show (see [29]) that in many cases sls DDLC has better numerical properties than even DDLC.

12.4 Factor Models for Time Series

Factor analysis has been developed by psychologists for measurement of intelligence in the beginning of the twentieth century. The motivations for the use of factor models are compression of the information contained in a high dimensional data vector into a small number of factors and the idea of underlying latent unobserved variables influencing the observations. Whereas the initial approach to factor analysis was oriented to data originating from independent, identically distributed random variables, the idea has been further generalized to the modeling of multivariate time series, thus compressing information in two dimensions, the cross-sectional and the time dimension. This idea has been pursued, rather independently, in a number of areas, such as econometrics [30–32] or signal processing [33]. This idea is of particular importance, if the cross-sectional dimension s is large (in relation to sample size T), where the so-called curse of dimensionality occurs. "Conventional" time series modeling by full AR, ARMA or "typical" state space model leads to a parameter space with dimension proportional to s^2; on the other hand the number of data points, for fixed T, is linear in s. Factor models are used to mitigate this curse of dimensionality. The basic, common equation for the different kinds of factor models considered here is of the form

$$y_t = \Lambda(z)\xi_t + u_t\,, \quad t \in \mathbb{Z}, \tag{12.57}$$

where y_t is the s-dimensional vector of observations, ξ_t are the $r < s$ typical dimensional factors, (u_t) is the noise and the transfer function $\Lambda(z) = \sum_{j=-\infty}^{\infty} \Lambda_j z^j$, $\Lambda_j \in \mathbb{R}^{s\times r}$, is called the factor loading matrix.

Throughout we assume $\mathbb{E}\xi_t = 0$, $\mathbb{E}u_t = 0$, (ξ_t) and (u_t) are stationary and regular with absolutely summable covariances and

$$\mathbb{E}\xi_t u_s' = 0 \quad \text{for all} \quad s, t\,. \tag{12.58}$$

For the spectral density f_y of (y_t) then we have an obvious notation

$$f_y(\lambda) = \Lambda(e^{-i\lambda}) f_\xi(\lambda) \Lambda(e^{-i\lambda})^* + f_u(\lambda)\,. \tag{12.59}$$

In addition we assume throughout that $\Lambda(e^{-i\lambda})$ and $f_\xi(\lambda)$ have rank r and that $f_y(\lambda)$ has rank s for all $\lambda \in [-\pi, \pi]$. A special case often considered occurs when $\Lambda(z) = \Lambda$ is constant. Then we have

$$\Sigma_y = \Lambda \Sigma_\xi \Lambda' + \Sigma_u\,, \tag{12.60}$$

where Σ_y denotes the variance matrix of y_t.

The assumptions imposed so far do not determine a reasonable model class in the sense that for given f_y or Σ_y too many models would be possible, see [34]. Thus, in order to obtain reasonable model classes, further assumptions have to be imposed. Three important cases are principal component analysis (PCA), linear factor models with idiosyncratic noise and generalized linear factor models which will be discussed below. For these model classes we are interested in

- Estimation of (a parametrized version of) $\Lambda(z)$, f_ξ, and f_u.
- Estimation of the factors ξ_t and of the latent variables $\hat{y}_t = \Lambda(z)\xi_t$.
- Forecasting.

Preceding to estimation in the narrow sense, problems of the structure of such models, in particular of identifiability, have to be discussed, see, e.g., [34]. Factor models (12.57) where Λ is constant, but where (ξ_t) and (u_t) are not necessarily white, are called *quasi static* [35] and *static* if (ξ_t) and (u_t) are white.

It should be noted that factor models (12.57) are closely related to errors-in-variables (EV) models

$$y_t = \hat{y}_t + u_t$$

where

$$\sum_{j=-\infty}^{\infty} w_j \hat{y}_{t-j} = 0 \, , \quad w_j \in \mathbb{R}^{(s-r)\times s} \, .$$

This is immediate for the quasi-static case, where the restriction that $\hat{y}_t$ has its image in a linear subspace of $\mathbb{R}^s$ is expressed in the factor formulation by the range of Λ and in the EV case by the kernel of w_0. For the dynamic case, this is explained in [34], see also [36].

The EV formulation emphasizes the point of view of "true" unobserved variables $\hat{y}_t$ satisfying the exact relation $w(z) = \sum_{j=-\infty}^{\infty} w_j z^j$ and that in principle all observed variables y_t may be contaminated by noise. In addition, both, the relation between the latent variables and the noise model are "symmetric," in the sense that no *a priori* assumption about the classification into inputs and outputs and even not about the number $s-r$ of equations has to be made.

For the quasi-static case, forecasting models are obtained by fitting AR(X) models to the estimated factors and by using these models for forecasting of factors and (using an estimate of Λ) of latent variables. For forecasting the observations y_t either the forecasts for the latent variables are used or these forecasts are combined with the individual forecasts for the one-dimensional noise components [35]. If $r \ll s$, then this gives a substantial reduction of dimension compared to full, e.g., AR(X) models.

12.4.1 Principal Component Analysis

In dynamic PCA [33] we commence from the canonical representation of the spectral density

$$f_y = O_1 \Omega_1 O_2 + O_2 \Omega_2 O_2 , \tag{12.61}$$

where $O_i(e^{-i\lambda})$, $\Omega_i(e^{-i\lambda})$, $i = 1, 2$ depend on frequency λ, $\Omega_1(e^{-i\lambda})$ is the diagonal matrix of the r largest eigenvalues of $f_y(\lambda)$, ordered according to decreasing size, $\Omega_2(e^{-i\lambda})$ is the diagonal matrix of the $s - r$ smallest eigenvalues of $f_y(\lambda)$, again ordered according to decreasing size and $O_1(e^{-i\lambda})$ and $O_2(e^{-i\lambda})$ are the matrices consisting of the corresponding eigenvectors. Under the assumptions given in [33], by

$$\begin{aligned} \xi_t &= O_1(z)^* y_t\,, & \Lambda(z) &= O_1(z)\,, \\ f_\xi(\lambda) &= \Omega_1(e^{-i\lambda})\,, & u_t &= O_2(z) O_2(z)^* y_t \\ f_u(\lambda) &= O_2(e^{-i\lambda})\Omega_2(e^{-i\lambda})O_2(e^{-i\lambda})^*\,, \end{aligned} \tag{12.62}$$

we obtain a special factor model of the type (12.57), the dynamic PCA model. PCA gives the best approximation of f_y by a rank r spectral density $\Lambda f_\xi \Lambda^*$ in the sense that $\operatorname{tr} \mathbb{E} u_t u_t'$ is minimal.

For estimation, f_y is replaced by an estimator of f_y and the estimators of ξ_t, $\Lambda(z)$, f_ξ and f_u are defined as in Eq. (12.62). Note that if all eigenvalues are assumed to be distinct, then the eigenvalues and the suitably normalized eigenvectors are continuous functions of the original matrix and thus consistent estimators of $f_y(\lambda)$ give consistent estimators of $\Lambda(e^{-i\lambda})$, $f_\xi(e^{-i\lambda})$, and $f_u(e^{-i\lambda})$. By the choice of r, the degree of dimension reduction in the cross-section, and, as a trade off, the quality of approximation are determined. Note that r is not intrinsic, in the sense that it is not a property of f_y. Dimension reduction in the time dimension may be performed by using finite-dimensional parameterizations. However note, that for rational f_y, the matrices on the right-hand side of Eq. (12.61) are not necessarily rational.

An important special case occurs if O_1 and O_2 in Eq. (12.61) do not depend on frequency λ, but Ω_1 and Ω_2 may depend on frequency λ; in this case the PCA is quasi static.

12.4.2 Factor Models with Idiosyncratic Noise

Here, in addition it is assumed that the noise components are uncorrelated, i.e., f_u is diagonal (or , in the static case, that Σ_u is diagonal). In other words the basic idea is not to look for the best approximation of observations y_t by the latent variables $\hat{y}_t$, but to separate the common components described by the factors, from the individual components, described by the noise. Note that the factors here have a splitting property in cross-section which is analogous to the property of the states in time: They make the components of (y_t) conditionally uncorrelated. Such models are commonly used, for, e.g., in finance, where for, e.g., returns in the stock market, the factors describe for instance the development of the market, and the "noise" describes the development of the individual firms.

The static model with idiosyncratic noise is the classical factor model, with a long history, dating back to the beginning of the twentieth century. Commencing from given Σ_y, we see from Eq. (12.60) that in the static case the following two identifiability problems arise:

- For given Σ_y, what is the set of all pairs $\Sigma_{\hat{y}} = \Lambda \Sigma_\xi \Lambda'$ and Σ_u, where $\Sigma_{\hat{y}}$ is positive semidefinite, singular and symmetric and Σ_u is positive semidefinite and diagonal, such that Eq. (12.60) is satisfied. In this set r may not be constant; we restrict ourselves to the subset, where r is minimal. Throughout r is used for the minimal r.

- How can we determine Λ and Σ_ξ from $\Sigma_{\hat{y}}$. Throughout $\Sigma_\xi = I$ is assumed, then Λ is unique up to multiplication by orthogonal matrices, corresponding to factor rotation.

As far as the first question is concerned, the answer is that $\Sigma_{\hat{y}}$ and Σ_u are, in general, not uniquely defined from Σ_y (see, e.g., [37]), but they are generically unique if r is smaller than or equal to the so-called Ledermann bound

$$\frac{2s+1}{2} - \sqrt{\frac{(2s+1)^2}{4} - s^2 + s}\,.$$

If uniqueness of $\Sigma_{\hat{y}}$ is obtained, Λ is made unique by a suitable normalization (see, e.g., [38]).

Estimators $\hat{\Lambda}$ and $\hat{\Sigma}_u$ of Λ and Σ_u are obtained from optimizing the (Gaussian log) likelihood function, which up to a constant, is given by

$$L_T(\Lambda, \Sigma_u \mid \hat{\Sigma}_{y,T}) = -\frac{T}{2} \log\det(\Lambda\Lambda' + \Sigma_u) - \frac{T}{2} \mathrm{tr}(\Lambda\Lambda' + \Sigma_u)^{-1} \hat{\Sigma}_{y,T} \tag{12.63}$$

subject to rank $\Lambda = r$, $\Sigma_u > 0$ and suitable normalization conditions on Λ. Here $\hat{\Sigma}_{y,T}$ denotes the sample variance

$$T^{-1} \sum_{t=1}^{T} y_t y_t'\,. \tag{12.64}$$

The corresponding ML estimators can be shown to be consistent. This holds even for the quasi-static case, where Λ is constant, but (ξ_t) and (u_t) may be correlated and thus Eq. (12.63) is no longer the likelihood.

As opposed to the PCA case, here, r or to be more precise, the minimal r in all decompositions (12.60) of Σ_y is intrinsic. Tests for determining r have been proposed in [39].

For the factor model with idiosyncratic noise, the factors, in general, are not functions of the observations and thus have to be approximated by the observations. One method for doing this is obtained from minimizing

$$\mathbb{E}(\xi_t - L y_t)(\xi_t - L y_t)'$$

over $L \in \mathbb{R}^{r \times n}$ in the ordering corresponding to positive semidefiniteness of matrices, giving $L = \Lambda' \Sigma_y^{-1}$, leading to a factor estimator (omitting T in the notation) of the form

$$\hat{\xi}_t = \hat{\Lambda}'(\hat{\Lambda}\hat{\Lambda}' + \hat{\Sigma}_u)^{-1} y_t\,. \tag{12.65}$$

For dynamic factor models, a rather complete structure theory, with focus on the relation $w(z)$ between the latent variables, has been developed in [34]. As far as the analogon to the first question above is concerned, namely the uniqueness of $\Lambda(e^{-i\lambda})f_\xi(\lambda)\Lambda(e^{-i\lambda})^*$ and $f_u(\lambda)$ from $f_y(\lambda)$, we have generic uniqueness for $r \leqslant s - \sqrt{s}$. More general, in [34], sets of observationally equivalent relations $w(z)$ are described and their continuous dependence on the spectral density f_y is shown. In addition, a description of the set of all spectral densities f_y corresponding to a given r is derived.

For estimation and model selection in the dynamic case we refer to [30] and [32]. In this area there is still a substantial number of unsolved problems.

12.4.3 Generalized Linear Dynamic Factor Models

In a number of applications, e.g., in asset pricing [40], in cross-country business cycle analysis or in monitoring and forecasting economic activity by estimation of common factors [41], the cross-sectional dimension may be very high, and may even exceed sample size. In addition, the assumption that f_u is diagonal turns out to be too restrictive for many applications, where, e.g., "local" dependency between noise components may occur.

Both the issue of weakening the assumption of uncorrelatedness on the noise components and the issue to exploit information contained in very high dimensional time series and to add information by adding an additional time series, led to the development of generalized linear quasi-static and dynamic factor models, see [42–44].

For the corresponding analysis the cross-sectional dimension s is not kept fixed. We consider a double sequence $(y_t{}^{(i)} \mid i \in \mathbb{N},\ t \in \mathbb{Z})$ of observations and assume that $(y_t^s = (y_t^{(i)})^{i=1,\dots,s} \mid t \in \mathbb{Z})$ has mean zero and is regular and stationary for every $s \in \mathbb{N}$. Using an obvious notation we write Eq. (12.57) as

$$y_t^s = \Lambda^s(z)\xi_t + u_t^s\,, \quad s \in \mathbb{N}\,. \tag{12.66}$$

Here, the factors ξ_t are assumed to be independent of s and, in particular, r is constant. A basic idea is to replace the assumption of uncorrelatedness of the noise components by a weak form of dependence which allows for an "averaging out" (for $s \to \infty$) of these components for certain linear combinations. For a complete set of assumption for Eq. (12.66) we refer to [42] for the quasi-static case and to [43] for the dynamic case. The main assumptions are:

Let $\omega_{s,k}^y$ denote the kth largest eigenvalue of the spectral density f_y^s of $(y_t^s \mid t \in \mathbb{Z})$; we use an analogous notation for $f_{\hat{y}}^s$ and f_u^s. Then we assume that all s eigenvalues of $f_y^s(\lambda)$ are distinct for all λ and that

- $\lim_{s\to\infty} \omega_{s,k}^y(\lambda) = \infty$ for all λ , $k = 1, \dots, r$ (i.e., the first r eigenvalues of f_y^s diverge for $s \to \infty$)
- there exists a $c > 0$ such that $\omega_{s,r+1}^y \leqslant c$, for all λ, $s \in \mathbb{N}$.

These assumptions are central for implying the existence of a sequence of models (12.66), where Λ^s, $\hat{y}_t^s$, and u_t^s are nested in the sense that, e.g., $u_t^{s+1} = (u_t^{s\prime}, u_t^{(s+1)})'$ and which satisfy the following assumptions:

- The spectral densities $f_{\hat{y}}^s$ of the latent variables $\hat{y}_t^s$ have rank r (assuming $s > r$) and the associated nonzero eigenvalues $\omega_{s,k}^{\hat{y}}(\lambda)$ diverge for $s \to \infty$, for all $\lambda, k = 1, \ldots, r$
- All eigenvalues of the spectral densities $f_u^s(\lambda)$ remain bounded for all λ and s.

The latter condition formalizes what we mean by weak dependence of the noise components.

These conditions do not guarantee identifiability for fixed s, but ensure asymptotic identifiability, e.g., in the sense that they allow to separate $\hat{y}_t^{(i)}$ and $u_t^{(i)}$ for $s \to \infty$.

Estimation for the quasi-static and the dynamic case, respectively, may be performed by quasi-static [42] or dynamic PCA [43], since PCA and generalized linear factor models are asymptotically equivalent in the sense that, e.g., the PCA latent variables and suitable estimators for these variables converge to the corresponding generalized factor model variables, see e.g., [43] for a more precise statement. The asymptotic analysis is performed for $T \to \infty$ and $s \to \infty$, again we refer to [42] for the quasi-static case and to [43] for the dynamic case for details.

For forecasting, dynamic PCA, in general, has a severe disadvantage, since in general

$$\hat{y}_t = O_1(z)O_1(z)^* y_t$$

is a non causal filtering operation and thus $\hat{y}_t$ may depend on y_s, $s > t$. One way to overcome this difficulty is to assume that $\Lambda^s(z)$ is polynomial with degree q, independent of s, and to write Eq. (12.66) as a quasi-static model

$$y_t^s = (\Lambda^s, \ldots, \Lambda_q^s)(\xi_t', \ldots, \xi_{t-q}') + u_t^s$$

on the cost of having higher dimensional factors. For estimation and forecasting in this context, we refer to [41, 44].

12.5 Summary and Outlook

In general terms, this contribution is concerned with data-driven modeling for multivariate, (equally spaced) discrete time, time series data. Despite that identification of nonlinear systems is of increasing importance, in most applications, for a number of reasons, linear systems still dominate. The author likes the statement that "nonlinear system identification" is word like "nonelephant-zoology." In particular for the multivariate case, there are still important open problems and accordingly intensive research in a number of sub-areas of linear system identification.

A basic assumption in this contribution is the use of stochastic models in a stationary context. This is a very common setting, however not a universally justified one. Recently linear system identification in a nonstochastic setting has been analyzed in detail (see [45, 46]) and there is a large body of literature on identification of unstable linear systems in a nonstationary context, in particular on cointegration, which is of great importance in economics.

Specifically, this contribution treats two topics. The first is identification of (multivariate) state space- and ARMA(X) systems. Despite the fact that such systems are more flexible compared to AR(X) systems, in many applications, the latter still prevail. The reasons for this are twofold. First, the least-squares type estimators of AR(X) systems are numerically fast and reliable and statistically asymptotically efficient at the same time, and second, there is no complicated structure theory involved in this case. We argue that powerful estimation procedures for (in particular multivariate) state space systems (and thus also for ARMA(X) systems) have to be based on a rather deep understanding of the underlying structure theory. A short account of structure theory is given. The intention here is not to give a survey on state space system identification; we focus on two important recent developments. The first is subspace, in particular CCA, estimation. The idea of subspace methods is to combine realization algorithms, which solve an idealized identification problem, e.g., commencing from the "true" transfer function, with a model reduction step (in most cases performed by SVD), leading to compression of information contained in the data. The second focus in the first topic is on a special parameterization, called DDLC, for MLE. The idea here is not to work with a finite number of prespecified coordinates, but to use the orthocomplement to the tangent space at a certain point in the equivalence class corresponding to a previous estimator as a parameter space. In this way the numerical performance of Gauss–Newton-type procedures for optimizing the likelihood is improved. Both, subspace identification and DDLC have been mainly developed in systems engineering and there is still little "technology transfer" to other areas.

A number of important areas such as "structural identification," taking into account "physical" *a priori* knowledge, identification for control or tracking time varying parameters have not been considered in the contribution.

The second topic treated in this contribution is factor models for time series. A main idea here is compression of information in cross-section and time, mainly in order to model high dimensional time series. Dynamic principal components, linear dynamic factor models with idiosyncratic noise and generalized linear dynamic factor models are considered. In particular for the two latter model classes, there is still a number of open problems, both in structure theory and in estimation. Generalized linear dynamic factor models have been developed in econometrics, and again there is little "technology transfer" to other areas.

Acknowledegments

I want to thank W. Scherrer, C. W. J. Granger, B. M. Pötscher, D. Bauer, T. Gneiting, H. Ledolter and M. Lippi for helpful comments.

References

[1] G. C. Tiao. In D. Pena et al., editors, *A Course in Time Series Analysis*. Wiley, New York, 2001.

[2] J. M. Dufour and D. Pelletier. Practical methods for modelling weak VARMA processes: Identification, estimation and specialization with macroeconomic application. Technical report, CIREQ, Université de Montréal.

[3] S. Gibson and B. Ninness. *Automatica*, 41:1667, 2005.

[4] Y. A. Rozanov. *Stationary Random Processes*. Holden Day, San Francisco, 1967.

[5] E. J. Hannan. *Multiple Time Series*. Wiley, New York, 1970.

[6] E. J. Hannan and M. Deistler. *The Statistical Theory of Linear Systems*. Wiley, New York, 1988.

[7] G. C. Reinsel. *Elements of Multivariate Time Series Analysis*. Springer, New York, 2003.

[8] M. Deistler. In D. Pena et al., editors, *A Course in Time Series Analysis*. Wiley, New York, 2001.

[9] H. Akaike. *SIAM J. Control*, 13:162, 1975.

[10] M. Deistler. In G. Goodwin, editor, *System Identification and Adaptive Control*. Springer, London, 2001.

[11] T. McKelvey and A. Helmersson. In *Proceedings 36th IEEE Conference on Decision and Control*, 1997.

[12] H. Akaike. In R. H. Mehra and D. G. Lainiotis, editors, *System Identification: Advances and Case Studies*. Academic Press, New York, 1976.

[13] W. E. Larimore. In H. S. Rao and P. Dorato, editors, *Proceedings of the 1983 American Control Conference*, page 445, 1983.

[14] P. Van Overschee and B. De Moor. *Subspace Identification for Linear Systems: Theory, Implementation, Applications*. Kluwer, Boston, 1996.

[15] D. Bauer. *Econometric Theory*, 21:181, 2005.

[16] T. Katayama. *Subspace Methods for System Identification*. Springer, London, 2005.

[17] M. Deistler, K. Peternell, and W. Scherrer. *Automatica*, 31:1865, 1995.

[18] K. Peternell, M. Deistler, and W. Scherrer. *Signal Proc.*, 52:161, 1996.

[19] D. Bauer, M. Deistler, and W. Scherrer. *Automatica*, 35:1243, 1999.

[20] D. Bauer, M. Deistler, and W. Scherrer. In *Proceedings of the IFAC Conference 'SYSID'*, 2000.

[21] A. Chiuso and G. Picci. *J. Econometrics*, 118:292, 2003.

[22] D. Bauer. *J. Time Series Anal.*, 26:631, 2005.

[23] T. McKelvey, A. Helmersson, and T. Ribarits. *Automatica*, 40:1629, 2004.

[24] J. A. Wills, B. Ninness, and S. Gibson. In *Proceedings of the IFAC World Congres*, 2005.

[25] T. Ribarits, M. Deistler, and T. McKelvey. *Automatica*, 40:789, 2004.

[26] M. Deistler, T. Ribarits, and B. Hanzon. *Compstat 2004*, page 137. Physika, Heidelberg, 2004.

[27] J. Bruls, C. T. Chou, and M. Verhaegen. In *Proceedings of the IFAC Conference 'SYSID'*, 1997.

[28] T. Ribarits, M. Deistler, and B. Hanzon. *Automatica*, 41:531, 2005.

[29] T. Ribarits, M. Deistler, and B. Hanzon. *Int. J. Adapt. Control Signal Proc.*, 18: 717, 2004.

[30] J. Geweke. *Latent Variables in Socio-Economic Model*. North Holland, Amsterdam, 1977.

[31] T. J. Sargent and C. A. Sims. In C. A. Sims, editor, *New Methods in Business Cycle Research*. Federal Reserve Bank of Minneapolis, Minneapolis, 1977.

[32] R. F. Engle and M. F. Watson. *J. Am. Stat. Assoc.*, 76:774, 1981.

[33] D. R. Brillinger. *Time Series: Data Analysis and Theory*. Holt, Rinehart and Winston, New York, 1981.

[34] W. Scherrer and M. Deistler. *SIAM J. Control Optimiz.*, 36:2418, 1998.

[35] M. Deistler and E. Hamann. *J. Financ. Econometrics*, 3:256, 2005.

[36] C. Heij, W. Scherrer, and M. Deistler. *SIAM J. Control Optimiz.*, 35:1924, 1997.

[37] R. E. Kalman. In R. Dauturay, editor, *Frontiers in Pure and Applied Mathematic*. North-Holland, Amsterdam, 1991.

[38] T. W. Anderson. *Ann. Stat*, 12:1, 1984.

[39] T. W. Anderson and H. Rubin. In *Proc. Third Berkeley Symposium on Math. Stat. Prob. V*, page 111, 1956.

[40] G. Chamberlain and M. Rothschild. *Econometrica*, 51:1281, 1983.

[41] J. H. Stock and M. W. Watson. *J. Business Economic Stat.*, 20:147, 2002.

[42] J. Bai. *Econometrica*, 71:135, 2003.

[43] M. Forni, M. Hallin, M. Lippi, and L. Reichlin. *J. Econometrics*, 119:231, 2004.

[44] M. Forni, M. Hallin, M. Lippi, and L. Reichlin. The generalized dynamic factor model, one sided estimation and forecasting. *J. Am. Stat. Assoc.*, 100: 830, 2005.

[45] I. Markovsky, J. C. Willems, S. Van Huffel, and B. De Moor. *Exact and Approximate Modeling of Linear Systems: A Behavioral Approach*. SIAM, Philadelphia, 2006.

[46] J. C. Willems. In B. A. Francis and J. C. Willems, editors, *Control of Uncertain Systems: Modelling, Approximation and Design (Festschrift for K. Glover)*. Springer, Berlin, 2006.

13 Spatio-Temporal Modeling for Biosurveillance Using a Spatially Constrained State Space Model

David S. Stoffer and Myron J. Katzoff

Real-time disease surveillance is an essential part of the detection of disease outbreaks. Although data are currently being collected in real-time, data analytic tools that support both temporal and spatial data analysis and visualization are lacking. In many cases, the analysis is accomplished by dropping either time or space. Here, we discuss a class of spatially constrained state space models, and we demonstrate its viability by analyzing weekly influenza and pneumonia mortality collected in the northeastern United States by the Centers for Disease Control. For biosurveillance, the main concern is whether the process has been tampered with by the introduction of an outside agent. For general disease surveillance, one is typically interested in whether or not an epidemic is imminent. Our idea is to develop an optimal method for the prediction of events using the available data in both space and time. If the number of events varies from its prediction in the next time period, this indicates the system should be investigated and monitored more closely.

13.1 Introduction

Real-time disease surveillance is essential in helping detect the presence of a disease outbreak, and in supporting the characterization of that outbreak by public health officials. Although data are being collected in real-time, for example, by the Centers for Disease Control (CDC) or by the Realtime Outbreak and Disease Surveillance Laboratory (RODS) at the University of Pittsburgh, data analytic tools that support both temporal and spatial data analysis and visualization are lacking. At the present time, most analyses are accomplished by dropping (by ignoring or by aggregating) either time or space.

We will present a viable method for monitoring such processes. The model is related to the STARMAX model first discussed in Stoffer [1, 2]. The STARMAX model is essentially a spatially constrained state space model, and we will demonstrate the benefits of the model for the general analysis of processes collected in both space and time.

Handbook of Time Series Analysis. Björn Schelter, Matthias Winterhalder, Jens Timmer

ISBN: 3-527-40623-9

For biosurveillance, one of the main concerns is whether the process has been tampered with by the introduction of an outside agent. Our idea is to develop optimal methods for the prediction of events using the available data (or history) in both space and time. This may be thought of as tracking an event (or disease). If the number of events varies from its prediction in the next time period, this produces a flag that indicates the process should be investigated and monitored more closely. Although various methods appear to be promising, it seems that a spatially constrained time series model is best suited for the job. The state space model was developed for tracking a space vehicle to make sure it remains on its orbit. In essence, we feel the biosurveillance problem has similarities to the tracking problem.

To fix ideas, suppose we observe several processes evolving in space over time, say

$$\{Y_t(s)\colon s = 1, \ldots, q; \quad t = 1, \ldots, n\}, \tag{13.1}$$

where s denotes the location of the process and t denotes time. For most biosurveillance problems, it is reasonable to assume the processes are observed regularly in time, but irregularly in space. Our goal is to predict $Y_{n+1}(s)$, for each s, using the data given in Eq. (13.1). If we let $\hat{Y}_{n+1}(s)$ denote the predicted value at location s, then our interest is in the innovations

$$\epsilon_{n+1}(s) = Y_{n+1}(s) - \hat{Y}_{n+1}(s) \tag{13.2}$$

for $s = 1, \ldots, q$. If, when $Y_{n+1}(s)$ is observed, $\epsilon_{n+1}(s)$ is unduly large in magnitude, at any location s, this produces a flag that indicates the events should be investigated and monitored by an expert (or experts). Hence, in biosurveillance, we see our goal as optimal prediction in time for each location, as opposed to optimal prediction of an unmonitored location (e.g., kriging).

13.2 Background

Although there has been a substantial amount of research in the area of spatio-temporal analysis, the area is not nearly as developed as purely spatial analysis or purely time series analysis. Much of the literature in the area of spatio-temporal analysis has been authored by spatial analysts, although some work has been done by time series analysts. For the most part, spatial analysts focus primarily on estimation and prediction in space using time as a nuisance dimension, whereas time series analysts focus on estimation and prediction in time using space as a nuisance dimension. In this section we give some background on the research in this area. We concern ourselves mainly with state space or dynamic linear models (DLMs) because that is the focus of this work.

Pfeifer and Deutsch [3, 4] developed estimation for space–time ARMA models for prediction in time using known spatial constraints. The model is a vector ARMA with (known) spatially weighted coefficient parameter matrices. The model was originally developed in [5]. As previously mentioned, this idea was generalized in Stoffer [1, 2], which is the basis of our investigation. In these papers, a spatially constrained ARMAX-type model was introduced. The state space form of the model was introduced for use in cases where there are possibly missing observations. Huang and Cressie [6] proposed a simple modeling technique using a vector autoregressive model with spatially dependent innovations.

Mardia et al. [7] defined the Kriged Kalman filter (KKF) as a particular type of state space model for the analysis of spatio-temporal data. In this method, the space–time field is decomposed into mean and error components. The mean component is expressed as a time-varying linear combination of a time-dependent parameter vector (state vector) and spatial fields (common fields). The state vector introduces a stochastic component into the mean structure. The spatial fields are selected from a basis of the space of all possible spatial kriging estimates for a given set of m sites and for a given second-order spatial structure (variogram). Then, the Kalman filter recursion is used. Maximum-likelihood estimation via Newton–Raphson or the expectation–maximization (EM) algorithm was suggested for parameter estimation.

Higdon [8, 9], used the state space form, but where the observation matrix has elements that are a kernel over space. The model consists of a state process whose dimension is the dimension of the number of locations of the underlying process of interest. The observation equation can be of smaller dimension, depending on which locations are actually being observed at a given time. The matrix that relates the observations to the states is assumed to be a spatial kernel.

Wikle and Cressie [10, 11] introduce the space–time Kalman filter by assuming that noisy observations are coming from an unobservable, latent, space–time process. The latent (or state) process is modeled as a sum of weighted lagged means and an error term. The weights from neighboring spatial locations form an orthonormal basis; the error term has temporal variability but no temporal dynamics. Then Kalman filter is used to obtain the value of the weighting parameters and the latent variable.

Sanso and Guenni [12, 13] used the Bayesian DLM, as described in [14], to analyze the Venezuelan rainfall data. Their framework accommodated time-varying seasonality, trends and dependent lagged values in its linear structure. Spatial correlations were handled by the parameters of the observation equation, by considering a completely unknown correlation matrix. Hence, their parameterization calls for informative priors. Then Monte Carlo methods were used to obtain posterior distributions for the parameters of the model.

A hierarchical DLM was introduced in Brown et al. [15], where a time-varying regression was used to find the relationship between gauge measurements and radar rainfall. The time-varying coefficients were stochastically modeled using a vector autoregressive (VAR) model, where the instantaneous covariance matrix has a component that influences the purely spatial covariance. This component was modeled in two ways, as separable and nonseparable correlation functions.

For a separable correlation function, a product of an exponential function and a Matern family correlation[1] was used. For nonseparable correlation functions, the method of blurring and smoothing was used. Then maximum likelihood was used to estimate the parameters.

Stroud et al. [16] modeled the mean function jointly in space and time by locally weighted mixtures of regression surfaces, i.e., the product of a weighting kernel and a set of basis functions, where the regression surfaces vary through time. Temporal trends and seasonal cycles and other exogenous variables can be included in the model. The authors used the Kalman filter and smoothing algorithms to obtain posterior predictive distributions in the closed form. This modeling technique does not resort to MCMC simulations.

13.3 The State Space Model

Because the state space model or DLM is the workhorse of our procedure, we present some of the basic ideas. In this case, we write the vector of observations at time t and for all q locations as a $q \times 1$ vector

$$\mathbf{Y}_t = \left(Y_t(1), \ldots, Y_t(q)\right)' . \tag{13.3}$$

In the DLM, we suppose that $\mathbf{Y}_t$ are observations of an unobserved latent, or state, process $\mathbf{X}_t$ that is p-dimensional. The state process is assumed to be observed with noise, say $\boldsymbol{v}_t$ through the *observation equation*

$$\mathbf{Y}_t = M_t \mathbf{X}_t + \boldsymbol{v}_t , \tag{13.4}$$

where M_t are $q \times p$ measurement matrices that describe how the states are observed at time t, and $\boldsymbol{v}_t$ is assumed to be q-dimensional white noise, with the variance–covariance matrix R. In its basic form, the *state equation*, which describes the dynamic behavior of the state, is given by a vector autoregression,

$$\mathbf{X}_{t+1} = F\mathbf{X}_t + G\boldsymbol{w}_t . \tag{13.5}$$

In Eq. (13.5), F is a $p \times p$ matrix of transition parameters, G is a $p \times r$ matrix of parameter coefficients that describe the relationship of the $r \times 1$ white noise process $\boldsymbol{w}_t$ to the states. We assume that the variance–covariance matrix of $\boldsymbol{w}_t$ is Q. In addition, we allow for the observation noise and state noise to be correlated at time t, that is,

$$\text{cov}\{\boldsymbol{w}_t, \boldsymbol{v}_u\} = S\delta_t^u$$

where S is an $r \times q$ matrix and $\delta_t^u = 1$ when $t = u$ and zero otherwise. Typically $\boldsymbol{v}_t$ and $\boldsymbol{w}_t$ are taken to be Gaussian processes, which are independent of the

[1] This is defined as $\rho(u; \phi, \kappa) = \left(2^{\kappa-1}\Gamma(\kappa)\right)^{-1}(u/\phi)^{\kappa}K_{\kappa}(u/\phi)$, where ϕ and κ are parameters and $K_{\kappa}(\cdot)$ denotes the modified Bessel function of the third kind, of order κ. The family is valid for $\phi > 0$ and $\kappa > 0$.

initial state vector, $\mathbf{X}_0 \sim N_p(\boldsymbol{\mu}_0, \Sigma_0)$. The coefficient matrices should be spatially constrained, of course, and we will discuss this problem in the next section.

In general, our goal is to predict $\mathbf{Y}_{t+1}$ from the data $\mathbf{Y}_1, \ldots \mathbf{Y}_t$. If the parameters are known and the process is Gaussian, then we are interested in calculating the minimum mean-square error predictor,

$$\mathbf{Y}_{t+1}^t \stackrel{\text{def}}{=} E\{\mathbf{Y}_{t+1} \mid \mathbf{Y}_t, \ldots, \mathbf{Y}_1\}, \tag{13.6}$$

and the mean-square prediction error (MSPE),

$$\Sigma_{t+1} \stackrel{\text{def}}{=} E\{(\mathbf{Y}_{t+1} - \mathbf{Y}_{t+1}^t)(\mathbf{Y}_{t+1} - \mathbf{Y}_{t+1}^t)'\}. \tag{13.7}$$

Noting, from Eq. (13.4), that

$$\mathbf{Y}_{t+1}^t = M_t \mathbf{X}_{t+1}^t,$$

where

$$\mathbf{X}_{t+1}^t \stackrel{\text{def}}{=} E\{\mathbf{X}_{t+1} \mid \mathbf{Y}_t, \ldots, \mathbf{Y}_1\},$$

our goal becomes obtaining $\mathbf{X}_{t+1}^t$ and its MSPE,

$$P_{t+1}^t \stackrel{\text{def}}{=} E\{(\mathbf{X}_{t+1} - \mathbf{X}_{t+1}^t)(\mathbf{X}_{t+1} - \mathbf{X}_{t+1}^t)'\}.$$

The results are contained in the famous Kalman filter, which we now state.

Theorem 13.1 (The Kalman filter). *For the state space model specified in Eq.* (13.4) *and Eq.* (13.5), *with initial conditions* $\mathbf{X}_1^0 = F\boldsymbol{\mu}_0$ *and* $P_1^0 = F\Sigma_0 F' + GQG'$, *for* $t = 1, \ldots, n$,

$$\mathbf{X}_{t+1}^t = F\mathbf{X}_t^{t-1} + K_t(\mathbf{Y}_t - M_t\mathbf{X}_t^{t-1}), \tag{13.8}$$

$$P_{t+1}^t = (F - K_t M_t)P_t^{t-1}(F - K_t M_t)' + GQG' + K_t R K_t' - GSK_t' - K_t S' G', \tag{13.9}$$

where the so-called gain matrix is given by

$$K_t = (FP_t^{t-1}M_t' + GS)(M_t P_t^{t-1} M_t' + R)^{-1}. \tag{13.10}$$

Proof. To establish Eq. (13.8), consider the innovations

$$\boldsymbol{\epsilon}_t = \mathbf{Y}_t - \mathbf{Y}_t^{t-1} = \mathbf{Y}_t - M_t \mathbf{X}_t^{t-1}, \tag{13.11}$$

and note that

$$\begin{aligned}\mathbf{X}_{t+1}^t &= E\{\mathbf{X}_{t+1} \mid \mathbf{Y}_1, \ldots, \mathbf{Y}_{t-1}, \mathbf{Y}_t\} = E\{\mathbf{X}_{t+1} \mid \mathbf{Y}_1, \ldots, \mathbf{Y}_{t-1}, \boldsymbol{\epsilon}_t\} \\ &= F\mathbf{X}_t^{t-1} + K_t \boldsymbol{\epsilon}_t,\end{aligned} \tag{13.12}$$

where

$$K_t = \text{cov}(\mathbf{X}_{t+1}, \boldsymbol{\epsilon}_t)[\text{var}(\boldsymbol{\epsilon}_t)]^{-1}.$$

The first part of the summand on the right-hand side of Eq. (13.12) follows because

$$E\{X_{t+1} \mid Y_1, \ldots, Y_{t-1}\} = E\{FX_t + Gw_t \mid Y_1, \ldots, Y_{t-1}\} = FX_t^{t-1} .$$

To evaluate P_{t+1}^t given by Eq. (13.9), first note that by using Eq. (13.4) and Eq. (13.12), we may write

$$X_{t+1} - X_{t+1}^t = (F - K_t M_t)(X_t - X_t^{t-1}) + Gw_t - K_t v_t .$$

Thus, Eq. (13.9) follows from straight forward calculations by noting $E(w_t v_t') = S$ and its transpose are the only cross-product terms that survive in the calculation.

The evaluation of K_t given in Eq. (13.10) also follows from straight forward calculations. To verify Eq. (13.10), we have

$$\text{cov}(X_{t+1}, \epsilon_t) = \text{cov}\{FX_t + Gw_t,\ M_t(X_t - X_t^{t-1}) + v_t\} = FP_t^{t-1}M_t' + GS ,$$

and

$$\Sigma_t = \text{var}(\epsilon_t) = \text{var}\{M_t(X_t - X_t^{t-1}) + v_t\} = M_t P_t^{t-1} M_t' + R . \quad (13.13)$$

Similarly, the initial conditions for the filter (13.8)–(13.10) are given by

$$X_1^0 = E(X_1) = F\mu_0$$

$$P_1^0 = \text{var}(X_1 - FX_1^0) = \text{var}(F[X_0 - \mu_0] + Gw_0) = F\Sigma_0 F' + GQG' .$$

□

We also remark that fixed inputs (exogenous variables) can enter into the model (13.4) and (13.5). The inclusion of inputs in the state and observation equations leads to simple and obvious adjustments to the predictions. For example, suppose u_t is an $\ell \times 1$ vector of fixed inputs, and the model is now

$$X_{t+1} = FX_t + Hu_t + Gw_t \quad (13.14)$$

$$Y_t = M_t X_t + \Gamma u_t + v_t \quad (13.15)$$

where H ($p \times \ell$) and Γ ($q \times \ell$) are the parameter matrices. Then, the only change to the filter is that Eq. (13.8) becomes

$$X_{t+1}^t = FX_t^{t-1} + Hu_t + K_t \epsilon_t$$

where the innovation is now

$$\epsilon_t = Y_t - M_t X_t^{t-1} - \Gamma u_t .$$

The values in Eq. (13.11) and Eq. (13.13) are important quantities that will be used for estimation. As previously mentioned, the prediction errors, ϵ_t, are called the *innovations*, with corresponding *innovation variance–covariance* matrices, Σ_t.

We may use Eq. (13.8) and Eq. (13.11) to write Eq. (13.4) and Eq. (13.5) in the *innovations form* of the model given by

$$\mathbf{X}_{t+1}^{t} = F\mathbf{X}_{t}^{t-1} + K_t\boldsymbol{\epsilon}_t \,, \tag{13.16}$$

$$\mathbf{Y}_t = M_t\mathbf{X}_{t}^{t-1} + \boldsymbol{\epsilon}_t \,. \tag{13.17}$$

If the process is not Gaussian, the Kalman filter yields best linear prediction. In this case, we may think of conditional expectation in the above arguments as projection onto the closed span of the space generated by the conditioning arguments.

For estimation, we can use the Gaussian form of the innovations likelihood. Let Θ denote the $k \times 1$ vector of parameters of interest, noting that in the model (13.4)–Eq. (13.5), we have $F = F(\Theta)$, $G = G(\Theta)$, $Q = Q(\Theta)$, and $R = R(\Theta)$. Letting $L_Y(\Theta)$ denote the likelihood of the data $\mathbf{Y}_1, \ldots, \mathbf{Y}_n$, note that we may write

$$L_Y(\Theta) = f_\Theta(\mathbf{Y}_1, \ldots, \mathbf{Y}_n) = \prod_{t=1}^{n} f_\Theta(\mathbf{Y}_t \mid \mathbf{Y}_{t-1}, \ldots, \mathbf{Y}_1) = \prod_{t=1}^{n} f_\Theta(\boldsymbol{\epsilon}_t) \,.$$

The innovations are Gaussian, hence, ignoring a constant, we may write the likelihood as

$$-\ln L_Y(\Theta) = \frac{1}{2}\sum_{t=1}^{n} \log|\Sigma_t(\Theta)| + \frac{1}{2}\sum_{t=1}^{n} \boldsymbol{\epsilon}_t(\Theta)'\Sigma_t(\Theta)^{-1}\boldsymbol{\epsilon}_t(\Theta) \,, \tag{13.18}$$

where we have emphasized the dependence of the innovations on the parameters Θ. Of course, Eq. (13.18) is a highly nonlinear and complicated function of the unknown parameters. A Newton–Raphson algorithm can be used to minimize Eq. (13.18) with respect to Θ. The steps involved in performing a Newton–Raphson estimation procedure are as follows:

1. Select initial values for the parameters, say, $\Theta^{(0)}$.

2. Run the Kalman filter, Eqs. (13.8)–(13.10), using the initial parameter values, $\Theta^{(0)}$, to obtain a set of innovations and error covariances, say, $\{\boldsymbol{\epsilon}_t^{(0)}, \Sigma_t^{(0)};\ t = 1, \ldots, n\}$.

3. Run one iteration of a Newton–Raphson procedure with $-\ln L_Y(\Theta)$ given in Eq. (13.18) as the criterion function to be minimized, to obtain a new set of estimates, say $\Theta^{(1)}$.

4. At iteration j ($j = 1, 2, \ldots$), repeat step 2 using $\Theta^{(j)}$ in place of $\Theta^{(j-1)}$ to obtain a new set of innovation values $\{\boldsymbol{\epsilon}_t^{(j)}, \Sigma_t^{(j)};\ t = 1, \ldots, n\}$. Then repeat step 3 to obtain a new estimate $\Theta^{(j+1)}$. Stop when the estimates or the likelihood stabilize; for example, stop when the values of $\Theta^{(j+1)}$ differ from $\Theta^{(j)}$, or when $L_Y(\Theta^{(j+1)})$ differs from $L_Y(\Theta^{(j)})$, by some predetermined, but small amount.

We stress the fact that it is not necessary for the data to be Gaussian to consider Eq. (13.18) as the criterion function to be used for parameter estimation. Furthermore, under certain rare conditions, the Gaussian MLE of Θ when the process is non-Gaussian is asymptotically optimal; details can be found in [17].

13.4 Spatially Constrained Models

To motivate our approach, first consider the problem of fitting individual ARMA models to each location. For ease, we will first concentrate on an ARMA(1,1) model. That is, suppose for each location $s = 1, \ldots, q$, we model the time series as

$$Y_t(s) = \phi(s)Y_{t-1}(s) + \nu_t(s) - \theta(s)\nu_{t-1}(s)\,, \tag{13.19}$$

where $\nu_t(s)$ is white noise. These models can be combined into a vector state space model as follows:

$$\begin{aligned} \mathbf{X}_{t+1} &= \begin{bmatrix} \phi(1) & \cdots & 0 \\ \vdots & \ddots & \vdots \\ 0 & \cdots & \phi(q) \end{bmatrix} \mathbf{X}_t + \begin{bmatrix} \phi(1)-\theta(1) & \cdots & 0 \\ \vdots & \ddots & \vdots \\ 0 & \cdots & \phi(q)-\theta(q) \end{bmatrix} \boldsymbol{\nu}_t\,, \\ \mathbf{Y}_t &= \mathbf{X}_t + \boldsymbol{\nu}_t\,, \end{aligned} \tag{13.20}$$

where $\mathbf{X}_t$, $\mathbf{Y}_t$ and $\boldsymbol{\nu}_t = \left(\nu_t(1), \ldots, \nu_t(q)\right)'$ are all $q \times 1$ vector processes, $\mathbf{Y}_t$ is as described in Eq. (13.3). To verify Eq. (13.20), for any $s = 1, \ldots, q$,

$$\begin{aligned} Y_t(s) &= X_t(s) + \nu_t(s) \\ &= \left\{\phi(s)X_{t-1}(s) + \phi(s)\nu_{t-1}(s) - \theta(s)\nu_{t-1}(s)\right\} + \nu_t(s) \\ &= \phi(s)Y_{t-1}(s) - \theta(s)\nu_{t-1}(s) + \nu_t(s)\,. \end{aligned}$$

Correlation among the locations can be introduced through R, the variance–covariance matrix of the $q \times 1$ noise process, $\boldsymbol{\nu}_t$. This technique, which was used in [18], will only be useful, however, if the processes are coherent equally across all frequencies. That is, for such a model, the squared coherency between any two locations, $Y_t(j)$ and $Y_t(k)$, is constant across all frequencies, ω, and will be $\rho^2_{jk}(\omega) = r^2_{jk}/(r_{jj}\, r_{kk})$, where r_{jk} is the (j,k)th element of the R matrix. An obvious extension of Eq. (13.20) is to write the state with general, rather than diagonal, coefficient parameters, say

$$\mathbf{X}_{t+1} = F\mathbf{X}_t + G\boldsymbol{\nu}_t \quad \text{and} \quad \mathbf{Y}_t = \mathbf{X}_t + \boldsymbol{\nu}_t\,,$$

where $\mathbf{X}_t$ and $\mathbf{Y}_t$ are $q \times 1$ vector processes, and F and G are both $q \times q$ parameter matrices. Thus, in full generality, there are $2q^2 + q(q+1)/2$ parameters (coefficient and variance–covariance components) to estimate, where we recall q is the number of locations. Hence, the estimation problem will become restrictive even for a relatively small number of sites.

Rather than use general F and G, it seems more appropriate to spatially constrain them using the knowledge of the spatial relationships among the sites. To fix ideas, we concentrate on the representation given in Eq. (13.20) and use some of the ideas presented in [2]. For example, consider the following model

$$\mathbf{X}_{t+1} = \mathrm{D}\Phi\mathbf{X}_t + \mathrm{D}(\Phi - \Theta)\boldsymbol{\nu}_t \quad t = 0, 1, \ldots, n \tag{13.21}$$

$$\mathbf{Y}_t = \mathbf{X}_t + \boldsymbol{\nu}_t \quad t = 1, \ldots, n \tag{13.22}$$

where $\mathbf{X}_t$ is the q-dimensional state vector, $\mathbf{Y}_t$ is the q-dimensional observation vector consisting of observations $Y_t(s)$ for $s = 1, \ldots, q$, taken at all locations at time t, and $\boldsymbol{\nu}_t$ is the q-dimensional noise vector with variance–covariance matrix R. The $q \times q$ parameter matrices Φ and Θ are, as in Eq. (13.20), *diagonal*, and D is a $q \times q$ matrix of spatial constraints with 1's along the diagonal. If D is the identity matrix, then there are no spatial constraints and Eqs. (13.21)–(13.22) become the model specified in Eq. (13.20).

Using the same arguments that showed Eq. (13.19) can be written as Eq. (13.20), the model (13.21)–(13.22) implies the model

$$\mathbf{Y}_t = \mathrm{D}\Phi\mathbf{Y}_{t-1} + \boldsymbol{\nu}_t - \mathrm{D}\Theta\boldsymbol{\nu}_{t-1} \, . \tag{13.23}$$

Of course, exogenous variables may be included, in which case the model becomes

$$\mathbf{Y}_t = \mathrm{D}\Phi\mathbf{Y}_{t-1} + \Gamma\mathbf{u}_t + \boldsymbol{\nu}_t - \mathrm{D}\Theta\boldsymbol{\nu}_{t-1} \, , \tag{13.24}$$

as previously discussed. We will refer to this model as a STARMAX(1, 1) model; this model can be compared with the less general model given in [2].

The STARMAX model is easily generalized to arbitrary orders and spatial constraints. For example,

$$\begin{aligned} \mathbf{X}_{t+1} &= \begin{bmatrix} \mathrm{D}_1\Phi_1 & \mathrm{D}_2\Phi_2 \\ \mathrm{I} & 0 \end{bmatrix} \mathbf{X}_t + \begin{bmatrix} \mathrm{D}_1(\Phi_1 - \Theta_1) \\ \mathrm{D}_2\Phi_2 \end{bmatrix} \boldsymbol{\nu}_t \\ \mathbf{Y}_t &= \begin{bmatrix} \mathrm{I}, 0 \end{bmatrix} \mathbf{X}_t + \Gamma\mathbf{u}_t + \boldsymbol{\nu}_t \end{aligned} \tag{13.25}$$

yields the STARMAX(2, 1) model

$$\mathbf{Y}_t = \mathrm{D}_1\Phi_1\mathbf{Y}_{t-1} + \mathrm{D}_2\Phi_2\mathbf{Y}_{t-2} + \Gamma\mathbf{u}_t + \boldsymbol{\nu}_t - \mathrm{D}_1\Theta_1\boldsymbol{\nu}_{t-1}, \tag{13.26}$$

where I is the $q \times q$ identity matrix, D_1 and D_2 are the $q \times q$ first-order and second-order spatial constraint matrices and Φ_1, Φ_2, Θ_1 are diagonal $q \times q$ matrices, as before. In this case, $\mathbf{X}_t$ is $2q \times 1$ and $\mathbf{Y}_t$ is $q \times 1$.

If we examine Eq. (13.23) for an individual site, $Y_t(s)$, we see that

$$\begin{aligned} Y_t(s) &= \sum_{k=1}^{q} d_{s,k}\phi_k Y_{t-1}(k) - \sum_{k=1}^{q} d_{s,k}\theta_k \nu_{t-1}(k) + \nu_t(s) \\ &= \phi_s Y_{t-1}(s) - \theta_s \nu_{t-1}(s) + \nu_t(s) \\ &\qquad + \sum_{k\neq s} d_{s,k}\{\phi_k Y_{t-1}(k) - \theta_k \nu_{t-1}(k)\}, \end{aligned} \tag{13.27}$$

for $s = 1, \ldots, q$, where we have written $D = \{d_{s,k}\}$, $\Phi = \text{diag}\{\phi_1, \ldots, \phi_q\}$ and $\Theta = \text{diag}\{\theta_1, \ldots, \theta_q\}$. From Eq. (13.27) we may deduce that forecasting the outcome at location s at time $n+1$, given the data $\mathbf{Y}_1, \ldots, \mathbf{Y}_n$, consists of two parts. The first part is based on the model for the individual site s, and the second part is a spatially weighted linear combination of the predicted outcomes from the other sites. We can write this predictor as

$$Y_{n+1}^n(s) = \phi_s Y_n(s) - \theta_s \nu_n^n(s) + \sum_{k\neq s} d_{s,k}\{\phi_k Y_n(k) - \theta_k \nu_n^n(k)\}, \tag{13.28}$$

where $\boldsymbol{\nu}_n^n = \left(\nu_n^n(1), \ldots, \nu_n^n(q)\right)' = \mathrm{E}(\boldsymbol{\nu}_n \mid \mathbf{Y}_1, \ldots, \mathbf{Y}_n)$.

An interesting aspect of this problem is that, in biosurveillance, the processes are typically evolving relatively slowly, over a week in our example. In that case, it may be of interest to include contemporaneously measured outcomes in the predictions. For example, we might consider changing Eq. (13.28) to

$$Y_{n+1}^n(s) = \phi_s Y_n(s) - \theta_s \nu_n^n(s) + \sum_{k\neq s} \delta_{s,k} Y_{n+1}(k) + d_{s,k}\{\phi_k Y_n(k) - \theta_k \nu_n^n(k)\}, \tag{13.29}$$

where the $\delta_{s,k}$ may be unknown parameters that can be spatially constrained. The prediction equations (13.29) can be obtained by rewriting the basic model given in Eq. (13.23) as

$$(I - \Delta)\mathbf{Y}_t = D\Phi\mathbf{Y}_{t-1} + \boldsymbol{\nu}_t - D\Theta\boldsymbol{\nu}_{t-1}, \tag{13.30}$$

where Δ has zeros along its diagonal and $\delta_{s,k}$ on the off-diagonals. Higher order models can be written analogously. There will be an identifiability problem here unless we assume that $\boldsymbol{\nu}_t$ has independent components; for further details, see Shumway and Stoffer ([19], pp. 397–400).

Another possibility that does not include adding more parameters would be to include contemporaneously measured outcomes in terms of their local innovations, namely,

$$Y_{n+1}^n(s) = \phi_s Y_n(s) - \theta_s \nu_n^n(s) + \sum_{k\neq s} \delta_{s,k}\{Y_{n+1}(k) - [\phi_k Y_n(k) - \theta_k \nu_n^n(k)]\}. \tag{13.31}$$

The prediction equation (13.31) may also be obtained from Eq. (13.30) by setting $d_{s,k} = -\delta_{s,k}$, where, as previously indicated, $\boldsymbol{v}_t$ has independent components.

We formally define a STARMAX(a, b) model for spatial data $\mathbf{Y}_t = \big(Y_t(1), \ldots, Y_t(q)\big)'$ collected (possibly irregularly) at q locations and regularly over time $t = 1, \ldots, n$, with inputs $\mathbf{u}_t$, as

$$\mathbf{Y}_t = \sum_{j=1}^{a} D_j \Phi_j \mathbf{Y}_{t-j} + \Gamma \mathbf{u}_t + \boldsymbol{v}_t - \sum_{j=1}^{b} D_j \Theta_j \boldsymbol{v}_{t-j}, \tag{13.32}$$

where Φ_j and Θ_j are the diagonal $q \times q$ matrices, $\boldsymbol{v}_t$ is white noise with the variance–covariance matrix R, and D_j are the $q \times q$ spatial constraint matrices with ones along the diagonal. The model may be put into a state space representation as was done in Eqs. (13.25) and (13.26), and the parameters may be estimated using the innovations likelihood given in Eq. (13.18).

Model identification proceeds as is typical for a multivariate ARMAX process. First, one would evaluate the dynamics of each univariate series and build the multivariate model from the univariate models. This approach is justified by considering fitting a first-order model, as discussed in Eq. (13.27), where one first proceeds by setting $d_{sk} = 0$ in Eq. (13.27). Then, the results of the individual model fits can be combined to construct the overall model; consideration of the distance matrices, D_j in Eq. (13.32), may be handled as described in [2]. We give a brief discussion here.

The specification of the spatial weighting matrices is left to the investigator of the space–time system so that as many of the physical characteristics and constraints of the map can be employed. For regularly spaced systems, equal scaled weighting is typically employed (see [20] or [4]). The weighting is a measure of inverse distance between neighbors in which the nearest neighbors have the most effect on each other. The weighting matrices adopted in the equal scaled scheme are of the form $W_{sk}^{(j)} = 1/n_s^{(j)}$ if locations s and $k = 1, \ldots, q$ are jth-order neighbors, and $W_{sk}^{(j)} = 0$ otherwise, where $W_{sk}^{(j)}$ is the skth element of a spatial weighting matrix $W^{(j)}$, and $n_s^{(j)}$ is the number of jth-order neighbors possessed by site s. Thus all nonzero weights of a given site for a particular spatial order are equal and scaled so that $\sum_k W_{sk}^{(j)} = 1$. To employ this idea in the STARMAX model, one could choose the spatial distance matrices in to be of the form

$$D_j = I + W^{(1)} + \cdots + W^{(\nu_j)}$$

where ν_j is the spatial order of the jth coefficient in the model.

For some irregularly spaced systems, a reasonable method of spatial weighting would be to base the weights on the Euclidean distance between each location. For example, if δ_{sk} is the distance between location s and location k, possible weighting functions might be $d_{j,sk} = c_j[\delta_{sk} + 1]^{-\alpha_j}$, $d_{j,sk} = c_j[\delta_{sk}^2 + 1]^{-\alpha_j}$ or $d_{j,sk} = c_j \exp[-\alpha_j \delta_{sk}]$ for some constants $c_j > 0$ and $\alpha_j \geqslant 0$, where $d_{j,sk}$

is the s, kth element of D_j. To include the effects of order on spatial weighting, one might choose $\alpha_j = j\alpha$, for example, where α is a constant. This approach, of course, may be modified by allowing nonsymmetric weighting schemes wherein $d_{j,sk} \neq d_{j,ks}$ when $s \neq k$.

An alternative to the use of weighting as a function of the distance between sites is to use the variogram to spatially weight the data. The variogram is currently used in kriging as a method for estimating the spatial variation of the map. Let δ_{sk} be the distance between site s and site k, and suppose that $E[Y_{t+h}(s) - Y_t(k)] = 0$ and

$$\mathrm{var}[Y_{t+h}(s) - Y_t(k)] = 2\eta_h(\delta_{sk}) .$$

The function $\eta_h(\delta_{sk})$ is called the variogram at lag h. These assumptions imply that the spatial variation is stationary in its increments and is weaker than the assumption of second-order spatial stationarity. The estimation of the variogram depends on the particular phenomenon being studied. If the sites are at regular spacing, one can estimate the variogram using the sample variance. If the experimental sites are irregularly spaced, they may be grouped by classes of distance (δ and angle ϕ), for example, all pairs of points less than one mile apart, from one to two miles apart, and so forth, separating the pairs oriented approximately north, south, east, and west. After estimating the variogram, one may wish to propose and fit a theoretical model. Possible models, whose behaviors are based on the sample variogram of actual data and are widely used, are the *generalized linear model*, the *spherical model*, the *exponential model* and the *Gaussian model*; see [2] for details. Once the experimenter has arrived at a suitable measure of spatial variation via the variogram, the measure may be used to create the spatial weighting matrices D based on inverse distance. For example, if $\hat{\eta}_1(\delta_{sk})$ is the estimated variogram between site s and site k at lag 1, one could choose D_1 to have elements of the form $d_{1,sk} = c[\hat{\eta}_1(\delta_{sk}) + 1]^{-\alpha}$ or $d_{1,sk} = c\,\exp\{-\alpha\hat{\eta}_1(\delta_{sk})\}$ for some constants $\alpha \geqslant 0$ and $c > 0$. Again note that one may choose asymmetric spatial weights.

13.5 Data Analysis

In this section we present an analysis of surveillance data. For background, the CDC receives weekly mortality reports from 122 cities and metropolitan areas in the United States within two, three weeks from the date of death. These reports summarize the total number of deaths occurring in these cities/areas each week, as well as the number due to pneumonia and influenza. This system consistently covers approximately one-third of the deaths in the United States and provides CDC epidemiologists with preliminary information with which to evaluate the impact of influenza on mortality in the United States and the severity of the currently circulating virus strains. The advantage of this system is that it provides

Fig. 13.1: Location of the three cities in our example; New York City in New York, Newark in New Jersey, and Philadelphia in Pennsylvania.

timely data two to three years before finalized mortality data are available from the National Center for Health Statistics (NCHS).

Most often, the data collected are counts. Although Poisson models have been developed for correlated data, we have found that the correlation structure and multivariate nature (considering both space and time) of surveillance data may be too complicated for analysis by Poisson time series models at this time. We do acknowledge the fact that models such as generalized linear ARMA models (see, e.g., [21]) and exponential family state space models (see, e.g., [22], Chapter 10) may be extended to include the spatial dimension. Also, the non-Gaussian models and corresponding methodology based on Markov Chain Monte Carlo methods presented in [23] might be able to be extended to the spatio-temporal problem. Although we may produce slightly better predictions developing models for count data, it is not clear whether the complexity of such models will render them useless for obtaining quick predictions that are needed for biosurveillance. We have, however, found that simple transformations such as differencing can accomplish much in simplifying the correlation structure of the data. The mortality data collected by the CDC exhibit long memory and level shifts.

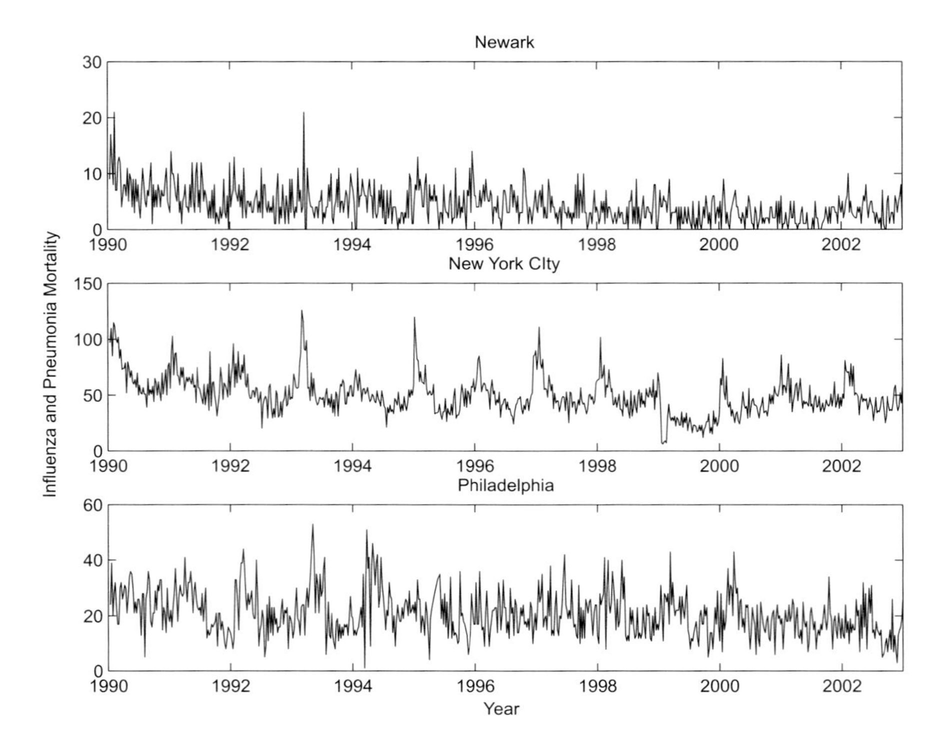

Fig. 13.2: Influenza and Pneumonia Mortality for Newark, New York City, and Philadelphia.

We discovered that a simple variance-stabilizing transformation (the square root transformation used for count data) followed by a differencing operation lead to a simplified correlation and distribution structure for the data in much the same way that one proceeds with financial data by considering percentage change (or returns) rather than the raw data.

For example, Fig. 13.1 displays the spatial relationship of three sites in the northeastern United States: Newark, New York City and Philadelphia. Figure 13.2 shows the combined influenza and pneumonia mortality series at these three CDC sites from 1990 to 2003. We will denote the mortality series by $m_t(s)$ for $s = 1$ (Newark), 2 (New York City), and 3 (Philadelphia). Figure 13.3 shows the autocorrelation function (ACF) of each series. The ACF for Newark shows classic long memory behavior. In addition to long memory, the ACF of the New York City and Philadelphia series show seasonal persistence. Also, it is clear from Fig. 13.2 that each series shows a slight negative trend over the approximate 13 year period and this may be accounting for the signs of long memory in the ACFs. The negative trend suggests that we might first difference each series prior to an analysis.

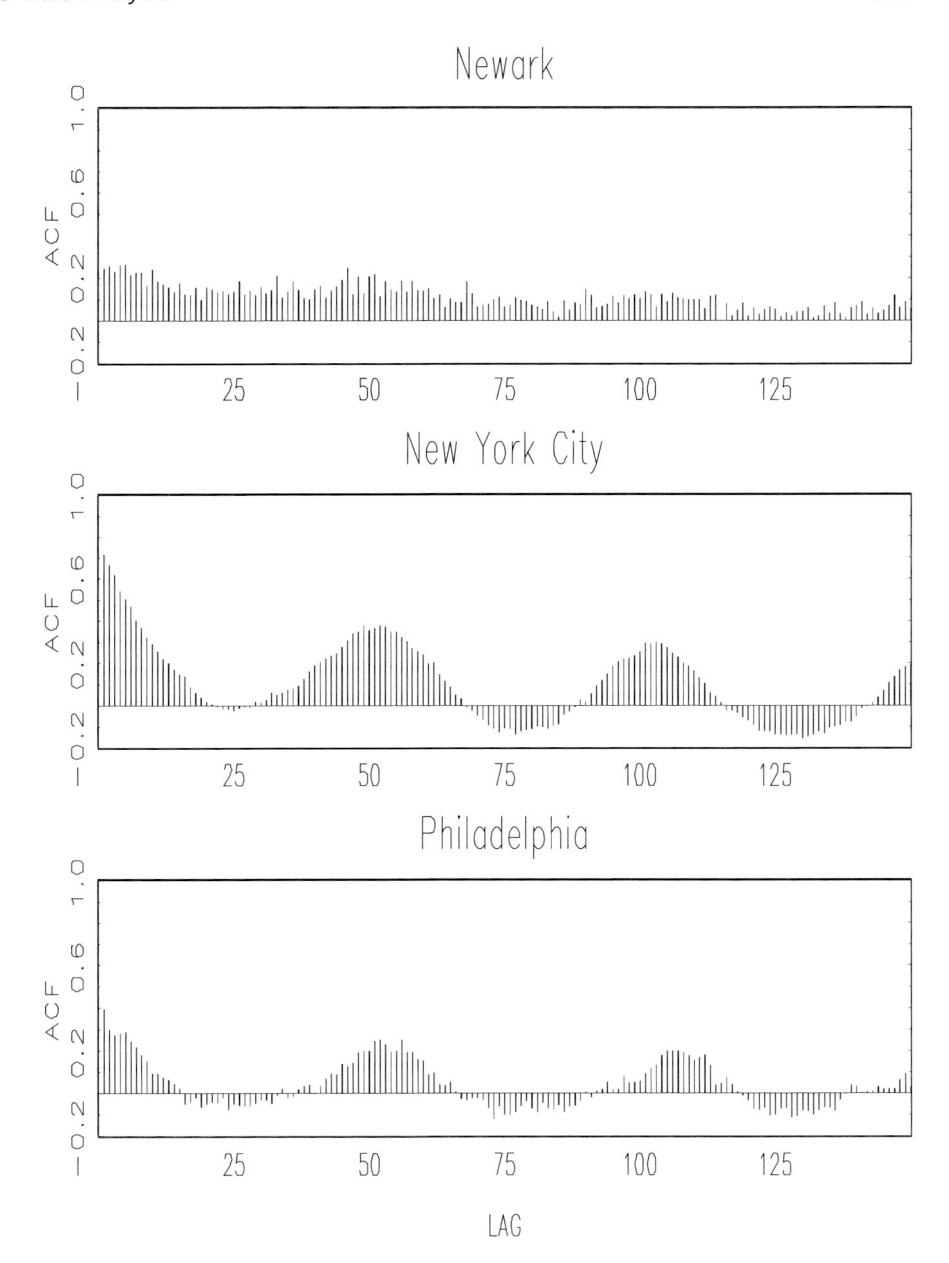

Fig. 13.3: ACFs of the Newark, New York City, and Philadelphia mortality series.

Because the data are counts, we first take a square root transformation for each site $s = 1, 2, 3$. Let

$$m_t^*(s) = \sqrt{m_t(s) + 1}\,,$$

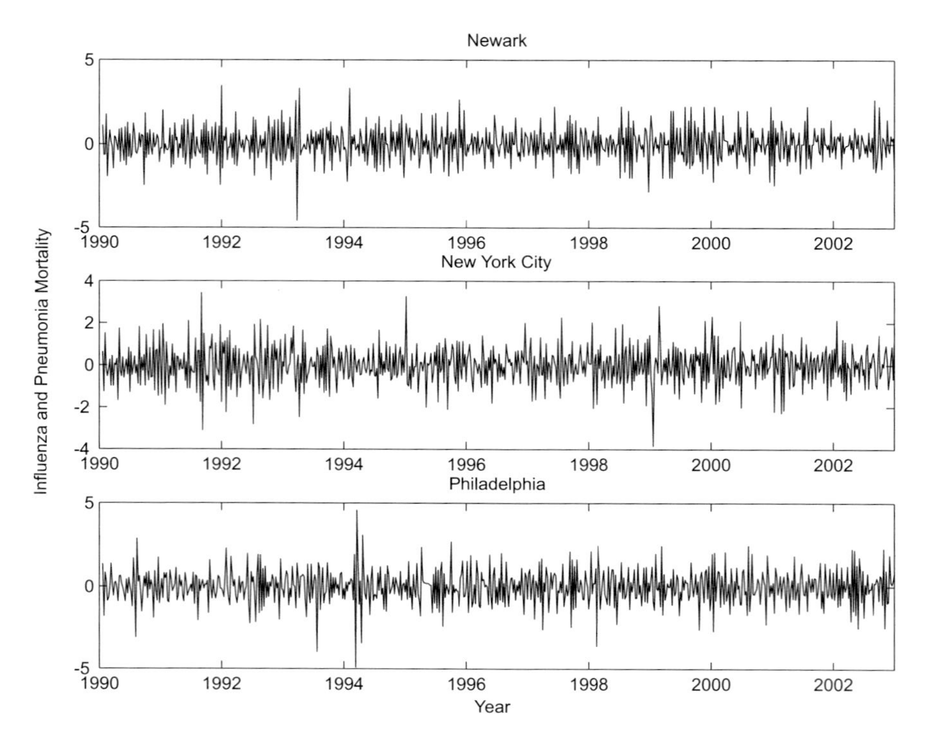

Fig. 13.4: Transformed Newark, New York City, and Philadelphia mortality series.

which may be considered a variance-stabilizing transformation. Figure 13.4 shows the first difference of each transformed series, that is,

$$\nabla m_t^*(s) = m_t^*(s) - m_{t-1}^*(s) .$$

It is clear that the transformed series are rather well behaved stationary series, with the possibility of a few outliers. Figure 13.5 compares the empirical distribution functions (EDFs) of each series, $\nabla m_t^*(s)$, with the corresponding Gaussian CDF. Clearly, except for the aforementioned outliers, a Gaussian model for the transformed series seems appropriate in this case.

Figures 13.6–13.8 show the ACFs and PACFs of each of the transformed series, $\nabla m_t^*(s)$, for $s = 1, 2, 3$. We note that each series $m_t^*(s)$ displays classic IMA$(1, 1)$ behavior. After preliminary fits of IMA$(1, 1)$ models to each separate series, it was clear that an AR term is needed for some of the series. Finally, we settle on a STARMAX$(1, 1)$ model for the transformed, difference data, $\mathbf{Y}_t = \left(\nabla m_t^*(1), \nabla m_t^*(2), \nabla m_t^*(3)\right)'$,

$$\mathbf{Y}_t = D\Phi\mathbf{Y}_{t-1} + \boldsymbol{\nu}_t - D\Theta\boldsymbol{\nu}_{t-1} , \tag{13.33}$$

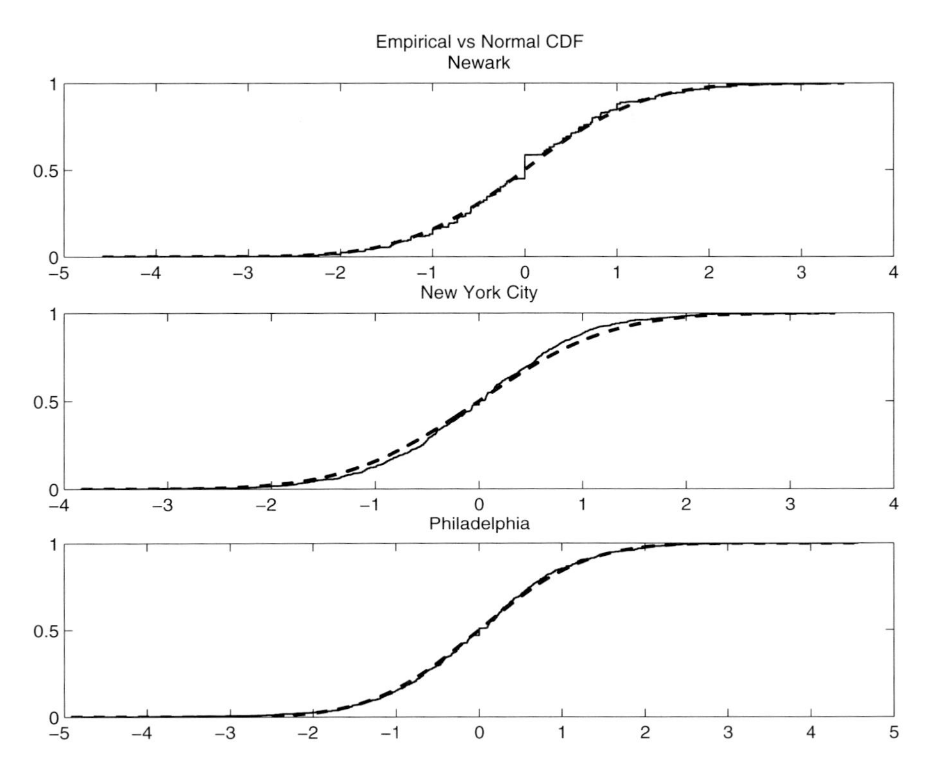

Fig. 13.5: EDFs (solid line) versus Normal CDFs (dashed line) for the transformed Newark, New York City, and Philadelphia mortality series.

where $\nu_t \sim$ iid $N_3(0, R)$. We note that forecasts of $m_t^*(s)$ can be easily obtained from the forecasts of $\mathbf{Y}_t$.

In our analysis, we used maximum-likelihood estimation to estimate the six off-diagonal elements of D, which we do not assume is symmetric, in addition to the three diagonal elements of Φ, the three diagonal elements of Θ, and the six elements of R. Thus, a total of 18 parameters are being estimated in contrast to the 24 parameters that would be needed for unconstrained model. In addition to reducing the number of parameters, of course, the D matrix helps in understanding the spatial relationships among the sites. The final estimates are as follows:

$$\hat{D} = \begin{bmatrix} 1 & -0.08 & 0.00 \\ -0.05 & 1 & 0.05 \\ -0.04 & -0.04 & 1 \end{bmatrix} \tag{13.34}$$

$$\hat{\Phi} = \text{diag}\{0.00_{(.04)}, 0.09_{(.08)}, 0.12_{(.06)}\} \tag{13.35}$$

$$\hat{\Theta} = \text{diag}\{0.94_{(.04)}, 0.63_{(.04)}, 0.85_{(.04)}\} \tag{13.36}$$

where the terms in parentheses are standard errors, and

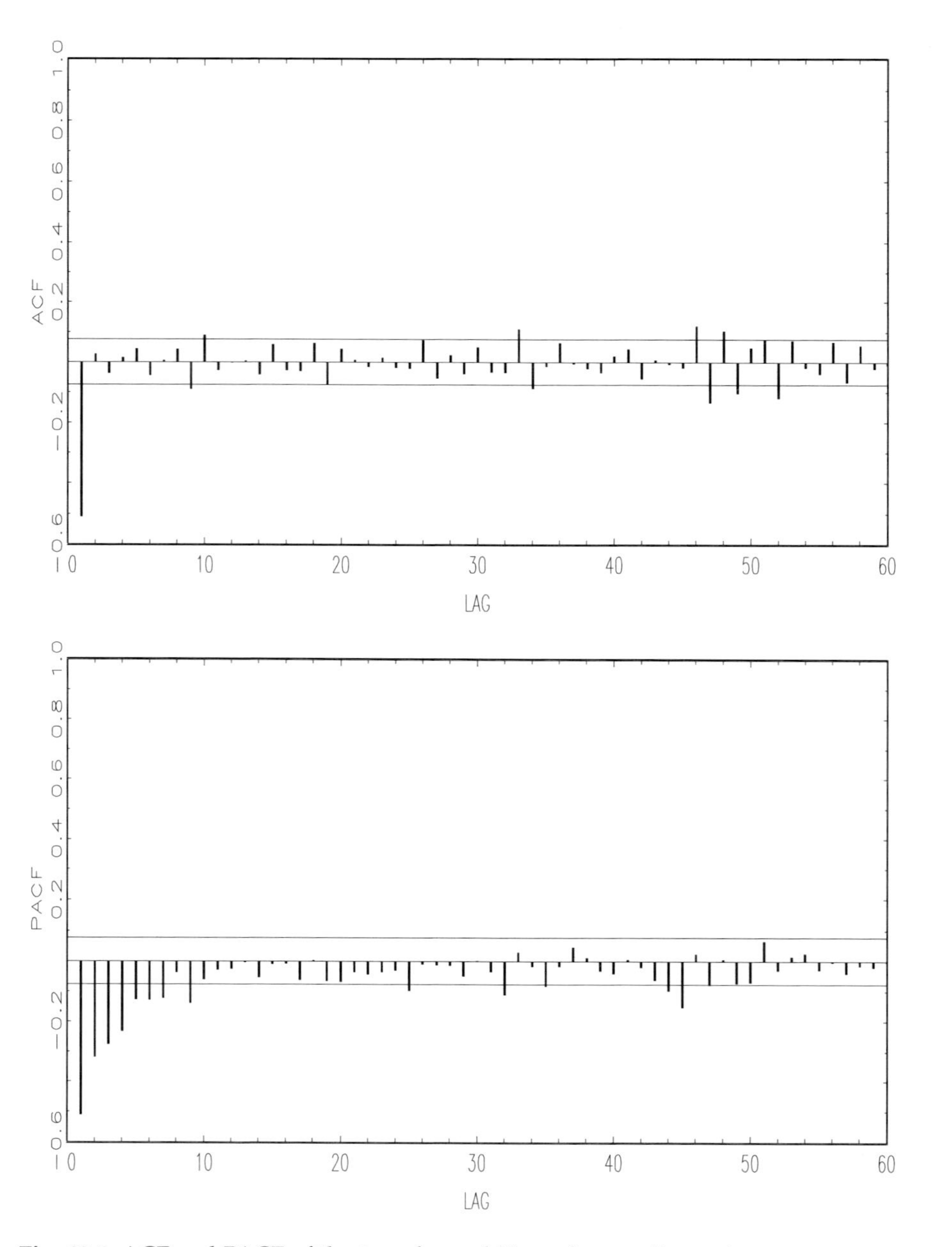

Fig. 13.6: ACF and PACF of the transformed Newark mortality series.

$$\text{chol}(\hat{R}) = \begin{bmatrix} 0.58 & 0.03 & 0.04 \\ & 0.79 & 0.01 \\ & & 0.80 \end{bmatrix} \tag{13.37}$$

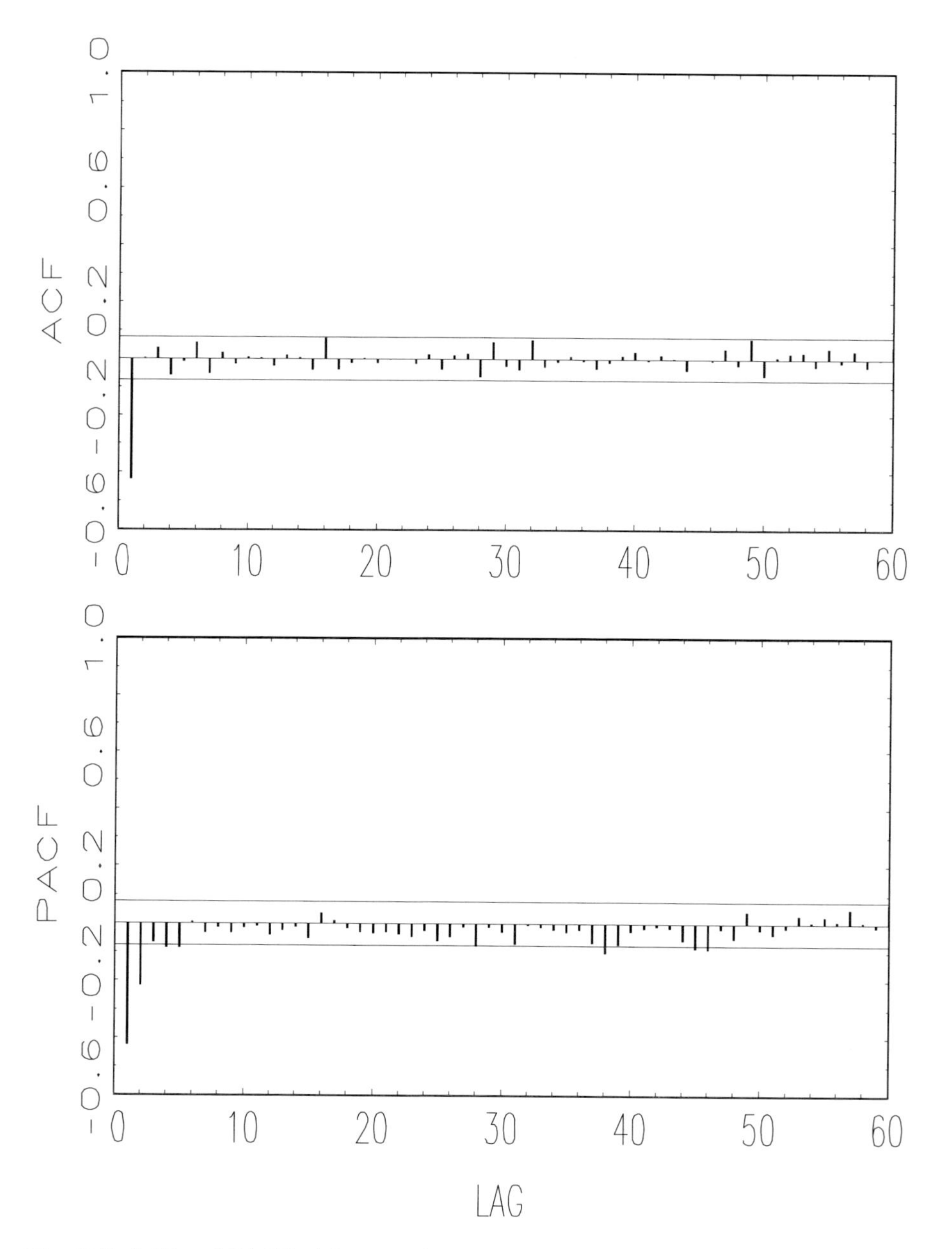

Fig. 13.7: ACF and PACF of the transformed New York City mortality series.

where $\text{chol}(\hat{R})$ is the Cholesky decomposition of the estimate of R (with the zeros deleted for ease of display).

From $\hat{\Phi}$ we note that the AR term is needed only for the Philadelphia mortality series, but not for Newark or New York City. As expected from the ACFs

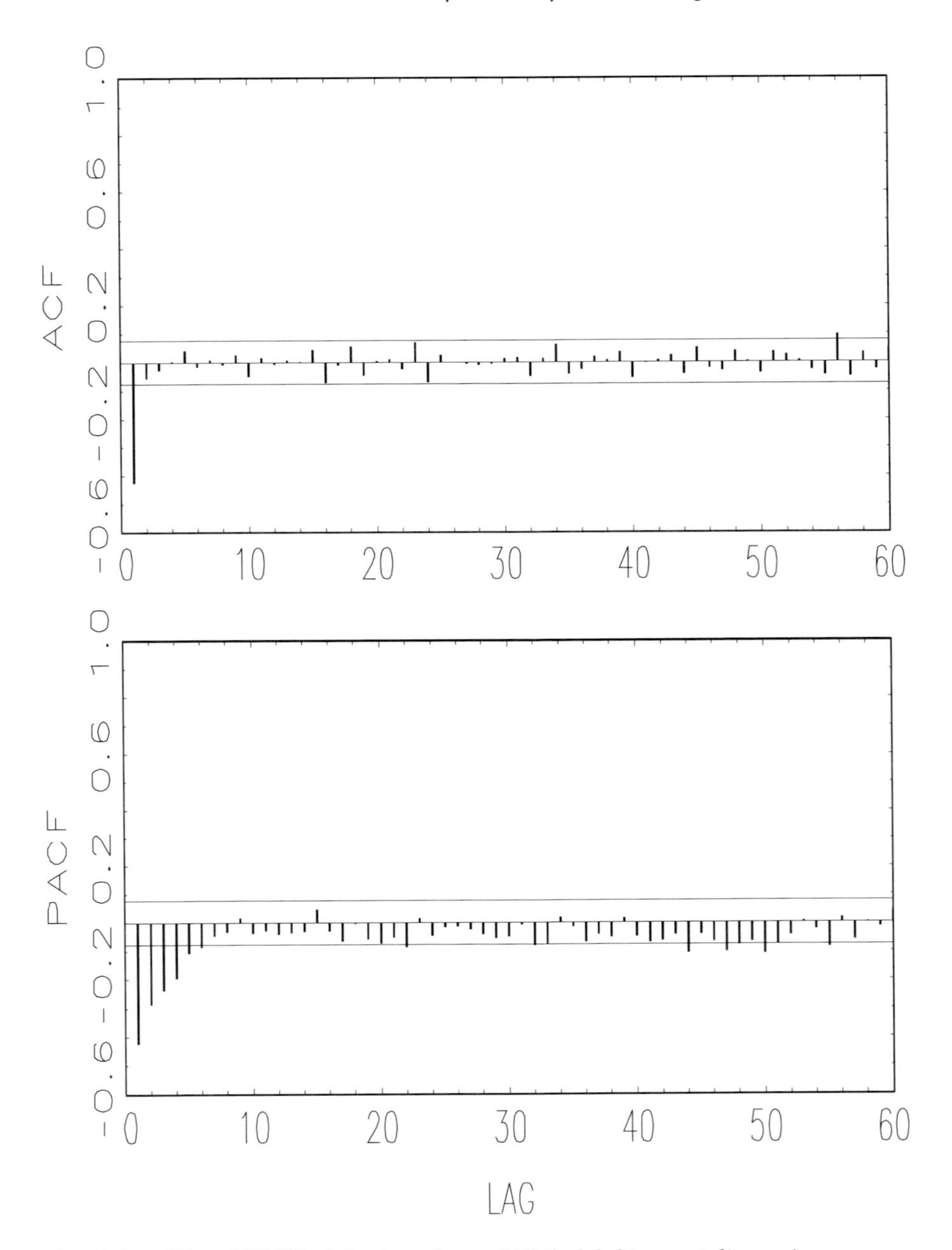

Fig. 13.8: ACF and PACF of the transformed Philadelphia mortality series.

and PACFs in Figs. 13.6–13.8, the terms in $\hat{\Theta}$ are highly significant, supporting the claim that the transformed mortality series are essentially IMA$(1, 1)$ series. It is interesting to consider the values of $\hat{\mathrm{D}}$. For example, the off-diagonal elements of $\hat{\mathrm{D}}$ are small, indicating that the transformed processes are nearly uncorrelated

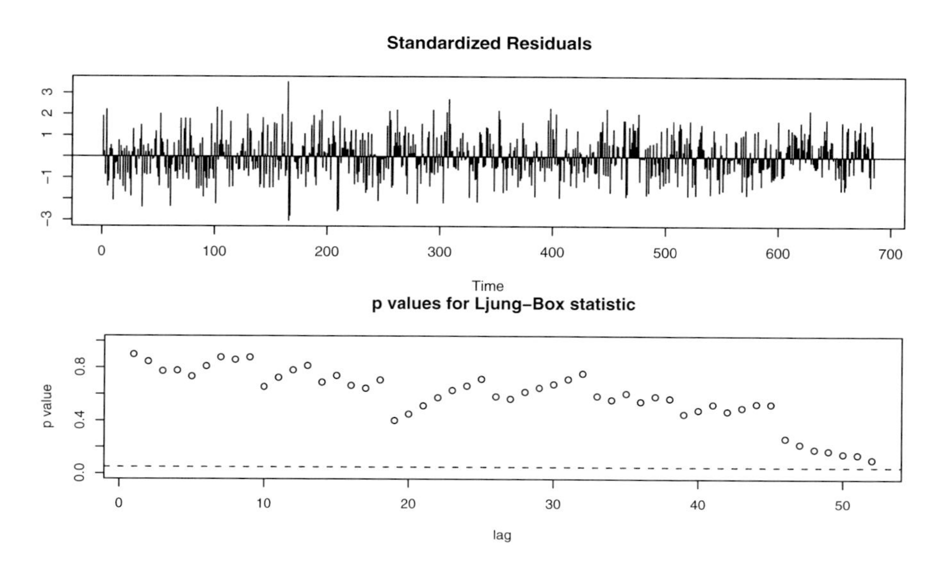

Fig. 13.9: Diagnostics for the Newark fit.

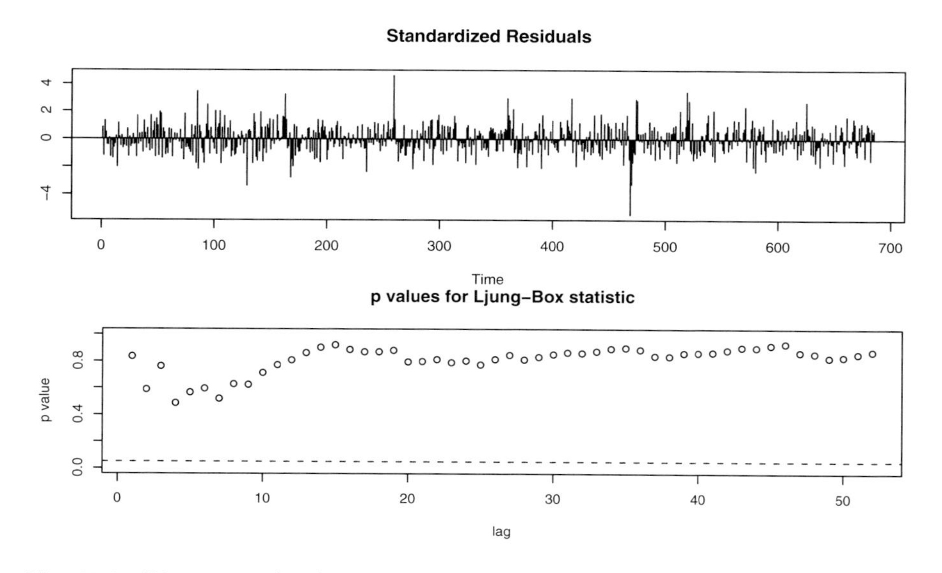

Fig. 13.10: Diagnostics for the New York City fit.

with each other. This fact is surprising when we consider how close the cities are to one another. We also note that the spatial weighting is not symmetric; the spatial relationship using Philadelphia mortality to predict Newark mortality is nonexistent ($\hat{d}_{13} = 0$), whereas the reverse is not true ($\hat{d}_{31} = -0.04$).

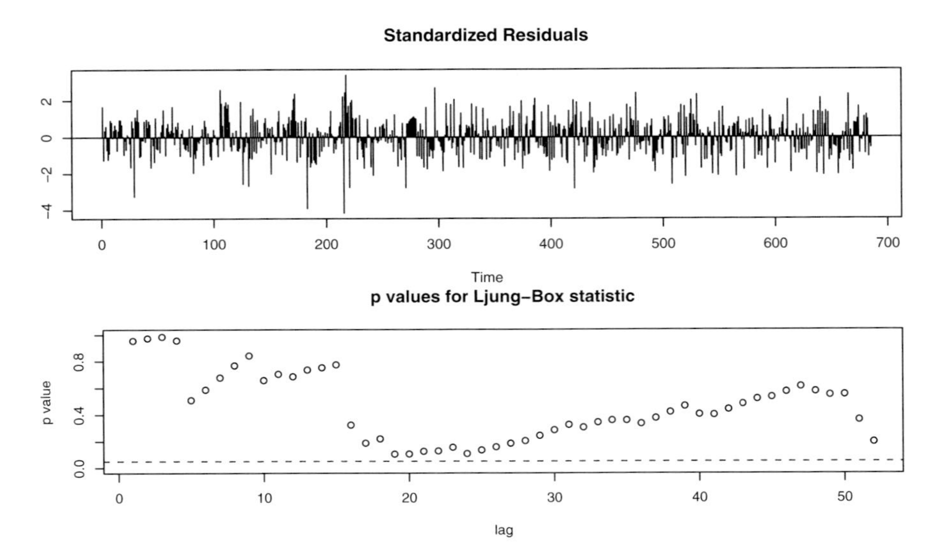

Fig. 13.11: Diagnostics for the Philadelphia fit.

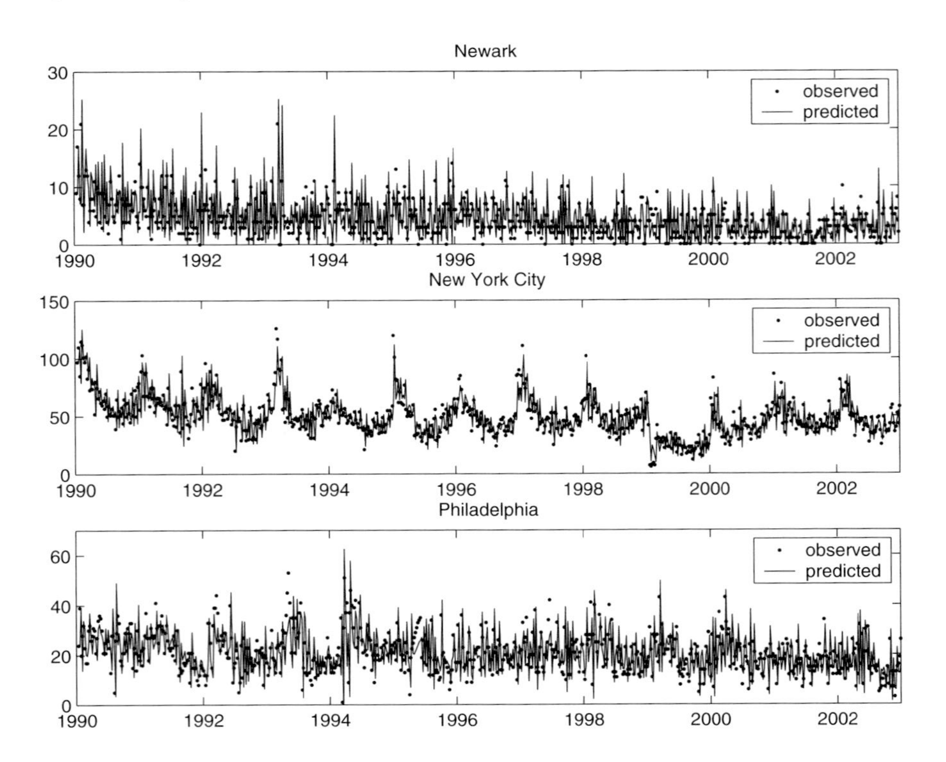

Fig. 13.12: Observed (as points) and predicted (as lines) values for each mortality series.

As previously mentioned, the Newark mortality series exhibit classic long memory behavior. Rather than fit a long memory model, we chose to difference the data, and this may be considered too severe an operation (as opposed to using fractional differencing) in some cases. While this does not seem to be a problem with the New York and Philadelphia series, there is some indication that the Newark series was over-differenced ($\hat{\theta}_1 = 0.94_{(.04)}$ is close to 1). Moreover, as seen in top of Fig. 13.3, the ACF of the Newark series is small, but significant, for large lag values, and it decreases linearly (rather than exponentially); this is a strong indication of long memory. Because our goal is short-term forecasting (one-step-ahead, in particular), we prefer to fit a simpler IMA$(1,1)$-type model than to fit a more complicated ARFIMA-type model to one of the series. As discussed in [24], if the interest is in short-term prediction, low-order ARMA models can produce competitive forecasts when predicting long memory time series with fractionally differenced ARFIMA structure.

Figures 13.9–13.11 show diagnostics for each fit by displaying the standardized innovations

$$\mathbf{e}_t(s) = \left(\mathbf{Y}_t(s) - \hat{\mathbf{Y}}_t^{t-1}(s)\right)/\sqrt{\hat{P}_t^{t-1}(s)}$$

where $\hat{\mathbf{Y}}_t^{t-1}(s)$ and $\hat{P}_t^{t-1}(s)$ are the estimated one-step-ahead prediction and the corresponding estimated MSPE at location s. The bottom halves of Figs. 13.9–13.11 show the p-value for the Ljung–Box–Pierce statistic for each lag, up to lag 52 (lag 52 corresponds to the yearly lag). It is apparent from the diagnostics that the innovations appear to be white noise, although there may be some small amount of correlation left at the one year lag in the Newark and Philadelphia residuals. Further investigation, however, showed that any correlation exhibited at the one year lag is insignificant.

In terms of biosurveillance, the standardized innovations can be used to set up some criteria for raising flags, indicating a series is out of control or has been tampered with. For example, we would suggest having an expert (or experts) closely monitor the series or related events, such as the number of emergency calls, if the standardized innovation exceeds 2.5 or 3. That is, if the observed mortality is larger than the predicted mortality by 2.5 or 3 times the standard prediction error, the process, or related events, should be scrutinized. Figure 13.12 shows each mortality series as points, with the predicted values as lines. We note that the predictions are very good, but there is a tendency to under-predict some peaks in the New York City series.

13.6 Discussion

Motivated by problems in biosurveillance, we extended the work of [1] and [2] for forecasting processes observed in both space and time. We presented the STARMAX model, which is a spatially constrained ARMAX model. The benefit of the model is that identification can be accomplished using well-known results from the fitting ARMAX models, while still reducing the numbers of unknown parameters. We showed the viability of the model by comparing one-week-ahead forecasts with their actual values for weekly influenza and pneumonia mortality in three cities in the northeastern portion of the USA.

Many possible extensions to the modeling technique exist. For example, while the STARMAX model uses the state space model for setting up the likelihood (for estimation) and the Kalman filter (for prediction), we did not fully use the idea of a latent process such as the state process $\mathbf{X}_t$. Recall the general model (without inputs) given in Eqs. (13.4)–(13.5). The data $Y_t(s)$, for $s = 1, \ldots, q$, are observations on a p-dimensional latent process, say $\mathbf{X}_t$, which is evolving as $\mathbf{X}_{t+1} = F\mathbf{X}_t + G\boldsymbol{w}_t$ where F is $p \times p$, G is $p \times r$, and $\boldsymbol{w}_t$ is the $r \times 1$ white noise with the variance–covariance matrix Q. The observations in this case can be written as $\mathbf{Y}_t = M_t\mathbf{X}_t + \boldsymbol{v}_t$ where M_t is a sequence of $q \times p$ measurement matrices, and $\boldsymbol{v}_t$ is white noise with the variance–covariance matrix R. The STARMAX model is only a special case of this general model, but other approaches to the space–time problem can be considered. As a simple example, suppose x_t represents a latent process that is the underlying cause of pneumonia mortality. In addition, for example, suppose that x_t follows an AR(2) process given by

$$x_t = \phi_1 x_{t-1} + \phi_2 x_{t-2} + w_t \, . \tag{13.38}$$

This equation can be written as

$$\mathbf{X}_t = \begin{bmatrix} \phi_1 & \phi_2 \\ 1 & 0 \end{bmatrix} \mathbf{X}_{t-1} + \begin{bmatrix} 1 \\ 0 \end{bmatrix} w_t \, , \tag{13.39}$$

where $\mathbf{X}_t = (x_t, x_{t-1})'$. Suppose further that, as in the data analysis section, we are taking observations at $q = 3$ locations, and let $\mathbf{Y}_t = \big(Y_t(1), Y_t(2), Y_t(3)\big)'$. The observation equation is then

$$\mathbf{Y}_t = M_t\mathbf{X}_t + \boldsymbol{v}_t \, , \tag{13.40}$$

where M_t is a 3×2 matrix (of possibly parameter values) and $\boldsymbol{v}_t$ is white noise that is independent of w_t. For example, if the first two sites are observing x_t, while the third site, which is far from the first two sites, is observing x_{t-1} (i.e., the process delayed by one time unit), then

$$M_t = \begin{bmatrix} 1 & 0 \\ 1 & 0 \\ 0 & 1 \end{bmatrix}$$

for all t. This, of course, assumes the relationship of the observations to the latent process x_t is known. If the relationship is unknown, M_t could consist of parameters to be estimated, with possible spatial constraints among the parameters.

Another problem that we have not focused on, is the case when observations are missing or are partially observed. For example, there were a few cases in the New York City series where no deaths were reported for a particular week, and where only a few deaths were reported in a week. This can be seen in the middle part of Fig. 13.12 around 1999, when there is a sudden drop in mortality. It is more than likely that this effect was caused by under-reporting, although we cannot be certain. In our example we filled in the few cases where zero deaths were reported and left the under-reporting alone. However, if it were a problem, we could try to fix it by using a technique discussed in Shumway and Stoffer ([19], § 4.4). The basic idea is that M_t in Eq. (13.40) would have a zero row for a time t in which data are missing or under-reported. The actual value could be estimated using Kalman smoothing, which we did not discuss here. We refer the reader to [19] for details.

Acknowledgements

This work was partially supported by the Centers for Disease Control and by a grant from the National Science Foundation.

References

[1] D. S. Stoffer. In O. D. Anderson, J. K. Ord, and E. A. Robinson, editors, *Time Series Analysis: Theory and Practice*. Elsevier/NorthHolland, New York, 1985.

[2] D. S. Stoffer. *J. Am. Stat. Assoc.*, 81:762, 1986.

[3] P. E. Pfeifer and S. J. Deutsch. *Technometrics*, 22:35, 1980.

[4] P. E. Pfeifer and S. J. Deutsch. *Technometrics*, 22:397, 1980.

[5] A. D. Cliff and J. K. Ord. *Trans. Inst. Brit. Geograph.*, 64:119, 1975.

[6] H. Huang and N. Cressie. *Comp. Stat. Data Anal.*, 22:159, 1996.

[7] K. V. Mardia, C. Goodall, E. J. Redfern, and F. J. Alonso. *Test*, 7:217, 1998.

[8] D. Higdon. *J. Environ. Ecolog. Stat.*, 5:173, 1998.

[9] D. Higdon. In C. Anderson, V. Barnett, P. C. Chatwin, and A. H. El-Shaarawi, editors, *Quantitative Methods for Current Environmental Issues*. Springer, NewYork, 2002.

[10] C. K. Wikle and N. Cressie. *Biometrika*, 86:815, 1999.

[11] C. K. Wikle and N. Cressie. In L. M. Berliner, D. Nychka, and T. Hoar, editors, *Studies in Atmospheric Sciences*. Springer, New York, 2000.

[12] B. Sanso and L. Guenni. *Appl. Stat.*, 48:345, 1999.

[13] B. Sanso and L. Guenni. *J. Am. Stat. Assoc.*, 95:1089, 2000.

[14] M. West and J. Harrison. *Bayesian Forecasting and Dynamic Models*. Springer, New York, 2nd edition, 1997.

[15] P. E. Brown, P. J. Diggle, M. E. Lord, and P. C. Young. *J. Roy. Stat. Soc. Ser. C*, 50:221, 2001.

[16] J. R. Stroud, P. Muller, and B. Sanso. *J. Roy. Stat. Soc. Ser. B*, 63:673, 2001.

[17] P. E. Caines. *Linear Stochastic Systems*. Wiley, New York, 1988.

[18] J. Haslett and A. E. Raftery. *Appl. Stat.*, 38:1, 1989.

[19] R. H. Shumway and D. S. Stoffer. *Time Series Analysis and Its Applications*. Springer, New York, 2000.

[20] J. Besag. *J. Roy. Stat. Soc. Ser. B*, 36:192, 1974.

[21] R. A. Davis, W. Dunsmuir, and S. Streett. *Biometrika*, 90:777, 2003.

[22] J. Durbin and S. J. Koopman. *Time Series Analysis by State Space Methods*. Oxford University Press, Oxford, 2001.

[23] B. P. Carlin, N. G. Polson, and D. S. Stoffer. *J. Am. Stat. Assoc.*, 87:493, 1992.

[24] K. S. Man. *Int. J. Forecast*, 19:477, 2003.

14 Graphical Modeling of Dynamic Relationships in Multivariate Time Series

Michael Eichler

The identification and analysis of interactions among multiple simultaneously recorded time series is an important problem in many scientific areas. Of particular interest are directed interactions that describe the dynamics of the systems and thus help to determine the causal driving mechanisms of the underlying system. The dynamic relationships among multiple series intuitively can be visualized by a path diagram (or graph), in which the variables are represented by vertices or nodes, and directed edges between the vertices indicate the dynamic or causal influences among the variables. In this chapter, we review recent results on the properties of such graphical representation, which show that path diagrams provide an ideal basis for discussing and investigating causal relationships in multivariate time series. The key role in this graphical approach is played by the so-called global Markov properties, which provide graphical conditions for the (in-)dependences that may be observed if only subprocesses instead of the full process are considered. Such considerations are, for example, central for the discussion of systems that may contain latent variables. The empirical analysis of dynamic interactions is commonly based on the concept of Granger causality. While this concept is well understood in the time domain, the time series of interest often are characterized in terms of their spectral properties. Therefore, particular emphasis will be given to the frequency-domain interpretation of Granger causality and the graphical concepts discussed in this chapter.

14.1 Introduction

The analysis of the interrelationships among multiple simultaneously recorded time series is an important problem in a variety of fields such as economics, engineering, the physical and the life sciences. Of particular interest are the dynamic relationships over time among the series, which help to determine the causal driving mechanisms of the underlying system. In neuroscience, for instance, signals reflecting neural activity such as electroencephalographic (EEG) or local field potentials (LFP) recordings have been used to learn patterns of interactions between brain areas that are activated during certain tasks and to improve thus our understanding of neural processing of information (e.g., [1, 2]).

Handbook of Time Series Analysis. Björn Schelter, Matthias Winterhalder, Jens Timmer

ISBN: 3-527-40623-9

The most commonly used approach for describing and inferring dynamic or causal relationships in multivariate time series is based on vector autoregressive models and the concept of Granger causality [3]. This probabilistic concept of causality is based on the common sense perception that causes always precede their effects in time: if one time series causes another series, knowledge of the former series should help to predict future values of the latter series after influences of other variables have been taken into account. Since the concept does not rely on an *a priori* specification of a causal model, it is particularly suited for empirical investigations of cause-effect relationships; being basically a measure of association, however, it can lead to the so-called spurious causalities if important relevant variables are not included in the analysis (e.g., [4]).

An intuitive approach to summarize the dynamic relationships in complex systems is to represent them in a graph, in which a set of vertices or nodes represents the variables and directed edges between the vertices indicate the dynamic or causal influences among the variables. The graphical representation of causal structures goes back to Wright [5, 6], who introduced path diagrams for the discussion of linear structural equation systems. More recently, graphs have been used to visualize and analyze the dependences among variables in multivariate data; for an introduction to the theory of graphical models we refer to the monographs of Whittaker [7], Cox and Wermuth [8], Lauritzen [9], and Edwards [10]. These theoretical advances and the introduction of Bayesian networks [11, 12] have stimulated new interest in graphical representations of causal structures and have led to the developments of concepts for a graph-theoretic analysis of causality (e.g., [13–16]).

For the analysis of the dynamic relationships in multivariate time series, Eichler [17, 18, 19] has introduced path diagrams that visualize the autoregressive structure of weakly stationary processes and, thus, encode the Granger-causal relationships among the variables of these processes. These graphs provide an ideal basis for discussing and investigating causal relationships in multivariate time series since, on the one hand, their Markov interpretation allows conclusions on which dependences may be observed in arbitrary subprocesses and, on the other hand, they have a natural causal interpretation if the observed process comprises all relevant variables. Thus, the graphs can be used, for instance, to examine whether the observed (in-)dependences in a vector time series are consistent with the theoretically predicted (in-)dependences derived from a hypothesized causal structure that possibly contains latent variables.

In this chapter we review the basic concepts for this graphical approach: Granger causality, path diagrams for vector autoregressions and their Markov properties, and statistical inference for such graphs. Since in many applications, especially in neuroscience, the time series of interest are characterized in terms of their spectral properties, particular emphasis will be given to the frequency-domain interpretation of Granger causality and the related graphical representations. We find that causal modeling in the frequency domain leads to linear structural equation systems for the frequency components of the process, whose structure is visualized by the path diagram associated with the autoregressive representation of the process.

14.2 Granger Causality in Multivariate Time Series

The concept of Granger causality is a fundamental tool for the empirical investigation of dynamic interactions in multivariate time series. This probabilistic concept of causality is based on the common sense conception that causes always precede their effects. Thus an event taking place in the future cannot cause another event in the past or present. This temporal ordering implies that the past and present values of a series X that influences another series Y should help to predict future values of this latter series Y. Furthermore, the improvement in the prediction of future values of Y should persist after any other relevant information for the prediction has been exploited. Suppose that the vector time series $\mathbf{Z}$ comprises all variables that might affect the dependence between X and Y such as confounding variables. Then we say that a series X Granger-causes another series Y with respect to the information given by the series $(X, Y, \mathbf{Z})$ if the value of $Y(t+1)$ can be better predicted by using the entire information available at time t than by using the same information apart from the past and present values of X. Here, "better" means a smaller variance of forecast error.

Because of the temporal ordering, it is clear that Granger causality can only capture functional relationships for which cause and effect are sufficiently separated in time. To describe causal dependences between variables at the same time point, Granger [3] proposed the notion of "instantaneous causality." In general, it is not possible to attribute a unique direction to such "instantaneous causalities" and we therefore will only speak of contemporaneous dependences.

In practice, the use of Granger causality mostly has been restricted to the investigation of linear relationships. This notion of linear Granger causality is closely related to the autoregressive representation of a weakly stationary process.

14.2.1 Granger Causality and Vector Autoregressions

Let $\mathbf{X}_V = \{\mathbf{X}_V(t), t \in \mathbb{Z}\}$ with $\mathbf{X}_V(t) = (X_v(t), v \in V)'$ be a weakly stationary vector time series with mean zero and covariances $\mathbf{c}(u) = \mathbb{E}\mathbf{X}_V(t)\mathbf{X}_V(t-u)'$. Throughout this chapter, we assume that the spectral density matrix

$$\mathbf{f}(\lambda) = \frac{1}{2\pi} \sum_{u \in \mathbb{Z}} \mathbf{c}(u) e^{-i\lambda u}$$

exists and that all its eigenvalues are bounded and bounded away from zero uniformly for all frequencies $\lambda \in [-\pi, \pi]$. Under these assumptions, the process $\mathbf{X}_V$ has an autoregressive representation of the form

$$\mathbf{X}_V(t) = \sum_{u \in \mathbb{N}} \mathbf{a}(u) \mathbf{X}_V(t-u) + \boldsymbol{\varepsilon}_V(t)\,, \tag{14.1}$$

where $\mathbf{a}(u)$ is a square summable sequence of $V \times V$ matrices and $\{\varepsilon_V(t)\}$ is a white noise process with mean zero and nonsingular covariance matrix $\boldsymbol{\Sigma}$. From the equation for $X_i(t)$, we obtain for the mean-square prediction error when $X_i(t)$ is predicted from the past values of $\mathbf{X}_V$

$$\operatorname{var}\big(X_i(t) \mid \bar{\mathbf{X}}_V(t-1)\big) = \operatorname{var}\big(\varepsilon_i(t)\big) = \sigma_{ii} . \tag{14.2}$$

Here, $\bar{\mathbf{X}}_V(t-1) = \{\mathbf{X}_V(t-u), u \in \mathbb{N}\}$ denotes the past values of $\mathbf{X}_V$ at time t and conditional variance is taken to be the variance about the linear projection.

Similarly, if we consider the subprocess $\mathbf{X}_{-j} = \mathbf{X}_{V\setminus\{j\}}$ consisting of all components but X_i, it follows from the above assumptions on the spectral matrix that $\mathbf{X}_{-j}$ has an autoregressive representation

$$\mathbf{X}_{-j}(t) = \sum_{u \in \mathbb{N}} \tilde{\mathbf{a}}(u)\mathbf{X}_{-j}(t-u) + \boldsymbol{\eta}_{-j}(t) , \tag{14.3}$$

where $\{\boldsymbol{\eta}_{-j}(t)\}$ is a white noise process with mean zero and covariance matrix $\tilde{\boldsymbol{\Sigma}}$. Thus, the mean square prediction error for predicting $X_i(t)$ from the past values of $\mathbf{X}_{-j}$ is given by

$$\operatorname{var}\big(X_i(t) \mid \bar{\mathbf{X}}_{-j}(t-1)\big) = \operatorname{var}\big(\eta_i(t)\big) = \tilde{\sigma}_{ii} . \tag{14.4}$$

In general, the mean square prediction error in Eq. (14.4) will be larger than that in Eq. (14.2), and the two variances will be equal if and only if the best linear predictor of $X_i(t)$ based on the full past $\bar{\mathbf{X}}_V(t-1)$ does not depend on the past values of X_j. This leads to the following definition of Granger noncausality, which we state more generally for vector subprocesses $\mathbf{X}_I$ and $\mathbf{X}_J$. Here, $|\mathbf{A}|$ denotes the determinant of a square matrix $\mathbf{A}$.

Definition 14.1. Let I and J be two disjoint subsets of V. Then $\mathbf{X}_J$ is *Granger-noncausal* for $\mathbf{X}_I$ with respect to $\mathbf{X}_V$ if the following two equivalent conditions hold:

(i) $\big|\operatorname{var}\big(\mathbf{X}_I(t) \mid \bar{\mathbf{X}}_{V\setminus J}(t-1)\big)\big| = \big|\operatorname{var}\big(\mathbf{X}_I(t) \mid \bar{\mathbf{X}}_V(t-1)\big)\big|$;

(ii) $\mathbf{a}_{IJ}(u) = 0$ for all $u \in \mathbb{N}$.

Furthermore, if $\boldsymbol{\Sigma}_{IJ} = 0$, we say that $\mathbf{X}_I$ and $\mathbf{X}_J$ are *contemporaneously uncorrelated* with respect to $\mathbf{X}_V$.

In other words, the variables $\mathbf{X}_I(t)$ and $\mathbf{X}_J(t)$ are contemporaneously uncorrelated with respect to $\mathbf{X}_V$ if they are uncorrelated after removing the linear effects of $\bar{\mathbf{X}}_V(t-1)$. We note that the autoregressive representations describe only linear relationships among the variables and thus, strictly speaking, relate to linear Granger noncausality. In the sequel, we will use the term Granger noncausality in this restricted meaning.

In practice, tests for Granger noncausality are mostly based on condition (ii) as it is formulated only in terms of the autoregressive coefficients in the full model and, thus, does not require fitting of multiple models (e.g., [4, 20–22]); the measure for conditional linear feedback proposed by Geweke [23], however, is based on condition (i).

From the definition of Granger noncausality in terms of the autoregressive parameters, it is clear that the notion of Granger noncausality depends on the multivariate time series $\mathbf{X}_V$ available for the analysis. If we consider only a subprocess $\mathbf{X}_{V'}$ with $V' \subseteq V$ instead of the full process $\mathbf{X}_V$, the vector time series $\mathbf{X}_{V'}$ has again an autoregressive representation

$$\mathbf{X}_{V'}(t) = \sum_{u \in \mathbb{N}} \tilde{\mathbf{a}}(u)\mathbf{X}_{V'}(t-u) + \tilde{\boldsymbol{\varepsilon}}_{V'}(t) ,$$

but the coefficients $\tilde{\mathbf{a}}(u)$ in general will differ from the coefficients $\mathbf{a}_{V'V'}(u)$ in the representation (14.1). To illustrate this dependence on the set of selected variables, we consider the four-dimensional vector autoregressive process $\mathbf{X}_V$ with components

$$\begin{aligned} X_1(t) &= \alpha X_4(t-2) + \varepsilon_1(t) , \\ X_2(t) &= \beta X_4(t-1) + \gamma X_3(t-1) + \varepsilon_2(t), \\ X_3(t) &= \varepsilon_3(t), \\ X_4(t) &= \varepsilon_4(t), \end{aligned} \tag{14.5}$$

where $\varepsilon_\nu(t)$, $\nu = 1, \ldots, 4$ are independent and identically normally distributed with mean zero and variance σ^2. From Eq. (14.5), we find that, for example, X_3 Granger-causes X_2 with respect to $\mathbf{X}_V$, but not X_1 or X_4. However, if we consider only the three-dimensional subprocess $\mathbf{X}_{\{1,2,3\}}$, simple calculations show that $\mathbf{X}_{\{1,2,3\}}$ is given by

$$\begin{aligned} X_1(t) &= \frac{\alpha\beta}{1+\beta^2} X_2(t-1) - \frac{\alpha\beta\gamma}{1+\beta^2} X_3(t-2) + \tilde{\varepsilon}_1(t) , \\ X_2(t) &= \gamma X_3(t-1) + \tilde{\varepsilon}_2(t) , \\ X_3(t) &= \tilde{\varepsilon}_3(t) , \end{aligned} \tag{14.6}$$

where $\tilde{\varepsilon}_3(t) = \varepsilon_3(t)$, $\tilde{\varepsilon}_2(t) = \varepsilon_2(t) + \beta X_4(t-1)$, and

$$\tilde{\varepsilon}_1(t) = \varepsilon_1(t) - \frac{\alpha\beta}{1+\beta^2} \varepsilon_2(t-1) + \frac{\alpha}{1+\beta^2} X_4(t-2) .$$

From this representation, it follows that X_3 Granger-causes not only X_2 but also X_1 with respect to $\mathbf{X}_{\{1,2,3\}}$. In contrast, if we restrict the information further and consider only the bivariate subprocess $\mathbf{X}_{\{1,3\}}$, we obtain from Eq. (14.6) that the two components X_1 and X_3 are two uncorrelated white noise processes; in particular, this implies that X_3 is Granger-noncausal for X_1 with respect to $\mathbf{X}_{\{1,3\}}$.

14.2.2 Granger Causality in the Frequency Domain

In many applications, the time series of interest are characterized in terms of their frequency properties; typical examples can be found in Chapters 15 and 16. It is therefore important to examine the relationships among multiple time series also in the frequency domain. The frequency-domain analysis of weakly stationary vector time series $\mathbf{X}_V$ is based on the spectral representation of $\mathbf{X}_V$, which is given by

$$\mathbf{X}_V(t) = \int_{-\pi}^{\pi} e^{i\lambda t}\, dZ_{\mathbf{X}_V}(\lambda), \tag{14.7}$$

where $dZ_{\mathbf{X}_V}(\lambda)$ is a random process on $[-\pi, \pi]$ that takes values in $\mathbb{C}^V$ and has mean zero and orthogonal increments (e.g., [24]). In this representation, the complex-valued random increments $dZ_{X_i}(\lambda)$ indicate the frequency components of the time series X_i at frequency λ. The increments are related to the spectral density matrix of $\mathbf{X}_V$ by

$$\mathbb{E}\big(dZ_{\mathbf{X}_V}(\lambda)\, dZ_{\mathbf{X}_V}(\mu)'\big) = \mathbf{f}(\lambda)\delta(\lambda - \mu)\, d\lambda\, d\mu\,,$$

where $\delta(u)$ is the Dirac-delta function. In other words, the spectral density matrix $\mathbf{f}(\lambda)$ can be viewed as the covariance matrix of the frequency components of $\mathbf{X}_V$ at frequency λ. Similarly, let

$$\varepsilon_V(t) = \int_{-\pi}^{\pi} e^{i\lambda t}\, dZ_{\varepsilon_V}(\lambda)$$

be the spectral representation of the error process $\varepsilon_V = \{\varepsilon_V(t)\}$ in the autoregressive representation of $\mathbf{X}_V$ in Eq. (14.1). Since ε_V is a white noise process with the covariance matrix $\mathbf{\Sigma}$, the increments $dZ_{\varepsilon_V}(\lambda)$ satisfy

$$\mathbb{E}\big(dZ_{\varepsilon_V}(\lambda)\, dZ_{\varepsilon_V}(\mu)'\big) = \mathbf{\Sigma}\delta(\lambda - \mu)\, d\lambda\, d\mu\,.$$

The autoregressive representation implies that the frequency components of the processes $\mathbf{X}_V$ and ε_V are related by the linear equation system

$$dZ_{\mathbf{X}_V}(\lambda) = \mathbf{A}(\lambda)\, dZ_{\mathbf{X}_V}(\lambda) + dZ_{\varepsilon_V}(\lambda)\,, \tag{14.8}$$

where

$$\mathbf{A}(\lambda) = \sum_{u \in \mathbb{N}} \mathbf{a}(u) e^{-i\lambda u} \tag{14.9}$$

is the Fourier transform of the autoregressive coefficients $\mathbf{a}(u)$. The coefficient $A_{ij}(\lambda)$ vanishes uniformly for all $\lambda \in [-\pi, \pi]$ if and only if X_j is Granger-noncausal for X_i with respect to $\mathbf{X}_V$. This suggests that the linear equation system Eq. (14.8) reflects the causal pathways by which the frequency components influence each other. More precisely, the complex-valued coefficient $A_{ij}(\lambda)$ indicates how a change in the frequency component of the series X_j affects the frequency component of X_i if all other components are held fixed, that is, $A_{ij}(\lambda)$ measures the direct causal effect of X_j on X_i at frequency λ.

As a coefficient in a linear equation system, $A_{ij}(\lambda)$ is not scale invariant, which makes it difficult to assess the strength of a directed relationship. Baccala and Sameshima [25, 26] used a factorization of the partial spectral coherence to derive a normalized frequency-domain measure for Granger causality, which they called the *partial directed coherence* (PDC). The PDC from X_j to X_i is defined as

$$\pi_{ij}(\lambda) = \frac{\bar{A}_{ij}(\lambda)}{\sqrt{\sum_{k \in V} |\bar{A}_{kj}(\lambda)|^2}},$$

where $\bar{\mathbf{A}}(\lambda) = \mathbf{I} - \mathbf{A}(\lambda)$ and $\mathbf{I}$ is the identity matrix. With this normalization, the PDC indicates the relative strength of the effect of X_j on X_i as compared to the strength of the effect of X_j on the other variables. Thus, partial directed coherence ranks the relative interaction strengths with respect to a given signal source. We note that other normalizations are possible; in Section 14.5, we propose an alternative rescaling based on an asymptotic significance level.

Linear equation systems have been widely used in economics and in the social sciences for simultaneously representing causal and statistical hypotheses relating a set of variables (e.g., [27–29]). In general, the structure of such systems is not uniquely determined by the distribution of the variables and, thus, cannot be determined empirically from data, but, on the contrary, must be determined from prior knowledge of the causal relations. In contrast, the coefficients in the above systems (14.9) are completely specified by the unique autoregressive representation of the process $\mathbf{X}_V$ and the implied requirements that $\mathbf{A}(\lambda)$ must be of the form (14.9) and that the covariance matrix of the error term $dZ_{\varepsilon_V}(\lambda)$ does not depend on the frequency λ.

Finally, we note that such causal interpretations should be treated with caution since they rely on the assumption that all relevant information has been included. The omission of important variables can lead to the so-called spurious causalities, which invalidate the causal interpretation of empirically determined Granger-causal relationships among the variables.

14.2.3 Bivariate Granger Causality

Although Granger [3, 30] always stressed the need to include all relevant information in an analysis to avoid spurious causalities, much of the literature on Granger causality has been concerned with the analysis of relationships between two time series or two vector time series (see, e.g., [31–34]). As a consequence, relationships among multiple time series are still quite frequently investigated using bivariate Granger causality, that is, analyzing pairs of time series separately (see, e.g., [35–38]). For a better understanding of this bivariate approach and its relation to a full multivariate analysis based on multivariate Granger causality, we will discuss in the sequel also the use of bivariate Granger causality for describing directed relationships among multiple time series.

Suppose that $\mathbf{X}_V$ is a weakly stationary process of the form Eq. (14.1). Then for $i, j \in V$ the bivariate subprocess $\mathbf{X}_{\{i,j\}}$ is again a weakly stationary process and has an autoregressive representation

$$\begin{aligned} X_i(t) &= \sum_{u \in \mathbb{N}} \tilde{a}_{ii}(u) X_i(t-u) + \sum_{u \in \mathbb{N}} \tilde{a}_{ij}(u) X_j(t-u) + \tilde{\varepsilon}_i(t)\,, \\ X_j(t) &= \sum_{u \in \mathbb{N}} \tilde{a}_{ji}(u) X_i(t-u) + \sum_{u \in \mathbb{N}} \tilde{a}_{jj}(u) X_j(t-u) + \tilde{\varepsilon}_j(t)\,, \end{aligned} \tag{14.10}$$

where $\tilde{\varepsilon}(t) = \big(\tilde{\varepsilon}_i(t), \tilde{\varepsilon}_j(t)\big)'$ is a white noise process with the covariance matrix $\tilde{\boldsymbol{\Sigma}}$. From this representation, it follows that X_j is bivariately Granger-causal for X_i if and only if the coefficients $\tilde{a}_{ij}(u)$ are zero for all lags $u \in \mathbb{N}$. Similarly, X_i and X_j are bivariately contemporaneously uncorrelated if $\tilde{\sigma}_{ij} = 0$.

14.3 Graphical Representations of Granger Causality

The causal relationships among the variables in complex multivariate systems are often visually summarized by graphs in which the nodes or vertices represent the variables and directed edges between the vertices indicate causal influences among the variables. In this section, we formally define such graphs for representing the multivariate or the bivariate Granger-causal relationships in multivariate time series; the properties of these graphs will then be discussed in Section 14.4.

14.3.1 Path Diagrams for Multivariate Time Series

Intuitively, the Granger-causal relationships in a weakly stationary vector time series $\mathbf{X}_V$ can be encoded and visualized by a path diagram in which the vertices $v \in V$ represent the components X_v of the process and directed edges ($\longrightarrow$) between the vertices indicate Granger-causal influences. To obtain a complete description of the dependence structure of $\mathbf{X}_V$, we additionally include undirected edges ($\dashrightarrow$) to depict contemporaneous correlations between the components of $\mathbf{X}_V$. Since the Granger-causal relationships of $\mathbf{X}_V$ are determined by the autoregressive representation of $\mathbf{X}_V$, we obtain the following definition of path diagrams associated with vector autoregressive processes [19, 39].

Definition 14.2. Let $\mathbf{X}_V$ be a weakly stationary time series with autoregressive representation Eq. (14.1). Then the *path diagram associated with* $\mathbf{X}_V$ is a graph $G = (V, E)$ with vertex set V and edge set E such that for $i, j \in V$ with $i \neq j$

(i) $j \longrightarrow i \notin E \iff a_{ij}(u) = 0 \quad \text{for } u \in \mathbb{N};$

(ii) $j \text{ - - - } i \notin E \iff \sigma_{ij} = 0.$

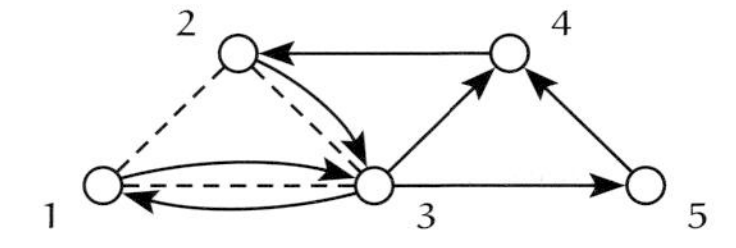

Fig. 14.1: Path diagram associated with a five-dimensional VAR(1) process that satisfies the parameter constraints in Eqs. (14.11) and (14.12).

In other words, the path diagram G contains a directed edge $j \longrightarrow i$ if and only if X_j Granger-causes X_i with respect to the full series $\mathbf{X}_V$; similarly, an undirected edge $i \dashrightarrow j$ is present in the path diagram if and only if X_i and X_j are contemporaneously correlated with respect to $\mathbf{X}_V$. For this reason, such path diagrams have also been called *Granger causality graphs* [17, 40].

The path diagram associated with a process $\mathbf{X}_V$ has also a natural interpretation in terms of the frequency components $dZ_{\mathbf{X}_V}(\lambda)$ of $\mathbf{X}_V$. As we have seen in Section 14.2.2 that the autoregressive representation of $\mathbf{X}_V$ corresponds to the linear equation systems

$$dZ_{\mathbf{X}_V}(\lambda) = \mathbf{A}(\lambda)\, dZ_{\mathbf{X}_V}(\lambda) + dZ_{\varepsilon_V}(\lambda)\,,$$

where the error component $dZ_{\varepsilon_V}(\lambda)$ has basically the covariance matrix $\mathbf{\Sigma}$. It follows that the path diagram G associated with $\mathbf{X}_V$ can also be viewed as the path diagram of the above linear equation systems[1] for all frequencies λ, and its edges equivalently are determined by the conditions

(i) $j \longrightarrow i \notin E \iff A_{ij}(\lambda) = 0 \quad \text{for } \lambda \in [-\pi, \pi];$

(ii) $j \text{ - - - } i \notin E \iff \sigma_{ij} = 0.$

We note that two vertices in a path diagram may be connected by up to three edges. As an example, we consider the five-dimensional vector autoregressive process

$$\mathbf{X}(t) = \mathbf{a}\mathbf{X}(t-1) + \varepsilon(t), \quad \operatorname{var}\big(\varepsilon(t)\big) = \mathbf{\Sigma}$$

with the coefficient matrix

$$\mathbf{a} = \begin{pmatrix} a_{11} & 0 & a_{13} & 0 & 0 \\ 0 & a_{22} & 0 & a_{24} & 0 \\ a_{31} & a_{32} & a_{33} & 0 & 0 \\ 0 & 0 & a_{43} & a_{44} & a_{45} \\ 0 & 0 & a_{53} & 0 & a_{55} \end{pmatrix} \tag{14.11}$$

[1] In path diagrams for structural equation systems, correlated errors commonly are represented by bi-directed edges ($\longleftrightarrow$) instead of dashed lines (- - -). Since in our approach directions are associated with temporal ordering, we prefer (dashed) undirected edges to indicate correlation between the error variables. Dashed edges with a similar connotation are used for covariance graphs (e.g., [8]), whereas undirected edges —— are commonly associated with nonzero entries in the inverse of the variance matrix (e.g., [9]).

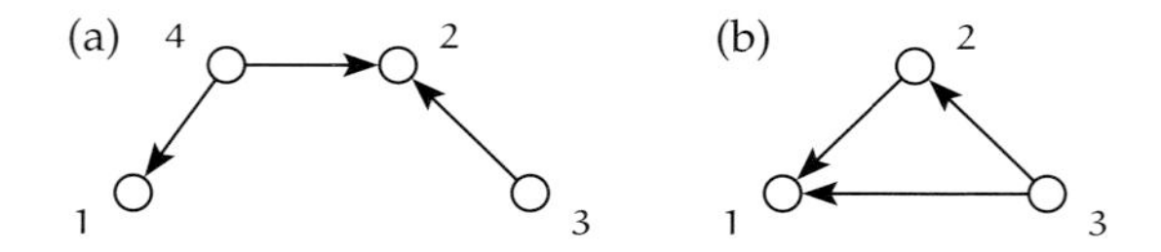

Fig. 14.2: Path diagrams associated with (a) four-dimensional process $\mathbf{X}_V$ given by Eq. (14.5) and with (b) three-dimensional subprocess $\mathbf{X}_{\{1,2,3\}}$.

and the covariance matrix

$$\boldsymbol{\Sigma} = \begin{pmatrix} \sigma_{11} & \sigma_{12} & \sigma_{13} & 0 & 0 \\ \sigma_{21} & \sigma_{22} & \sigma_{23} & 0 & 0 \\ \sigma_{31} & \sigma_{32} & \sigma_{33} & 0 & 0 \\ 0 & 0 & 0 & \sigma_{44} & 0 \\ 0 & 0 & 0 & 0 & \sigma_{55} \end{pmatrix} . \tag{14.12}$$

The autoregressive structure of $\mathbf{X}_V$ is visualized by the associated path diagram shown in Fig. 14.1. The diagram indicates, for example, that there is a feedback loop between variables X_1 and X_3, or that X_1 affects X_4 indirectly with X_3 as mediating variable.

From our discussion in Section 14.2.1, it is clear that the path diagram depends on the set of variables included in the process $\mathbf{X}_V$. To illustrate this dependence, let us again consider the four-dimensional process in Eq. (14.5). Its associated path diagram is depicted in Fig. 14.2(a), which, for example, shows that X_3 is Granger-noncausal for X_1 with respect to $\mathbf{X}_V$. In contrast, if we consider only variables X_1, X_2, and X_3, the corresponding autoregressive representation in Eq. (14.6) yields the path diagram in Fig. 14.2(b); in this graph, there is a directed edge from vertex 3 to vertex 1, which implies that X_3 Granger-causes X_1 with respect to the subprocess $\mathbf{X}_{\{1,2,3\}}$.

We note that more detailed graphical descriptions of the dependences among the components of $\mathbf{X}_V$ are possible by representing each variable $X_\nu(t)$ for all time points t by a separate node (see, e.g., [40–42]). However, identification of such graphs easily becomes infeasible due to the large number of possible edges. Moreover, such a level of detail is not always wanted; in particular, graphs of this type have no direct interpretation in terms of the frequency components of the process.

14.3.2 Bivariate Granger Causality Graphs

When the directed relationships in a vector time series $\mathbf{X}_V$ are described in terms of bivariate Granger causality, the results of such bivariate analyzes again can be graphically represented by a path diagram. In these graphs, bivariate Granger-causal relationships will be indicated by the dashed directed edges (- - →) in order to distinguish these edges from the directed edges in multivariate path diagrams,

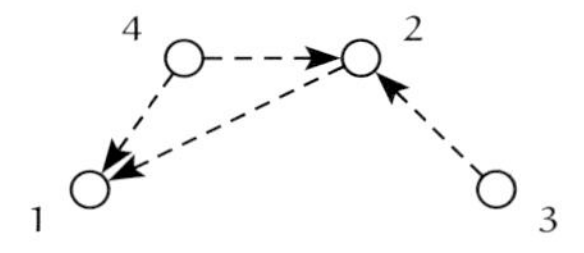

Fig. 14.3: Bivariate Granger causality graph associated with four-dimensional process $\mathbf{X}_V$ given by Eq. (14.5).

which represent Granger causal influences with respect to the full multivariate process $\mathbf{X}_V$. This leads to the following definition of *bivariate path diagrams* or *bivariate Granger causality graphs*, which visualizes the bivariate connectivities in vector time series.

Definition 14.3. Let $\mathbf{X}_V$ be a weakly stationary time series of the form Eq. (14.1). Then the *bivariate path diagram associated with* $\mathbf{X}_V$ is a graph $G = (V, E)$ with vertex set V and edge set E such that for all $i, j \in V$ with $i \neq j$

(i) $\quad j \dashrightarrow i \notin E \iff \tilde{a}_{ij}(u) = 0 \quad \text{for } u \in \mathbb{N},$

(ii) $\quad j \text{ --- } i \notin E \iff \tilde{\sigma}_{ij} = 0,$

where $\tilde{a}_{ij}(u)$, $u \in \mathbb{N}$ and $\tilde{\sigma}_{ij}$ are the parameters in the autoregressive representation (14.10) of the bivariate subprocess $\mathbf{X}_{\{i,j\}}$.

From the above definition, it is clear that, for any subprocess $\mathbf{X}_S$ of $\mathbf{X}_V$, the bivariate Granger causality graph of $\mathbf{X}_S$ is given by the subgraph G_S that is obtained from the bivariate causality graph G by removing all vertices that are not in S and all edges—directed or undirected—that do not have both endpoints in S.

As an example, we again consider the four-dimensional process in Eq. (14.5). For the bivariate Granger causality graph, we have to determine the bivariate autoregressive representations for all pairs X_i and X_j. Simple calculations show, for example, that $\mathbf{X}_{\{1,2\}}$ is given by

$$\begin{aligned} X_1(t) &= \frac{\alpha\beta}{1+\beta^2+\gamma^2} X_2(t-1) + \tilde{\varepsilon}_1(t), \\ X_2(t) &= \tilde{\varepsilon}_2(t). \end{aligned}$$

Furthermore, we have already shown in Section 14.2.1 that the components X_1 and X_3 are completely uncorrelated in a bivariate analysis. Evaluating similarly the autoregressive representations for all other bivariate subprocesses, we obtain the bivariate path diagram in Fig. 14.3 as a visualization of the bivariate Granger-causal relationships among the variables. In this graph, the directed edge $2 \dashrightarrow 1$ suggests a causal influence of X_2 on X_1. Comparison with the corresponding multivariate path diagram in Fig. 14.2(a) shows that this "causal influence" is spurious as it is only induced by the common influence from variable X_4.

In general, the relationship between the two notions of multivariate and bivariate Granger causality is more complicated than in this simple example, and, in most cases, an analytic derivation of the bivariate representation would be very difficult to obtain. In the following section, we discuss graphical conditions that allow drawing conclusions about one graph from the other.

14.4 Markov Interpretation of Path Diagrams

The edges in the path diagrams discussed in this chapter represent pairwise Granger-causal relationships with respect to either the complete process in the case of multivariate path diagrams or with respect to bivariate subprocesses in the case of path diagrams depicting bivariate connectivity structures. The results in this section show that both types of path diagrams more generally provide sufficient conditions for Granger-causal relationships with respect to subprocesses X_S for arbitrary subsets S of V.

14.4.1 Separation in Graphs and the Global Markov Property

The basic idea of graphical modeling is to represent the Markov properties of a set of random variables in a graph by relating certain separation properties of the graph to statements about conditional independence or, in the linear case, partial noncorrelation between the variables. To this end, we firstly review a path-oriented concept of separating subsets of vertices in a mixed graph that has been used to represent the Markov properties of linear structural equation systems (e.g., [43, 44]). Following Richardson [45] we will call this notion of separation in mixed graphs as m-separation.

Let $G = (V, E)$ be a mixed graph with directed edges ($\longrightarrow$) and undirected edges (- - -). A *path* in G is a sequence $\pi = \langle e_1, \ldots, e_n \rangle$ of edges $e_i \in E$ with an associated sequence of vertices $v_0, \ldots, v_n$ such that edge e_i connects vertices v_{i-1} and v_i. We say that v_0 and v_n are the *endpoints* of the path, while the vertices $v_1, \ldots, v_{n-1}$ are the *intermediate vertices* on the path. Note that the vertices v_i in the sequence do not need to be distinct and that therefore the paths considered in this chapter may be self-intersecting.

Furthermore, an intermediate vertex c on a path π is said to be a *collider* on the path if the edges preceding and succeeding c on the path both have an arrowhead or a dashed tail at c, i.e., $\longrightarrow c \longleftarrow$, $\longrightarrow c$ - - -, - - - $c \longleftarrow$, - - - c - - -; otherwise the vertex c is said to be a *noncollider* on the path. Next, let S be a subset of V and let i and j be two vertices that are not in S. Then a path π between the vertices i and j is said to be m-*connecting* given the set S if

(i) every noncollider on the path is not in S and

(ii) every collider on the path is in S,

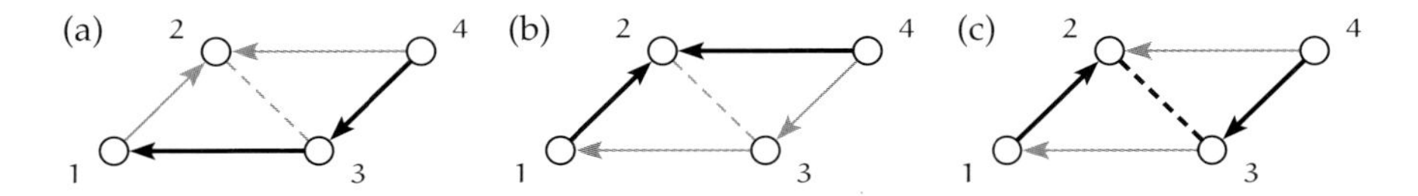

Fig. 14.4: Illustration of m-separation in mixed graphs: Vertices 1 and 4 are m-separated given $S = \{3\}$ since all paths between 1 and 4 are m-blocked given S. (a) path $4 \longrightarrow 3 \longrightarrow 1$ is m-blocked by noncollider $3 \in S$; (b) path $4 \longrightarrow 2 \longleftarrow 1$ is m-blocked by collider $2 \notin S$; (c) path $4 \longrightarrow 3 \dashrightarrow 2 \longleftarrow 1$ is m-blocked by collider $2 \notin S$

otherwise we say the path is m-*blocked* given S. If all paths between i and j are m-blocked given S, then i and j are said to be m-*separated* given S. Similarly, two disjoint subsets I and J are said to be m-separated given S if for every pair $i \in I$ and $j \in J$, the vertices i and j are m-separated given S.

To illustrate these graph-theoretic concepts, we consider the graph in Fig. 14.4. In this graph, vertices 1 and 4 are m-separated given $S = \{3\}$. To show this, we have to examine all paths between the two vertices:

- We note that every path that passes through vertex 2 contains this vertex as a collider. Two examples of such paths are given in Fig. 14.4(b) and (c). Since 2 is not contained in $S = \{3\}$, all these paths are m-blocked given S.

- The only path between vertices 1 and 4 that does not pass through vertex 2 is the path $4 \longrightarrow 3 \longrightarrow 1$ (Fig. 14.4(a)). The intermediate vertex 3 on this path is a noncollider and, thus, the path is m-blocked given $\{3\}$.

It follows that there exists no path between 1 and 4 that is m-connecting given $S = \{3\}$, and the vertices 1 and 4 are consequently m-separated given S.

For linear structural equation systems, Koster [44] has shown that the associated path diagrams have indeed a Markov interpretation, namely, if two sets I and J of vertices are m-separated given a third set S, the corresponding variables $\mathbf{X}_I$ and $\mathbf{X}_J$ are independent conditionally on $\mathbf{X}_S$. The linear equation system (14.8) for the frequency components $dZ_{X_V}(\lambda)$ suggests that a similar result also holds for the frequency components in the time series case. Moreover, since the frequency components at different frequencies are uncorrelated—or independent in the Gaussian case—the separation statements should also translate into noncorrelation between complete subprocesses.

To make this precise, let $\mathbf{X}_V$ be a weakly stationary time series with autoregressive representation (14.1), and let G be its associated multivariate path diagram. Furthermore, suppose that I, J, and S are disjoint subsets of V, and let $\mathbf{Y}_{I|S}$ and $\mathbf{Y}_{J|S}$ be the residual time series of $\mathbf{X}_I$ and $\mathbf{X}_J$, respectively, after the linear effects of the components in $\mathbf{X}_S$ have been removed (see [46], Section 8.3). Then the two subprocesses $\mathbf{X}_I$ and $\mathbf{X}_J$ are partially uncorrelated given $\mathbf{X}_S$ if

$$\operatorname{corr}\big(\mathbf{X}_I(t), \mathbf{X}_J(s) \mid \mathbf{X}_S\big) = \operatorname{corr}\big(\mathbf{Y}_{I|S}(t), \mathbf{Y}_{J|S}(s)\big) = 0 \tag{14.13}$$

for all $t, s \in \mathbb{Z}$; this will be denoted by $\mathbf{X}_I \perp \mathbf{X}_J \mid \mathbf{X}_S$. For an alternative formulation in the frequency domain, let

$$\mathbf{f}_{IJ|S}(\lambda) = \mathbf{f}_{IJ}(\lambda) - \mathbf{f}_{IS}(\lambda)\mathbf{f}_{SS}(\lambda)^{-1}\mathbf{f}_{SJ}(\lambda) = \mathbf{f}_{\mathbf{Y}_{I|S}\mathbf{Y}_{J|S}}(\lambda)$$

be the partial cross-spectrum between $\mathbf{X}_I$ and $\mathbf{X}_J$ given $\mathbf{X}_S$, and let $\mathbf{R}_{IJ|S}(\lambda)$ be the partial spectral coherency given by

$$R_{ij|S}(\lambda) = \frac{f_{ij|S}(\lambda)}{\sqrt{f_{ii|S}(\lambda) f_{jj|S}(\lambda)}} \tag{14.14}$$

for $i \in I$ and $j \in J$ (see [46], Section 8.3). Then condition (14.13) is equivalent to

$$\mathbf{R}_{IJ|S}(\lambda) = 0 \quad \text{for all } \lambda \in [-\pi, \pi] \,. \tag{14.15}$$

Since the partial spectral coherency can be viewed as the partial correlation between frequency components, this implies that $\mathrm{d}\mathbf{Z}_{\mathbf{X}_I}(\lambda)$ and $\mathrm{d}\mathbf{Z}_{\mathbf{X}_J}(\lambda)$ are partially uncorrelated given $\mathrm{d}\mathbf{Z}_{\mathbf{X}_S}(\lambda)$ for all frequencies $\lambda \in [-\pi, \pi]$. With these definitions, it can be shown (e.g., [19]) that path diagrams associated with vector time series have a Markov interpretation both in the time and the frequency domain.

Theorem 14.1. *Suppose $\mathbf{X}_V$ is a weakly stationary time series with autoregressive representation (14.1), and let G be the path diagram associated with $\mathbf{X}_V$. Furthermore, let I, J, and S be disjoint subsets of V. Then, if I and J are m-separated given S, the process $\mathbf{X}_V$ satisfies*

(i) $\mathbf{X}_I \perp \mathbf{X}_J \mid \mathbf{X}_S$;

(ii) $\mathrm{d}\mathbf{Z}_{\mathbf{X}_I}(\lambda) \perp \mathrm{d}\mathbf{Z}_{\mathbf{X}_J}(\lambda) \mid \mathrm{d}\mathbf{Z}_{\mathbf{X}_S}(\lambda)$ *for all* $\lambda \in [-\pi, \pi]$.

This property is called the global Markov property with respect to G.

As an example, we again consider the four-dimensional process in Eq. (14.5) and its associated path diagram in Fig. 14.2(a). Here, vertices 1 and 3 are linked by the path $1 \longleftarrow 4 \longrightarrow 2 \longleftarrow 3$. Obviously, the path is m-connecting given S only if $S = \{2\}$ since 2 is a collider and 4 is a noncollider on this path. It follows from Theorem 14.1 that the two processes X_1 and X_3 are uncorrelated in a bivariate analysis, but not in a trivariate analysis that also includes X_2.

14.4.2 The Global Granger-Causal Markov Property

Next, we discuss how the graph-theoretic concepts presented in the previous section can be used for deriving Granger noncausality relations from path diagrams. For a better understanding of the problem, we firstly consider the autoregressive process $\mathbf{X}_V$ given by

$$\begin{aligned} X_1(t) &= \alpha X_2(t-1) + \varepsilon_1(t)\,, \\ X_2(t) &= \beta X_3(t-1) + \varepsilon_2(t)\,, \\ X_3(t) &= \varepsilon_3(t) \end{aligned} \tag{14.16}$$

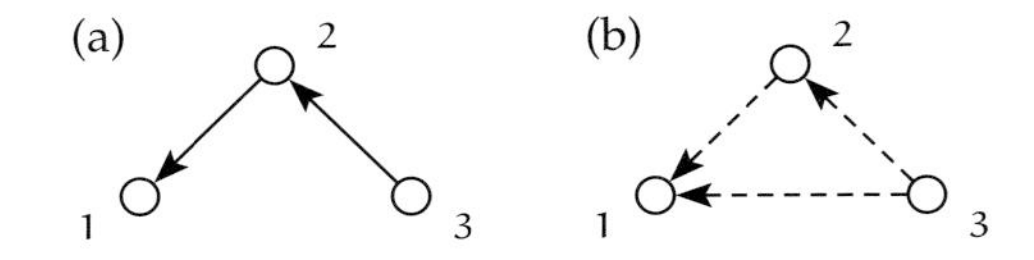

Fig. 14.5: (a) Multivariate path diagram associated with the trivariate process $\mathbf{X}_V$ given by Eq. (14.16); (b) bivariate path diagram associated with $\mathbf{X}_V$

with $\mathrm{var}\big(\varepsilon(t)\big) = \mathbf{I}$. The associated path diagram is shown in Fig. 14.5(a). The diagram shows a directed path from vertex 3 to 1, which suggests an indirect causal influence of X_3 on X_1. Indeed, noting that the autoregressive representation of the subprocess $\mathbf{X}_{\{1,3\}}$ is given by

$$X_1(t) = \alpha\beta X_3(t-2) + \tilde{\varepsilon}_1(t)\,,$$
$$X_3(t) = \tilde{\varepsilon}_3(t)$$

with $\tilde{\varepsilon}_1(t) = \varepsilon_1(t) + \beta\varepsilon_2(t-1)$, $\tilde{\varepsilon}_3(t) = \varepsilon_3(t)$, and associated bivariate path diagram as shown in Fig. 14.5(b), we find that X_3 bivariately Granger-causes X_1, whereas X_1 is bivariately Granger-noncausal for X_3. Obviously, the notion of m-separation is too strong for the derivation of such Granger noncausality relations from multivariate path diagrams: the definition of m-separation requires that all paths between vertices 1 and 3 are m-blocked whereas the path $3 \longrightarrow 2 \longrightarrow 1$ intuitively is interpreted as a causal link from X_3 to X_1. Consequently, the path should not be considered when discussing Granger noncausality from X_1 to X_3.

The example suggests the following definition. A path π between vertices j and i is said to be *i-pointing* if it has an arrowhead at the endpoint i. More generally, a path π between J and I is said to be *I-pointing* if it is i-pointing for some $i \in I$. In order to establish Granger noncausality from X_J to X_I, it is sufficient to consider only all I-pointing paths between I and J (cf. [19]).

Theorem 14.2. *Suppose $\mathbf{X}_V$ is a weakly stationary time series with autoregressive representation* (14.1) *and let G be the path diagram associated with $\mathbf{X}_V$. Furthermore, suppose that $S \subset V$ and let I and J be the two disjoint subsets of S. If every I-pointing path between J and I is m-blocked given $S \setminus J$, then $\mathbf{X}_J$ is Granger-noncausal for $\mathbf{X}_I$ with respect to $\mathbf{X}_S$.*

Similarly, a graphical condition for contemporaneous correlation can be obtained. Intuitively, two variables X_i and X_j are contemporaneously uncorrelated with respect to $\mathbf{X}_S$ if they are contemporaneously uncorrelated with respect to $\mathbf{X}_V$ and, furthermore, the variables are not jointly affected by past values of the omitted variables $\mathbf{X}_{V\setminus S}$. For a precise formulation of the condition, we need the following definition. A path π between vertices i and j is said to be *bi-pointing* if it has an arrowhead at both endpoints i and j. Then the sufficient condition for contemporaneous correlation can be stated as follows (cf. [19]):

Theorem 14.3. *Suppose* $\mathbf{X}_V$ *is a weakly stationary time series with autoregressive representation* (14.1), *and let* $G = (V, E)$ *be the path diagram associated with* $\mathbf{X}_V$*. Furthermore, suppose that* $S \subset V$ *and let* I *and* J *be the two disjoint subsets of* S*. If*

(i) $i \text{ --- } j \notin E$ *for all* $i \in I$ *and* $j \in J$*, and*

(ii) every bi-pointing path between I *and* J *is* m*-blocked given* S*,*

then $\mathbf{X}_I$ *and* $\mathbf{X}_J$ *are contemporaneously uncorrelated with respect to* $\mathbf{X}_S$*.*

In other words, if two variables X_i and X_j are contemporaneously correlated in the subprocess $\mathbf{X}_S$, then they are also contemporaneously correlated in the full process $\mathbf{X}_V$ or the contemporaneous correlation is due to confounding through the variables along an m-connecting path between i and j.

As an example, consider the four-dimensional process $\mathbf{X}_V$ given by Eq. (14.5). The path diagram associated with $\mathbf{X}_V$ is shown in Fig. 14.6(a). Suppose that we are interested in the Granger-causal relationships that hold for the three-dimensional subprocess $\mathbf{X}_{\{1,2,3\}}$.

- The directed edge $3 \longrightarrow 2$ implies that X_3 Granger-causes X_2 also with respect to $\mathbf{X}_{\{1,2,3\}}$.
- Vertices 1 and 3 are connected by the path $3 \longrightarrow 2 \longleftarrow 4 \longrightarrow 1$. Of the two intermediate vertices 2 and 4 on this path, the former is an m-collider, whereas the latter is an m-noncollider. Thus the path is m-blocked given the set $\{2\}$, which implies by Theorem 14.2 that X_3 is Granger-noncausal for X_1 in a bivariate analysis but not in a trivariate analysis including X_2.
- Vertices 1 and 2 are connected by the bi-pointing path $1 \longleftarrow 4 \longrightarrow 2$, which is m-blocked only given vertex 4. Therefore, it follows by Theorems 14.2 and 14.3 that X_1 and X_2 Granger-cause each other and additionally are contemporaneously correlated regardless whether X_3 is included in the analysis or not.

The Granger-causal relationships with respect to $\mathbf{X}_{\{1,2,3\}}$ that can be inferred from the path diagram in Fig. 14.6(a) can be summarized by the graph in Figure 14.6(b).

More generally, if a mixed graph G encodes certain Granger noncausality relations of a process $\mathbf{X}_V$, we say that $\mathbf{X}_V$ satisfies a Markov property with respect to the graph G.

Definition 14.4. We say that a weakly stationary time series $\mathbf{X}_V$ satisfies the *global Granger-causal Markov property* with respect to a mixed graph G if for all $S \subseteq V$ and all disjoint subsets I and J of S the following conditions hold:

(i) $\mathbf{X}_J$ is Granger-noncausal for $\mathbf{X}_I$ with respect to $\mathbf{X}_S$ whenever in the graph G every I-pointing path between J and I is m-blocked given $S \setminus J$.

(ii) $\mathbf{X}_I$ and $\mathbf{X}_J$ are contemporaneously uncorrelated with respect to $\mathbf{X}_S$ whenever in the graph G the sets I and J are not connected by an undirected edge (---) and every bi-pointing path between I and J is m-blocked given S.

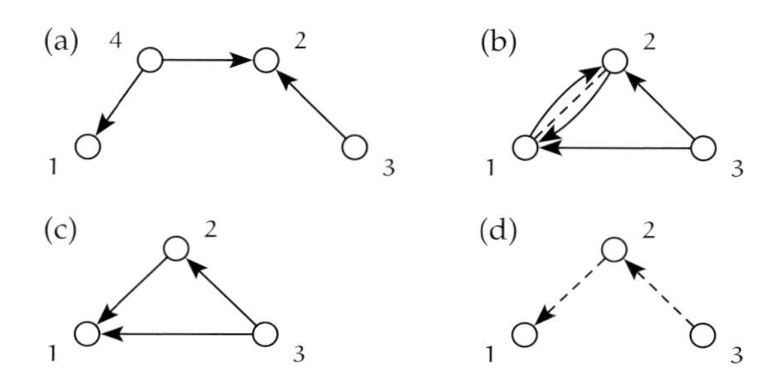

Fig. 14.6: (a) Path diagram of four-dimensional process X_V; (b) derived path diagram of $X_{\{1,2,3\}}$ obtained from the graph in (a); (c) path diagram of $X_{\{1,2,3\}}$; (d) bivariate path diagram of $X_{\{1,2,3\}}$

With this definition, Theorems 14.2 and 14.3 state that a weakly stationary process X_V with autoregressive representation (14.1) satisfies the global Granger-causal Markov property with respect to its multivariate path diagram G.

For the four-dimensional vector time series X_V in Eq. (14.5), we have shown above that the Granger-causal relationships with respect to the subprocess $X_{\{1,2,3\}}$ that can be derived from the multivariate path diagram (Fig. 14.6(a)) are encoded by the graph in Fig. 14.6(b). It follows from Theorems 14.2 and 14.3 that the trivariate subprocess $X_{\{1,2,3\}}$ satisfies the global Granger-causal Markov property with respect to the graph in Fig. 14.6(b). On the other hand, the autoregressive representation of the subprocess $X_{\{1,2,3\}}$ is given in Eq. (14.6); the corresponding path diagram is depicted in Fig. 14.6(c). We note that this path diagram is a subgraph of the graph in Fig. 14.6(b), which has been derived from the multivariate path diagram of the complete series X_V. This demonstrates that Theorems 14.2 and 14.3 provide only sufficient, not necessary conditions for Granger noncausality with respect to subprocesses.

14.4.3 Markov Properties for Bivariate Path Diagrams

Next, we discuss the properties of the bivariate path diagrams introduced in Section 14.3.2. Recall that these path diagrams may have two kind of edges, namely dashed directed edges ($\dashrightarrow$) and undirected edges ($---$). The representation of bivariate Granger-causal relationships by dashed directed edges allows applying the concept of m-separation without further modifications. More precisely, let G be a mixed graph with directed edges ($\dashrightarrow$) and undirected edges ($---$) and let π be a path in G. Then the intermediate vertices on π can be characterized as colliders and noncolliders as in the previous section, that is, an intermediate vertex c on the path π is said to be a collider if the edges preceding and succeeding c on the path both have an arrowhead or a dashed tail at c. However, since G contains only edges of the form $\dashrightarrow$ or $---$, it follows that all paths π in G are pure-collider paths, that is, all intermediate vertices are colliders. Consequently, a path π between vertices i and j is m-connecting given a set S if and only if all intermediate vertices are contained in S.

In the previous section, we have shown that the concepts of m-separation and of pointing paths can be used to derive Granger noncausality relations with respect to subprocesses $\mathbf{X}_S$ from multivariate path diagrams. The same is also true for bivariate path diagrams. More precisely, we have the following result (cf. [39]):

Theorem 14.4. *Let $\mathbf{X}_V$ be a weakly stationary time series with autoregressive representation* (14.1) *and let* G *be the bivariate path diagram of $\mathbf{X}_V$. Then $\mathbf{X}_V$ satisfies the global Granger-causal Markov property with respect to* G.

For an illustration of the Markov interpretation of bivariate path diagrams, we consider again the four-dimensional process $\mathbf{X}_V$ in Eq. (14.5) and suppose that variable X_4 has not been observed. The bivariate path diagram associated with the subprocess $\mathbf{X}_{\{1,2,3\}}$ is depicted in Fig. 14.6(d); as noted before it can be obtained as subgraph of the bivariate path diagram associated with the complete process $\mathbf{X}_V$ (Fig. 14.3). What can we learn from this diagram about the Granger-causal relationships with respect to $\mathbf{X}_S = \mathbf{X}_{\{1,2,3\}}$?

- Since there is no 3-pointing path in the graph, it follows that the components X_1 and X_2 are Granger-noncausal for X_3 with respect to $\mathbf{X}_S$. Similarly, the absence of an undirected edge or a bi-pointing path between vertex 3 and the other two vertices implies that $\mathbf{X}_{\{1,2\}}$ and X_3 are contemporaneously uncorrelated with respect to $\mathbf{X}_S$.

- Vertices 1 and 3 are connected by the 1-pointing path $3 \dashrightarrow 2 \dashrightarrow 1$. This suggests that in a trivariate analysis based on $\mathbf{X}_S$ the series X_3 Granger-causes X_1.

- Similarly, because of the 2-pointing path $1 \dashleftarrow 2 \dashleftarrow 3 \dashrightarrow 2$, we cannot conclude that X_1 is Granger-noncausal for X_2 with respect to $\mathbf{X}_S$. Since the path is also bi-pointing, we additionally cannot rule out that X_1 and X_2 are contemporaneously correlated with respect to $\mathbf{X}_S$.

Summarizing the results, we find that the bivariate path diagram associated with $\mathbf{X}_S$ encodes the same statements about Granger noncausality or contemporaneous noncorrelation with respect to $\mathbf{X}_S$ as the graph in Fig. 14.6(b).

14.4.4 Comparison of Bivariate and Multivariate Granger Causality

The notion of Granger causality is based on the idea that a correlation between two variables that cannot be explained otherwise must be a causal influence; the temporal ordering then determines the direction of the causal link. This approach requires that all relevant information is included in the analysis. Given data from a multivariate time series $\mathbf{X}_V$, it therefore seems plausible to discuss Granger causality with respect to the full multivariate process $\mathbf{X}_V$.

Fig. 14.7: (a) Multivariate path diagram associated with the process $\mathbf{X}_V$ in Eq. (14.17); (b) bivariate path diagram associated with $\mathbf{X}_V$.

As an example, we consider the vector time series $\mathbf{X}_V$ given by

$$\begin{aligned} X_1(t) &= \alpha X_2(t-2) + \varepsilon_1(t)\,, \\ X_2(t) &= \varepsilon_2(t)\,, \\ X_3(t) &= \beta X_2(t-1) + \varepsilon_3(t)\,, \end{aligned} \tag{14.17}$$

where $\{\varepsilon(t)\}$ is a white noise process with $\mathrm{var}(\varepsilon(t)) = \mathbf{I}$. Simple calculations show that the bivariate path diagram of $\mathbf{X}_V$ is given by the graph in Fig. 14.7(b). Here, the bivariate analyses suggest a causal link from X_3 to X_1 although the observed correlation between X_1 and X_3 is only due to confounding by X_2. In contrast, the multivariate path diagram in Fig. 14.7(a) correctly shows neither direct connections nor a causal pathway between X_1 and X_3. This inability of the bivariate approach to discriminate between causal influences and confounded relationships has been noted by several authors (e.g., [47–49]).

One serious problem that arises in practice is that relevant variables are omitted from the analysis, for example, because they could not be measured. For an illustration, we consider again the four-dimensional process $\mathbf{X}_V$ in Eq. (14.5). As in the previous section, we assume that only the subprocess $\mathbf{X}_S = \mathbf{X}_{\{1,2,3\}}$ is available for an analysis of interrelationships. The multivariate path diagram in Fig. 14.6(c) indicates the presence of a direct causal link from X_3 to X_1, whereas in a bivariate analysis of $\mathbf{X}_{\{1,3\}}$ this Granger-causal influence vanishes. In this situation, the bivariate path diagram in Fig. 14.6(d) clearly provides a better graphical description of the relationships among the variables than the multivariate path diagram.

More generally, it can be shown that systems in which all relationships between the observed variables are due to confounding by latent variables can be best represented by bivariate path diagrams. In contrast, multivariate path diagrams are best suited for the representation of causal structures that do not involve confounding by latent variables. In practice, however, causal structures may be a combination of both situations with only a part of the Granger-causal relationships being due to confounding by latent variables. In such cases neither graphical representation would provide an optimal description of the dependences among the observed variables. Eichler [39] presented a graphical approach for evaluating the connectivity of such systems based on general mixed graphs that generalize both multivariate and bivariate path diagrams.

14.5 Statistical Inference

In practice, the autoregressive structure of the processes of interest typically is unknown and must be identified from data. One straightforward approach is to test for the presence of edges in the path diagram; this approach can be used for both types of path diagrams. In the case of multivariate path diagrams, the path diagram can be identified alternatively by model selection based on fitting graphical vector autoregressive models that are constrained according to a path diagram (e.g., [50, 51]).

14.5.1 Inference in the Time Domain

For the analysis of empirical data, VAR(p) models can be fitted using least-squares estimation. For observations $\mathbf{X}_V(1), \ldots, \mathbf{X}_V(T)$ from a d-dimensional multiple time series $\mathbf{X}_V$, let $\hat{\mathbf{R}}_p = \big(\hat{\mathbf{R}}_p(u,v)\big)_{u,v=1,\ldots,p}$ be the $pd \times pd$ matrix composed by submatrices

$$\hat{\mathbf{R}}_p(u,v) = \frac{1}{T-p} \sum_{t=p+1}^{T} \mathbf{X}(t-u)\mathbf{X}(t-v)' .$$

Similarly, we set $\hat{\mathbf{r}}_p = \big(\hat{\mathbf{R}}_p(0,1), \ldots, \hat{\mathbf{R}}_p(0,p)\big)$. Then the least-squares estimates of the autoregressive coefficients are given by

$$\hat{\mathbf{a}}(u) = \sum_{v=1}^{p} (\hat{\mathbf{R}}_p)^{-1}(u,v)\hat{\mathbf{r}}_p(v) \tag{14.18}$$

for $u = 1, \ldots, p$, while the covariance matrix $\mathbf{\Sigma}$ is estimated by

$$\hat{\mathbf{\Sigma}} = \frac{1}{T} \sum_{t=p+1}^{T} \hat{\varepsilon}(t)\hat{\varepsilon}(t)' ,$$

where $\hat{\varepsilon}(t) = \mathbf{X}(t) - \sum_{u=1}^{p} \hat{\mathbf{a}}(u)\mathbf{X}(t-u)$ are the least-squares residuals. The estimates $\hat{a}_{ij}(u)$ are asymptotically jointly normally distributed with mean $a_{ij}(u)$ and covariances satisfying

$$\lim_{T\to\infty} T \operatorname{cov}\big(\hat{a}_{ij}(u), \hat{a}_{kl}(v)\big) = H_{jl}(u,v)\sigma_{ik} ,$$

where $H_{jl}(u,v)$ are entries in the inverse $\mathbf{H}_p = \mathbf{R}_p^{-1}$ of the covariance matrix $\mathbf{R}_p$. For details, we refer to Lütkepohl [52].

The coefficients $a_{ij}(u)$ depend like any regression coefficient on the unit of measurement of X_i and X_j and thus are not suited for comparisons of the strength of causal relationships between different pairs of variables. Therefore, Dahlhaus and Eichler [40] proposed partial directed correlations as a measure of the strength of causal effects. For $u > 0$, the partial directed correlation $\pi_{ij}(u)$ is defined as the correlation between $X_i(t)$ and $X_j(t-u)$ after removing the linear effects of $\mathbf{X}_{V\setminus\{b\}}(t-u)$, $u \in \mathbb{N}$. Similarly, we define $\pi_{ij}(0)$ as the correlation between $X_i(t)$ and $X_j(t)$ after removing the linear effects of $\mathbf{X}_V(t-u)$, $u \in \mathbb{N}$, while for $u < 0$ we have $\pi_{ij}(u) = \pi_{ji}(-u)$. It has been shown (see [53]) that estimates for the partial directed correlations $\pi_{ij}(u)$ with $u > 0$ can be obtained from the parameter estimates of a VAR(p) model by rescaling the coefficients $\hat{a}_{ij}(u)$,

$$\hat{\pi}_{ij}(u) = \frac{\hat{a}_{ij}(u)}{\sqrt{\hat{\sigma}_{ii}\hat{\tau}_{ij}(u)}}$$

where

$$\hat{\tau}_{ij}(u) = \hat{K}_{jj} + \sum_{v=1}^{u-1} \sum_{k,l \in V} \hat{a}_{kj}(v)\hat{K}_{kl}\hat{a}_{lj}(v) + \frac{\hat{a}_{ij}(u)^2}{\hat{\sigma}_{ii}}$$

with $\hat{\mathbf{K}} = \hat{\mathbf{\Sigma}}^{-1}$. For $u = 0$, we obviously have

$$\hat{\pi}_{ij}(0) = \frac{\hat{\sigma}_{ij}}{\sqrt{\hat{\sigma}_{ii}\hat{\sigma}_{jj}}}.$$

For large sample length T, the partial directed correlations are approximately normally distributed with mean $\pi_{ij}(u)$ and variance $1/T$.

Tests for Granger-causal relationships among the variables can be derived from the asymptotic distribution of the parameters of the VAR(p) model. More precisely, let $\hat{V}(u,v) = \hat{H}_{jj}(u,v)\hat{\sigma}_{ii}$ be the estimate of the asymptotic covariance between $\hat{a}_{ij}(u)$ and $\hat{a}_{ij}(v)$, let $\hat{\mathbf{V}}$ be the corresponding $p \times p$ matrix and set $\hat{\mathbf{W}} = \hat{\mathbf{V}}^{-1}$ with entries $\hat{W}(u,v)$. Then the existence of a Granger-causal effect of X_j on X_i can be tested by evaluating the test statistic

$$S_{ij} = T \sum_{u,v=1}^{p} \hat{a}_{ij}(u)\hat{W}(u,v)\hat{a}_{ij}(v)\,.$$

Under the null hypothesis that X_j is Granger-noncausal for X_i with respect to $\mathbf{X}_V$, the test statistic S_{ij} is asymptotically χ^2-distributed with p degrees of freedom.

14.5.2 Inference in the Frequency Domain

In the frequency domain, the Granger-causal relationships in a multivariate time series $\mathbf{X}_V$ can be evaluated by the Fourier transform

$$\hat{\mathbf{A}}(\lambda) = \sum_{u=1}^{p} \hat{\mathbf{a}}(u)e^{-i\lambda u}\,,$$

where $\hat{\mathbf{a}}(u)$, $u = 1, \ldots, p$, are the autoregressive estimates given by Eq. (14.18). From this, estimates for the partial directed coherence can be obtained by suitable normalization. We note that because of the asymptotic normality of the estimates $\hat{a}_{ij}(u)$ the real and imaginary parts of $\hat{A}_{ij}(\lambda)$ are also jointly asymptotically normally distributed. Furthermore, it has been shown (see [54]) that, if $A_{ij}(\lambda) = 0$, then the asymptotic distribution of

$$T\frac{|\hat{A}_{ij}(\lambda)|^2}{C_{ij}(\lambda)}, \tag{14.19}$$

where

$$C_{ij}(\lambda) = \sigma_{ii}\Big(\sum_{k,l=1}^{p} H_{jj}(k,l)\big[\cos(k\lambda)\cos(l\lambda) + \sin(k\lambda)\sin(l\lambda)\big]\Big) \tag{14.20}$$

is that of a weighted average of two independent χ^2-distributed random variables each with one degree of freedom. Noting that the $1-\alpha$ quantiles of this asymptotic distribution can be bounded by the $1-\alpha$ quantile $\chi^2_{1,1-\alpha}$ of a χ^2-distribution with one degree of freedom, we can use

$$\frac{1}{T}\hat{C}_{ij}(\lambda)\chi^2_{1,1-\alpha},$$

where $\hat{C}_{ij}(\lambda)$ is an estimate of $C_{ij}(\lambda)$ in Eq. (14.20), as an approximate α-significance level for testing whether $A_{ij}(\lambda) = 0$. Similarly, a significance level for the partial directed coherence can be derived [54].

We note that the functions $A_{ij}(\lambda)$ like the coefficients $a_{ij}(u)$ depend on the unit of measurement of X_i and X_j and thus are unsuitable for comparing the strength of Granger-causal relationships between different pairs of variables. As noted before, the partial directed coherence does not provide a complete solution to this problem as it measures the relative strength with respect to a given signal source. Instead, we will consider for the examples in Section 14.6 the statistic

$$\hat{\alpha}^2_{ij}(\lambda) = \frac{|\hat{A}_{ij}(\lambda)|^2}{\hat{C}_{ij}(\lambda)},$$

which allows the use of the same significance level $\chi^2_{1,1-\alpha}/T$ for all frequencies λ and all pairs $i, j \in V$. We will call the statistic the *rescaled partial directed coherence* (PDC) from X_j to X_i.

14.5.3 Graphical Modeling

An alternative approach for inference on causal structures in multivariate time series is based on fitting graphical vector autoregressive models. For given graph $G = (V, E)$ and order p, we consider vector autoregressive (VAR) models of the form

$$\mathbf{X}_V(t) = \sum_{u=1}^{p} \mathbf{a}(u)\mathbf{X}_V(t-u) + \varepsilon_V(t), \quad \mathrm{var}\big(\varepsilon(t)\big) = \mathbf{\Sigma},$$

where the parameters $\mathbf{a}(u)$, $u = 1, \ldots, p$, and $\mathbf{\Sigma}$ satisfy the constraints

(i) $a_{ij}(u) = 0$ for $u = 1, \ldots, p$, whenever $j \longrightarrow i \notin E$ and

(ii) $\sigma_{ij} = 0$, whenever $i \text{ --- } j \notin E$.

It follows that the processes $\mathbf{X}_V$ satisfy the global Granger-causal Markov property with respect to the graph G, and we therefore call the VAR model with these constraints on the parameters the graphical vector autoregressive model of the order p with respect to graph G or short the VAR(p,G) model.

Given observations $\mathbf{X}_V(1), \ldots, \mathbf{X}_V(T)$, the unconstrained parameters in a VAR(p,G) model can be estimated iteratively by the following two steps.

(i) Let the estimate $\hat{\mathbf{\Sigma}}$ be fixed. Then the estimates $\hat{\mathbf{a}}(u)$, $u = 1, \ldots, p$ are determined as the solution of the linear equations

$$\left(\sum_{v=1}^{p} \hat{\mathbf{\Sigma}}^{-1} \mathbf{a}(v) \hat{\mathbf{R}}_p(u, v)\right)_{ij} = \left(\hat{\mathbf{\Sigma}}^{-1} \hat{\mathbf{R}}_p(0, v)\right)_{ij}$$

for $u = 1, \ldots, p$ and all $i, j \in V$ such that $j \longrightarrow i \in E$ under the constraints that $a_{ij}(u) = 0$, whenever the directed edge $j \longrightarrow i$ is absent in the graph G.

(ii) Let $\hat{\mathbf{a}}(u)$, $u = 1, \ldots, p$ be fixed and let $\hat{\varepsilon}(t)$ be the corresponding residuals. Then the estimate $\hat{\mathbf{\Sigma}}$ is obtained by solving the nonlinear equations

$$(\mathbf{\Sigma}^{-1})_{ij} = (\mathbf{\Sigma}^{-1} \hat{\mathbf{\Sigma}}_0 \mathbf{\Sigma}^{-1})_{ij}$$

for all $i, j \in V$ such that $i \text{ --- } j \in E$, where $\hat{\mathbf{\Sigma}}_0 = \frac{1}{T} \sum_{t=p+1}^{T} \hat{\varepsilon}(t)\hat{\varepsilon}(t)'$ is an unconstrained estimate of $\mathbf{\Sigma}$.

The second step corresponds to fitting a covariance model to the residuals $\hat{\varepsilon}(t)$, which is determined by the above zero constraints on the covariance matrix $\mathbf{\Sigma}$. An iterative algorithm for fitting such covariance models has been introduced by Drton and Richardson [55]. Since the solution of both sets of equations are not independent, an iteration of the two steps is needed to obtain a joint solution. For details on fitting graphical vector autoregressive models, we refer to Eichler [53].

Graphical vector autoregressive models can be used to determine the Granger-causal relationships among multiple time series by minimizing model selection criteria like AIC [56] or BIC [57]. The AIC for the VAR(p,G) model is given by

$$\mathrm{AIC}(p, G) = \frac{1}{2} \log|\hat{\mathbf{\Sigma}}| + \frac{r}{T},$$

where $\hat{\mathbf{\Sigma}}$ is the estimate for $\mathbf{\Sigma}$ in the VAR(p,G) model and r is the number of unconstrained parameters in the model.

14.6 Applications

In this section, we present three examples to demonstrate how graphical representations facilitate our understanding of interrelationships in multivariate time series.

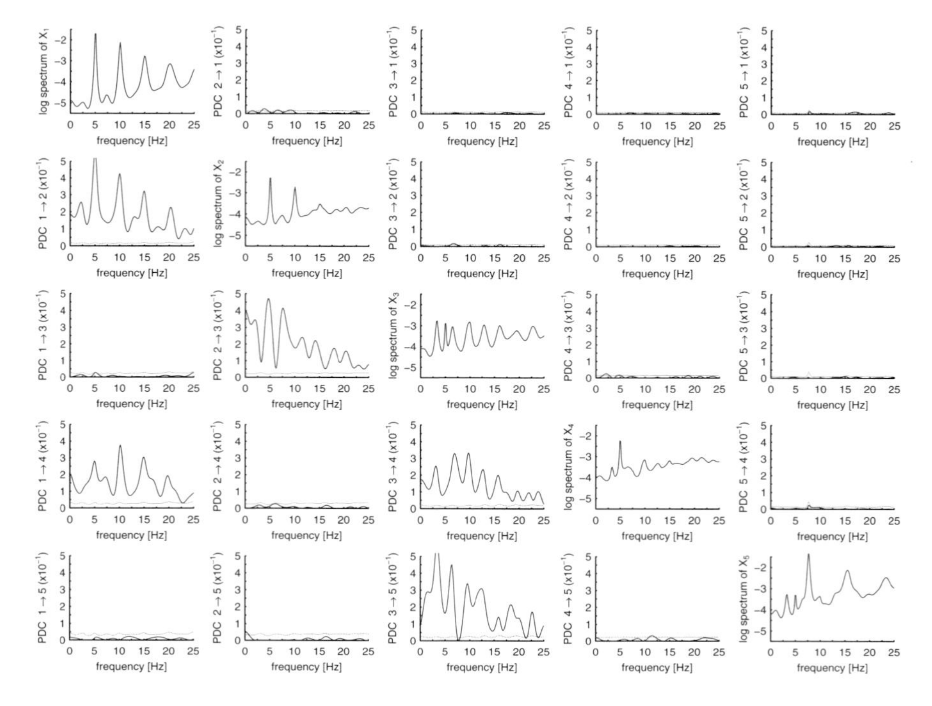

Fig. 14.8: Results for neuronal spike train data: estimates of log-spectral densities (on diagonal) and nonnormalized PDC $|A_{ij}(\lambda)|^2$ (off-diagonals). The dotted lines signify pointwise 95 % test bounds for the hypothesis that the PDC is zero.

14.6.1 Frequency-Domain Analysis of Multivariate Time Series

In our first example, we review various frequency-domain-based methods for the description of interrelations among multiple time series and discuss their relations to each other. To illustrate the theoretical results, we apply the methods to neuronal spike train data recorded from the lumbar spinal dorsal horn of a pentobarbital-anaesthetized rat during noxious stimulation. The firing times of ten neurons were recorded simultaneously by a single electrode with an observation time of 100 s. The data have been described in detail in Sandkühler and Eblen-Zajjur [58]; the connectivity among the recorded neurons has been analyzed previously by partial correlation analysis [59] and partial directed correlations [60].

For the analysis, we converted the spike trains of five neurons to binary time series and fitted a VAR model of the order $p = 100$. Figure 14.8 displays the estimated spectra for these five neurons. The strong peaks in the spectra for neurons 1 and 2 indicate that these neurons show rhythmic discharges at 5 Hz; similarly, neuron 5 fires rhythmically at 7.5 Hz.

For the identification of the effective connectivity among these five neurons,

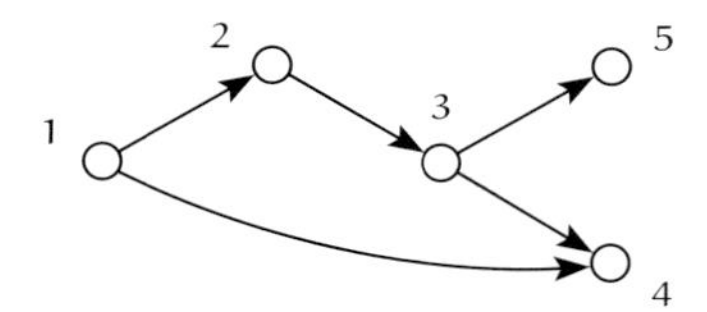

Fig. 14.9: Results for neuronal spike train data: multivariate path diagram identified from the PDCs in Fig. 14.8.

we have estimated the nonnormalized PDC $|A_{ij}(\lambda)|^2$ (Fig. 14.8). The PDC detects strongly significant directed relationships for five pairs of neurons. Additionally, tests for contemporaneous noncorrelation yielded no significant links between the neurons. Thus, the dependences between the five neurons can be represented by the path diagram in Fig. 14.9.

One nondirectional measure for the direct interdependences between the frequency components of a process $\mathbf{X}_V$ is the *partial spectral coherence* $|R_{ij|V\setminus\{i,j\}}(\lambda)|^2$ with $R_{ij|V\setminus\{i,j\}}(\lambda)$ defined as in Eq. (14.14) (see, e.g., [46, 61]). As we have seen in Section 14.4.1, it is closely related to the Markov interpretation of multivariate path diagrams in the frequency domain. In particular, Theorem 14.1 implies that the partial spectral coherence $|R_{ij|V\setminus\{i,j\}}(\lambda)|^2$ vanishes uniformly for all frequencies λ, whenever the vertices i and j are m-separated given $V \setminus \{i, j\}$.

Figure 14.10 shows nonparametric and parametric estimates of the partial spectral coherence for the neuronal spike train data. Here, the partial spectral coherence between neurons i and j shows a strong association between the corresponding frequency components, whenever i and j are connected by an edge. Additionally, we also find a small, but significant partial spectral coherence between neurons 1 and 3, which corresponds with the graphical characterization since in the path diagram in Fig. 14.9 vertices 1 and 3 are linked by the m-connecting path $1 \longrightarrow 4 \longleftarrow 3$.

Another important measure for directed information flow in multivariate systems is the *directed transfer function* (DTF), which has been proposed by Kamiński and Blinowska [62] and is based on the transfer function $\mathbf{B}(\lambda) = \big(\mathbf{I} - \mathbf{A}(\lambda)\big)^{-1}$. The transfer function relates the frequency components of $\mathbf{X}$ and ε_V by the linear system

$$dZ_{X_V}(\lambda) = \mathbf{B}(\lambda)\, dZ_{\varepsilon_V}(\lambda)$$

and thus describes how the frequency components of the input process ε_V are transformed by the linear system to the frequency components of the output process X. In particular, the entry $B_{ij}(\lambda)$ measures the response of variable X_i to sinusoidal random shocks of frequency λ at variable X_j. The DTF is a normalized version of the transfer function given by

$$\gamma_{ij}^2(\lambda) = \frac{|B_{ij}(\lambda)|^2}{\sum_k \big|B_{ik}(\lambda)\big|^2} \tag{14.21}$$

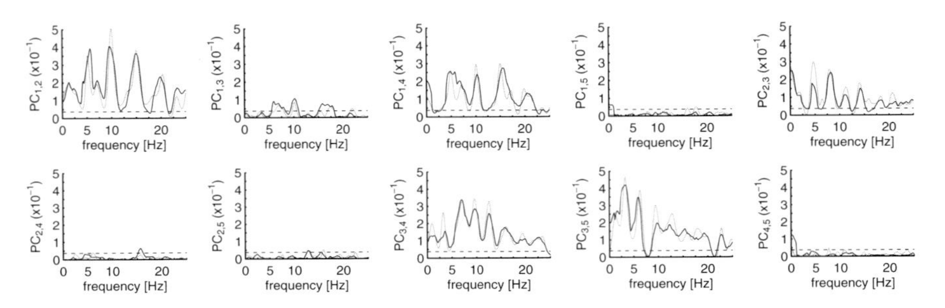

Fig. 14.10: Nonparametric (solid lines) and parametric (dotted lines) estimates of partial spectral coherence for the neuronal spike train data. For the nonparametric estimates, the horizontal dashed lines signify pointwise 95 % test bounds for the hypothesis that the partial spectral coherence is zero.

and describes the ratio of the influence of component X_j on component X_i to all the influences on component X_i. Due to the normalization, the DTF takes values in $[0, 1]$. For the comparison of the information flow for different target processes or between different experiments, also a nonnormalized version of the DTF given by

$$\theta_{ij}^2(\lambda) = |B_{ij}(\lambda)|^2 \tag{14.22}$$

has been suggested [35, 49]. Expanding the inverse $\big(\mathbf{I} - \mathbf{A}(\lambda)\big)^{-1}$ as a geometric series, we find that

$$B_{ij}(\lambda) = A_{ij}(\lambda) + \sum_{k=1}^{d} A_{ik}(\lambda)A_{kj}(\lambda) + \sum_{k_1,k_2=1}^{d} A_{ik_1}(\lambda)A_{k_1k_2}(\lambda)A_{k_2j}(\lambda) + \cdots . \tag{14.23}$$

It follows that the DTF accumulates the information flow from direct pathways—measured by $A_{ij}(\lambda)$—as well as from indirect pathways via components $X_{k_1}, \ldots, X_{k_r}$. In particular, this implies that the DTF from X_j to X_i vanishes uniformly for all frequencies, whenever there exists no directed path $j \longrightarrow \cdots \longrightarrow i$ in the multivariate path diagram associated with $\mathbf{X}_V$. To illustrate this fact, we estimated the DTF for the neuronal spike train data (Fig. 14.11) with pointwise significance levels as described in Eichler [63]. Comparing the results with the path diagram in Fig. 14.9, we find that the DTF indeed identifies information flow from neuron j to neuron i, whenever there is a directed path from j to i in the path diagram, which is in line with the graph theoretical predictions.

We conclude that the DTF can be used to describe the propagation of information in multivariate systems, but cannot be used for the detection of the pathways by which the information is propagated, which would entail discrimination between direct and indirect interactions. This also implies that the DTF cannot be used as a measure for Granger causality as defined in Definition 14.1 (see [63]).

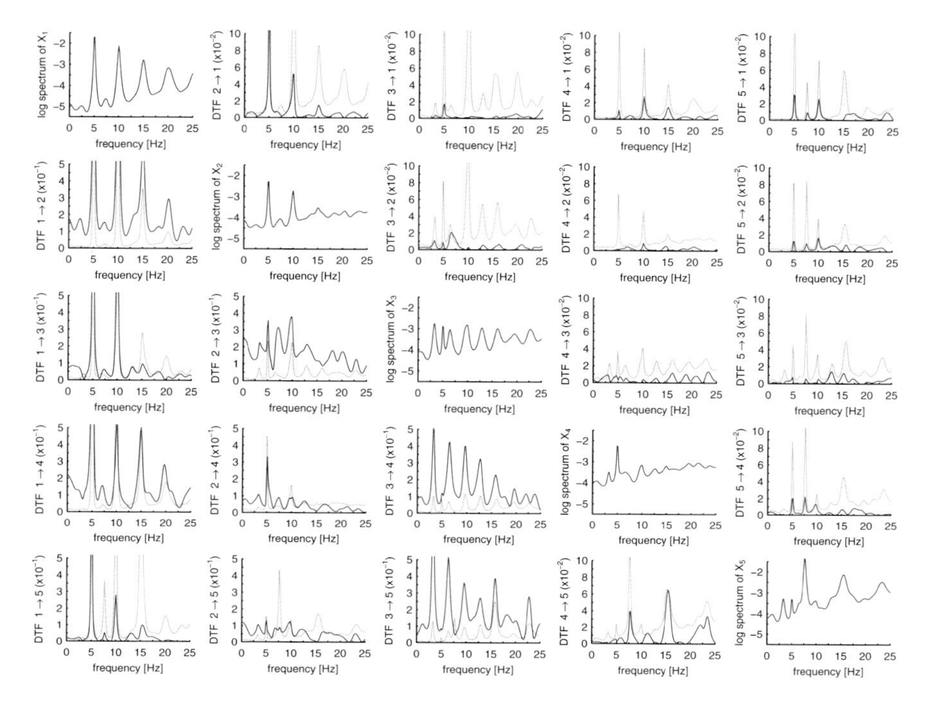

Fig. 14.11: Estimates of log-spectral densities (on diagonal) and normalized DTF $\gamma_{ij}^2(\lambda)$ (off-diagonals) for the neuronal spike train data. The dotted lines signify pointwise 95 % test bounds for the hypothesis that the DTF is zero.

To resolve the problem of indirect information flow, Korzeniewska et al. [64] proposed a modification of the DTF, which combines the DTF and the partial spectral coherence. This *direct DTF* (dDTF) is defined as the product

$$\delta_{ij}(\lambda) = \gamma_{ij}(\lambda)|R_{ij|V\setminus\{i,j\}}(\lambda)|\,.$$

The motivation behind this definition is that the DTF $\gamma_{ij}(\lambda)$ measures the propagation of information within a system and, in particular, identifies the direction of the information flow—both direct and indirect—while the partial spectral coherence vanishes if there is no direct interaction between the corresponding frequency components [65]. From the graphical conditions for the partial spectral coherence and the DTF, we immediately find that the dDTF $\delta_{ij}(\lambda)$ vanishes at all frequencies λ whenever in the path diagram

- i and j are m-separated given $V \setminus \{i, j\}$ or
- there exists no directed path $j \longrightarrow \ldots \longrightarrow i$.

Since the second condition determines only if there is information flow from j to i, the discrimination of direct and indirect information flow must be accom-

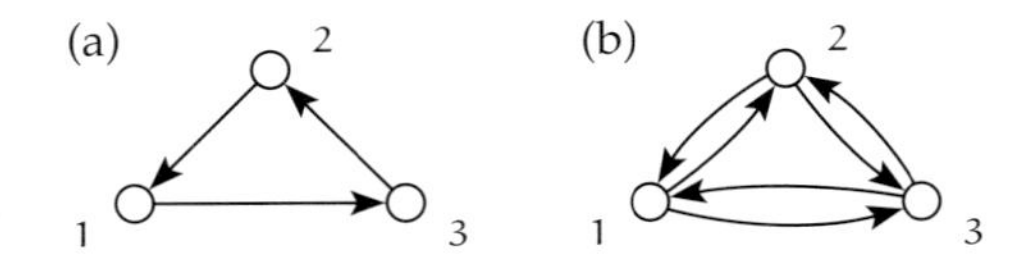

Fig. 14.12: (a) Path diagram with one cycle; (b) direct information flow as identified by the dDTF.

plished by the first condition. This, however, is obviously not the case since two vertices i and j are not m-separated given all other vertices $V \setminus \{i, j\}$ if and only if

(i) they are linked by an edge (regardless of its direction or type) or

(ii) connected by a path of the form $i \longrightarrow k \longleftarrow j$.

In particular, this implies that the discrimination fails whenever the path diagram contains a directed cycle, that is, a path of the form $v \longrightarrow \ldots \longrightarrow v$. As an example, we consider the path diagram in Fig. 14.12(a): in this graph, any two vertices i and j are connected by a directed path from j to i (either $j \longrightarrow i$ or $j \longrightarrow k \longrightarrow i$) and are linked by an edge (either $j \longrightarrow i$ or $i \longrightarrow j$), which means that the dDTF $\delta_{ij}(\lambda)$ is nonzero for all i and j. Clearly, in this case, the dDTF cannot distinguish between direct and indirect information flow (Fig. 14.12(b)).

The effect in (ii) that two independent variables become conditionally dependent if they both affect a third variable that is included in the conditioning set is well known in graphical modeling theory and is called the marrying parents effect (see, e.g., [7, 66]). For an illustration of this effect and how it affects the dDTF, we consider again the neuronal spike train data. In the path diagram in Fig. 14.9 showing the identified connectivity for the five neurons, we find that the two vertices 1 and 3 are linked by both a directed path ($1 \longrightarrow 2 \longrightarrow 3$) and an m-connecting path ($1 \longrightarrow 4 \longleftarrow 3$). According to the above characterization, this implies that the dDTF from X_1 to X_3 is nonzero, and indeed the estimates in Fig. 14.13 show two small peaks at frequencies 5 Hz and 10 Hz in the dDTF from neuron 1 to neuron 3. The assessment of the significance of these peaks is difficult since the statistical properties of the dDTF have not been investigated so far. However, we note that the path $1 \longrightarrow 4 \longleftarrow 3$ is only m-connecting if vertex 4 is included in the separating set. In other words, if neuron 4 is omitted from the analysis, the dDTF should become zero. The corresponding estimates of the dDTF obtained from the process $\mathbf{X}_{\{1,2,3,5\}}$ are also shown in Fig. 14.13 (dotted curves). Comparing these estimates with those obtained from the full process, we find that the dDTF from neuron 1 to neuron 3 is reduced considerably, while for all other pairs the omission of neuron 4 leaves the estimates basically unchanged. This indicates that the peaks in the former estimate of the dDTF from neuron 1 to neuron 3 were indeed induced by the combination of an m-connecting and a directed pathway from X_1 to X_3.

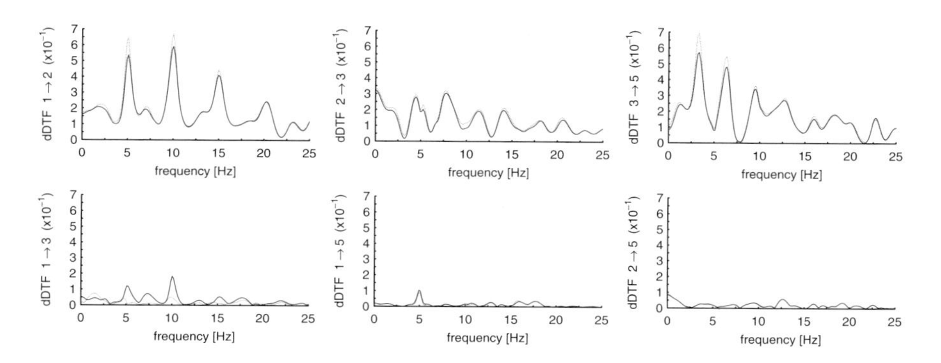

Fig. 14.13: Direct DTF (dDTF) for the neuronal spike train data: dDTF obtained from the five-dimensional process $\mathbf{X}_{\{1,\ldots,5\}}$ (solid lines) and dDTF obtained from the four-dimensional process $\mathbf{X}_{\{1,2,3,5\}}$ (dotted lines).

If the true path diagram is a directed acyclic graph, that is, it does not contain any undirected edges or directed cycles, then the iterative algorithm presented in Dahlhaus et al. [66] can be applied to identify direct information flow among the components of $\mathbf{X}_V$ by the dDTF. However, in general, identification based on the dDTF can lead to wrongly detected relationships. Therefore, analysis of the information flow and the connectivity in multivariate systems should be based on the PDC or the DTF, which both have a clear interpretation as direct and as total information flow, respectively.

14.6.2 Identification of Tremor-Related Pathways

The second example is concerned with the analysis of simultaneous electroencephalographic (EEG) and electromyographic (EMG) recordings from patients suffering from essential tremor. This neurological disease manifests itself by an involuntary, oscillatory movement of parts of the body, mainly the upper limbs, with a typical trembling frequency of 4 Hz to 10 Hz. In previous studies based on coherence analysis, tremor correlated cortical activity has been observed in the EEG [67, 68], but the direction of the relationship remained unclear.

The analyzed data consist of the EMG from the left-wrist extensor measuring the movement of that hand and the recordings from EEG channels C4 and PZA, which both showed a strong correlation with the EMG at the tremor frequency of about 5 Hz. The EMG signal was band-pass filtered to avoid aliasing effects and undesired slow drifts. Additionally, the signal was digitally full wave rectified. The resulting time series reflects the muscle activity encoded in the envelope of the originally measured signal.

Figure 14.14 shows estimates of the log-spectral densities and the PDC for the data. Furthermore, Table 14.1 shows the significant contemporaneous correlations between the series. This leads to the path diagram in Fig. 14.15(a). We

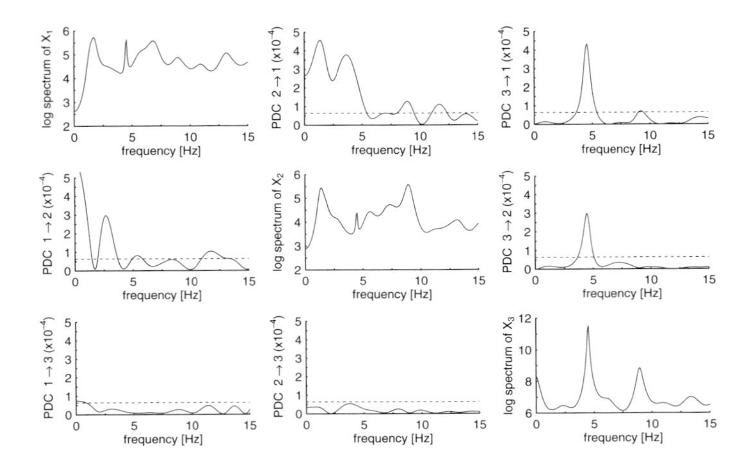

Fig. 14.14: Results for tremor-related EEG channels C4 (X_1) and PZA (X_2) and EMG channel (X_3): estimates of log-spectral densities (*on diagonal*) and rescaled PDC $\alpha^2_{ij}(\lambda)$ (*off-diagonals*). The horizontal dashed lines signify pointwise 95 % test bounds for the hypothesis that the PDC is zero.

Tab. 14.1: p-values for testing for contemporaneous noncorrelation in the tremor-related EEG/EMG signals.

C4---PZA	C4---EMG	PZA---EMG
0.000	0.011	0.103

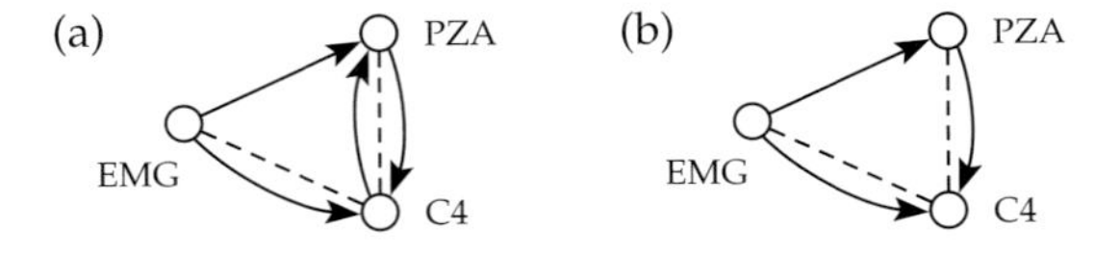

Fig. 14.15: Path diagrams for tremor-related EEG/EMG data: (a) path diagram for dependences over frequency range 0 Hz to 25 Hz; (a) path diagram for dependences at tremor frequency $\lambda \approx 5$ Hz.

note that the EMG signal Granger-causes the EEG signals of both channels C4 and PZA, which suggests that the muscle activity is reflected in the cortex via proprioceptive afferences. Additionally, we find a significant contemporaneous correlation between the EMG signal and channel C4. Since we cannot identify a direction for this association, it remains an open question whether the oscillatory cortical activity reflected in the signal in channel C4 is involved in the generation of the tremor.

Alternatively, we could restrict ourselves to the dependences at the tremor frequency, which leads to the omission of the edge C4 $\longrightarrow$ PZA (Fig. 14.15(b)). The conclusions concerning the relationship between the EMG signal and the cortical activity, however, remain the same.

14.6.3 Causal Inference

In the last example, we apply the graphical approach to concurrent recordings from EEG and functional magnetic resonance imaging (fMRI) for the investigation of the interrelations between the alpha rhythm in the EEG and blood oxygenation level dependent (BOLD) responses in the fMRI. The data and their requisition are described in detail in Goldman et al. [69].

The EEG was sampled at 200 Hz from an array of 16 bipolar pairs, with an additional channel for the EKG and scan trigger. For the analysis, the time-varying spectrum of the EEG has been decomposed by parallel factor (PARAFAC) analysis into trilinear components (called atoms), each being the product of a spatial, spectral, and temporal factors [70]. The PARAFAC analysis extracted three significant atoms characterized by their spectral signature. Only the temporal factor of the alpha atom corresponding to a frequency range 8 Hz to 12 Hz was included in the effective connectivity analysis.

The fMRI series were measured with a time resolution of 2.5 s. Here, we consider two time series of length $T = 108$ for two regions in the brain, namely visual cortex and thalamus, whose activation seemed directly related with the EEG alpha atom, namely visual cortex and thalamus. For each region, the time series was obtained by averaging the time series of all voxels in that region.

Tab. 14.2: p-values for testing for multivariate and bivariate contemporaneous noncorrelation in the fMRI/EEG data.

	VC --- TH	VC --- EEG	TH --- EEG
Bivariate	0.08	0.70	0.22
Multivariate	0.00	0.49	0.10

For the analysis of the effective connectivity, we have fitted a VAR model of order 2 to the data; the order has been determined by minimizing the AIC. Figure 14.16 shows the estimates of the PDC obtained by a trivariate analysis (solid lines) and by bivariate analyses (dotted lines). Additionally we have tested for contemporaneous noncorrelation; the results are given in Table 14.2. The results of the analyses are summarized by the multivariate and bivariate path diagrams $G^{(m)}$ and $G^{(b)}$ in Fig. 14.17(a) and (b), respectively. Here, the multivariate path diagram $G^{(m)}$ implies that thalamus and visual cortex neither Granger-cause the EEG alpha atom nor are they contemporaneously correlated with the EEG component, while the bivariate path diagram $G^{(b)}$ additionally encodes that, firstly, the EEG alpha atom does not bivariately Granger-cause the thalamus and, sec-

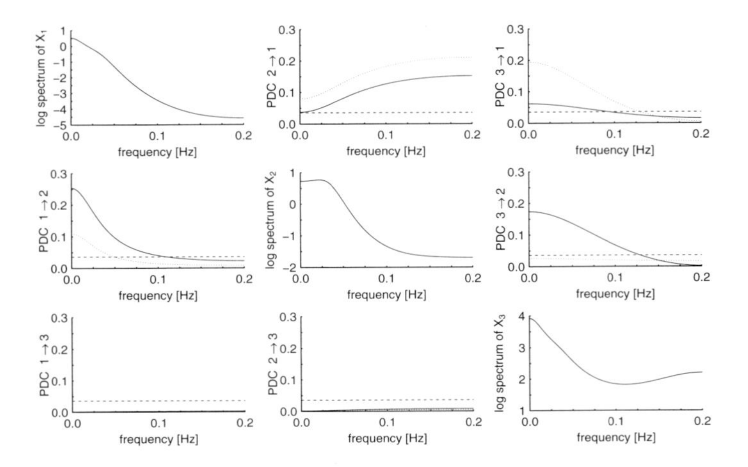

Fig. 14.16: Results for fMRI time series from visual cortex (X_1) and thalamus (X_2) and EEG alpha atom (X_3): estimates of log-spectral densities (on diagonal) and rescaled PDC $\alpha_{ij}^2(\lambda)$ (off-diagonals). The dotted lines represent the rescaled PDCs obtained from bivariate analysis of the corresponding pairs X_i and X_j. The horizontal dashed lines signify pointwise 95 % test bounds for the hypothesis that the PDC is zero.

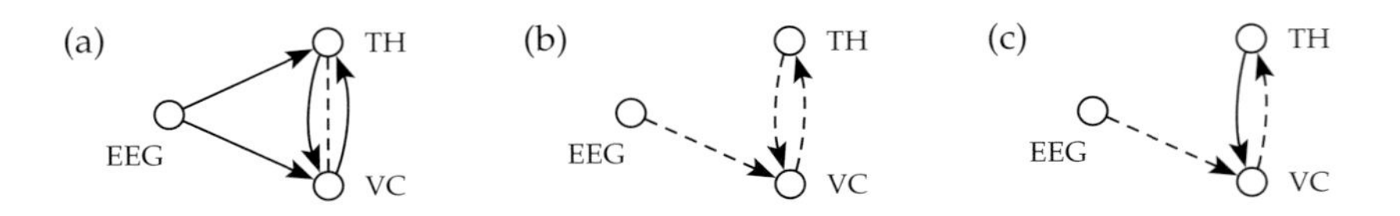

Fig. 14.17: Identification of effective connectivity between the EEG alpha atom, the visual cortex, and the thalamus: (a) multivariate path diagram; (b) bivariate path diagram; (c) alternative path diagram that is Markov equivalent to the graph in (b).

ondly, visual cortex and thalamus are bivariately contemporaneously uncorrelated. Thus the bivariate Granger causality graph encodes more Granger noncausality relations than the multivariate path diagram, which suggests that at least part of the directed relationships shown in the latter are induced by latent variables.

To describe systems that are partly affected by latent variables, Eichler [39] considered more general graphical representations that combine features of bivariate and multivariate path diagrams. In these graphs, ordinary directed edges ($\longrightarrow$) represent causal links while the dashed directed edges ($\dashrightarrow$) indicate spurious causalities induced by latent variables. An example of such a graph is shown in Fig. 14.17(c). In contrast to $G^{(b)}$, this graph indicates a causal influence from the thalamus to the visual cortex. Simple evaluations show that the graph is Markov equivalent to the bivariate path diagram, that is, it encodes the same relationships among the variables. This implies that we cannot decide empirically between the two graphs as possible descriptions of the connectivity among the variables. We note that in both the graphs the correlation between EEG alpha atom and thalamic BOLD responses that is observed in a multivariate analysis is attributed to the indirect link EEG ⇢ VC ⇢ TH mediated by the visual cortex. This is in line with previous results [70], which identified the visual cortex as the source of the "EEG alpha rhythm." Similarly, we note that the contemporaneous correlation between thalamus and visual cortex in a multivariate analysis is attributed to the pathway TH ⇠ VC ⇠ EEG ⇢ VC.

14.7 Conclusion

In this chapter, we have described a graphical approach for visualizing and analyzing the causal relationships in multivariate time series based on the concept of Granger causality. We have seen that by the global (Granger-causal) Markov property certain pathways in a graph can be related to dependences between the variables. This can be exploited for determining whether a given causal structure that possibly contains unmeasured latent variables is consistent with the dynamic dependences that have been found empirically between the observed variables. The graphical analysis shows in particular that the causal structure of systems that may be affected by latent variables in general cannot be resolved by multivariate and bivariate analyses alone, but only by examination of Granger noncausality relations with respect to all possible subseries.

In Section 14.6.3, we have briefly touched general Granger causality graphs for the representation of causal structures with latent variables. Unlike bivariate or multivariate path diagrams, which can be specified by pairwise Granger causality relations, these graphs are determined solely through the global Granger-causal Markov property. This holds a number of problems for the empirical identification of causal structures. First, such general Granger causality graphs are not uniquely determined by the Granger noncausality relations that they encode; Fig. 14.17 has shown an example of two such Markov equivalent graphs. Secondly, the identification of such graphical representations is based on a multistep procedure where each step requires the fitting of a new autoregressive model to a subseries. As a consequence, it is impossible to compare two graphical representations of the effective connectivity and to test between them. Moreover, the statistical errors in different steps may lead to contradictory results. To avoid these problems associated with this multistep identification, future research aims at the development of new graphical time series models that satisfy the global Granger-causal Markov property with respect to such general Granger causality graphs; the identification of the causal structure could then be achieved by model selection.

Acknowledgements

The data on essential tremor in Section 14.6.2 were recorded by B. Hellwig and B. Guschlbauer at the Department of Neurology of the University of Freiburg, and the EEG-fMRI experiments discussed in Section 14.6.3 were conducted by Robin Goldman and Mark Cohen, which is gratefully acknowledged. Furthermore, the author wishes to thank Pedro Valdéz-Sosa and Eduardo Martínez Montes for many helpful comments on the EEG-fMRI data set.

References

[1] B. Schack, P. Rappelsberger, S. Weiss, and E. Möller. Adaptive phase estimation and its application in EEG analysis of word processing. *J. Neurosci. Methods*, 93:49–59, 1999.

[2] H. Liang, M. Ding, R. Nakamura, and S. L. Bressler. Causal influences in primate cerebral cortex during visual pattern discrimination. *NeuroReport*, 11:2875–2880, 2000.

[3] C. W. J. Granger. Investigating causal relations by econometric models and cross-spectral methods. *Econometrica*, 37:424–438, 1969.

[4] C. Hsiao. Autoregressive modeling and causal ordering of econometric variables. *J. Econ. Dyn. Control*, 4:243–259, 1982.

[5] S. Wright. Correlation and causation. *J. Agric. Res.*, 20:557–585, 1921.

[6] S. Wright. The method of path coefficients. *Ann. Math. Stat.*, 5:161–215, 1934.

[7] J. Whittaker. *Graphical Models in Applied Multivariate Statistics*. Wiley, Chichester, 1990.

[8] D. R. Cox and N. Wermuth. *Multivariate Dependences—Models, Analysis and Interpretation*. Chapman and Hall, London, 1996.

[9] S. L. Lauritzen. *Graphical Models*. Oxford University Press, Oxford, 1996.

[10] D. Edwards. *Introduction to Graphical Modelling*. Springer, New York, 2nd edition, 2000.

[11] J. Pearl. Fusion, propagation and structuring in belief networks. *Artif. Intell.*, 29:241–288, 1986.

[12] J. Pearl. *Probabilistic Inference in Intelligent Systems*. Morgan Kaufmann, San Mateo, CA, 1988.

[13] J. Pearl. Causal diagrams for empirical research (with discussion). *Biometrika*, 82:669–710, 1995.

[14] J. Pearl. *Causality*. Cambridge University Press, Cambridge, UK, 2000.

[15] P. Spirtes, C. Glymour, and R. Scheines. *Causation, Prediction, and Search*. MIT Press, Cambridge, MA, 2nd edition, 2001. With additional material by David Heckerman, Christopher Meek, Gregory F. Cooper and Thomas Richardson.

[16] S. L. Lauritzen. Causal inference from graphical models. In O. E. Barndorff-Nielsen, D. R. Cox, and C. Klüppelberg, editors, *Complex Stochastic Systems*, pages 63–107. CRC Press, London, 2001.

[17] M. Eichler. *Graphical Models in Time Series Analysis*. PhD thesis, Universität Heidelberg, 1999.

[18] M. Eichler. Graphical modelling of time series. Technical report, Universität Heidelberg, 2001.

[19] M. Eichler. Granger-causality and path diagrams for multivariate time series. *J. Econ.*, 2006. In press.

[20] C. A. Sims. Macroeconomics and reality. *Econometrica*, 48:1–4, 1980.

[21] H. Y. Toda and P. C. B. Philipps. Vector autoregressions and causality. *Econometrica*, 61:1367–1393, 1993.

[22] B. Hayo. Money-output Granger causality revisited: an empirical analysis of EU countries. *Appl. Econ.*, 31:1489–1501, 1999.

[23] J. F. Geweke. Measures of conditional linear dependence and feedback between time series. *J. Am. Stat. Assoc.*, 79:907–915, 1984.

[24] P. J. Brockwell and R. A. Davis. *Time Series: Theory and Methods*. Springer, New York, 2nd edition, 1991.

[25] K. Sameshima and L. A. Baccalá. Using partial directed coherence to describe neuronal ensemble interactions. *J. Neurosci. Methods*, 94:93–103, 1999.

[26] L. A. Baccalá and K. Sameshima. Partial directed coherence: a new concept in neural structure determination. *Biol. Cybern.*, 84:463–474, 2001.

[27] T. Haavelmo. The statistical implications of a system f simultaneous equations. *Econometrica*, 11:1–12, 1943.

[28] A. S. Goldberger. Structural equation models in the social sciences. *Econometrica*, 40:979–1001, 1972.

[29] K. A. Bollen. *Structural Equations with Latent Variables*. Wiley, New York, 1989.

[30] C. W. J. Granger. Testing for causality, a personal viewpoint. *J. Econ. Dyn. Control*, 2:329–352, 1980.

[31] J. P. Florens and M. Mouchart. A linear theory for noncausality. *Econometrica*, 53:157–175, 1985.

[32] Y. Hosoya. On the Granger condition for non-causality. *Econometrica*, 45: 1735–1736, 1977.

[33] D. Tjøstheim. Granger-causality in multiple time series. *J. Econ.*, 17:157–176, 1981.

[34] J. F. Geweke. Measurement of linear dependence and feedback between multiple time series. *J. Am. Stat. Assoc.*, 77:304–313, 1982.

[35] M. Kamiński, M. Ding, W. A. Truccolo, and S. L. Bressler. Evaluating causal relations in neural systems: Granger causality, directed transfer function and statistical assessment of significance. *Biol. Cybern.*, 85:145–157, 2001.

[36] R. Goebel, A. Roebroeck, D.-S. Kim, and E. Formisano. Investigating directed cortical interactions in time-resolved fMRI data using vector autoregressive modeling and Granger causality mapping. *Magn. Reson. Imaging*, 21:1251–1261, 2003.

[37] W. Hesse, E. Möller, M. Arnold, and B. Schack. The use of time-variant EEG Granger causality for inspecting directed interdependencies of neural assemblies. *J. Neurosci. Methods*, 124:27–44, 2003.

[38] A. Brovelli, M. Ding, A. Ledberg, Y. Chen, R. Nakamura, and S. L. Bressler. Beta oscillations in a large-scale sensorimotor cortical network: directional influences revealed by granger causality. *Proc. Natl. Acad. Sci. India A., Phys. Sci. USA*, 101:9849–9854, 2004.

[39] M. Eichler. A graphical approach for evaluating effective connectivity in neural systems. *Philos. Trans. R. Soc. A, Phys. Sci. B*, 360:953–967, 2005.

[40] R. Dahlhaus and M. Eichler. Causality and graphical models in time series analysis. In P. Green, N. Hjort, and S. Richardson, editors, *Highly Structured Stochastic Systems*. University Press, Oxford, 2003.

[41] C. M. Queen and J. Q. Smith. Multiregression dynamic models. *J. Roy. Stat. Soc. B*, 55:849–870, 1993.

[42] M. Reale and G. Tunnicliffe Wilson. Identification of vector AR models with recursive structural errors using conditional independence graphs. *Stat. Methods Appl.*, 10:49–65, 2001.

[43] P. Spirtes, T. S. Richardson, C. Meek, R. Scheines, and C. Glymour. Using path diagrams as a structural equation modelling tool. *Soc. Methods Res.*, 27: 182–225, 1998.

[44] J. T. A. Koster. On the validity of the Markov interpretation of path diagrams of Gaussian structural equations systems with correlated errors. *Scand. J. Stat.*, 26:413–431, 1999.

[45] T. Richardson. Markov properties for acyclic directed mixed graphs. *Scand. J. Stat.*, 30:145–157, 2003.

[46] D. R. Brillinger. *Time Series: Data Analysis and Theory*. McGraw-Hill, New York, 1981.

[47] K. J. Blinowska, R. Kuś, and M. Kamiński. Granger causality and information flow in multivariate processes. *Phys. Rev., E*, 70:050902(R), 2004.

[48] R. Kuś, M. Kamiński, and K. J. Blinowska. Determination of EEG activity propagation: Pair-wise versus multichannel estimate. *IEEE Trans. Biomed. Eng.*, 51:1501–1510, 2004.

[49] M. Kamiński. Determination of transmission patterns in multichannel data. *Philos. Trans. R. Soc. A, Phys. Sci. B*, 360:947–952, 2005.

[50] J. Corander and M. Villani. A Bayesian approach to modelling graphical vector autoregressions. *J. Time Ser. Anal.*, 27:141–156, 2006.

[51] M. Eichler. *Maximum Likelihood Estimation for Graphical Autoregressions*. University of Maastricht, 2006. Preprint.

[52] H. Lütkepohl. *Introduction to Multiple Time Series Analysis*. Springer, New York, 1993.

[53] M. Eichler. *Graphical Modelling for Multivariate Time Series Using Chain Graphs*. University of Maastricht, 2006. Preprint.

[54] B. Schelter, M. Winterhalder, M. Eichler, M. Peifer, B. Hellwig, B. Guschlbauer, C. H. Lücking, R. Dahlhaus, and J. Timmer. Testing for directed influences among neural signals using partial directed coherence. *J. Neurosci. Methods*, 152:210–219, 2006.

[55] M. Drton and T. S. Richardson. Iterative conditional fitting for estimation of a covariance matrix with zeros. Technical report, University of Washington, 2004.

[56] H. Akaike. Fitting autoregressive models for prediction. *Ann. Inst. Stat. Math.*, 21:243–247, 1969.

[57] G. Schwarz. Estimation the dimension of a model. *Ann. Stat.*, 6:461–464, 1978.

[58] J. Sandkühler and A. A. Eblen-Zajjur. Identification and characterization of rhythmic nociceptive and non-reciceptive spinal dorsal horn neurons in the rat. *Neurosci.*, 61:991–1006, 1994.

[59] M. Eichler, R. Dahlhaus, and J. Sandkühler. Partial correlation analysis for the identification of synaptic connections. *Biol. Cybern.*, 89:289–302, 2003. doi: 10.1007/s00422-003-0400-3.

[60] R. Dahlhaus and M. Eichler. Causality and graphical models for multivariate time series and point processes. In H. Hutten and P. Kroesl, editors, *IFMBE Proceedings EMBEC 2002*, volume 3(2), pages 1430–1431, 2002.

[61] J. R. Rosenberg, A. M. Amjad, P. Breeze, D. R. Brillinger, and D. M. Halliday. The Fourier approach to the identification of functional coupling between neuronal spike trains. *Prog. Biophysics Mol. Biol.*, 53:1–31, 1989.

[62] M. J. Kamiński and K. J. Blinowska. A new method of the description of the information flow in the brain structures. *Biol. Cybern.*, 65:203–210, 1991.

[63] M. Eichler. On the evaluation of information flow in multivariate systems by the directed transfer function. Technical report, University of Heidelberg, 2005.

[64] A. Korzeniewska, M. Manczak, M. Kaminski, K. Blinowska, and S. Kasicki. Determination of information flow direction between brain structures by a modified Directed Transfer Function method (dDTF). *J. Neurosci. Methods*, 125:195–207, 2003.

[65] R. Dahlhaus. Graphical interaction models for multivariate time series. *Metrika*, 51:157–172, 2000.

[66] R. Dahlhaus, M. Eichler, and J. Sandkühler. Identification of synaptic connections in neural ensembles by graphical models. *J. Neurosci. Meth.*, 77: 93–107, 1997.

[67] J. Timmer, M. Lauk, B. Köster, B. Hellwig, S. Häußler, B. Guschlbauer, V. Radt, M. Eichler, G. Deuschl, and C. H. Lücking. Cross-spectral analysis of tremor time series. *Int. J. Bif. Chaos*, 10:2595–2610, 2000.

[68] B. Hellwig, S. Häußler, B. Schelter, M. Lauk, B. Guschlbaur, J. Timmer, and C. H. Lücking. Tremor correlated cortical activity in essential tremor. *Lancet*, 357:519–523, 2001.

[69] R. I. Goldman, J. M. Stern, J. Engel, and M. S. Cohen. Simultaneous EEG and fMRI of the alpha rhythm. *NeuroReport*, 13:2487–2492, 2002.

[70] E. Martínez-Montes, P. A. Valdés-Sosa, F. Miwakeichi, R. I. Goldman, and M. S. Cohen. Concurrent EEG/fMRI analysis by multiway partial least squares. *NeuroImage*, 22:1023–1034, 2004.

15 Multivariate Signal Analysis by Parametric Models

Katarzyna J. Blinowska and Maciej Kamiński

Multivariate time series analysis by parametric models finds a broad range of applications: in biomedical research, economics, geophysics or industry. To fully utilize the information contained in the recorded signals, methods describing the relations in the whole data set are needed. Parametric methods extract a meaningful description of the data and then the signal properties are derived from the parameters of the model, not from the data themselves. Particularly useful in respect of multivariate time series analysis are autoregressive models (MVAR) which fulfill the maximum entropy property.

In the following the formalism and the method of the estimation of the model is presented. The basic statistical measures used in multivariate data analysis are described. The concept of coherence and ordinary (bivariate), partial and multiple coherences are introduced and their properties are characterized.

Linear modeling allows for description of causal relations between data channels within a multichannel set. The basic concepts, e.g., Granger causality are introduced. Then an extension of the formalism for a multivariate case is presented. The Directed Transfer Function (DTF) is described and its properties are discussed.

Different measures involving the causality relations in time series or direction of signal propagation are considered. Multivariate methods such as DTF and partial directed coherence (PDC) are compared with a bivariate approach by means of simulations. Their performance is also tested on experimental signals.

Of particular interest, especially in the field of biomedical applications, is the dynamical propagation of signals. The short-time directed transfer function (SDTF) allows for estimation of propagation of signals in time and frequency when multiple realizations of the investigated process are available. The formalism of SDTF estimation is described and then the performance of the function is characterized.

Examples of application of the presented formalism to experimental data are presented. In particular, the propagation of electroencephalography (EEG) and local field potentials (LFP) signals in time and frequency is considered.

Handbook of Time Series Analysis. Björn Schelter, Matthias Winterhalder, Jens Timmer

ISBN: 3-527-40623-9

15.1 Introduction

Currently, state-of-the-art measurements in a variety of fields offer large batteries of data recorded by multiple sensors. Multichannel data analysis inherently includes analysis of interrelations between channels. When a single data channel is considered, we may calculate measures describing only that channel, the so-called *auto*-quantities. A multivariate set of data, e.g., a multichannel EEG recorded simultaneously, contain auto-quantities for every data channel of the set and, moreover, it contains information about interrelations between data channels of the set, called *cross*-quantities. Cross-correlation in the time domain or coherence in the frequency domain are typical examples of such quantities. It should be noted that (i) cross-quantities are independent of and are not directly related to auto-quantities of the same set and (ii) they are functions of two (or more) channels. If the measured signals come from the same system or interconnected systems, they are usually correlated. In tracking causality between these signals methods are needed which consider the system as a whole entity and take into account mutual dependencies between full set of signals.

Data analysis methods can be divided in two groups: nonparametric and parametric. The nonparametric approach relies on estimating the desired quantities directly from the data. For instance, Fourier analysis is a nonparametric method and spectral estimates are obtained by calculations performed on the data samples. On the other hand, the parametric approach is based on another idea: a data-generation model is assumed and fitted to the signals. The signals are then represented by a set of model parameters. All further analysis is performed on the fitted model parameters. When the data are of random character, a stochastic model of data generation should be assumed. Such signals, containing a random component, are often encountered in biomedical recordings or industrial processes. A good example of such data could be EEG signals. Although the parametric analysis can be applied to a wide range of time series, from economics to dendrology, the main issues connected with that technique will be exemplified in this chapter on biomedical signals.

15.2 Parametric Modeling

The model approximating the time series should be chosen with care in order to describe the data appropriately. The problem of proper selection of the model to the given data can be considered a drawback of the parametric approach. On the other hand, that approach has many advantages. The description of the process is simplified and its properties can be estimated from the model itself. Based on that property, the parametric analysis in the frequency domain can overcome the window problem, which is always present in the nonparametric approach. Fourier analysis theory assumes operations on infinite or periodic signals. Finite data sets (that means: all real data sets) are considered as multiplied by a finite time window function. The transform of the data is then always convolved with the transform of the window function, which distorts the spectral estimate. Parametric modeling assumes validity of the model inside and outside the window of observation, which is more realistic than assuming the signal to be zero when we do not measure it. The model-based spectra have no sidelobes and are smooth, since they are typically described by an analytical function. Moreover, parametric modeling is especially suitable for the consideration of multichannel data, since it allows for defining truly multichannel estimators of causal relations between channels; this important property will be discussed later in this chapter.

The origins of the linear modeling lie in economical and social sciences, yet it is now a popular technique in many fields of science and engineering. In biomedical data analysis from a wide class of possible models the autoregressive (AR) and autoregressive-moving average (ARMA) models are of primary importance. Such models can describe a wide class of signals commonly appearing in practice and there are numerous examples of their successful applications. The theoretical foundation of multichannel AR model can be found as early as in the 1960s. In the paper from 1965, Akaike [1] considered frequency characteristics of a system having multiple inputs. Later, in the 1970s, several authors considered such linear models in data analysis, e.g., [2–7]. Measures of dependencies between channels such as correlation, coherence or causality were first introduced for pairs of channels. Granger (Nobel prize winner in 2003) defined the causality principle for two time series and applied it to economic problems [8]. However, even in the early attempts of identification of interrelation between signals a reservation was made concerning the validity of information drawn from bivariate measures in a case when more than two channels are involved in a given process (Granger, Gersch 1972). The three-channel AR model was elaborated by Gersch [9–11] and tested on epileptic EEG signals, with the indication concerning the extension of the model to the arbitrary number of channels. The formalism for the estimation of the MVAR model for the arbitrary number of channels and calculation of ordinary, partial and multiple coherences and application of that formalism to biological signals was given by Franaszczuk et al. 1985. In [12] coherences were calculated for electrocorticogram data (ECoG) registered from four electrodes. The MVAR model, besides its wide range of applications in electroencephalography (EEG) analysis, has been used in functional magneto-resonance imaging (fMRI) data processing as well [13]. The formalism of MVAR coefficient estimation developed in this paper was later used in designing the Directed Transfer Function (DTF) and for the calculation of partial and multiple coherences and DTFs, e.g., for 21 channels of EEG in Kamiński et al. [14]. Moreover, the DTF method has been applied to localize epileptic foci [15], to determine LFP propagation between brain structures in different behavioral states of animals [16] to investigate EEG activity propagation in different sleep stages [14] and to study epileptogenesis [17].

15.3 Linear Models

The AR model assumes that

$$\mathbf{X}(t) = \left(X_1(t), X_2(t), \ldots, X_k(t)\right)^T \tag{15.1}$$

—a sample of data at a time t—can be expressed as a sum of its p previous values weighted by model coefficients $\mathbf{A}$ plus a random value $\mathbf{E}(t)$

$$\mathbf{X}(t) = \sum_{j=1}^{p} \mathbf{A}(j)\mathbf{X}(t-j) + \mathbf{E}(t)\,. \tag{15.2}$$

The p is called the model order. For a k-channel process $\mathbf{X}(t)$ and $\mathbf{E}(t)$ are vectors of size k and the coefficients $\mathbf{A}$ are $k \times k$-sized matrices.

Equation (15.2) can be easily transformed to describe relations in the frequency domain. After rewriting Eq. (15.2) in the form (sign of $\mathbf{A}$ changed)

$$\mathbf{E}(t) = \sum_{j=0}^{p} \mathbf{A}(j)\mathbf{X}(t-j) \tag{15.3}$$

the application of Z transform yields

$$\begin{aligned} \mathbf{E}(f) &= \mathbf{A}(f)\mathbf{X}(f) \\ \mathbf{X}(f) &= \mathbf{A}^{-1}(f)\mathbf{E}(f) = \mathbf{H}(f)\mathbf{E}(f) \\ \mathbf{H}(f) &= \left(\sum_{m=0}^{p} \mathbf{A}(m)\exp(-2\pi i m f \Delta t)\right)^{-1}. \end{aligned} \tag{15.4}$$

Details of the procedure can be found in various signal analysis textbooks and papers [18–23]. From the form of that equation we see that the model can be considered as a linear filter with white noises $\mathbf{E}(f)$ on its input (flat dependence on frequency) and the signals $\mathbf{X}(f)$ on its output. The matrix of filter coefficients $\mathbf{H}(f)$ is called the transfer matrix of the system. It contains information about all relations between data channels in the given set. It easily follows that the spectral matrix is given by

$$\mathbf{S}(f) = \mathbf{X}(f)\mathbf{X}^*(f) = \mathbf{H}(f)\mathbf{E}(f)\mathbf{E}^*(f)\mathbf{H}^*(f) = \mathbf{H}(f)\mathbf{V}\mathbf{H}^*(f), \tag{15.5}$$

where the asterisk denotes a transpose and complex conjugate operation. The matrix $\mathbf{S}(f)$ contains auto-spectra of each channel on the diagonal and cross-spectra off the diagonal.

The moving average (MA) model is defined by

$$\sum_{i=0}^{q} \mathbf{B}(i)\mathbf{E}(t-i) = \mathbf{X}(t)\,. \tag{15.6}$$

The data sample $\mathbf{X}(t)$ is generated as a weighted (with coefficients $\mathbf{B}$) sum of q previous white noise values $\mathbf{E}(t)$. Although this type of linear model is not directly applicable in biomedical data analysis, it can be shown that a finite order MA model can be expressed as an AR model (possibly of infinite order) and vice versa. Moreover, MA model can be combined with the previously described AR model producing an autoregressive-moving average (ARMA) model, commonly used in parametric data analysis. It is defined as

$$\sum_{i=0}^{q} \mathbf{B}(i)\mathbf{E}(t-i) = \sum_{j=0}^{p} \mathbf{A}(j)\mathbf{X}(t-j) . \tag{15.7}$$

ARMA models can describe a more general class of processes than AR models. It can be shown that a spectrum of an AR process has the form of a constant over a polynomial (of $\mathbf{A}$ coefficients) while a spectrum of an ARMA process has the form of a ratio of polynomials (of $\mathbf{B}$ and $\mathbf{A}$ coefficients). Roots of polynomials in the denominator correspond to maxima (peaks) in the spectrum and roots of the polynomial in the numerator correspond to dips in the spectrum. Therefore, AR models can describe well a signal containing a set of distinct rhythms responsible for peaks in the spectrum. Additionally, ARMA models can handle well a process with dips (together with peaks) in its spectral power. However, dips in spectral power are a rather rare feature in biomedical data. Moreover, ARMA models, although similar to AR models, require nonlinear algorithms for the estimation of parameters. Procedures are more complicated, typically iterative, in contrast to AR modeling algorithms which are rather straightforward. These facts may explain lower popularity of ARMA applications in the field of biomedical data analysis.

15.4 Model Estimation

The parametric analysis starts with fitting a model to the data. We will present the main issues of the fitting procedure using AR model as an example. Each type of model requires a different algorithm for estimating its parameters [18–27]. There is an abundance of publications concerning the estimation of AR model parameters. Although computational speed is no longer a key issue when choosing an algorithm, small differences in properties of the estimates may favor the application of a particular algorithm to a certain type of data.

Since typically signals of stochastic nature are investigated, estimation procedures rely on statistical properties of the available data. One must make sure that the analyzed data segment is stationary, i.e., the statistical properties of the data do not vary in time, and long enough to get reliable estimates. It is hard to give any precise limits; however, we must assume that the number of available data points is several times bigger than the number of data channels. In the case of short data windows or nonstationary signals, there are special techniques to deal with the data, which will be described later in this chapter. It is worth noting that typically spectral estimates of short data segments obtained by means of parametric modeling perform better than similar estimates obtained using a nonparametric approach.

Before starting a fitting procedure certain preprocessing steps are needed. First of all, the temporal mean should be subtracted for every channel. Equation (15.2) is written assuming that the data are of zero mean. Additionally, in most cases normalization of the data is recommended by dividing each channel by its temporal variance. This is especially useful when data channels have different amplification ratio.

Another problem is the choice of the model order p. An order too low may not allow to describe the data to its full extent while too big an order may introduce spurious artifacts to the estimates. Sometimes it is possible to evaluate the optimal model order directly. For instance, the spectrum of an AR model is given by a rational function with a polynomial of order p in the denominator. So, the number of maxima (peaks) in the spectrum cannot exceed the number of roots of the polynomial in the denominator. Because the roots are always in conjugate pairs, we can expect $p/2$ (or $(p-1)/2$ for an odd p) peaks in the spectrum. If we know that data would contain more rhythmic components we should extend the model order accordingly. Unfortunately, such a simple deduction is not possible for multichannel models where the spectrum in each channel is given by a more complicated formula. Certain statistical criteria have been proposed to deal with the problem of optimal model order selection, like Akaike's final prediction error FPE or Akaike information criterion AIC [19, 28]. We calculate the value of a criterion for every model order within a certain range. The criterion value takes its minimum for the optimal model order. Such criteria are designed to find a balance between the tendency to increase the accuracy of the fit by increasing the model order and a penalty function designed to decrease the order value.

The classical technique of AR model parameters estimation is the Yule–Walker algorithm which will be presented below. It requires calculating the correlation matrix $\mathbf{R}$ of the system up to lag p

$$R_{ij}(s) = \frac{1}{N_S} \sum_{t=1}^{N_S} X_i(t) X_j(t+s) \quad \text{for} \quad s = 0, \ldots, p \,. \tag{15.8}$$

In the next step the model equation (15.1) is multiplied by $\mathbf{X}^T(t-s)$, for $s = 0, \ldots, p$ and expectations of both sides of each equation are taken. Assuming that the noise component is not correlated with the signals, we get a set of linear equations to solve, the Yule–Walker equations

$$\begin{pmatrix} \mathbf{R}(0) & \mathbf{R}(-1) & \dots & \mathbf{R}(p-1) \\ \mathbf{R}(1) & \mathbf{R}(0) & \cdots & \mathbf{R}(p-2) \\ \vdots & \vdots & \ddots & \vdots \\ \mathbf{R}(1-p) & \mathbf{R}(2-p) & \dots & \mathbf{R}(0) \end{pmatrix} \begin{pmatrix} \mathbf{A}(1) \\ \mathbf{A}(2) \\ \vdots \\ \mathbf{A}(p) \end{pmatrix} = \begin{pmatrix} \mathbf{R}(-1) \\ \mathbf{R}(-2) \\ \vdots \\ \mathbf{R}(-p) \end{pmatrix} \tag{15.9}$$

$$\mathbf{V} = \sum_{j=0}^{p} \mathbf{A}(j)\mathbf{R}(j) .$$

Another popular method is the Burg (LWR) algorithm [29]. It is a recursive procedure, where the matrix $\mathbf{R}$ is not calculated. The Burg algorithm produces high-resolution spectra and is preferred when closely spaced spectral components are to be distinguished. Sinusoidal components in a spectrum are better described by the covariance algorithm or its modification. Recently, a Bayesian approach has been proposed for estimating the optimal model order and model parameters as well [30, 31]. In most cases, however, the spectra produced by different algorithms are very similar to each other.

15.5 Cross Measures

In order to evaluate relations between channels of a multivariate dataset, cross-quantities, depending on two or more time series simultaneously, are used. The commonly known cross quantity is coherence. The ordinary coherence between signals i and j is defined as the normalized cross-spectral element S_{ij}

$$K_{ij}(f) = \frac{S_{ij}(f)}{\sqrt{S_{ii}(f)S_{jj}(f)}} . \tag{15.10}$$

Coherence is a complex number, having an amplitude and phase. The normalization assures that the modulus of the function takes values within the range $[0, 1]$. The $K_{ij}(f)$ describes which part of both signals is common and coherent in phase in the channels i and j at frequency f.

If a data set contains more than two channels, the signals can be related with each other in a more complicated way. Namely, two (or more) signals may simultaneously have a common component. Depending on the character of relations between channels, some of them may be connected directly with each other and some connections can be indirect (through other channels). To distinguish between these situations partial and multiple coherences were introduced.

Partial coherence is defined using minors of the spectral matrix $\mathbf{S}$, in the following way:

$$C_{ij}(f) = \frac{\mathbf{M}_{ij}(f)}{\sqrt{\mathbf{M}_{ii}(f)\mathbf{M}_{jj}(f)}} , \tag{15.11}$$

where $\mathbf{M}_{ij}$ is a minor of $\mathbf{S}$ with the ith row and jth column removed. Its properties are similar to ordinary coherence, however, it is nonzero only when the given relation between channels is direct. If a signal in a given channel can be explained by a linear combination of some other signals of the set, the partial coherence between them will be low.

Multiple coherence is given by

$$G_i(f) = \sqrt{1 - \frac{\det(\mathbf{S}(f))}{S_{ii}(f)\mathbf{M}_{ii}(f)}} \,. \tag{15.12}$$

Its value describes the amount of common components in the given channel and the rest of the set. If the value of multiple coherence is low then the channel has no common signal with any other channel of the set.

All coherence functions can be calculated from the spectral matrix **S** by means of nonparametric methods as well. However, by application of parametric modeling we get the spectral matrix for the whole multichannel system. This property is very important in multichannel data analysis and will be discussed later in Section 15.8.

15.6 Causal Estimators

Although at a first glance it seems that the phase of the coherence function can be utilized to estimate the direction of influence between signals, in practice this is rarely possible and often leads to ambiguous results as will be demonstrated below. Therefore, other reliable measures of causal relations were proposed. In order to precisely describe causal relations, a basic definition of causality should be adopted. The definition given by Granger [8] received big popularity because it can be easily transformed to time series modeling. Granger defined causality in terms of predictability of time series which was based on previous works of Wiener [32]. Let us consider two time series X and Y. If we try to predict a value of $X(t)$ using p previous values of the series X only, we get a prediction error e_1

$$X(t) = \sum_{j=1}^{p} A_{11}(j)X(t-j) + e_1(t) \,. \tag{15.13}$$

If we try to predict a value of $X(t)$ using p previous values of the series X and q previous values of Y we get another prediction error e_2

$$X(t) = \sum_{j=1}^{p} A'_{11}(j)X(t-j) + \sum_{j=1}^{p} A_{12}(j)Y(t-j) + e_2(t) \,. \tag{15.14}$$

If the variance of e_2 (after including series Y to the prediction) is lower than the variance of e_1 we say that Y causes X in the sense of Granger causality.

Parametric analysis of time series provides a natural tool to describe causal relations. When considering Eq. (15.3) we see that all the relations between data channels are contained in the transfer matrix **H**. We may define directed transfer function (DTF) which describes causal influence of channel j on channel i at frequency f (Kamiński and Blinowska [33])

$$\gamma_{ij}^2(f) = \frac{|H_{ij}(f)|^2}{\sum_{m=1}^{k} |H_{im}(f)|^2} . \tag{15.15}$$

The above equation defines a normalized version of DTF, which takes values from zero to one producing a ratio between the inflow from channel j to channel i to all the inflows to channel i. Sometimes it is easier to abandon the normalization property and use values of elements of transfer matrix which are related to causal connection strength [34]. The nonnormalized DTF can be defined as

$$\theta_{ij}^2(f) = |H_{ij}(f)|^2 . \tag{15.16}$$

The DTF method, although is based on the Granger causality idea modeled by a MVAR model, describes rather a joint effect of transmission between channels than direct relations. The original definition of the Granger causality, which (in terms of linear models) was given for a pair of channels, can be extended for the multichannel case. Then we predict signal $X_1(t)$ using all available signals. If we are interested in a specific relation, say, between channels X_1 and X_m, we compare prediction errors in a situation when channel m is included or not included into the prediction

$$X_1(t) = \sum_{i=1}^{k} \sum_{j=1}^{p_i} A_{1i}(j) X(t-j) + e_3(t) \quad i \neq m . \tag{15.17}$$

Similar to the case of coherences, there is still a problem of identifying direct and indirect causal relations in the frequency domain. DTF does not discriminate between these two types of relations. Several functions were proposed to solve the problem. The partial directed coherence (PDC)
indexpartial directed coherence (PDC) was defined by Baccala and Sameshima [35] in the following form:

$$P_{ij}(f) = \frac{A_{ij}(f)}{\sqrt{\mathbf{a}_j^*(f)\mathbf{a}_j(f)}} . \tag{15.18}$$

In the above equation $A_{ij}(f)$ is an element of $\mathbf{A}(f)$—a Fourier transform of model coefficients $\mathbf{A}(t)$, where $\mathbf{a}_j(f)$ is jth column of $\mathbf{A}(f)$ and the asterisk denotes the transpose and complex conjugate operation. Although it is a function operating in the frequency domain, the dependence of $\mathbf{A}(f)$ on the frequency has not a direct correspondence to the power spectrum.

Another function—the direct Directed Transfer Function (dDTF)—was proposed in Korzeniewska et al. [36]. The dDTF is defined as a multiplication of a modified DTF by partial coherence. The modification of DTF concerned normalization of the function in such a way as to make the denominator independent of frequency. The dDTF ($\chi ij(f)$) showing direct propagation from channel j to i is defined as

$$\chi_{ij}^2(f) = F_{ij}^2(f) C_{ij}^2(f)$$
$$F_{ij}^2(f) = \frac{|H_{ij}(f)|^2}{\sum_f \sum_{m=1}^{k} |H_{im}(f)|^2} . \quad (15.19)$$

$\chi_{ij}(f)$ has a nonzero value when both functions $F_{ij}^2(f)$ and $C_{ij}^2(f)$ are nonzero, in that case there exists a causal relation between channels $j \to i$ and that relation is direct.

Because of different normalizations, the results of PDC and dDTF may differ in specific situations. This point will be illustrated by means of simulations in the next chapter.

15.7 Modeling of Dynamic Processes

In order to fit a linear model to a dataset the data segment must be long enough to fulfill the requirement that the number of fitted parameters must not exceed the number of data points. In practice, to assure correct statistical properties of the model, we need several times more data points than model parameters. The number of MVAR parameters is pk^2, where p is the model order and k is the number of channels, whereas the number of data points is given by kn, where n is the data segment length in each channel. The number of MVAR parameters increases strongly with the number of channels and sometimes it is difficult to obtain stationary data of appropriate length to fit the model well. This is especially the case for dynamic phenomena, e.g., evoked potentials. Several techniques have been proposed to deal with the problem of nonstationary data modeling.

Some approaches extend the fixed-parameter linear models by including adaptive changes of the model parameters in time. Besides allowing for periodic (or quasi-periodic) components modeled by a set of seasonal parameters in the models, continuous changes of the parameters in time are calculated. This can be accomplished by estimating the parameters over a short time window and successively including new points in the estimate to update the set of model parameters. Another approach is the recursive Kalman [37, 38] filter algorithm. We will not present these methods in a more detailed way; theory and examples can be found in [39–42]. Instead, we will focus on the short sliding window idea, proposed in [43].

When multiple repetitions of an experiment are available, another approach can be proposed. We may repeat the experiment and treat data from each repetition as a realization of the same stochastic process. Then the number of data points is nkN_T (where N_T is the number of realizations) and their ratio to the number of parameters effectively increases. Based on this observation, we can divide a nonstationary recording into shorter time windows, short enough to treat the data within a window as quasi stationary. In practice, due to random jitter effects, it is impossible to obtain perfect synchronization of trials in time. Instead, we use the property that auto- and cross-correlations within each trial are preserved and do not depend on jitter. We calculate the correlation matrix for each trial separately. The resulting model coefficients are based on the correlation matrix averaged over trials. The correlation matrix has the form

$$\tilde{R}_{ij}(s) = \frac{1}{N_T} \sum_{r=1}^{N_T} R_{ij}^{(r)}(s) = \frac{1}{N_T} \sum_{r=1}^{N_T} \frac{1}{N_S} \sum_{t=1}^{N_S} X_i^{(r)}(t) X_j^{(r)}(t+s) . \tag{15.20}$$

The averaging concerns correlation matrices for short data windows—data are not averaged in the process. The details of the procedure involve specific preprocessing in order to avoid problems with model fitting. Besides data normalization in each channel, it is recommended to subtract the ensemble average from the data and divide them by the ensemble variance in each channel to reduce the risk of instability of the models in certain data windows. The choice of window size depends on the investigated problem and it is always a compromise between quality of fit (the ratio between the number of data points and the number of model parameters) and time resolution. Discussion of the preprocessing steps was presented in [34, 43].

By application of the above-described procedure the MVAR coefficients are obtained for each short data window and, subsequently, the estimators characterizing the signals (power spectra, coherences, DTFs) are found. By means of a sliding window the evolution in time is determined. In this way multivariate estimators may be expressed as functions of time and frequency. The Short-Time Directed Transfer Function (SDTF, STDTF) obtained in this way creates a possibility to follow the dynamics of transmissions between data channels.

To estimate the variance of evaluated SDTF functions, the bootstrap approach [44–46] can be utilized. In this technique we simulate multiple experiments by repeatedly randomly selecting a set of input data trials from the pool of experiment repetitions. For each trial the calculation of the model parameters and the estimators is performed. That procedure allows to evaluate the distribution of the results.

There remains a problem of estimation of the admissible level of flows. This can be accomplished by means of the surrogate data technique given by Theiler et al. [47]. The idea is to construct a dataset similar to the given one, but with all causal relations between channels removed. To accomplish this, the data are Fourier transformed, their phases are replaced by random numbers from a flat distribution and then they are transformed back to the time domain by inverse Fourier transform. Such surrogate signal has the same amplitude spectrum as the original one but phases are random in each channel. Modeling and analysis performed on repeatedly generated surrogate datasets provides the baseline distribution for the given estimator of directedness.

15.8 Simulations

The performance of different methods of multivariate data analysis can be illustrated by means of numerical experiments simulating patterns of flows [48–50]. In order to make simulations similar to real experimental situations, as an input time series a human EEG signal recorded from a scalp electrode was used. The signal of 20 s duration (2560 points) was highpass filtered with cut-off frequency of 3 Hz. In construction of the flow pattern the signal was in each step successively delayed; also in each step a random Gaussian noise (each time drawn from a different generator) was added.

In the following simulations where differences between multivariate and bivariate approach are presented, certain functions will be applied twice to the same set of data. The results will be estimated for the whole system of channels simultaneously and for all pairs of channels from the given set. The cases will be referred as "multichannel" or "bivariate," respectively, but one should remember that the same function was used (differently) to obtain both results.

15.8.1 Common Source in Three Channels System

The first simulation, including only three channels, illustrates a basic property of causal relations in the multichannel systems. In this case the delay in channel two was one sample and in channel three two samples. The results are shown in Fig. 15.1. It is easy to notice that the correct pattern of flows is obtained—we get DTF functions indicating propagation from channel one only. If the same system would be analyzed pairwise, additionally the two $\rightarrow$ three transmission would be found (see the next simulation). However, there is no such transmission in the system. In this case the application of a multichannel measure can help to avoid confusing results.

Note that the correct pattern was found for very noisy signals. In the simulation the variance of the noise in channels two and three was nine times bigger than the input EEG signal in channel one. This robustness to noise of the DTF function is especially important for biological time series, where the noise component is usually very strong.

15.8.2 Activity Sink in Five Channels System

Based on the observation from the above simulation, we may construct a more complicated system. Quite common a situation in biological systems is the case of propagation of activity from a source to locations situated at different distances, where recording electrodes are placed. In the following extended simulation the signal from a source channel one was transmitted with delays of one to four samples to the channels two to five. The signals in the destination channels were embedded in noise twice as big as the input signal.

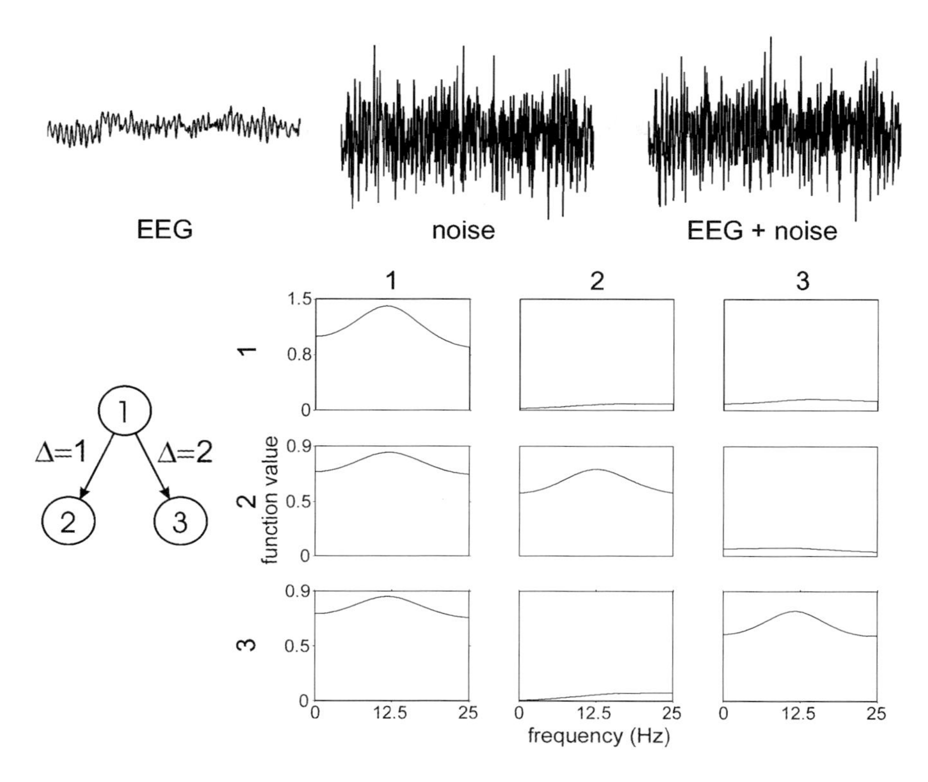

Fig. 15.1: Top: simulated signals. Bottom left: simulation scheme. Bottom right: in each box DTF as a function of frequency (0 Hz to 25 Hz); the numbers above the columns indicate output channels, the numbers on the left of the rows indicate destination channels. Reprinted with permission from [48] (© (2004) by the American Physical Society).

This pattern of flows was used as a model for testing the performance of bivariate versus multivariate estimates of directionality. The first of the tested methods was the bivariate Granger causality. In this approach the MVAR model was fitted to two channels at a time and the Granger causality estimate was calculated. The bivariate results (Fig. 15.2) show propagation not only from source channel 1 but from other channels which were not sources of signal in this simulation. Propagation was found always when a coherent phase difference existed between a pair of channels.

Similar results are obtained from the consideration of phases of bivariate coherences (Fig. 15.3). In Fig. 15.3 the ordinary coherences calculated pairwise are shown, their amplitude spectra are presented at the upper triangle and phases at the lower triangle of the picture matrix. From the phase spectrum of coherences, values corresponding to the frequency of the maximum of amplitude spectrum, namely 11 Hz, were estimated. Subsequently we have found the corresponding

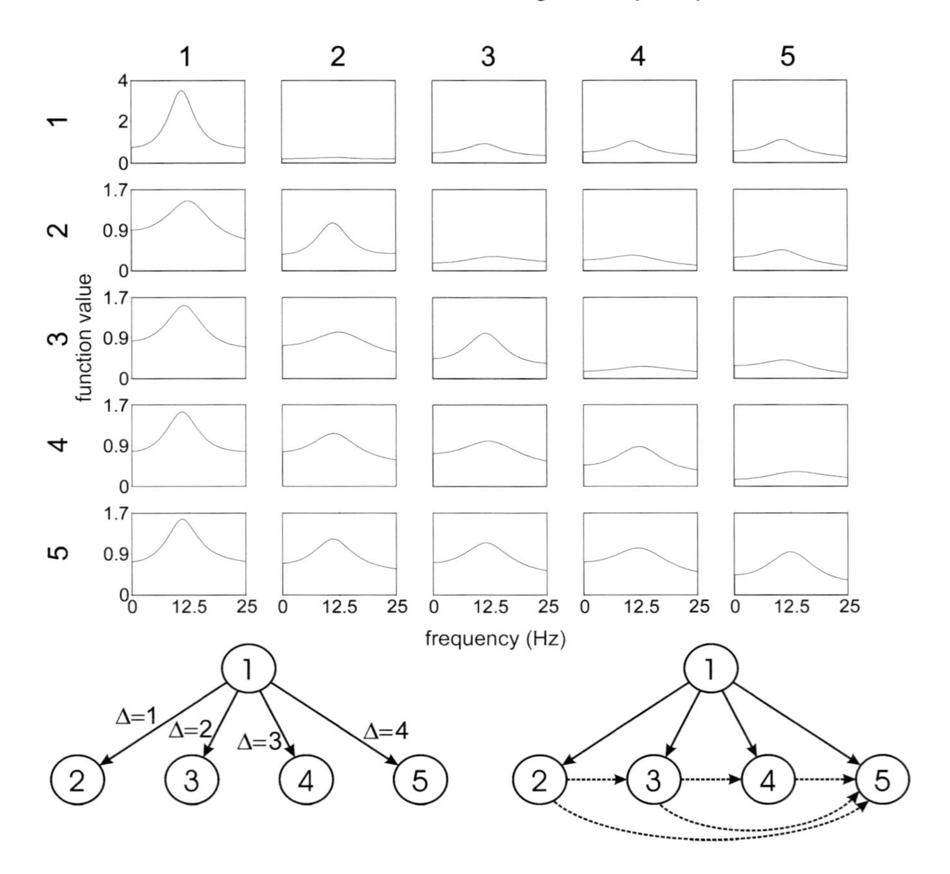

Fig. 15.2: Top: Granger causality calculated pairwise, each graph represents the function describing transmission from the channel marked above the column to the channel marked on the left of the row. Granger causality in arbitrary units on vertical axes; graphs on the diagonal contain power spectra; frequency on horizontal axes (0 Hz to 25 Hz range); Bottom left: simulation scheme. Bottom right: resulting flow scheme. Black arrows represent true (simulated) flows, dotted arrows represent false flows found by the applied method. Reprinted with permission from [48] (© (2004) by the American Physical Society).

delays (in samples). The obtained effective pattern of propagations together with the input diagram of flows are illustrated at the bottom of the figure. In the picture showing phases one can observe discontinuities connected with the fact that phases are determined modulo 2π. This ambiguity makes determination of flows from coherences even more doubtful.

The DTF functions are obtained by fitting the MVAR model simultaneously to all channels of the simulated process. The resulting flow scheme is presented in Fig. 15.4. In this case the pattern of flows is reproduced correctly. One can see only

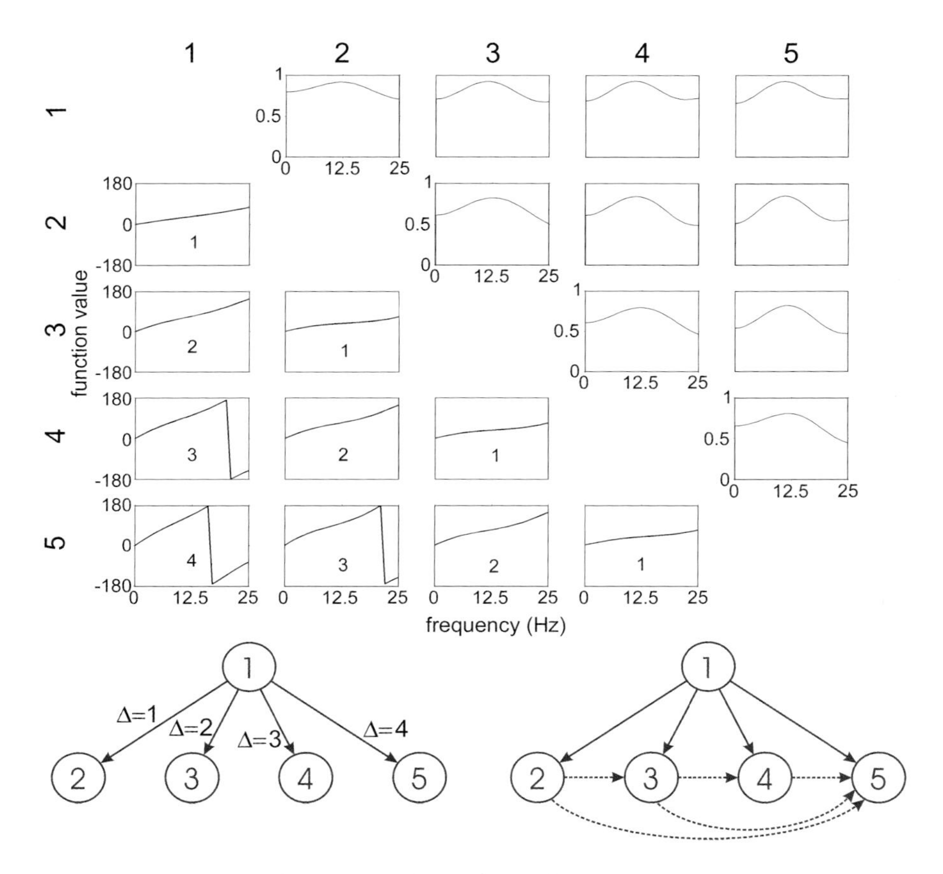

Fig. 15.3: Pair-wise coherences and resulting flows. Top: coherence amplitude (black graphs above diagonal) and coherence phase (graphs below diagonal); each graph represents the function for the pair of channels marked on the left of the row and above the column; on the horizontal axes frequency (0 Hz to 25 Hz); on vertical axes coherence amplitudes (0 to 1 range) or phases ($-180°$ to $180°$ range); delay values (in samples) estimated from phases, marked by the numbers shown over the phase graphs. Bottom left: simulation scheme. Bottom right: resulting flow scheme. The same convention in drawing arrows as in Fig. 15.2. Reprinted with permission from [48] (© (2004) by the American Physical Society).

small "leak flows" in the direction opposite to the designed one. The question concerning an admissible level of weak flows may be resolved by means of the surrogate data test or by the bootstrap method. The advantage of the surrogate data test is that the shapes of the spectra are preserved. The maximal levels of flows obtained from surrogate data (illustrated at the bottom of Fig. 15.4) are similar to the "leak flows" of the results obtained by DTF (upper part of Fig. 15.4).

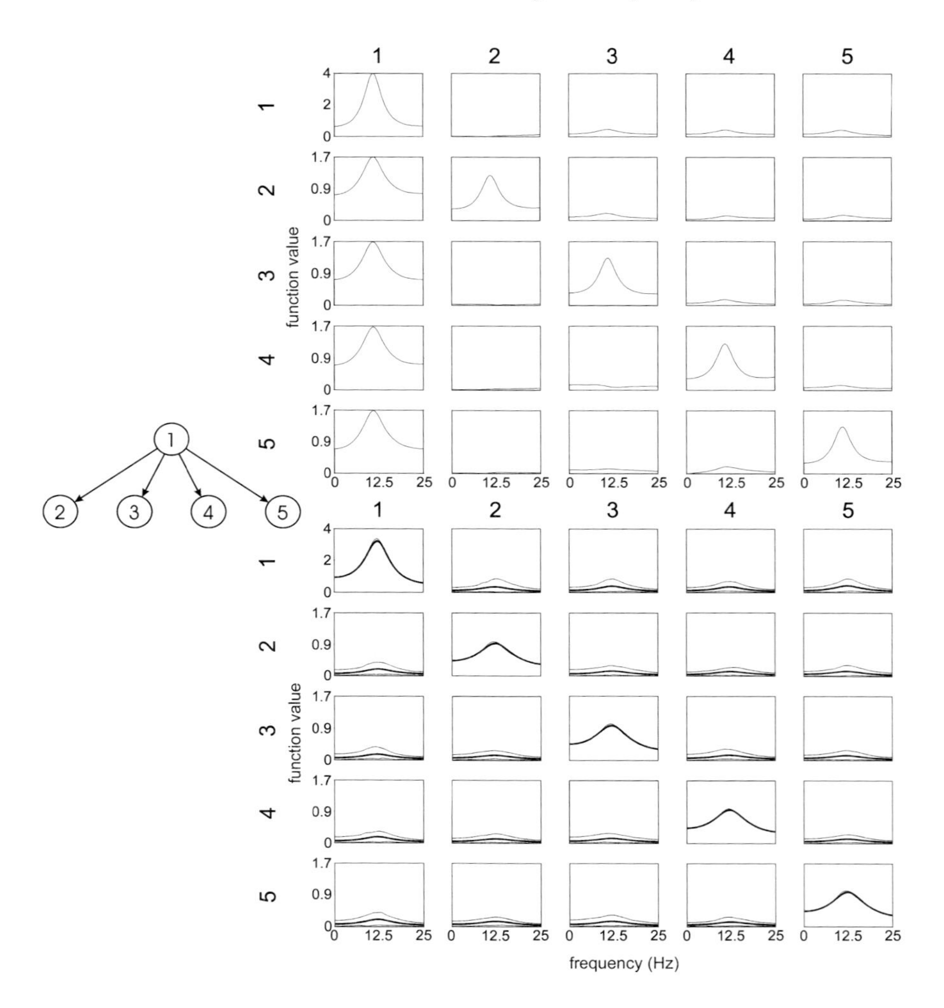

Fig. 15.4: Top: nonnormalized multichannel DTFs for the same simulation as in Figs. 15.2 and 15.3. Each graph represents the function describing transmission from the channel marked above the column to the channel marked on the left of the row (on the diagonal power spectra). Bottom: DTFs obtained from surrogate data. Thick line: the average obtained from 100 surrogates. 95 % of surrogate realizations are contained between the thin lines. Plots in both panels in the same scale in the arbitrary units. Frequency on horizontal axes, 0 Hz to 25 Hz range. At left the resulting flow pattern. Reprinted with permission from [48] (© (2004) by the American Physical Society).

15.8.3 Cascade Flows

The next simulations will concern a more complicated situation encountered, e.g., in case of signals measured from electrodes implanted in different brain

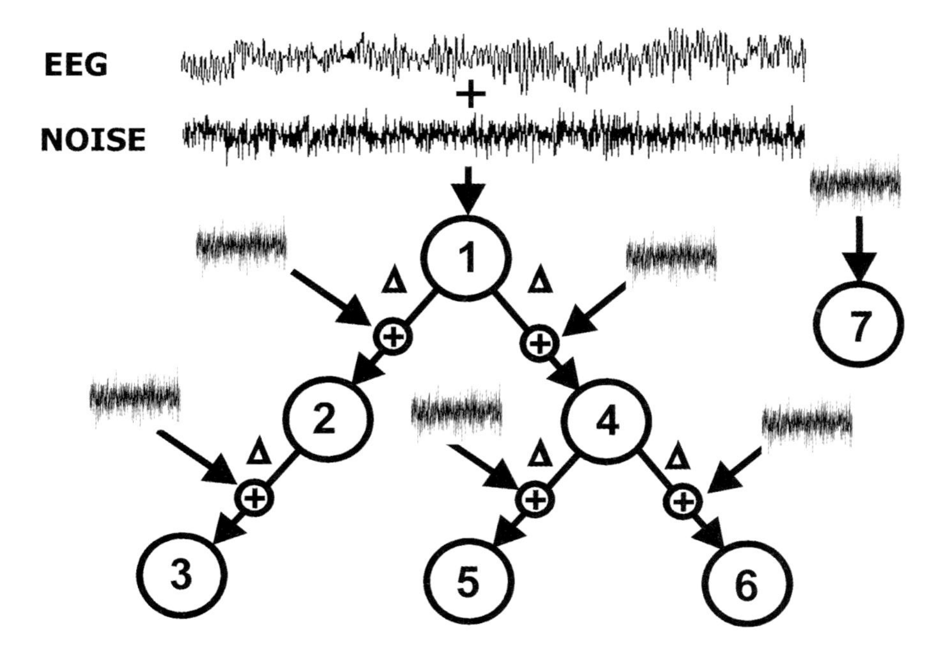

Fig. 15.5: Scheme of simulation. Signal in channel 1 is generated by addition of white noise to the experimental EEG signal. Δ denotes time delay of one sample. Reprinted with permission from [49] (© IEEE 2005).

structures. The scheme of the pattern of flows is shown in Fig. 15.5. The signal in the input channel was the same as in the previous simulations. In each step the signal was successively delayed by one sample; also in each step a random Gaussian noise was added and the time series obtained in this way were transmitted to another channel with a weight 0.8. The amplitude of noise, added in each step, was 0.5 of the amplitude of the original EEG signal. The signal from channel one propagated to channel three through channel two and to channels five and six through channel four. Channel seven was uncoupled to the other channels in any way.

Figure 15.6 presents the results obtained by means of the Granger causality measure calculated pairwise. We can observe that besides the simulated flows some additional propagations are obtained —e.g., from channel two to six, from two to five and from four to three. This result comes from the fact that in case of a difference in delays for bivariate estimates we obtain a flow from the less delayed channel to the more delayed channel, even if they are not connected. This effect is absent for multivariate estimates, when all sources of the signal are included in the calculations.

In the next picture (Fig. 15.7) the ordinary coherences calculated pairwise are shown. The delays between channels and the resulting flow scheme were

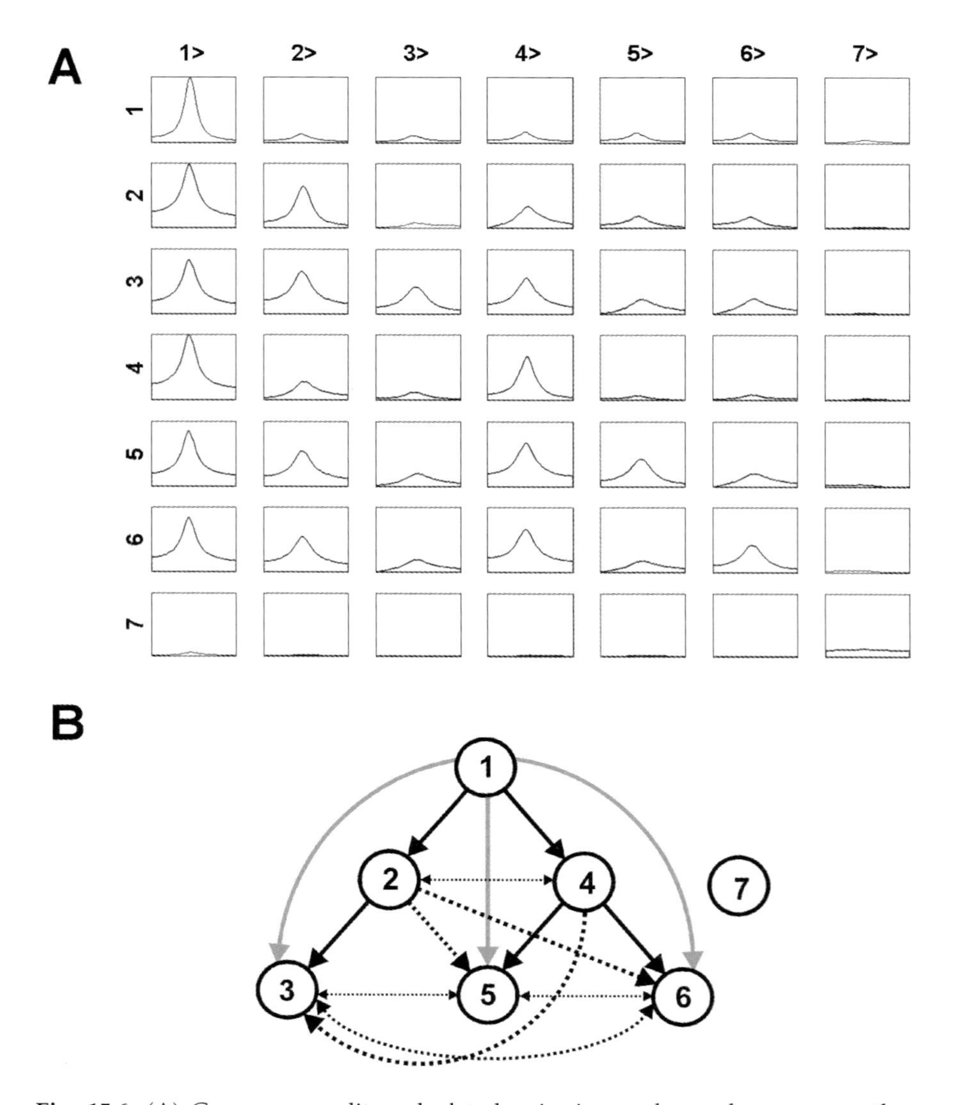

Fig. 15.6: (A) Granger causality calculated pairwise, each graph represents the function describing transmission from the channel marked above the row to the channel marked on the left of the row, frequency on horizontal axes (0 Hz to 25 Hz range); Granger causality in arbitrary units on vertical axes; graphs on the diagonal contain power spectra. (B) the resulting flow scheme. Black arrows represent true (simulated) flows, gray arrows represent indirect flows revealed by the applied method, dotted arrows represent false flows found by the applied method. Reprinted with permission from [49] (© IEEE 2005).

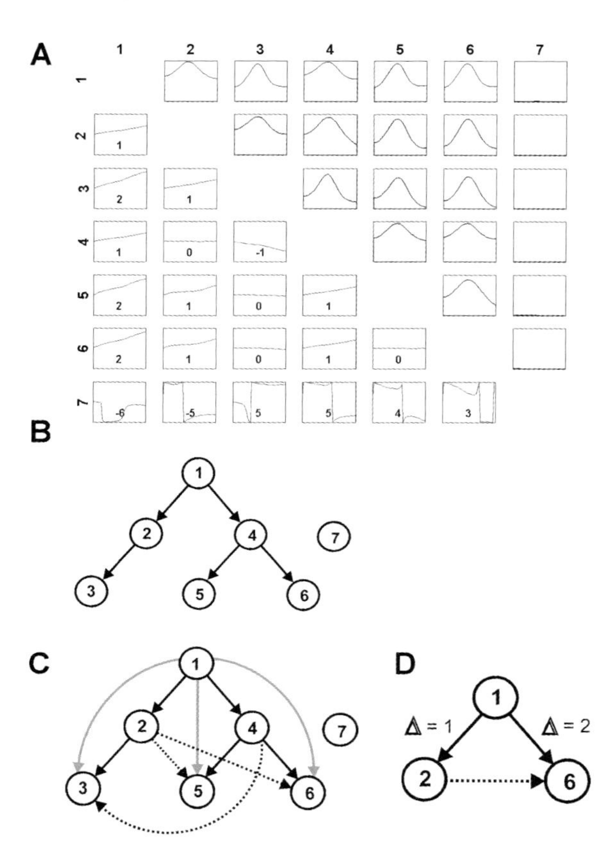

Fig. 15.7: Pair-wise coherences and resulting flow scheme for simulation I shown in Fig. 15.5. (A) coherence amplitude (black graphs above diagonal) and coherence phase (graphs below diagonal); each graph represents the function for pair of channels marked on the left of the row and above the column; on the horizontal axes frequency (0 Hz to 25 Hz); on vertical axes coherence amplitudes (0 to 1 range) or phases ($-\pi$ to π range); delay values (in samples) estimated from phases, marked by the numbers shown over the phase graphs; (B) simulated pattern of flows; (C) pattern of flows estimated from coherence values; (D) pattern of flows obtained from bivariate coherence estimate for different delays between channels. Convention of drawing arrows the same as in Fig. 15.6. Reprinted with permission from [49] (© IEEE 2005).

obtained in the same way as in the example shown in Fig. 15.2. Again we obtain the flows for each pair of electrodes differing in the delay value between them.

The results of application of a multivariate estimator are shown in Fig. 15.8. The pattern of flows is almost correct, except that in the case of cascade, indirect flows are present.

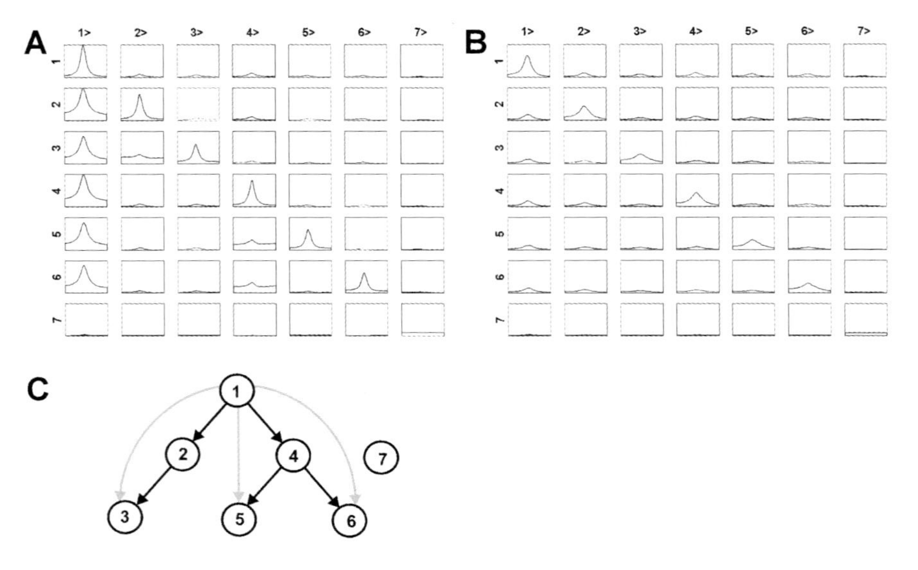

Fig. 15.8: (A) Nonnormalized multichannel DTFs for the simulation illustrated in Fig. 15.5.; (B) DTFs obtained from surrogate data; organization of pictures A and B similar to Fig. 15.2 (on the diagonal power spectra). (C) the resulting flow pattern. Plots in A and B in the same scale in arbitrary units. Reprinted with permission from [49] (© IEEE 2005).

15.8.4 Comparison between DTF and PDC

In order to distinguish direct from indirect flows the direct Directed Transfer Function (dDTF) may be used. It was designed especially for evaluation of experimental results from electrodes implanted in the brain structures [36]. This function is constructed by multiplication of a modification of DTF by partial coherence (Eq. (15.19)). Figure 15.9 shows partial coherences and dDTF found for the system of signals illustrated in Fig. 15.5. In this case the pattern of flows is determined correctly.

Another multivariate method, which makes the distinction between indirect and direct flows possible is Partial Directed Coherence—PDC (Baccala and Sameshima [35]). The application of PDC to the pattern of flows shown in Fig. 15.5 is illustrated in Fig. 15.10. The application of PDC gives similar results as dDTF, except that PDC estimators depend on frequency in a way different from power spectra, which might sometimes cause difficulties in interpretation of the results.

In simple situations results obtained by dDTF and PDC agree with each other; however, there are cases where the results of PDC and DTF give different patterns of flows. This is the case when more than one source emits the activity to the destination channels. In order to clarify the differences between DTF and PDC a series of simulations were performed. In the following simulations again the signal in source channel one is the same as in the preceding numerical exper-

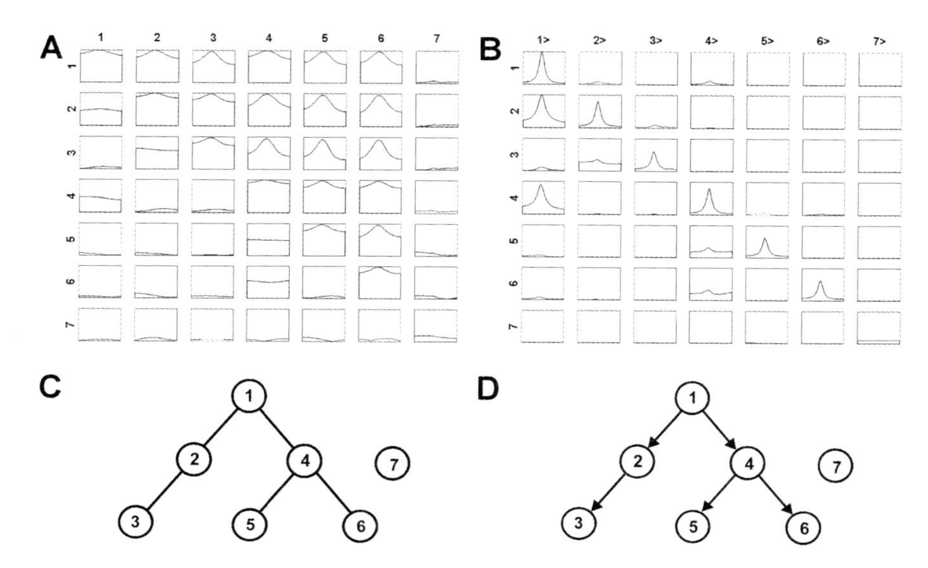

Fig. 15.9: (A) Ordinary (graphs above diagonal), partial (graphs below diagonal) and multiple coherences (graphs on the diagonal) for the simulation shown in Fig. 15.5; in each panel: vertical axis—amplitude in 0 to 1 range, horizontal axis—frequency in 0 Hz to 25 Hz range; (B) dDTFs for the simulated data (power spectra shown on the diagonal); (C) pattern of direct connections estimated from partial coherences; (D) pattern of direct flows estimated from dDTFs. Reprinted with permission from [49] (© IEEE 2005).

iments. This signal is transmitted with weight 0.8 and the delay of one sample to channels two, three and four, with the noise components drawn from different distributions. Time series in channel five is constructed in the same way as signal one, but the input EEG comes from a different subject. This signal plus the noise component is transmitted with delay of one sample and variance four times smaller than the variance of signal one: in simulation II to channel four, in simulation III to channels two, three and four (with different noise components). In simulation IV the scheme is similar to simulation III, except that the strengths of all the transmitted activities are equal. The results of these simulations are illustrated in Fig. 15.11.

It is easy to see that DTF and PDC show the same correct directions of flows, however, there are differences in their intensities. For simulation II the pattern of flows is well reproduced by DTF; however, for PDC weak propagation from channel five becomes predominant in the absence of other flows from that channel. In simulation III, PDC shows similar intensities of flows from electrodes one to five, although originally those from electrode five are much weaker. In the case of simulation IV, when the intensities of flows are same for both sources, the results for DTF and PDC are very similar. These results follow from different

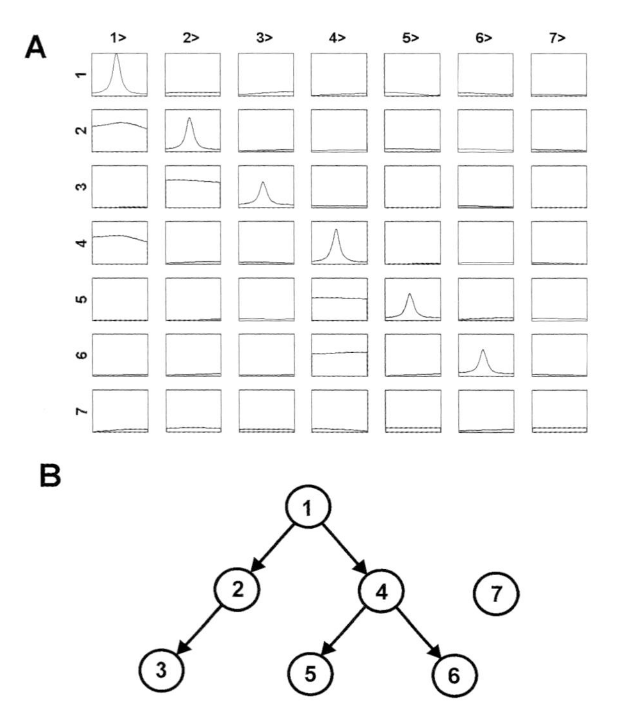

Fig. 15.10: (A) PDC functions for simulation I (Fig. 15.5); (B) resulting pattern of flows. Organization of the picture the same as in Fig. 15.2. Reprinted with permission from [49] (© IEEE 2005).

normalizations in DTF and PDC. DTF is normalized with respect to the inflows to the destination channel and PDF with respect to the outflows from the given channel, therefore for PDC it is difficult to estimate the strengths of the flows. As the authors of the PDC method [35, 51, 52] admit, PDC portrays the relative strengths of direct pairwise structure interactions, while DTF represents a balance of signal power that spreads from one structure to different destinations. Simulations II, III, and IV will help to understand some discrepancies obtained by application of different methods to the same experimental data.

15.9 Multivariate Analysis of Experimental Data

15.9.1 Human Sleep Data

The performance of different estimators can be best explained by their application to real experimental data. The meaning of the ordinary, partial, and multiple coherences will be illustrated by human EEG recorded during sleep and relaxed state, since for these behavioral states the main features of EEG activity are known and relatively long stationary epochs can be recorded. The data

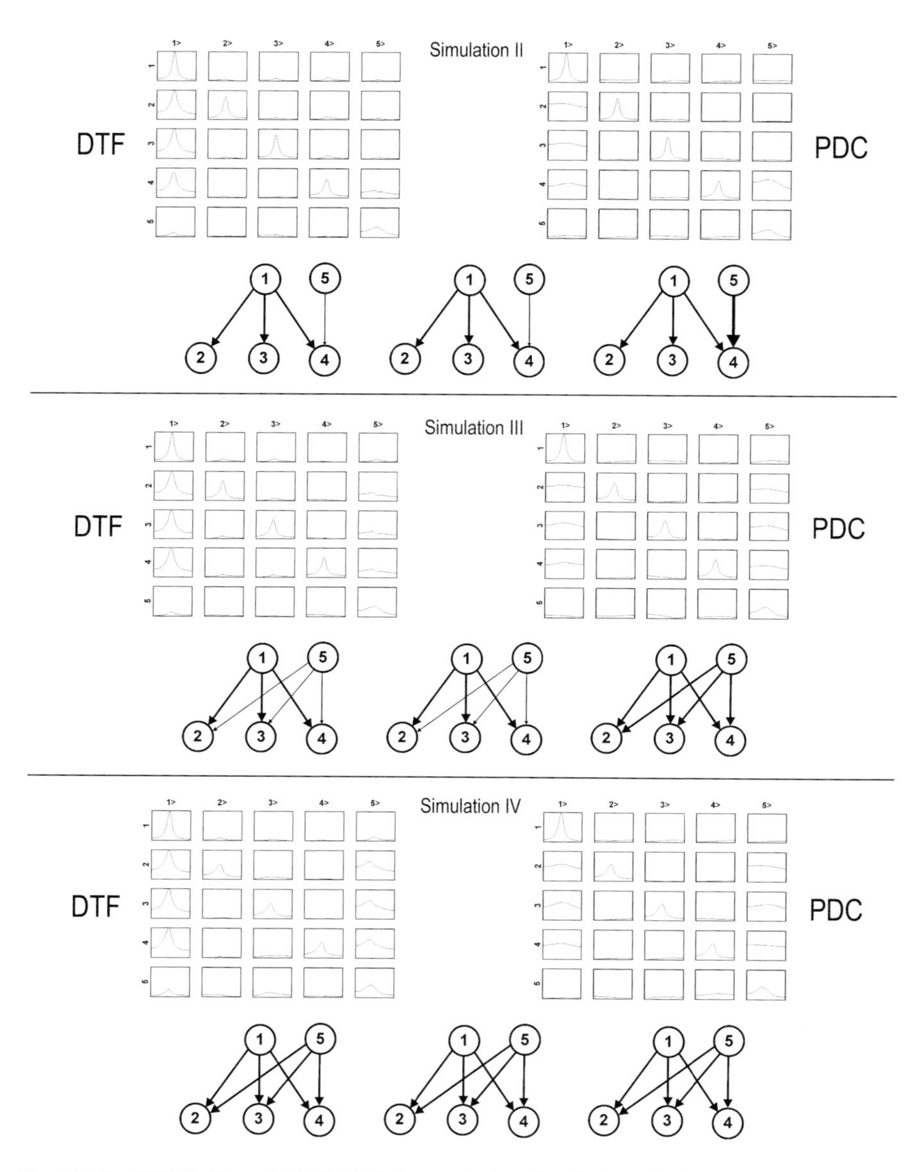

Fig. 15.11: DTF (left) and PDC (right) results for simulations II, III, IV (described in the text). Below the pictures representing DTF and PDC deduced flows are shown for each simulation. The schemes of simulations are shown in the middle column. Reprinted from with permission [49] (© IEEE 2005).

analysis was done by fitting MVAR models to continuous artifact free stationary epochs. The typical graph of coherences for sleep stage two is presented in Fig. 15.12. The multiple coherences are all high, indicating a close relation be-

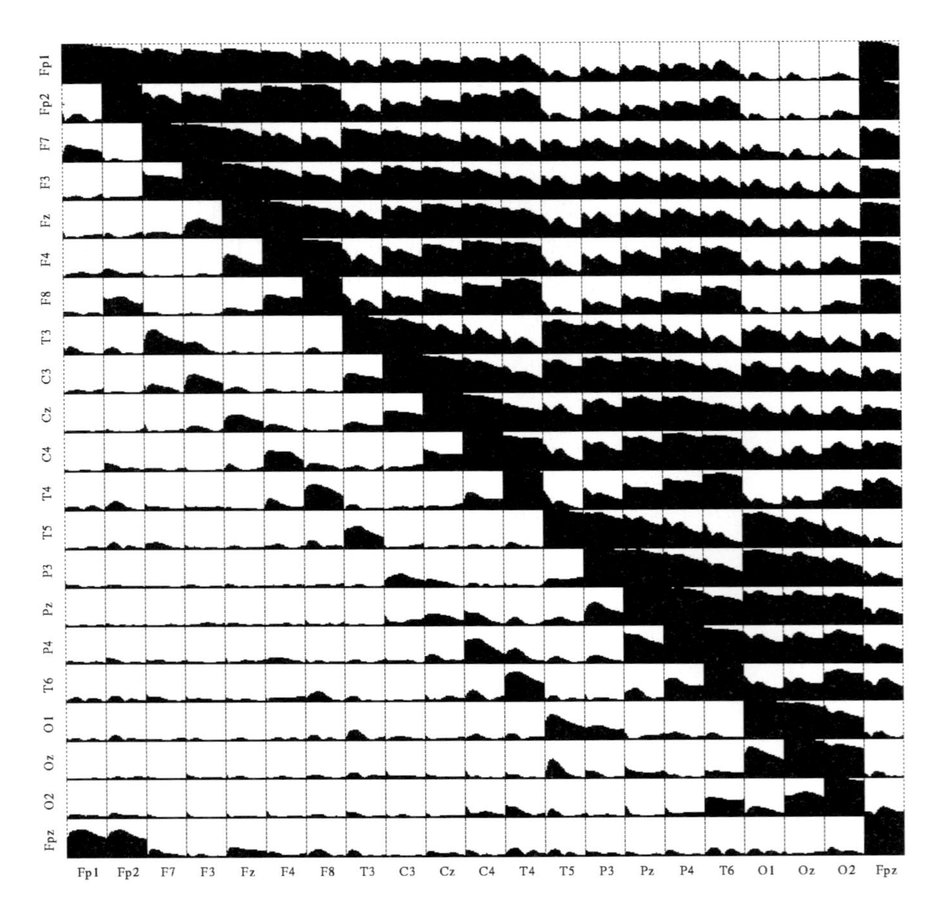

Fig. 15.12: Example of ordinary, partial, and multiple coherences for a set of 21 channels EEG recorded in sleep stage two. Each box in the presented matrix is a graph of a coherence function connecting channel marked below the relevant column and channel marked at left of the relevant row. Frequency on horizontal axes (0 Hz to 30 Hz), function value on vertical axes (0–1). Ordinary coherences above the diagonal, multiple coherences on the diagonal and partial coherences below the diagonal of the matrix of graphs. Reprinted from [14], © (1997), with permission from Elsevier.

tween all the channels in the set. In fact, multiple coherences are usually high for an EEG recorded by scalp electrodes. A scalp EEG shows strong dependencies between all channels in a large frequency range. In consequence, when we consider only two channels at a time, neglecting all the others, we do not know if the relation between them comes from their mutual dependencies or if they are due to feeding from other channels. Closer inspection of ordinary coherences (upper triangle) shows that they depend mainly on the distance between the respective electrodes.

The partial coherences, showing only directly coupled channels, are mostly very low, with little change in frequency. The averaged values of partial coherences were nonzero practically only for neighboring electrodes. This finding is in agreement with the observations of Bullock et al. [53] who analyzed coherences for electrodes implanted in the cortex at different distances, indicating that coherence in the cortex is within a range of 10 mm to 20 mm. In [53] a weak dependence of coherences on frequency was observed, which is also the case for partial coherences (Fig. 15.12). Low values of partial coherences may seem contradictory to the results of many papers showing high coherence between distant electrodes, however, usually the authors analyzed ordinary coherences. Ordinary coherences, especially for scalp electrodes indicate a sum of many indirect relations between channels, therefore it is difficult to draw firm conclusions from ordinary coherences about the interactions between channels.

Partial coherences are mostly observed for neighboring electrodes, although interesting conclusions can be drawn from the pattern of their strength which changes depending on the behavioral state. In the sleep study significant changes were found for partial coherences for different sleep stages (Kamiński et al. [14]).

In order to visualize the performance of the estimators of directedness the best way is to find an example where the sources of activity are known. This is the case for alpha rhythm. With the eyes closed the sources of the EEG activity are placed in the visual cortex at the back of the head and also some sources may be placed frontally.

The calculations were performed on a 21 EEG channels (10–20) system recorded in awake state with eyes closed. The signal was highpass filtered above 3 Hz and lowpass filtered below 50 Hz. The evaluated epoch length was 20 s. In the case of the bivariate Granger causality the MVAR model was fitted to two channels at a time. Since in our simulations the pattern of flows found by means of coherence phase analysis and by the Granger causality measure were identical, we made calculations only for pairwise Granger causality.

For the multivariate measures DTF, dDTF, multivariate Granger causality and PDC, the MVAR model was fitted simultaneously to 21 channels. The model order found by means of AIC criterion was four. For all estimators the calculated transmissions between all channels were integrated in the 7 Hz to 15 Hz range in order to represent alpha activity. The results showing the direction of propagation and intensity of flow are illustrated in Fig. 15.13.

The multivariate estimates of Granger causality and DTF are quite similar and they show a very consistent pattern of flows, directed mainly from the posterior parts of the head toward the front. The difference between both estimates can be observed for the Fz electrode. It comes from the fact that the Granger causality is not normalized, it is simply an element of the transfer matrix. DTF is normalized with respect to inflows. The Fz electrode sends the activity, but there is also a

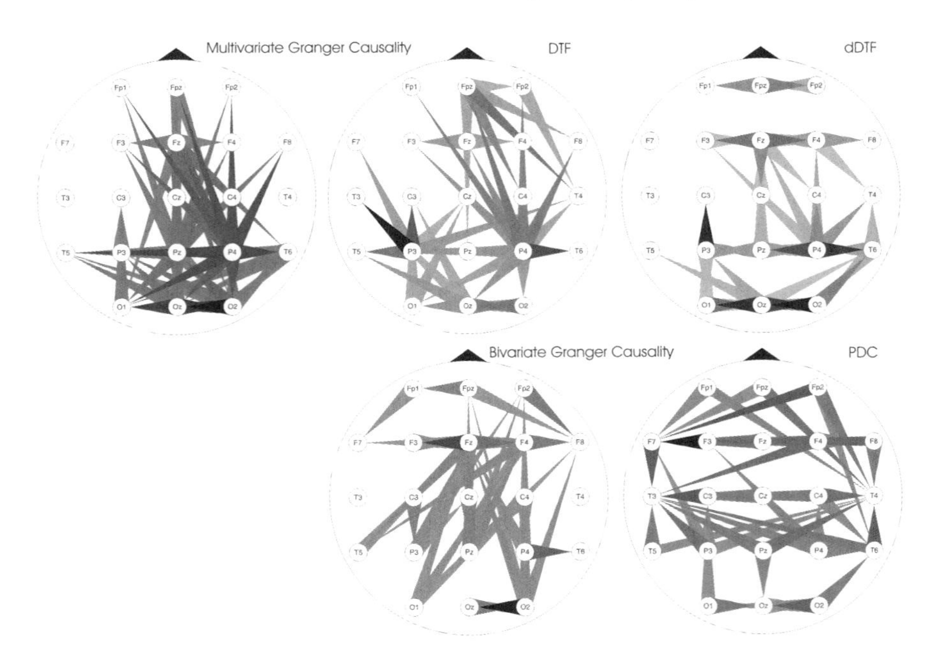

Fig. 15.13: Direction of flows for 21-channel EEG (awake state eyes closed) obtained by means of different methods. The shade of gray of the arrow represents the strength of the connection (black = the strongest), for each method 40 strongest flows are shown. Reprinted from with permission [49] (© IEEE 2005).

lot of activity flowing to the destination channels from the posterior electrodes, so the denominator in Eq. (15.6) is quite large, which diminishes the values of DTFs showing outflows from Fz. For Granger causality and DTF there is no propagation from the temporal electrodes. This is practically also the case for dDTF. The dDTF shows only direct flows, we can see that in this case the pattern of flows is consistent with anatomy, e.g., a lack of direct connection between Oz and Pz, Fz, and Fpz—locations where hemispheres are partitioned. The main sources of the activity—namely, electrodes P3, P4, O2, Oz, O1—are the same as for the other multivariate estimates.

Inspecting the results of application of the PDC function to the same data epoch we observe a different picture. One can notice that, unlike the results of dDTF, some channels became sinks. This is due to the normalization of PDC. In fact, we do not see the transmission, as is the case for dDTF, but the ratio between the flow to a given channel with respect to all the outflows from the considered channel. In this way, a channel propagating activity in all directions will show weaker flows than a channel propagating only in one direction. Therefore, the method is not suitable for identification of sources of EEG activity, but it may be useful when the destination channel is of primary interest.

The pattern of propagations obtained for the bivariate estimates of the Granger

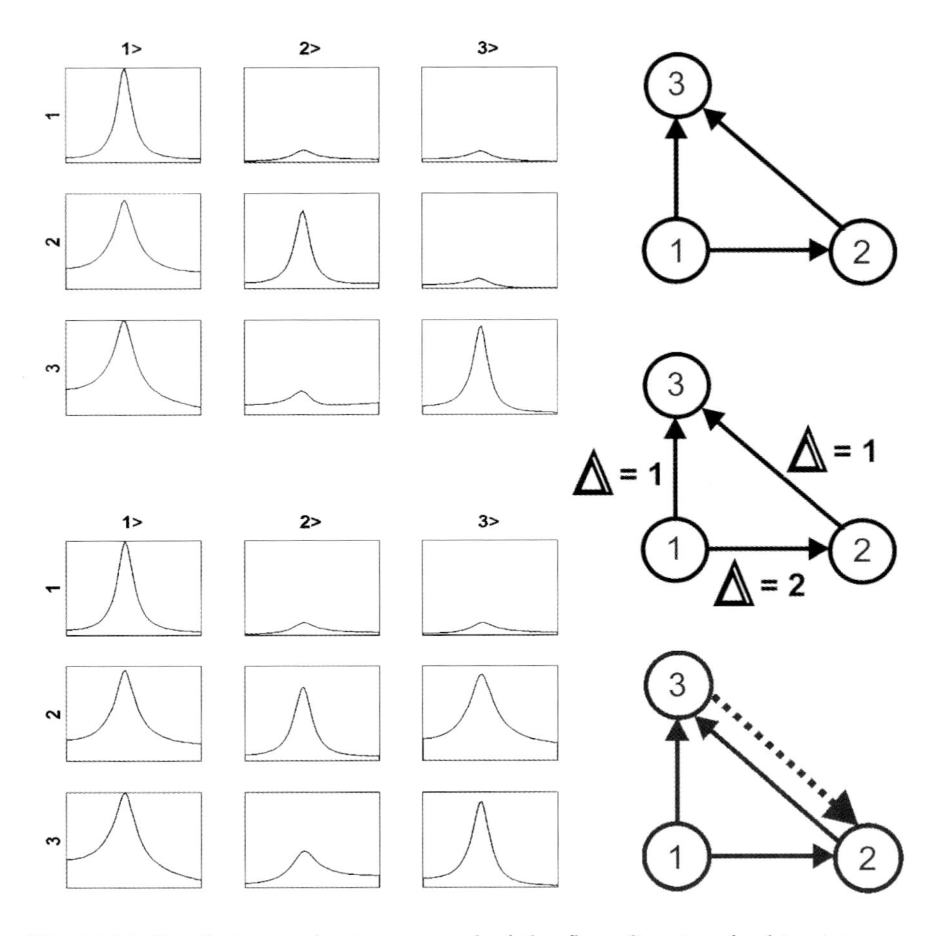

Fig. 15.14: Simulation explaining reversal of the flow direction for bivariate estimate of Granger causality. From the top:—DTF and resulting flow pattern, bottom—bivariate Granger causality and deduced flow pattern. In the middle the simulation scheme is shown (Δ—represents delay value in samples). Dotted line shows nonexisting flow found by bivariate estimate. Reprinted with permission from [49] (© IEEE 2005).

causality does not reveal any clear tendencies. The strong outflow from Fpz to Fp1 may perhaps be explained by the fact that probably the delays of the EEG signals coming to Fpz are smaller than the delays for Fp1 and F3. In this case we observe for electrode C3 a flow in a different direction than shown by all three multivariate estimates discussed above. In order to explain better this phenomenon we have made a simple simulation shown in Fig. 15.14. If the delay between channels two and three is bigger than between one and three, we have inversion of flow direction for the bivariate estimate. Such a situation, as shown in our simulation example, can easily happen for experimental EEG. The above considerations demonstrate that drawing any physiological conclusions from pairwise estimates of causality is very risky, if not impossible.

15.9.2 Application of a Time-Varying Estimator of Directedness

Quite often the dynamics of a multivariate process is of primary interest. This is especially the case for many biological systems. A good example may be information processing by the brain which takes place in a fairly short time scale. Therefore it is important to be able to estimate the topographic pattern of flows not only as a function of frequency, but also follow its dynamics. The SDTF offers an opportunity to trace propagation changes in time when multiple realizations of an experiment are available. The performance of SDTF will be illustrated on the example of the voluntary finger movement. The experiment involved the lifting of the right- or left-hand index finger (or the imagination of this task) after presentation of a cue indicating right/left direction. Eight seconds long epochs were considered with a cue appearing in the fifth second. The description of the experiment may be found in [54].

In order to follow the time evolution of transmissions a compromise has to be found between the window length and the number of channels. Two separate sets of nine electrodes were taken into account, one located over the left hemisphere sensorimotor area and another at the opposite positions over the right hemisphere; the middle electrodes were shared by both sets. Signals from distinct hemispheres could be treated separately because of little coherence and weak flow found between them. A 50 points (400 ms) long window was chosen, which resulted in a ratio of data points to number of parameters of about 50. In order to calculate SDTF as a function of time the window position was consecutively shifted by 10 points (80 ms). In order to better recognize artifacts generated by contraction of neck muscles, for each subject a certain threshold was established for energy cumulated in the 15 Hz to 40 Hz band during an epoch, and trials surpassing this threshold were rejected. Special measures to eliminate high frequency artifacts were taken because in this experiment changes of flows in the beta and gamma bands were of primary interest.

The SDTFs as functions of time and frequency are shown in Fig. 15.15. When analyzing the matrix of SDTFs reflecting the propagation between the channels, one can observe a characteristic gap in propagation in the beta band around time zero (cue presentation), especially in the locations connected with a motor task (electrodes C3, C1).

The analysis of graphs such as presented in Fig. 15.15 may be difficult. In order to follow better the time course of propagation of particular rhythms the SDTF values may be integrated in the frequency bands of interest. In this way the time course of propagation in the given band may be followed. In Fig. 15.16 the propagation in the beta band (17 Hz to 23 Hz) as a function of time is shown

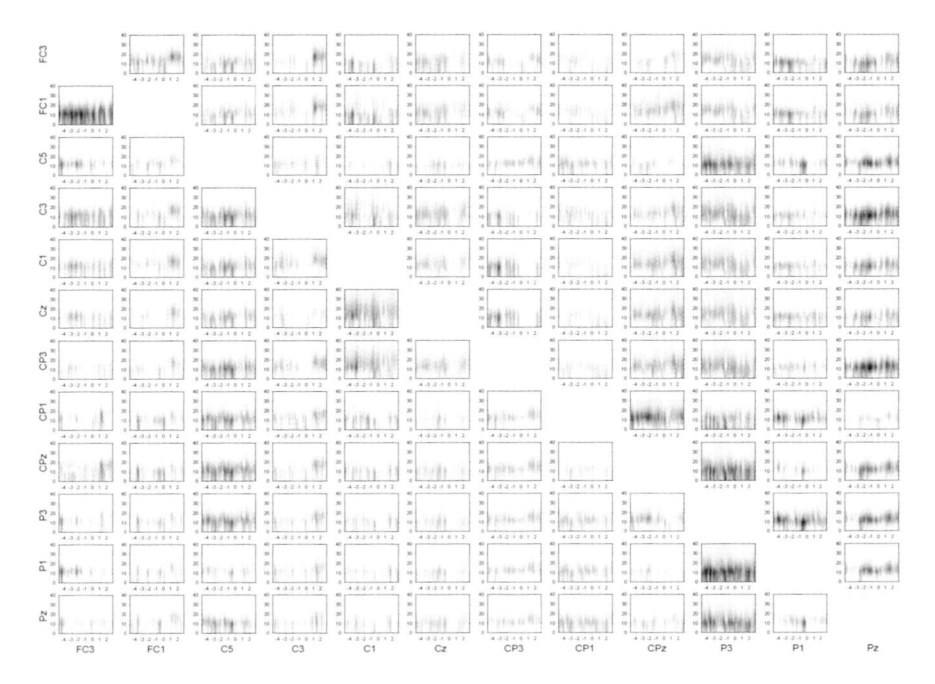

Fig. 15.15: (a) Example of propagation of EEG activity in the left hemisphere during right-hand movement imagination. In each small panel SDTF as a function of time (horizontal axis in seconds) and frequency (vertical axis in hertz) is presented. Intensity of flow coded by shades of gray (black = the strongest). Intensity scale is the same for all panels. The flow of activity is from the electrode marked above the column to the electrode marked at the relevant row. On the diagonal power spectra are shown. Reprinted from [54], © (2001), with permission from Elsevier.

together with the corridors of errors determined by the bootstrap method. One can see the similarities in the time evolution for the same pairs of electrodes for both subjects and the characteristic decrease of activity during the movement and increase after the completion of the task. By examining the curves it is possible to determine from which electrode the propagation starts first.

In the investigation of cognitive and control processes gamma activity is of special interest since it is connected with the information processing by the brain [55]. Gamma activity is hard to detect by means of scalp electrodes. It is observed for some subjects only. Thanks to its selectivity in respect of phase dependencies and robustness in respect to noise, the SDTF method makes it possible to follow the dynamics of gamma activity propagation. The best way to present the abundant information supplied by DTF is to present it in the form of a movie. Animations of gamma activity propagation are accessible at web page `http://brain.fuw.edu.pl/~kjbli/DTF_MOV.html`. The animations illustrate characteristic features of gamma rhythm propagation in case of a real

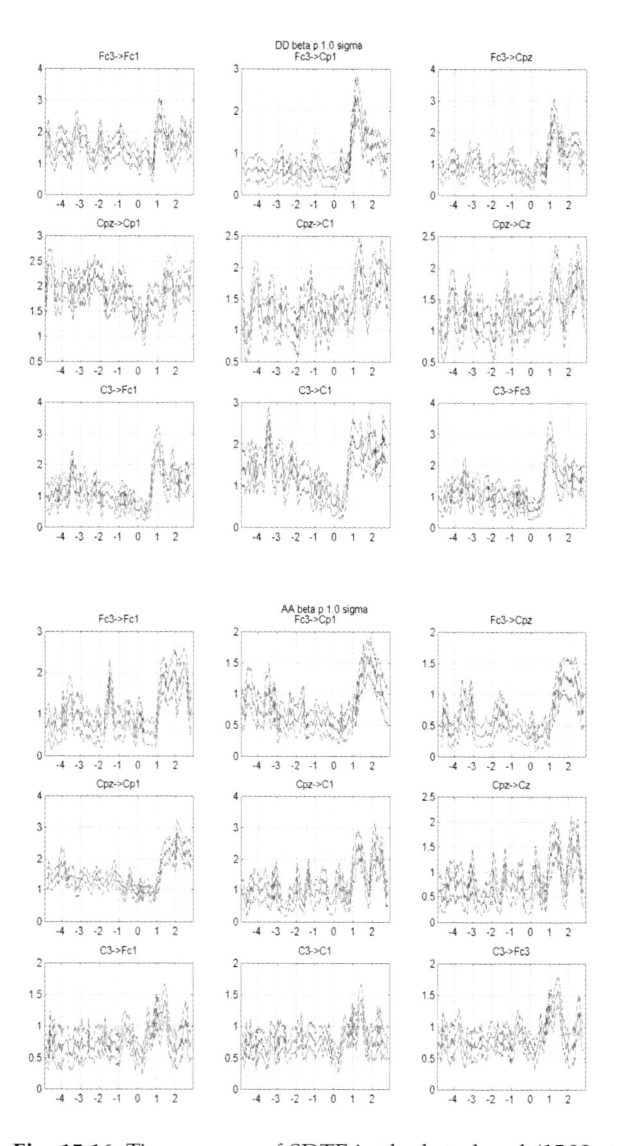

Fig. 15.16: Time course of SDTF in the beta band (15 Hz to 30 Hz) for two subjects. The corridors of errors at the level of one standard deviation are shown. Reprinted from [54], © (2001), with permission from Elsevier.

or imaginary movement. The interpretation of the observed pattern of flows is straightforward: during the real movement a short burst of activity signaling the command to perform the task is emitted from the corresponding motor areas. It is followed by a flow from the more frontal areas, which may be interpreted as recognition of the performance. In case of imagination the process is much

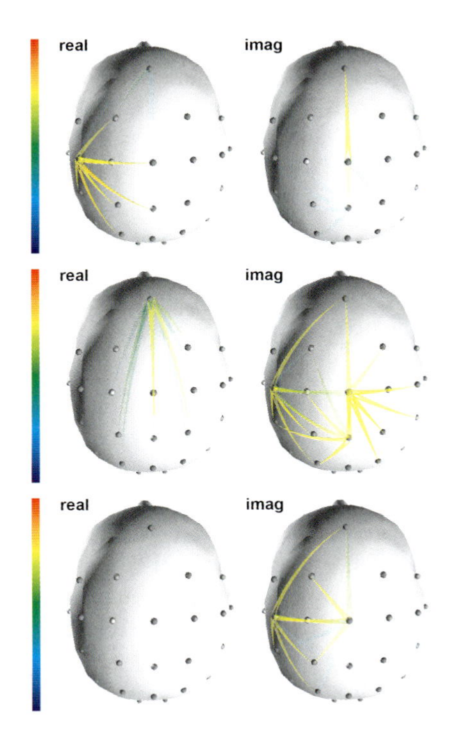

Fig. 15.17: Example of transmissions in gamma band for real (left column) and imaginary (right column) finger movement. Upper row presents the propagation at 0.3 s, middle row at 1.1 s, and bottom row at 1.4 s after the cue presentation. Arrows represent increase of the transmission in respect to the reference period (between 80 % and 100 % percentiles, color scale at the left).

longer and several structures of sensorimotor cortex are involved, it especially concerns Supplementary Motor Area (located mainly in beneath the Fz electrode). Screenshots of one of the animations illustrating a typical situation are shown in Fig 15.17. The above example of the Short-time Directed Transfer Function application shows that SDTF gives a coherent and detailed description of information processing by brains connected with motor control. The obtained evidence is in agreement with the known topographic and spectral features of the investigated task and at the same time new information is obtained that is not accessible by standard methods.

15.10 Discussion

In this chapter we have considered the possibilities offered by a parametric MVAR model in respect of finding measures describing the relations between multichannel data sets. MVAR offers a comprehensive means of depicting the properties of the related signals in the frequency domain, which is especially important in the case of time series characterized by rhythmic components. Once an MVAR model is adequately estimated many powerful spectral quantities can be derived such as spectra, coherences, and causal influence measures.

The information contained in coherences is usually interpreted as a measure of coupling between signals in a given frequency range; however, in drawing conclusions about the coupling strength not only ordinary (bivariate) coherences should be considered. Quite often the set of signals is strongly interrelated, especially if the signals are produced by the same process. Multiple coherences show the relation between a given channel and all the others channels of the set. They give an indication about the strength of the interactions in the considered system. In the example given above concerning the EEG measured from scalp electrodes, each of the channels revealed a very strong coupling with the system. It may be taken as an indication that in consideration of relations between channels partial coherences should be taken into account, since they are a measure of direct interaction between channels discriminating against influences of other channels of the set. Complete information about coherence in the system may be obtained by not only estimating ordinary, but also multiple and partial coherences. Most papers, particularly in the field of neuroscience, address exclusively ordinary coherences, which describe only part of the information contained in the multivariate data structure and can hardly be interpreted in terms of a real coupling between given channels.

The problem of determining directionality and finding causal relationships between time series is at present at the center of interest in many different fields, e.g., neuroscience, geophysics, economy, and sociology. Information about causality is coded in the phases between the channels of a process, although correct procedures are needed to reveal this information. We have demonstrated by simulations and by examples of experimental signals, why attempts to find directions from pairwise measures failed when more than two time series were involved. The phases of ordinary coherences give very little information, since they are blurred by multiple relationships between channels and, moreover, coherence phases are ambiguous by definition (determined modulo 2π). For that reason the conclusions, which can be drawn from pairwise phase calculations, are usually very weak.

The DTF function is a measure, which makes it possible to find causality relations for an arbitrary number of channels, with a reservation imposed by the statistical requirements concerning the number of model parameters in relation to the data points. DTF advantages are robustness with respect to noise or constant phase disturbances. Multivariate estimators of directedness DTF, nonnormalized DTF and dDTF, show slightly different aspects of propagation, depending on normalization, but their results are consistent and compatible with the physiological and anatomical evidence. All the results obtained by simulations also hold for SDTF since it is practically the same estimate, only the MVAR coefficients are calculated in a different way. By revealing the temporal changes of pattern of transmissions SDTF opens a way to elucidate dynamical evolution of nonstationary processes.

An interesting feature of DTF is the fact that it allows for extraction of weak components from the noise background, if they reveal definite phase dependence. An example may be the observation of gamma activity propagation during the motor task. In this case gamma activity was hardly observed in the spectra, but thanks to the selectivity of DTF a pattern of gamma activity propagation was determined.

In this chapter we have demonstrated that correct causal relationships and directions of signal propagation can only be found when all the interacting channels are evaluated simultaneously. The issue of the completeness of information was pointed out already by Granger [56], he stated namely that a correct causality measure can only be assessed, if the signal set contains all the possible relevant information of the problem. How can we be sure that our battery of time series forms a complete set? It is a difficult question, but usually we have some *a priori* knowledge on the process generating signals and we can expect which channels are mutually interdependent. The best solution is to take full sets of signals; however, that is not always possible, since the number of channels is connected to the number of model parameters, which cannot be too high in comparison with the number of points, as was pointed out above. In finding the balance between the number of channels and the data window, partial coherences may be helpful. If the partial coherence between two sets of channels is low we can assume low coupling between these sets and consider them separately. This was the case for our evoked response data where the signals from both the brain hemispheres were considered separately because of low partial coherences between electrodes belonging to the different hemispheres.

Some pitfalls should be mentioned in the context of the application of multivariate parametric models. The calculation of the MVAR model coefficients is based on an estimation of the correlation matrix, therefore no preprocessing involving the introduction of correlation between signals should be used. Introduction of any additional correlation ruins the causal estimate completely. We stress this point, since e.g., in the field of brain signals analysis in order to discriminate against the volume conduction the Laplacian or Hjorth transforms are used. Such preprocessing introduces additional correlations and moreover it is not necessary, since DTF is insensitive to zero phase disturbance, hence it discriminates against volume conduction.

In the recent literature there are many papers devoted to the nonlinear measures of dependencies between channels. These measures are exclusively bivariate, since the design of multivariate nonlinear estimators of causality is very difficult and the problem is not resolved yet. In each particular case the question arises if we commit a bigger error assuming linearity or by using bivariate measures. As it was pointed out, pairwise estimates can lead to ambiguous or even wrong conclusions. Therefore it is worth testing if the assumption of linearity does not hold. For EEG and LFP it was demonstrated by numerous studies based on linear versus nonlinear forecasting [57] or surrogate data techniques [58, 59] that EEG and LFP can be considered a colored noise time series and that one can trace nonlinear behavior only in certain phases of epileptic seizure [60]. However, even in this case linear techniques perform well: e.g., in [15] the power of the multivariate DTF function in epileptic focus localization was demonstrated. Another aspect, which can not be neglected is the fact that nonlinear methods are usually much more sensitive to noise. Therefore, if there is no strong evidence of nonlinearity the linear approximation may be recommended.

In this study we have emphasized the importance of a multivariate approach, which merits more attention, since the pitfalls in evaluating the direction of causal relations in EEG or LFP connected with the use of bivariate instead of multivariate techniques are much more serious than the limitations connected with the assumption of linearity of time series.

15.11 Acknowledgements

This work was partly supported by the KBN grant to the Institute of Experimental Physics.

References

[1] H. Akaike. *Ann. Inst. Stat. Math.*, 20:425, 1968.

[2] P. B. C. Fenwick, P. Mitchie, J. Dollimore, and G. W. Fenton. *Agressologie*, 10: 553, 1969.

[3] P. B. C. Fenwick, P. Mitchie, J. Dollimore, and G. W. Fenton. *Int. J. Bio. Med. Comput.*, 2:281, 1971.

[4] A. Isaksson, A. Wennberg, and L. H. Zetterberg. *Proc. IEEE*, 69:451, 1981.

[5] R. Oguma. *J. Nucl. Sci. Technol.*, 17:677, 1980.

[6] P. Whittle. *Biometrika*, 50:129, 1963.

[7] L. Zetterberg. *Math. Biosci.*, 5:227, 1969.

[8] C. W. J. Granger. *Econometrica*, 37:424, 1969.

[9] W. Gersch. *Math. Biosci.*, 14:177, 1972.

[10] W. Gersch. *Math. Biosci.*, 7:205, 1970.

[11] W. Gersch and J. Yonemoto. *Comput. Biomed. Res.*, 10:113, 1977.

[12] P. J. Franaszczuk, K. J. Blinowska, and M. Kowalczyk. *Biol. Cybern.*, 51:239, 1985.

[13] L. Harrison, W. D. Penny, and K. Friston. *NeuroImage*, 19:1477, 2003.

[14] M. Kamiński, K. J. Blinowska, and W. Szelenberger. *Electroenceph. Clin. Neurophys.*, 102:216, 1997.

[15] P. J. Franaszczuk, G. K. Bergey, and M. Kamiński. *Electroenceph. Clin. Neurophys.*, 91:413, 1994.

[16] A. Korzeniewska, S. Kasicki, M. Kamiński, and K. J. Blinowska. *J. Neurosci. Meth.*, 73:49, 1997.

[17] J. Medvedev and O. Willoughby. *Int. J. Neurosci.*, 97:149–67, 1999.

[18] S. M. Kay. *Modern Spectral Estimation*. Prentice-Hall, Englewood Cliffs, NJ, 1988.

[19] S. L. Marple. *Digital Spectral Analysis with Applications*. Prentice-Hall Signal Processing Series. Simon & Schuster, New Jersey, 1987.

[20] J. D. Hamilton. *Time Series Analysis*. Princeton University Press, Princeton, 1994.

[21] G. Box, G. M. Jenkins, and G. Reinsel. *Time Series Analysis: Forecasting and Control*. Prentice-Hall, Englewood Cliffs, NJ, 1994.

[22] M. B. Priestley. *Spectral Analysis and Time Series*. Academic, New York, 1981.

[23] B. H. Jansen. *CRC Crit. Rev. Biomed. Eng.*, 12:343, 1985.

[24] G. M. Jenkins and D. G. Watts. *Spectral Analysis and Its Applications*. Holden-Day, San Francisco, 1968.

[25] A. V. Oppenheim and R. W. Schafer. *Discrete-Time Signal Processing*. Prentice Hall, Englewood Cliffs, NJ, 1989.

[26] J. G. Proakis and D. G. Manolakis. *Digital Signal Processing: Principles, Algorithms and Applications*. Prentice Hall, Englewood Cliffs, NJ, 1996.

[27] B. Kemp and F. H. Lopes da Silva. In R. Weitkunat, editor, *Digital Biosignal Processing*, pages 129–156. Elsevier, Amsterdam, 1991.

[28] H. Akaike. *IEEE Trans. Autom. Control*, 19:716, 1974.

[29] M. Morf, A. Vieira, D. Lee, and T. Kailath. *IEEE Trans. Geosci. Electronics*, 16: 85, 1978.

[30] G. E. P. Box and G. C. Tiao. *Bayesian Inference in Statistical Analysis*. Wiley, New York, 1992.

[31] W. D. Penny and S. J. Roberts. *IEE Proc. Vision Image Sig. Proc.*, 149:33, 2002.

[32] N. Wiener. In E. F. Beckenbach, editor, *Modern Mathematics for Engineers*, chapter 8. McGraw-Hill, New York, 1956.

[33] M. Kamiński and K. J. Blinowska. *Biol. Cybern.*, 65:203, 1991.

[34] M. Kamiński, M. Ding, W. Truccolo, and S. Bressler. *Biol. Cybern.*, 85:145, 2001.

[35] L. A. Baccalá and K. Sameshima. *Biol. Cybern.*, 84:463, 2001.

[36] A. Korzeniewska, M. Maczak, M. Kamiński, K. J. Blinowska, and S. Kasicki. *J. Neurosci. Meth.*, 125:195, 2003.

[37] M. Arnold, W. H. R. Miltner, R. Bauer, and C. Braun. *IEEE Trans. Biomed. Eng.*, 45:553, 1998.

[38] R. E. Kalman. *Trans. ASME J. Basic Eng.*, 82:35, 1960.

[39] W. Gersch. Methods of analysis of brain electrical and magnetic signal. In *Handbook of Electroencephalography and Clinical Neurophysiology*, volume 1. Elsevier, Amsterdam, 1987.

[40] A. Benveniste, M. Metivier, and P. Priouret. *Adaptive Algorithms and Stochastic Approximations*. Springer, Berlin, 1990.

[41] W. Hesse, E. Möller, M. Arnold, and B. Schack. *J. Neurosci. Meth.*, 124:27, 2003.

[42] E. Möller, B. Schack, M. Arnold, and H. Witte. *J. Neurosci. Meth.*, 105:143, 2001.

[43] M. Ding, S. L. Bressler, W. Yang, and H. Liang. *Biol. Cybern.*, 83:35, 2000.

[44] B. Efron and R. J. Tibshirani. *An Introduction to the Bootstrap*. Chapman and Hall, London, 1993.

[45] B. Efron. *Ann. Stat.*, 7:1, 1979.

[46] A. M. Zoubir and B. Boashash. *IEEE Sig. Proc. Mag.*, 15:56, 1998.

[47] J. Theiler, S. Eubank, A. Longtin, B. Galdrikian, and D. Farmer. *Physica D*, 58:77, 1992.

[48] K. J. Blinowska, R. Kuś, and M. Kamiński. *Phys. Rev. E*, 70:050902, 2004.

[49] R. Kuś, M. Kamiński, and K. J. Blinowska. *IEEE Trans. Biomed. Eng.*, 51:1501, 2004.

[50] K. J. Blinowska, R. Kuś, and M. Kamiński. *Virt. J. Biol. Phys. Res.*, 8:1–4, 2004.

[51] L. A. Baccalá, K. Sameshima, G. Ballester, A. C. Do Valle, and C. Timo-Iaria. *App. Sig. Proc.*, 5:40, 1998.

[52] K. Sameshima and L. A. Baccalá. *J. Neurosci. Meth.*, 94:93, 1999.

[53] T. H. Bullock, M. C. McClune, J. Z. Achimowicz, V. J. Iragui-Madoz, R. B. Duckrow, and S. S. Spencer. *Electroenceph. Clin. Neurophys.*, 95:161, 1995.

[54] J. Ginter Jr., K. J. Blinowska, M. Kamiński, and P. J. Durka. *J. Neurosci. Meth.*, 110:113, 2001.

[55] S. L. Bressler. *Trends Neurosci.*, 13:161, 1990.

[56] C. W. J. Granger. *J. Econ. Dyn. Control*, 2:329, 1980.

[57] K. J. Blinowska and M. Malinowski. *Biol. Cybern.*, 66:159, 1991.

[58] P. Achermann, R. Hartmann, A. Gunzinger, W. Guggenbühl, and A. A. Borbély. *Electroenceph. Clin. Neurophys.*, 90:384, 1994.

[59] C. Stam, J. P. M. Pijn, P. Suffczyński, and F. H. Lopes da Silva. *Clin. Neurophysiol.*, 110:1801, 1999.

[60] J. P. M. Pijn, D. N. Velis, M. J. van der Heyden, J. DeGoede, C. W. M. van Veelen, and F. H. Lopes da Silva. *Brain Topogr.*, 9:249, 1997.

16 Computer Intensive Testing for the Influence Between Time Series

Luiz Antonio Baccalá, Daniel Y. Takahashi, and Koichi Sameshima

Recent years have seen several different quantitative approaches to gauging the mutual influence between multiple simultaneously measured time series with applications that range from physiology to economics. Some of them, specially those that portray that influence in the frequency domain, like *partial directed coherence*, in connection to the parametric modeling of jointly stationary time series lead to estimators whose asymptotic behavior, even if known, is of limited practical value, as many time series of interest can only be often considered stationary over very limited time spans. This chapter examines how to use the actually observed data itself to: (a) set limits on the significance of the null hypothesis of absence of relationship between the observed time series and (b) to produce confidence interval estimates when the mutual influence is of significance. Two different strategies to produce bootstrapped estimates are considered. The first one is based on random model residual resampling and the second one on spectral phase shuffling. Their relative merits are examined and examples of their application to both real and simulated data are considered.

16.1 Introduction

Cost reductions in data acquisition technology have produced an overwhelming increase in the available information in many research areas. For instance, electroneurophysiology which went from the simultaneous measurement of a handful of signals to that of hundreds of electrodes in both high resolution digital electroencephalography (EEG) [1] and in multisingle unit recordings [2] is allowing a more systematic treatment of the crucial question of functional connectivity: i.e., finding how and when brain areas communicate among themselves both under normal and pathological situations. For example, one such important connectivity question is seizure focus determination in epilepsy [3].

To address questions like these, some authors have recently proposed a number of techniques (see Table 16.1 for a partial summary) based on frequency domain representations of multivariate autoregressive models (of order p)

Handbook of Time Series Analysis. Björn Schelter, Matthias Winterhalder, Jens Timmer

ISBN: 3-527-40623-9

$$\begin{bmatrix} x_1(k) \\ \vdots \\ x_N(k) \end{bmatrix} = \sum_{r=1}^{p} \mathbf{A}_r \begin{bmatrix} x_1(k-r) \\ \vdots \\ x_N(k-r) \end{bmatrix} + \begin{bmatrix} w_1(k) \\ \vdots \\ w_N(k) \end{bmatrix} \tag{16.1}$$

of simultaneously measured time series $x_i(n)$, $(1 \leqslant i \leqslant N)$ where the coefficients matrix $\mathbf{A}_r$ whose i, jth entry $a_{ij}(r)$ describes the linear relationship between time series $x_i(k)$ and $x_j(k)$ at the rth past lag and where $w_i(k)$ represent the driving innovations.

Underlying Table 16.1 proposals is the idea of Granger causality [4][1], whereby a time series $x_j(n)$ is said to *Granger cause* $x_i(n)$ if it is possible to significantly improve prediction of the latter from the (exclusive) knowledge of $x_j(n)$'s past. For example, consider $N = 2$ and first equation in Eq. (16.1)

$$\begin{aligned} x_1(k) = a_{11}(1)x_1(k-1) + a_{12}(2)x_1(k-2) + \cdots \\ + a_{12}(0)x_2(k) + a_{12}(1)x_2(k-1) + \cdots . \end{aligned} \tag{16.2}$$

If $a_{12}(r) = 0$ for $r > 0$, this implies that $x_2(k)$'s past has no bearing on $x_1(k)$ values, i.e., $x_2(k)$ does not *Granger cause* $x_1(k)$.

From the second equation in Eq. (16.1)

$$\begin{aligned} x_2(k) = a_{22}(1)x_2(k-1) + a_{21}(2)x_2(k-2) + \cdots \\ + a_{21}(0)x_1(k) + a_{21}(1)x_1(k-1) + \cdots \end{aligned} \tag{16.3}$$

Granger causality's unreciprocal nature becomes clear since even if all $a_{12}(r) = 0$ this does not mean that $a_{21}(r) = 0$. This fact leads to the possibility of deducing the direction of information flow from $x_1(k)$ to $x_2(k)$ if $a_{21}(r) \neq 0$ for some $r \leqslant p$.

Thus functional connectivity inference can be reduced to hypothesis testing for

$$H_0 : a_{ij}(r) = 0 \tag{16.4}$$

for all r between 1 and p.

Whereas many approaches [5] exist for directly testing Eq. (16.4) in the time domain, tests like

$$H_0 : |\pi_{ij}(\lambda)|^2 = 0 \tag{16.5}$$

based on the allied frequency domain representations of Granger causality in Table 16.1, which hold if and only if Eq. (16.4) hold, are either scarce or nonexistent. In fact, former neuroscience applications [6–10], where frequency-band information is specially relevant, were either made by extensive simulation [6, 11], via

[1] In economics it has become a tool in the empirical investigation of systematic time precedence/feedback questions between time series like employment and inflation.

the choice of arbitrary thresholds [9] or more recently through the use of bootstrap related ideas [12] similar to one of the approaches examined in the present chapter (Section 16.3.2).

In fact, data-aided means like bootstrap of testing Eq. (16.5) are specially important because practical time series are usually short or non-Gaussian and violate the commonest assumptions used in developing asymptotic tests for Eq. (16.4) or Eq. (16.5) which at best amount only to rough indicators of the actual connectivity.

Tab. 16.1: Partial list of time series connectivity inference methods based on multivariate autoregressive models (16.1) in the frequency domain whose description is based on

$$\mathbf{A}(\lambda) = \mathbf{I} - \sum_{r=1}^{p} A_r e^{-2\pi i r \lambda}$$

for the normalized frequency $|\lambda| \leqslant 0.5$ (The normalized sampling frequency is given by the ratio the frequency of interest with respect to the sampling rate, both in hertz) where $\mathbf{I}$ is an $N \times N$ identity matrix, $\mathbf{i} = \sqrt{-1}$. This allows one to define $\mathbf{a}_k(\lambda) = [A_{1k}(\lambda) \cdots A_{Nk}(\lambda)]^T$ out of $\mathbf{A}(\lambda)$'s columns. Also let $\mathbf{H}(\lambda) = \mathbf{A}^{-1}(\lambda)$ and $\mathbf{h}_k(\lambda) = [H_{k1}(\lambda) \ldots H_{kN}(\lambda)]^T$ built from $\mathbf{H}(\lambda)$'s lines. Further let $\mathbf{\Sigma}_w$ be the innovations covariance matrix and σ_{ii}^2 the variance of $w_i(k)$.

Method	Expression	Reference
Cross spectrum (CS)	$S_{ij}(\lambda) = \mathbf{h}_i^T(\lambda)\mathbf{\Sigma}_w \mathbf{h}_j(\lambda)$	[13]
Coherence (C)	$C_{ij}(\lambda) = \dfrac{\mathbf{h}_i^T(\lambda)\mathbf{\Sigma}_w \mathbf{h}_j(\lambda)}{\sqrt{(\mathbf{h}_i^T(\lambda)\mathbf{\Sigma}_w \mathbf{h}_j(\lambda))\,(\mathbf{h}_j^T(\lambda)\mathbf{\Sigma}_w \mathbf{h}_j(\lambda))}}$	[13]
Partial coherence (PC)	$\kappa_{ij}(\lambda) = \dfrac{\mathbf{a}_i^T(\lambda)\mathbf{\Sigma}_w^{-1}\mathbf{a}_j(\lambda)}{\sqrt{(\mathbf{a}_i^T(\lambda)\mathbf{\Sigma}_w^{-1}\mathbf{a}_i(\lambda))\,(\mathbf{a}_j^T(\lambda)\mathbf{\Sigma}_w^{-1}\mathbf{a}_j(\lambda))}}$	[14]
Directed coherence (DC)	$\gamma_{ij}(\lambda) = \dfrac{\sigma_{jj} H_{ij}(\lambda)}{\sqrt{\sum_{j=1}^{N} \sigma_{jj}^2 \lvert H_{ij}(\lambda)\rvert^2}}$	[15]
Partial directed coherence (PDC)	$\pi_{ij}(\lambda) = \dfrac{A_{ij}(\lambda)}{\sqrt{\sum_{j=1}^{N} \lvert A_{ij}(\lambda)\rvert^2}}$	[16]
Directed transfer function (DTF)	$\mathrm{DTF}_{ij}(\lambda) = \dfrac{H_{ij}(\lambda)}{\sqrt{\sum_{j=1}^{N} \lvert H_{ij}(\lambda)\rvert^2}}$	[7]
Generalized PDC (GPDC)	$\bar{\pi}_{ij}(\lambda) = \dfrac{\frac{1}{\sigma_{ii}} A_{ij}(\lambda)}{\sqrt{\sum_{i=1}^{N} \frac{1}{\sigma_{ii}^2} \lvert A_{ij}(\lambda)\rvert^2}}$	[17]

Rather than full generality, after a bootstrap refresher (Section 16.2), this chapter concentrates (Section 16.3) on two of the many possible data-aided approaches for testing Eq. (16.5). This is followed by some numerical illustrations (Section 16.4) and a brief discussion (Section 16.5).

16.2 Basic Resampling Concepts

Statistical estimation is concerned with making the best possible use of measured quantities affected by random perturbations. To improve the reliability of computing a quantity $\theta(u)$ that depends on the measurement of u, one employs an estimator $\hat{\theta}$ that combines randomly perturbed measurements (samples) $u(1), \ldots, u(K)$ of u. The statistical problem then is to describe the reliability of the values for θ that are produced by $\hat{\theta}$ in the form of quantities like bias $(\beta(\hat{\theta}))$, mean-squared error (MSE$(\hat{\theta})$), confidence intervals, hypothesis test threshold values and so on. The latter are often called *level 2* statistics whereas the computed variable of interest $\hat{\theta}$ is a case of *level 1* statistic.

An elementary example of this is the measurement of some constant value θ_0 in random noise w. The samples produced are described by

$$u(k) = \theta_0 + w(k), \tag{16.6}$$

where $w(k)$ are assumed independent and identically distributed (iid) for simplicity.

An estimator for θ_0 is

$$\hat{\theta}_0 = \frac{1}{K} \sum_{k=1}^{K} u(k) \tag{16.7}$$

and its statistical performance (given by the level 2 statistics) depends on what is known about the description of w's randomness; for example if w's mean is zero, Eq. (16.7) is unbiased (i.e., $\beta(\hat{\theta}_0) = 0$).

Much of mathematical statistics is concerned with explicitly describing the level 2 statistics as a function of one's knowledge of $G\,(w(1), \ldots, w(K))$, w's sampling distribution. The available answers are usually asymptotic (hold for $K \to \infty$) and depend on how much is known about G *a priori*. For example, if in addition to being iid, $w(k)$ are also Gaussian, the explicit asymptotic confidence intervals for θ_0 depend on whether w's variance σ_w^2 is known or whether it must also be estimated from the observations.

In many cases of interest, neither is the sample size K large nor much is known about G. This apparent dead end was overcome in many important cases by Efron's proposal of the idea of bootstrap [18, 19] which consisted of using the actually observed data to infer the level 2 statistics.

In the case of Eq. (16.7), bootstrap proceeds as

Procedure 16.1. *Bootstrap for iid data:*

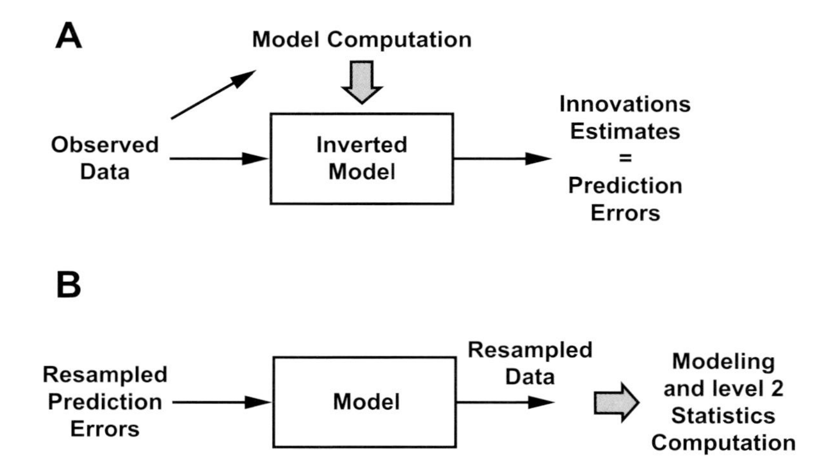

Fig. 16.1: (A) In the model-based approach, the observed data are used to generate a model whose inversion leads to estimates of the innovations time series $w_i(\cdot)$ in the form of prediction errors (residuals). (B) If the model is adequate, passing residual whiteness tests, one can sample the residuals with replacement and use the estimated model to generate resampled time series whose modeling, in turn, leads to the computation of the level 2 statistics of interest.

1. *Randomly draw data with replacement and equal probability from the original sample set* $U = \{u(1), \ldots, u(K)\}$ *to produce* N^* *bootstrapped sets* $U^{(l)} = \{u_1^{(l)}, \ldots, u_{K^*}^{(l)}\}$ *where* $u_{k^*}^{(l)}$ *are the resampled data.*

2. *Obtain the empirical distribution* F^* *for* $\hat{\theta}_0^*$ *via*

$$\{\hat{\theta}_0(U^{(1)}), \ldots, \hat{\theta}_0(U^{(N^*)})\} \tag{16.8}$$

3. *Use* F^* *to compute the level 2 statistics of interest.*

The resampling literature [19] tells us that as N^* increases (for large enough K) through Procedure 16.1, $F^* \to F$, the actual sampling distribution of $\hat{\theta}_0$. Whereas bootstrap convergence holds for Eq. (16.7) and leads to statistically efficient results for small K, the same method may fail for other estimators; it fails for order statistics such as $\min U$ or $\max U$ even if w's distributions are defined on bounded intervals [20, 21].

A major hindrance to the applicability of Procedure 16.1 to general time series is the iid requirement as time series data samples usually exhibit some form of interdependence which needs to be considered or circumvented (Section 16.3) [22].

16.3 Time Series Resampling

Observed time series, univariate or multivariate, correspond to samples

$$\mathbf{x}(\cdot) = \{\mathbf{x}(k), k = 1, \ldots, K\} \tag{16.9}$$

gathered sequentially from some random data generating mechanism (random process)[2]. Time series analysis aims at describing the underlying sequential data generating mechanisms. Usually the description is partial and concerns just some aspects of the data dependence pattern. For stationary mechanisms, one may be interested in the spectral representation of the data, or equivalently in the autocovariance $\gamma_{\mathbf{x}}(m) = E[\mathbf{x}(n)\mathbf{x}^T(n+m)]$ to be inferred from a single observed process *realization*. Parameters from models like Eq. (16.1) and derived quantities in Table 16.1 represent possible parameters of interest (level 1 statistics) whose statistical accuracy needs to be assessed (via the level 2 statistics).

To carry out this process, as for Eq. (16.6), one must use the observed data, Eq. (16.9), to generate other time sequences which preserve some (or all) properties of the parameters of interest. This is represented by the transformation

$$\mathbf{x}^{(*)}(\cdot) = \mathcal{T}(\mathbf{x}(\cdot), \xi) \tag{16.10}$$

that generates the sequence $x^{(*)}(\cdot)$ randomly as represented by the random variable ξ in Eq. (16.10). Iterated application of Eq. (16.10) leads to resampled realizations

$$\{\mathbf{x}^{(1)}(\cdot), \ldots, \mathbf{x}^{(N^*)}(\cdot)\}, \tag{16.11}$$

wherefrom the resampled parameters

$$\{\hat{\theta}\left(\mathbf{x}^{(1)}(\cdot)\right), \ldots, \hat{\theta}\left(\mathbf{x}^{(N^*)}(\cdot)\right)\} \tag{16.12}$$

are generated and whose empirical distribution must approximate $F(\hat{\theta})$ for a good choice of $\mathcal{T}$. Clearly, Procedure 16.1 is a special case of $\mathcal{T}(\mathbf{x}(\cdot), \xi)$ that is only good for time series without time dependence (white noise as the samples in Eq. (16.6) may be interpreted). Note that ξ's role is represented by step 1 of Procedure 16.1.

Hence in time series resampling, one must choose $\mathcal{T}$ not only to match the estimator of interest but also to handle the time dependence adequately.

In the remainder of this chapter, adequate description by models like Eq. (16.1) that encode the time dependence in its parameters is assumed. As a result, if this encoding is successful, the time dependence between modeling residues is abolished and leads to $\mathcal{T}(\mathbf{x}(\cdot), \xi)$ based on prior data modeling as described in Section 16.3.1.

Model independent resampling can be achieved in the case of stationary time series by taking into account their spectral representation in terms of independent random increments [23]. This is covered in Section 16.3.2 whereas other more general time series resampling strategies are briefly discussed in Section 16.3.3.

[2] Boldface quantities like $\mathbf{x}(\cdot) = [x_1(\cdot), \ldots, x_N(\cdot)]^T$ denote multivariate time series.

Due to the emphasis on providing answers to both confidence intervals and to performing null hypothesis tests like Eq. (16.5), the forthcoming sections distinguish between correlation preserving $\mathcal{T}_P$ and correlation abolishing $\mathcal{T}_R$ transformations.

16.3.1 Residue Resampling

Because of the central importance of the iid requirement, the first idea that comes to mind is to use $\mathcal{T}$ to reduce the original observations equation (16.9) to an equivalent iid data set as an intermediary step.

Models like Eq. (16.1), when adequately fitted, readily lead to a representation of $\mathbf{x}(\cdot)$ that is composed of the model parameters and its residuals[3]

$$\mathbf{w}(\cdot) = \{\mathbf{w}(k), k = 1, \ldots, K\}. \tag{16.13}$$

Model parameters encode the time relationship between $\mathbf{x}(\cdot)$ components and their respective time samples whereas the residuals $\mathbf{w}(\cdot)$ represent what cannot be predicted from past $\mathbf{x}(\cdot)$ based on the fitted model. When the models are in addition *invertible* [24], Eq. (16.13) may be used directly to recompose the observed time series (see Fig. 16.1). This is the key to:

Procedure 16.2. *Model-based crosscorrelation preserving resampling ($\mathcal{T}_P^{\mathcal{M}}$):*

1. *Use the data (16.9) to fit a model $\mathcal{M}$, say Eq. (16.1), and ensure that the residual Eq. (16.13) cannot be distinguished from iid (uncorrelated) time series (modeling diagnostics). This step also produces $\theta(\mathcal{M})$, the level 1 statistics of interest.*
2. *Produce $\mathbf{w}^{(*)}(\cdot)$ time series by resampling Eq. (16.13) with replacement and equal probability (the ξ step).*
3. *Use $\mathbf{w}^{(*)}(\cdot)$ to generate $\mathbf{x}^{(*)}(\cdot)$ from $\mathcal{M}$.*
4. *Analysis of $\mathbf{x}^{(*)}(\cdot)$ leads to model $\mathcal{M}^{(*)}$ wherefrom θ^* is generated.*
5. *Repeat the steps 2–4, N^* times and use the resulting θ^*'s to compute F_θ^* to approximate F_θ.*

While Procedure 16.2 preserves the interaction between the time series and can be used to produce confidence intervals, to test the null hypothesis of lack of interaction, i.e., null hypothesis like Eq. (16.4) or Eq. (16.5), one must break the relationship between the time series by modeling them separately. Thus, each resampled component time series preserves all of its spectral distributions and at

[3] If $\mathbf{x}(\cdot)$ are Gaussian and the model is adequately fitted, $\mathbf{w}(\cdot)$ cannot be distinguished from iid time series. For linear models like Eq. (16.1), even if gaussianity does not hold, $\mathbf{w}(\cdot)$ are just uncorrelated in time and this suffices. Model fitting adequacy is ensured by whiteness tests [5, 23, 24] on Eq. (16.13).

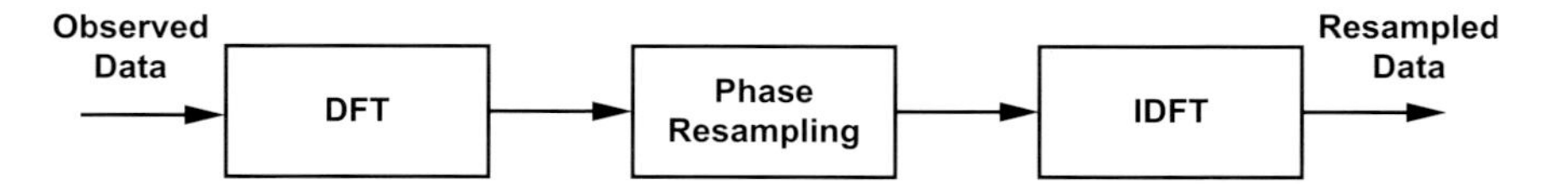

Fig. 16.2: Schematic diagram of the phase-resampling approach to bootstrap. The original time series are transformed to the frequency domain where the phase is randomly altered. Inverse Fourier transformation to the time domain produces the resampled time series whose analysis provides the level 2 statistics information.

the same time obliterates any existing mutual interactions. Hence, any joint multivariate analysis on the separately resampled time series mimics the variability one would observe if the underlying time series were unrelated. This allows establishing null hypothesis test decision thresholds as summarized in the next procedure:

Procedure 16.3. *Model-based resampling without cross-correlation preservation ($\mathcal{T}_R^{\mathcal{M}}$): It is identical to Procedure 16.2, except that*

1. *It is applied to each component time series $x_i(\cdot)$ separately and independently from the other time series and generates $\mathcal{M}_i$ models whose residuals $w_i(\cdot)$ are used for resampling and reconstituting $x_i^{(*)}(\cdot)$.*
2. *The reassembled joint time series $\mathbf{x}^{(l)}(\cdot) = [x_1^{(l)}(\cdot), \ldots, x_N^{(l)}(\cdot)]^T$ are then used to generate $\mathcal{M}^{(l)}$ models wherefrom the distribution of $\hat{\theta}(\mathcal{M}^{(l)})$ approximates $F_{\hat{\theta}}$ under the null hypothesis of lacking interaction between the component time series.*

Ordinarily, these approaches are limited by outliers or when model poles are close to the unit circle [13] calling for alternative approaches (see Examples 16.2 and 16.3).

16.3.2 Phase Resampling

The essence of this approach is to realize that for frequencies given by $\nu = k/K$, where k is an integer ($0 \leqslant k \leqslant K-1$), the Discrete Fourier Transform, DFT of $x_i(k)$:

$$X_i(\nu) = \frac{1}{\sqrt{K}} \sum_{m=0}^{K-1} x_i(m) e^{-2\pi i m \nu} \tag{16.14}$$

are approximately iid complex random variables [21, 25]. As is well known, the periodogram of $x_i(k)$

$$I_X(k) = \left| X_i\left(\frac{k}{K}\right) \right|^2 \tag{16.15}$$

is the basis for nonparametric estimates of the spectrum of $x_i(k)$ after adequate local averaging [23, 25]. This means that time series whose DFT values differ by a random phase $\xi(\nu)$ essentially have the same estimated spectrum. Thus, different waveshapes $x_i^{(*)}(k)$ with the same estimated spectrum may be created.

By rewriting Eq. (16.14) in polar form, $X_i(\nu) = |X_i(\nu)|e^{j\Phi_{X_i}(\nu)}$, one may randomize the phase either by writing:

Procedure 16.4. *Phase randomization with correlation nullification ($\mathcal{T}_R^{\Phi}$) [26]:*

$$\Phi_{X_i}^{(*)}(\nu) = \xi_i(\nu), \tag{16.16}$$

where $\xi_i(\nu)$ are uniform mutually independent real random variables in $[-\pi, \pi]$ or

Procedure 16.5. *Phase randomization with correlation preservation ($\mathcal{T}_P^{\Phi}$) [27]:*

$$\Phi_{X_i}^{(*)}(\nu) = \Phi_{X_i}(\nu) + \xi(\nu), \tag{16.17}$$

where $\xi(\nu)$ are real uniform random variables in $[-\pi, \pi]$ and are produced independently at each frequency ν.

so that phase randomized $x_i^{(*)}(k)$ are obtained from computing the inverse DFT of $X_i(\nu)$

$$x_i^{(*)}(k) = \frac{1}{\sqrt{K}} \sum_{m=0}^{K-1} X_i^{(*)}\left(\frac{m}{K}\right) e^{\frac{2\pi imk}{K}} \tag{16.18}$$

for $k = 0, \ldots, K-1$. This is summarized in Fig. 16.2.

What distinguishes Eq. (16.17) is that the same random phase is used in perturbing all time series for a given ν and this choice leaves the cross-spectra invariant [27] whereas in Eq. (16.16) the phases differ for each time series as well so that on average unrelated time series are produced.

It should be remarked that the suitability of $\mathcal{T}_P^{\Phi}$ is proved in [28] who point out the necessity of extracting the sample mean before using the method for adequate convergence. The inadequacy of this method for non-Gaussian time series is also discussed.

Also more general spectral resampling than the choice of random phases is possible if adequate spectral estimation is carried out [23–25, 29] whereby the full χ_2^2 spectral statistic of $\left|X_i\left(\frac{k}{K}\right)\right|^2$ can be used (Gaussian data) or if adequate (nonparametric) spectral estimates are available (see Section 16.3.3).

16.3.2.1 Some Computational Issues

The residue resampling methods are very easy to program. The resampled residue data are given by

$$w_{(\cdot)}^{*}(1 : L_B + K) = w_{(\cdot)}(\lceil K\,\mathrm{rand}(1 : L_B + K)\rceil), \tag{16.19}$$

where rand stands for a random number generator function with uniform distribution in $[0, 1]$ and where L_B is the number of burn-in data points used to obtain stationary output from simulating Eq. (16.1) with a total computation cost of $O\left((L_B + K)N^2p + Np\right)$ floating point operations versus $O(K \log_2 K)$ operations for the phase resampling methods if a fast algorithm is used [13]. Usually L_B is increased until the estimates become stable.

16.3.3 Other Resampling Methods

The methods of the previous sections are by no means the only ones. The first proposals for time series bootstrapping (or for other dependent data) was made via the so-called *block* bootstraping methods [30] which admit a variety of "flavors" like moving block bootstrap (MBB), nonoverlapping block bootstrap (NBB), circular block bootstrap (CBB), and stationary bootstrap (SB) [22].

The essential feature of these methods is to randomly select data blocks of an appropriate length l from the original time series data set. When rejoined, the blocks produce resampled time series $\mathbf{x}^*(\cdot)$ for analysis. These different block methods differ in how block selection takes place: for example CBB uses periodically extended time series [22].

It is important to note, however, that for the purpose of inferring quantities that depend on *second order statistics* as those in Table 16.1, one must use 'block of blocks' methods which consist of picking up blocks themselves made up by other blocks whose length reflects the lag structure dependence of the data [22].

As in Section 16.3.1, the block procedures may be applied to the time series vector $\mathbf{x}(\cdot)$ as whole ($\mathcal{T}_P$) or to each component time series $x_i(\cdot)$ independently ($\mathcal{T}_R$).

Other bootstrap methods include subsampling (a generalized form of 'jackknife') [31], sieve bootstrap [32] and transformation-based bootstrap (TBB) of which the method in Section 16.3.2 is an example. In fact, more general use of the frequency domain (FDB) for bootstrap is discussed in [22, 33, 34].

All of these methods differ in their statistical efficiency and applicability. Lengthy theoretical details can be found in [22].

16.4 Numerical Examples and Applications

In this section, we examine the resampling strategies adopted in Sections 16.3.1 and 16.3.2, first via simple toy models to illustrate the statistical behavior that should be expected under controlled situations and then apply the methods to some experimentally observed data.

16.4.1 Simulated Data

When the innovations $w_i(k)$ have identical variances and the model is bivariate, DC, DTF, PDC, GPDC in Table 16.1 amount to basically the same quantities. As such, consider

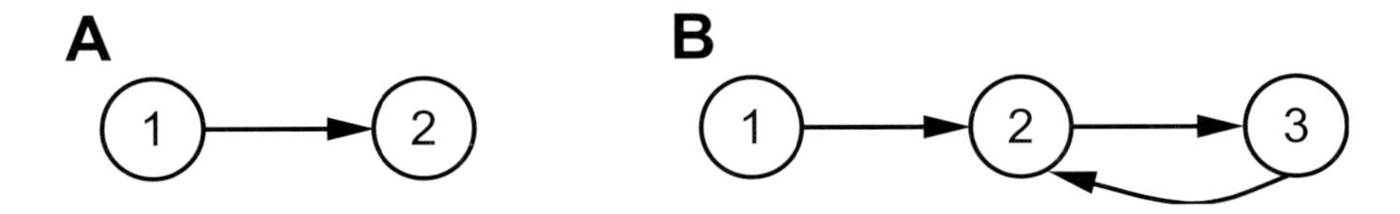

Fig. 16.3: (A) Connectivity patterns for Examples 16.1 and 16.2 and (B) for Example 16.3.

Example 16.1. The simplest possible case of analyzing data generated by

$$\begin{bmatrix} x_1(k) \\ x_2(k) \end{bmatrix} = \begin{bmatrix} 0.5 & 0 \\ 0.5 & 0.5 \end{bmatrix} \begin{bmatrix} x_1(k-1) \\ x_2(k-1) \end{bmatrix} + \begin{bmatrix} w_1(k) \\ w_2(k), \end{bmatrix} \tag{16.20}$$

where $w_i(k)$ are iid mutually independent random zero mean unit variance Gaussian sequences. The connectivity between $x_1(k)$ and $x_2(k)$ is summarized in Fig. 16.3(A).

Null connectivity hypothesis tests using PDC's estimator for $\mathcal{T}_R$ resampling strategies for the absent connection $x_2(k) \to x_1(k)$ are compared in Fig. 16.4. It is seen that less strict thresholds are provided by $\mathcal{T}_R^{\mathcal{M}}$ than for $\mathcal{T}_R^{\Phi}$ which is closer to the asymptotic thresholds defined in [35] when just $K = 100$ points are used.

For $x_1(k) \to x_2(k)$, connectivity cannot be rejected as its PDC value is well above the null hypothesis thresholds (Fig. 16.5). In this case PDC confidence limits obtained via bootstrap can be computed and the $\mathcal{T}_P^{\mathcal{M}}$ $(2.5\,\%, 97.5\,\%)$ limits are slightly larger than those computed using $\mathcal{T}_P^{\Phi}$.

Further insight on PDC's bootstrap distributions under the null connection hypothesis for $\lambda = 0.1088$ is provided in Fig. 16.6 where one readily sees in this case that $\mathcal{T}_R^{\Phi}$ is less likely to generate false positives than $\mathcal{T}_R^{\mathcal{M}}$. For comparison, numerical approximations of the asymptotic PDC distributions are also depicted [35].

Now consider an example that differs from the previous one in that its poles are close to the unit circle.

Example 16.2. Two unidirectionally coupled linear oscillator structures with the same connectivity pattern as that of Example 16.1 (Fig. 16.3(A)) whose data generating model is described by the following matrices:

$$\mathbf{A}_1 = \begin{bmatrix} 1.8266 & 0 \\ 0.3 & 1.7537 \end{bmatrix} \quad \mathbf{A}_2 = \begin{bmatrix} -0.9409 & 0 \\ 0 & -0.9409 \end{bmatrix} \tag{16.21}$$

and where $w_i(k)$ are defined as in Example 16.1.

This model was chosen because extensive simulations [9, 11, 36, 37] have shown the high degree of false positive detections in the reverse unconnected direction using frequency independent thresholds like $|\pi_{ij}(\lambda)|^2 > 0.1$ for coupled oscillators when little energy dissipation is involved. In this case, only $K = 100$ points (roughly just two observed cycles) are used in the inferences.

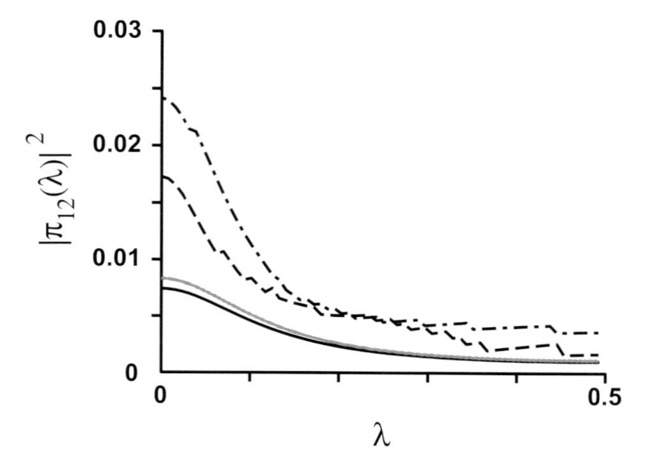

Fig. 16.4: Under the null hypothesis, the bootstrap thresholds for $\mathcal{T}_R^{\mathcal{M}}$ (dot dashed), $\mathcal{T}_R^{\Phi}$ (dashed) and the theoretical asymptotic value (gray) for PDC estimates (solid) of Example 16.1 for the nonexisting $x_2(k) \to x_1(k)$ connection by using $K = 100$ data points as a function of the normalized frequency λ.

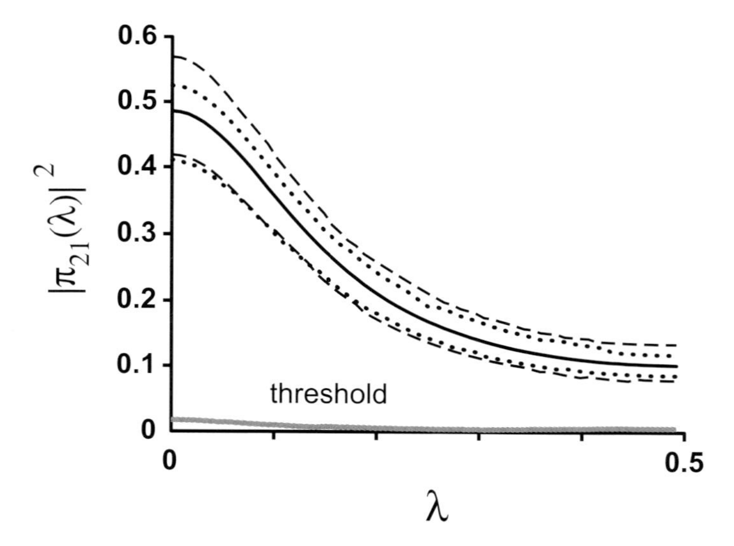

Fig. 16.5: Confidence intervals $(2.5\,\%, 97.5\,\%)$ for the existing $x_1(k) \to x_2(k)$ ($\mathcal{T}_P^{\Phi}$ – dots $\mathcal{T}_P^{\mathcal{M}}$ – dashed) in Example 16.1 using $K = 100$ points to estimate PDC (solid) as a function of the normalized frequency λ. Hypothesis tests 95 % thresholds are also shown at the bottom (gray) and reflect the connection's high significance.

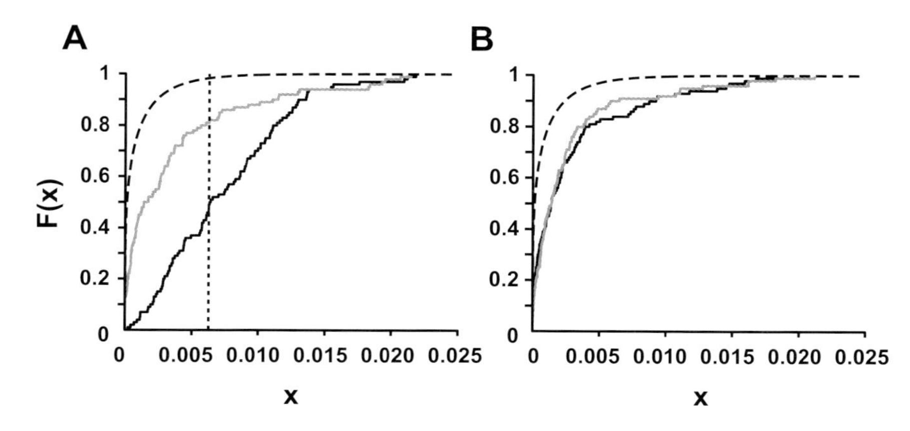

Fig. 16.6: Comparative cumulative resampling distributions ($\mathcal{T}_R^{\Phi}$ in black, $\mathcal{T}_R^{\mathcal{M}}$ in gray and asymptotic PDC from [35] – dashed) under the null hypothesis of connection, respectively, $x_2(k) \to x_1(k)$ (A) and $x_1(k) \to x_2(k)$ (B) for the normalized frequency $\lambda = 0.109$ using $K = 100$ in Example 16.1. The vertical dashed line corresponds to estimated PDC value which is absent from (B) as its value is above the scale.

The inference results are shown in Fig. 16.7. Decision threshold dependence on K is portrayed in Fig. 16.8 for $\lambda = 0.09$ that corresponds to the frequency where the estimated PDC is maximum. Note that in accord with theory, residual resampling furnishes slacker thresholds in this case. A flavor of the small dependence of the decision threshold value on the number of resamples for $\mathcal{T}_R^{\Phi}$ can be appreciated in Fig. 16.9.

Illustration of the reversed roles of the methods in regard to confidence limit estimation for $x_1(n) \to x_2(n)$ can be appreciated in Fig. 16.10 where the estimated PDC value lies within computed $\mathcal{T}_P^{\mathcal{M}}$ methods as opposed to $\mathcal{T}_P^{\Phi}$ methods which generate evidently biased limits. Compare this situation with that of Fig. 16.5.

The next example covers more than two time series. In this case, the connectivity estimators differ in what they conceptually mean [38]. As opposed to the case of PDC, whose asymptotic frequency domain characteristics have been recently worked out [35], most of the other estimators in Table 16.1 have unknown precise asymptotic behavior for parametric estimation methods. As such, bootstrap methods provide the only approximate guidelines to the statistical variability that should be expected.

Example 16.3. Consider the data generating model described by:

$$\mathbf{A}_1 = \begin{bmatrix} 1.6498 & 0 & 0 \\ 0.1 & 1.663 & -0.81 \\ 0 & 1 & 0 \end{bmatrix} \quad \mathbf{A}_2 = \begin{bmatrix} -0.81 & 0 & 0 \\ 0 & 0 & 0 \\ 0 & 0 & 0 \end{bmatrix} \tag{16.22}$$

(see Fig. 16.3(B)) for $w_i(k)$ statistics defined as before. This model is also one of the two coupled oscillators where access is now available to the internal variables

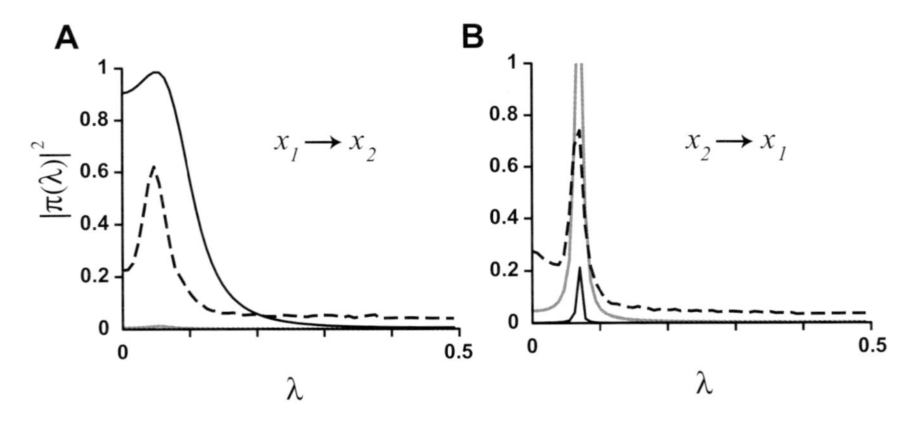

Fig. 16.7: Null hypothesis test results for the existing (A) $x_1(k) \rightarrow x_2(k)$ and the nonexisting connection (B) $x_2(k) \rightarrow x_1(k)$ in Example 16.2 for $K = 100$ points using $\mathcal{T}_R^{\Phi}$ (dashed) thresholds against the actual PDC estimate (solid). The gray lines refer to thresholds obtained via the approximation adopted in [35] and mean, that for this case, asymptotic reliable decisions are not possible as the latter thresholds are above the $|\pi_{12}(\lambda)|^2 = 1$ theoretical upper bound.

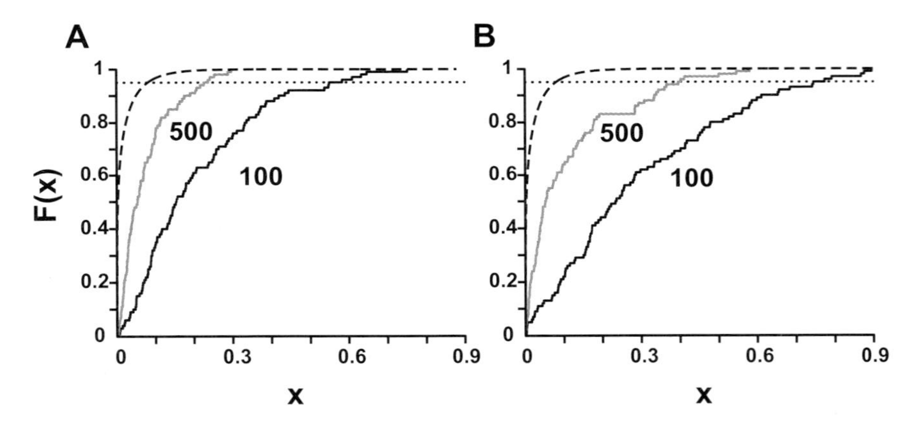

Fig. 16.8: Comparison between the null hypothesis test distributions for $N^* = 100$ resamples, respectively, for $\mathcal{T}_R^{\Phi}$ (A) and $\mathcal{T}_R^{\mathcal{M}}$ (B) using $K = 100$ (black) and $K = 500$ (gray) data points at $\lambda = 0.09$ showing convergence to the theoretical asymptotic distribution (dashed). The horizontal dotted line refers to the 95 % probability threshold.

of the second receiving oscillator whose oscillation, if it were alone, would be achieved by the feedback in Fig. 16.3(B).

As before, a short data segment ($K = 100$) is used in the illustrations using $N^* = 100$ resampled time series.

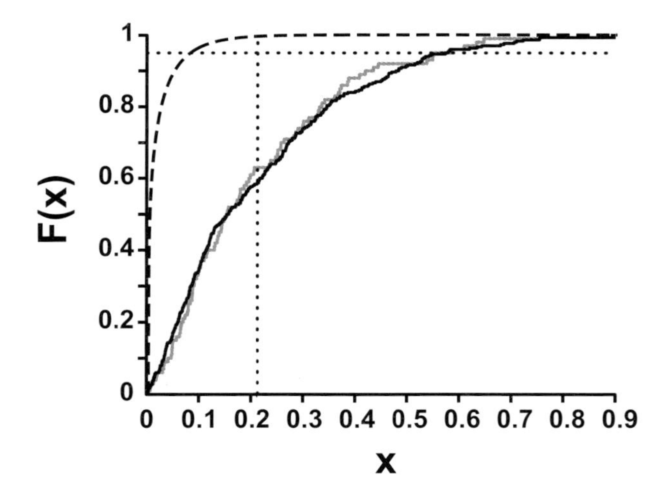

Fig. 16.9: Changes in null hypothesis distribution for $\mathcal{T}_R^{\Phi}$ applied to Example 16.2 at $\lambda = 0.1$ for the $x_2(k) \to x_1(k)$ as a function the number of resampled time series N^*. It is readily seen that $N^* = 100$ (black) resampled series is not markedly different from $N^* = 500$ (gray). The estimated PDC value (vertical dotted line) is well below the 95 % $\mathcal{T}_R^{\Phi}$ threshold value (signalled by the horizontal dotted line).

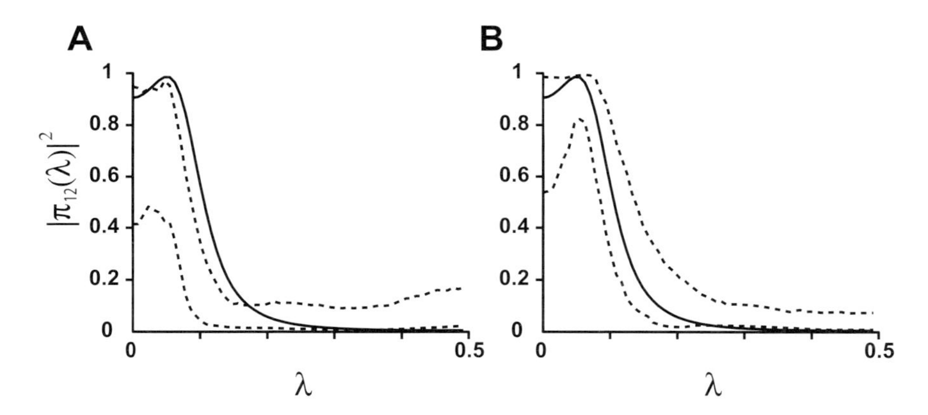

Fig. 16.10: Confidence interval (2.5 %, 97.5 %) results (dashed lines) against the estimated PDC (solid) for $\mathcal{T}_P^{\Phi}$ (A) and $\mathcal{T}_P^{\mathcal{M}}$ (B) for the existing $x_1(n) \to x_2(n)$ connection portraying the interval bias of each method.

From theoretical considerations, since DTF involves a matrix inversion dispensed by PDC, one should expect that its bootstrap results will be subject to larger variability. This is confirmed by comparing the results in Figs. 16.11 and

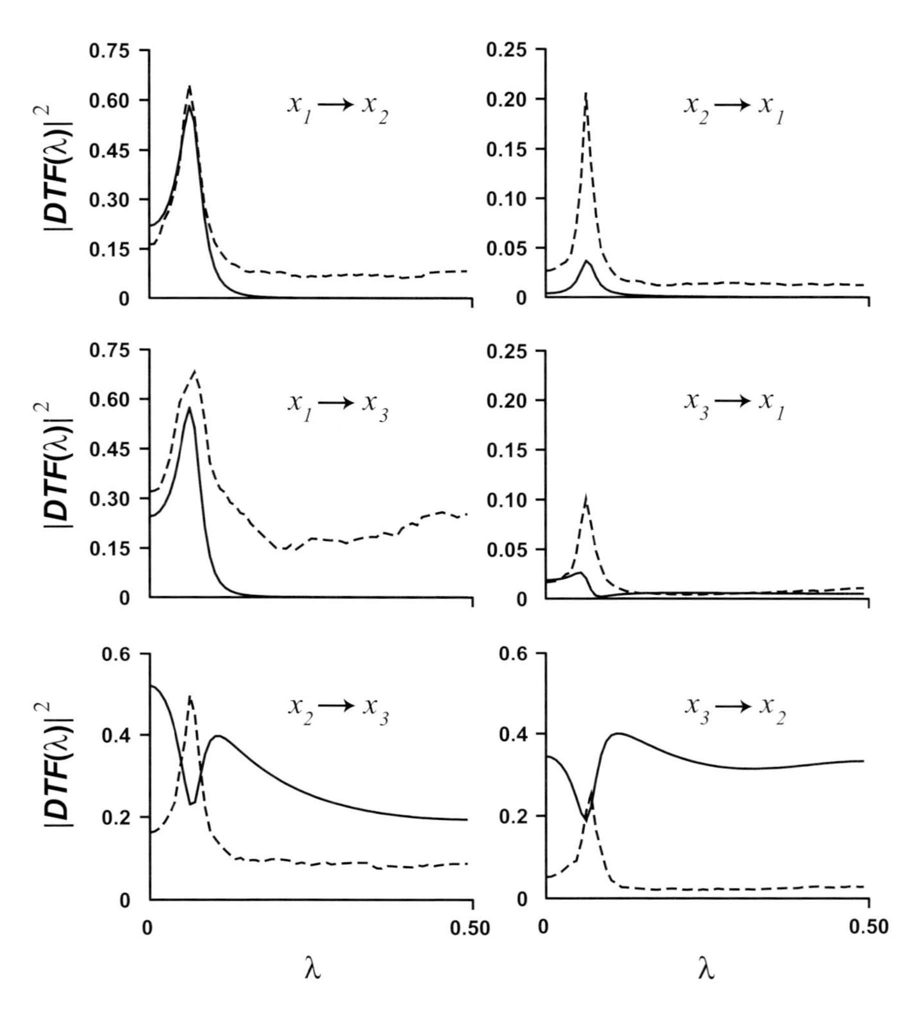

Fig. 16.11: DTF null hypothesis results for the connections in Example 16.3 using $K = 100$ (and $N^* = 100$) made by comparing the estimated DTF (solid) against the 95 % thresholds via $\mathcal{T}_R^{\Phi}$ (dashed). Only the $x_2(n) \to x_1(n)$ and $x_3(n) \to x_1(n)$ witness correct decisions for all frequency bands. The fact that power emanates from $x_1(n)$ to $x_3(n)$ is incorrectly identified as is the mutual feedback between $x_2(n)$ and $x_3(n)$ at the resonance frequency.

16.12. Improvement in DTF descriptions for this model is attained for $K = 300$ as can be appreciated in Fig. 16.13.

16.4.2 Real Data

The first real data example is taken from Andrews and Herzberg [39].

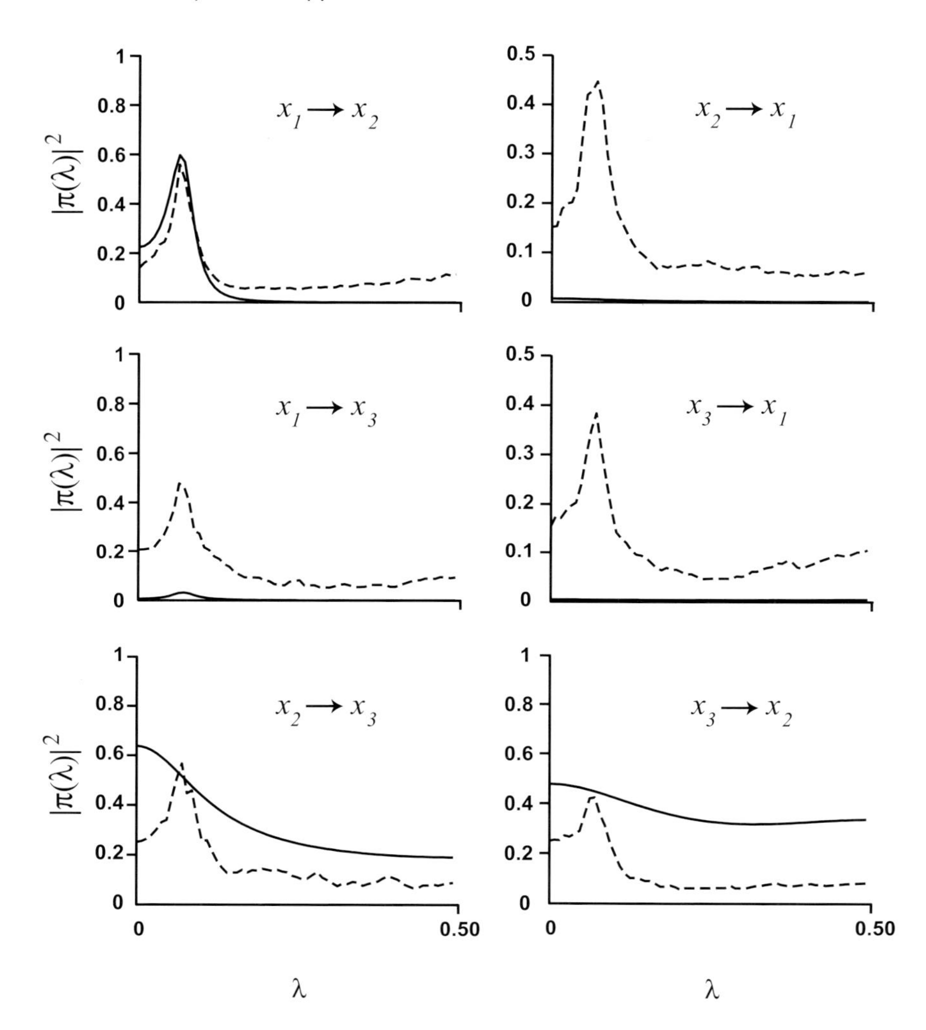

Fig. 16.12: PDC null hypothesis test results reproduce much more closely the theoretical structure in Fig. 16.3(B) (Example 16.3) as the PDC estimates (solid) are above the $\mathcal{T}_R^{\Phi}$ 95 % thresholds (dashed) for almost all frequencies when the connection exists.

Example 16.4. The data relate the time series of melanoma incidence in Connecticut after trend removal $(x_1(k))$ and Wölfer sunspot data $(x_2(n))$ between 1936 and 1972 in a total of $K = 37$ data points.

Naturally, the true causal relation can only be that of solar activity induced melanomas. The resulting model is severely unbalanced with respect to $w_i(k)$ innovation time series variances. Results of using $\mathcal{T}_R^{\Phi}$ for PDC connectivity significance analysis in Fig. 16.14 with[4] and without data normalization prior to

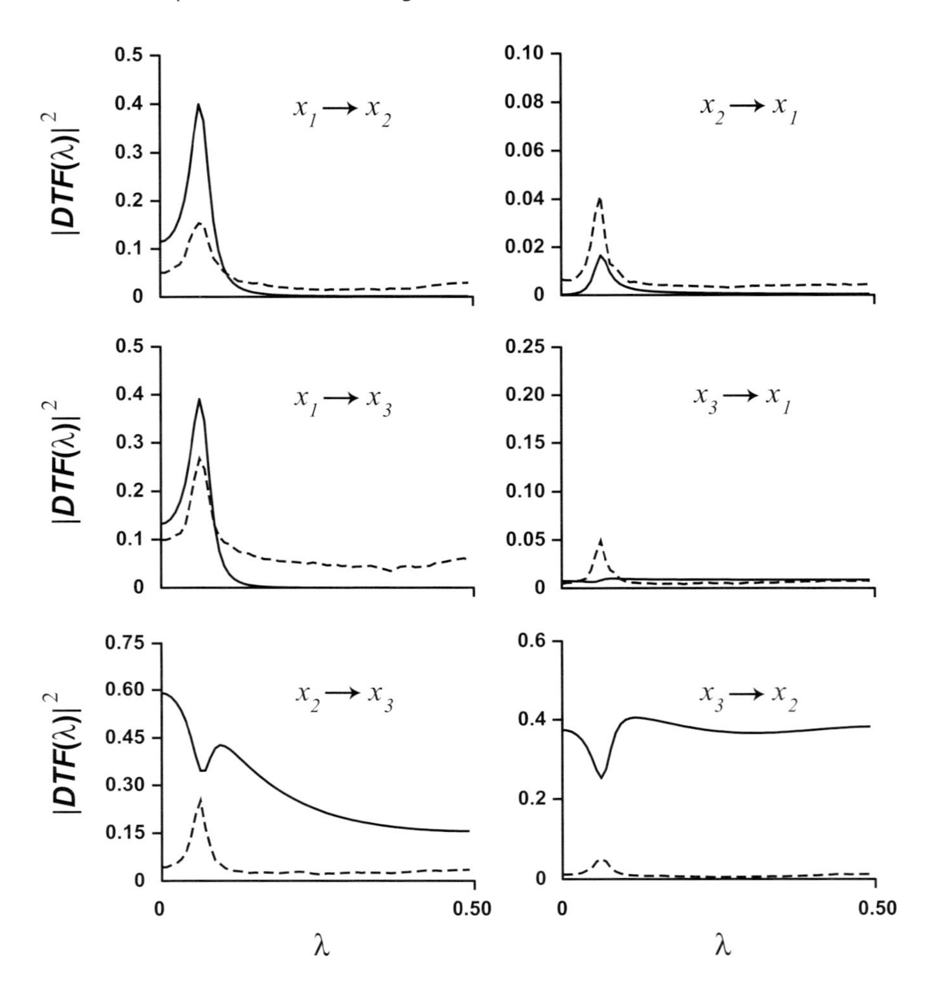

Fig. 16.13: An example of the same type of DTF (solid) results under null hypothesis tests for $K = 300$ show much closer agreement to the actual theoretically DTF connectivity description as thresholds (dashed) are correctly crossed (solid) for the structure in Example 16.3. Compare with Fig. 16.11.

modeling leads to false conclusions, whereas the use of GPDC, because of its variance stabilization properties [17], leads to correct conclusions both with and without data normalization (Fig. 16.15). It is worth noting that the usual time domain Granger causality tests agree with those obtained visually through GPDC.

Experience has, in fact, shown that lack of variance stabilization can lead to unacceptably high decision error rates. This is what induced the introduction of GPDC in the first place [17].

[4] By fitting $x_1(k)/\sigma_{x_1}$ and $x_2(k)/\sigma_{x_2}$ instead of the actual series $x_1(k)$ and $x_2(k)$, where σ_{x_i} stand for the estimated standard deviations of the time series.

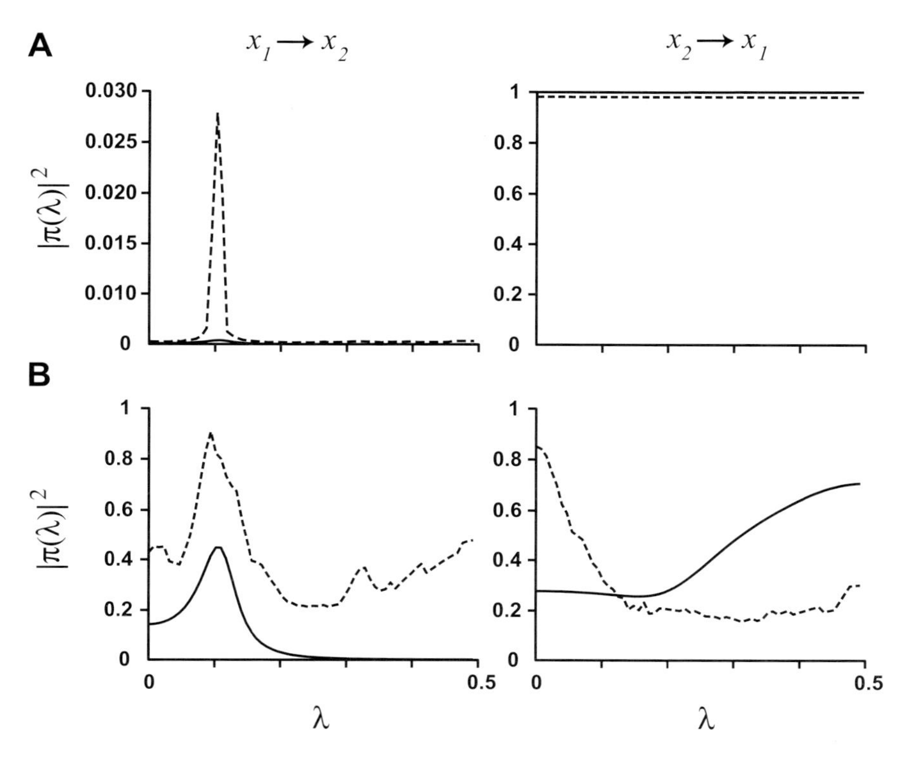

Fig. 16.14: PDC results (solid) and the allied decision thresholds (dashed) using $\mathcal{T}_R^{\Phi}$ for unnormalized (A) and normalized (B) data showing reversed causality estimates between the detrended melanoma $(x_1(k))$ and Wölfer sunspot time series $(x_2(k))$.

Example 16.5. Three time series (T3, T4, and O1) obtained via a standard international 10–20 EEG system and sampled at 200 Hz were derived from a patient with left mesial temporal lobe epilepsy, with a seizure focus roughly localized at the T3 channel area as clinically diagnosed with post-surgical confirmation from the Neurological Division of Hospital das Clínicas from the University of São Paulo. Two distinct data segments (1000 data points, i.e., 5 s), during and immediately before a seizure onset, separated by 20 s to exclude the transition period, were used in characterizing the relationship between these brain areas.

Three-variate models were estimated for each segment with model orders obtained via AIC (Akaike's Information Criterion) leading to $p = 4$ and $p = 5$, respectively. Model adequacy, in addition, was ensured by a Portmanteau test on estimated $w_i(k)$ autocorrelations whose whiteness could not be rejected at 5 % [5].

After PDC computation, null hypothesis tests were performed at each frequency for each channel pair at 5 %. When Eq. (16.5) could be rejected using $\mathcal{T}_R^{\Phi}$,

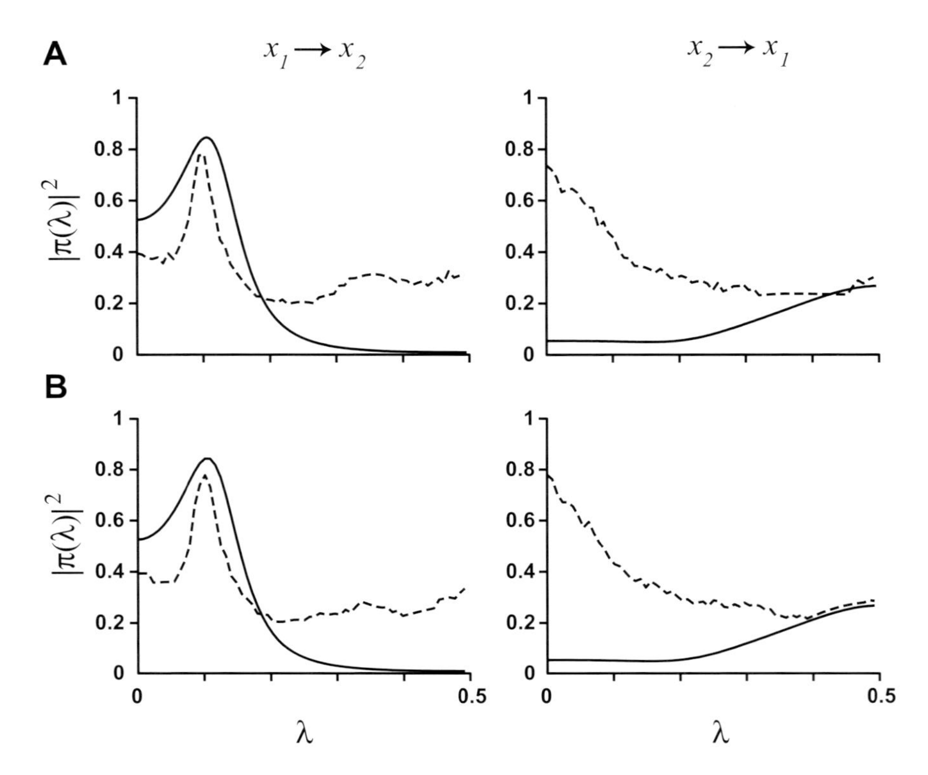

Fig. 16.15: GPDC results (solid) and the allied decision thresholds (dashed) using $\mathcal{T}_R^{\Phi}$ for unnormalized (A) and normalized (B) data showing correct causality estimates. Detrended Melanoma time series $(x_1(k))$ and Wölfer sunspot time series $(x_2(k))$. Compare with Fig. 16.14.

confidence intervals were computed under the normal approximation, leading to Figs. 16.16 and 16.17 via $\mathcal{T}_P^{\Phi}$.

Before seizure onset, $\mathcal{T}_R^{\Phi}$ indicates lack of significant interactions below 5 Hz, even though PDC is significant for higher frequencies (Fig. 16.16). During the seizure, lower frequencies (Fig. 16.17) become significant in agreement with physiological information since temporal lobe seizures are characterized by both low-frequency oscillations ($\approx$ 3 Hz) and channel synchronization [1]. Confidence intervals were plotted based on $\mathcal{T}_P^{\Phi}$ using $N^* = 800$ resampled time series.

The data in this example was used to illustrate the asymptotic PDC results in [35] and the present results using bootstrap methods agree both with time domain asymptotic Granger Causality tests [5] and with those in [35].

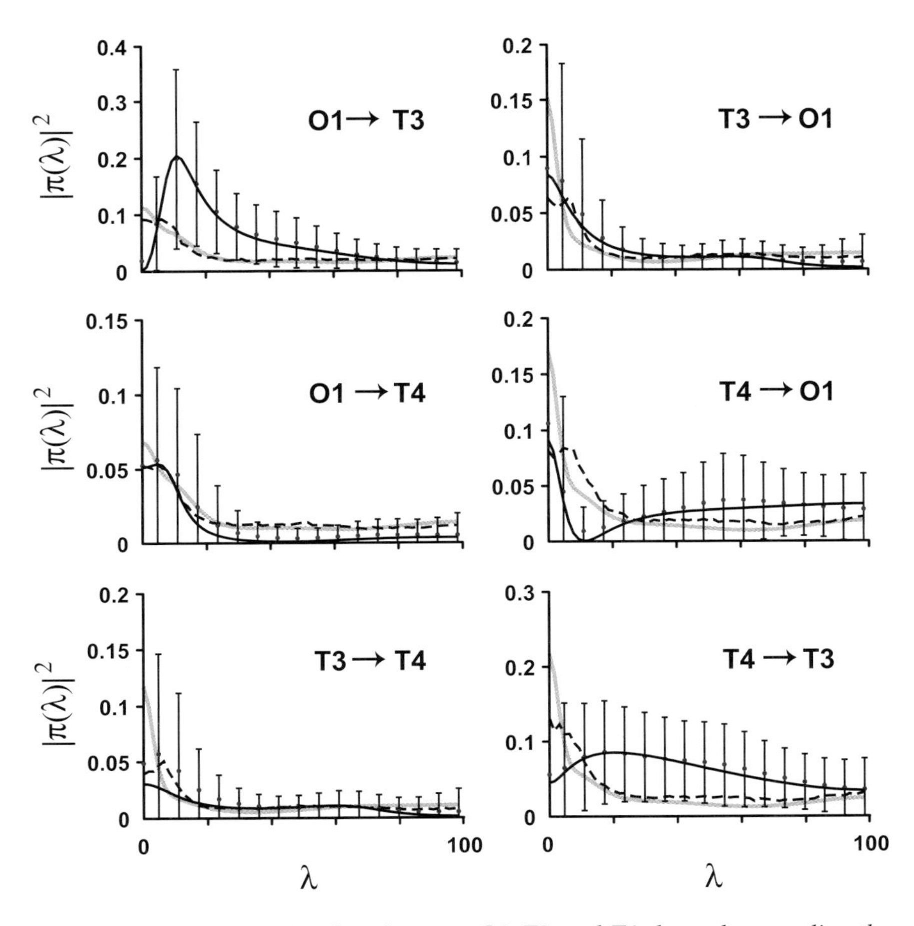

Fig. 16.16: Estimated PDC values between O1, T3, and T4 channels preceding the seizure onset. Black solid lines represent the estimated PDC value for each frequency in hertz. Bootstrap thresholds ($\mathcal{T}_{R}^{\Phi}$) at 5 % for each frequency (dashed lines) are contrasted to asymptotic ones (gray lines). Between both limits are in fair agreement. The bootstrap (2.5 %, 97.5 %) confidence intervals ($\mathcal{T}_{P}^{\Phi}$) for the estimated PDC are also plotted.

16.5 Discussion

A large variety of resampling methods is available. Only two such methods were considered here and it was possible to illustrate the convenience, if not the need for all these methods as they manage to adequately capture the level 2 statistics with different degrees of reliability depending on what is intended. This is clearly shown in Fig. 16.10, where $\mathcal{T}_{P}^{\Phi}$ fails to capture reasonable confidence interval limits as opposed to $\mathcal{T}_{P}^{\mathcal{M}}$. In fact, $\mathcal{T}_{P}^{\Phi}$ inadequacy is immediately apparent calling for the use of other resampling methods to estimate the confidence limits. The rea-

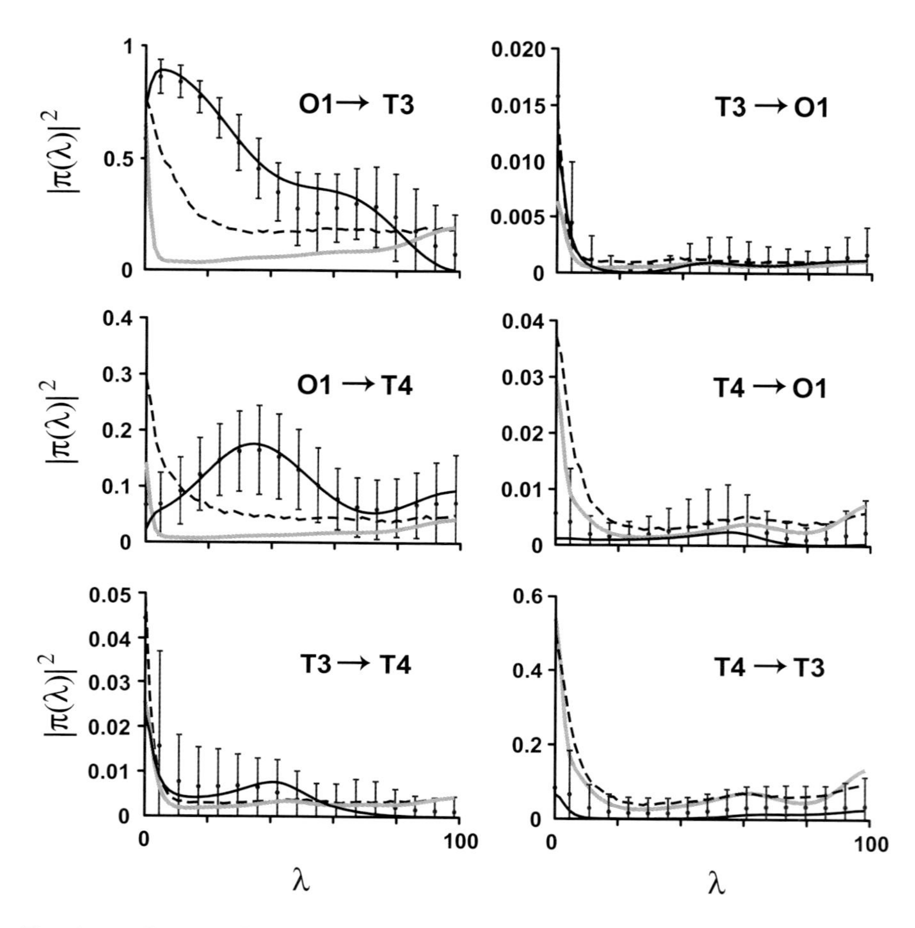

Fig. 16.17: Estimated PDC values between O1, T3, and T4 channels during the seizure onset. See legend to Fig. 16.16 for further explanation.

son for $\mathcal{T}_P^{\Phi}$'s failure is associated with the limited cross-spectral range explored by resampling through the addition of just a common random phase factor for each resample as opposed to perturbing the spectral amplitudes as well.

It is interesting to note that the number of resamples N^*, if large enough, does not seem to be too much of an issue (see Fig. 16.9).

The list of resampling methods presented here is by no means exhaustive and many more ideas can be used as done for example in [36], where neuronal spike time series were randomly circularly rotated before reconstruction and analysis [40] to generate phase scrambled data with identical spectral properties for each resample. The development of other application ready resampling tools is therefore welcome.

The applicability of these methods is not restricted to the quantities in Table 16.1, on the contrary, the empirical resampling distribution of model variables like $a_{ij}(r)$ or even of time domain statistics used in conventional Granger causality tests could also have been computed.

What is of even greater importance is that these methods provide a clearer picture of how much variability should be expected from a model derived from a given data set, which is most often just what is needed in practice to convey a sense of how reliable one's conclusion about connectivity should be.

It is perhaps reassuring that specially for null hypothesis testing the resampling methods, given limitations peculiar to each, converge to the asymptotic decision threshold limits of quantities with known statistics (Fig. 16.8). In fact, in practical situations, resampling provides relatively large limits that prevent the false positives that were common in this case [9, 36, 37] when arbitrary frequency independent thresholds were employed.

Finally, the importance of using variance stabilized estimators is shown in Example 16.4. One should note, however, that it is possible to go one step ahead and "bootstrap the bootstrap" to achieve variance stabilization [41, 42] so that even in this case, this problem that often leads to high rates of incorrect connectivity inferences can be circumvented.

16.6 Conclusions

Both methods discussed here show that the actually observed data can provide insight on the amount of variability that can be expected from modeling an observed data set and lead to reasonable criteria for connectivity analysis.

Each method has its domain of applicability that needs to be considered *vis-à-vis* what is desired such as hypothesis testing or confidence interval determination or some other *level 2* statistic of interest.

Acknowledgements

Supported by FAPESP 02/06925-1 and CNPq 305576/2003-8.

References

[1] D. D. Daly and T. A. Pedley. *Current Practice of Clinical Electroencephalography*. Raven Press, New York, 2nd edition, 1990.

[2] M. A. L. Nicolelis. *Methods in Neural Ensemble Recordings*. CRC Press, Boca Raton, London, 1999.

[3] L. A. Baccalá, M. Y. Alvarenga, K. Sameshima, C. L. Jorge, and L. H. Castro. *J. Int. Neurosci.*, 3:379, 2004.

[4] C. W. J. Granger. *Econometrica*, 37:424, 1969.

[5] H. Lütkepohl. *Introduction to Multiple Time Series Analysis*. Springer, Berlin, 2nd edition, 1993.

[6] S. M. Schnider, R. H. Kwong, F. A. Lenz, and H. C. Kwan. *Biol. Cybern.*, 60: 203, 1989.

[7] M. J. Kaminski and K. J. Blinowska. *Biol. Cybern.*, 65:203, 1991.

[8] M. Kaminski, K. Blinowska, and W. Szelenberger. *Electroencephal. Clin. Neurophysiol.*, 102:216, 1997.

[9] K. Sameshima and L. A. Baccalá. *J. Neurosci. Methods*, 94:93, 1999.

[10] E. E. Fanselow, K. Sameshima, L. A. Baccalá, and M. A. L. Nicolelis. *Proc. Natl. Acad. Sci.*, 98:15330, 2001.

[11] L. A. Baccalá and K. Sameshima. In F. Rattay, editor, *World Congress on Neuroinformatics*, volume 1, pages 546–553. 2001.

[12] M. Kaminski, M. Z. Ding, W. A. Truccolo, and S. L. Bressler. *Biol. Cybern.*, 85:145, 2001.

[13] S. M. Kay. *Modern Spectral Estimation: Theory and Application*. Prentice-Hall, Englewood Cliffs, NJ, 1988.

[14] L. A. Baccalá. In N. Callaos, D. Rosario, and B. Sanches, editors, *World Multiconference on Systemics, Cybernetics and Informatics*, volume 6, pages 10–14, Orlando, 2001. International Institute of Informatics and Systemics.

[15] L. A. Baccalá, K. Sameshima, G. Ballester, A. C. Valle, and C. Timo-Iaria. *Appl. Sig. Proc.*, 5:40, 1998.

[16] L. A. Baccalá and K. Sameshima. *Biol. Cybern.*, 84:463, 2001.

[17] L. A. Baccalá, D. Y. Takahashi, and K. Sameshima. Generalized partial directed coherence. 2006. Submitted to XVI Congresso Brasileiro de Automática, Salvador, Bahia.

[18] B. Efron. *The Jackknife, the Bootstrap and other Resampling Plans*. Number 38 in CBMS-NSF Regional Conference Series in Applied Mathematics. SIAM, Philadelphia, 1982.

[19] P. Hall. *The Bootstrap and Edgeworth Expansion*. Springer, Berlin, 1992.

[20] D. N. Politis. *IEEE Sig. Proc. Mag.*, 15:39, 1998.

[21] A. C. Davison and D. V. Hinkley. *Bootstrap Methods and their Application*. Cambridge University Press, Cambridge, 1997.

[22] S. N. Lahiri. *Resampling Methods for Dependent Data*. Springer, New York, London, 2003.

[23] M. B. Priestley. *Spectral Analysis and Time Series*. Academic Press, London, 1982.

[24] P. J. Brockwell and R. A. Davis. *Time Series: Theory and Methods*. Springer, New York, 2nd edition, 1991.

[25] D. B. Percival and A. T. Walden. *Spectral Analysis for Physical Applications*. Cambridge University Press, Cambridge, 1993.

[26] J. Theiler, S. Eubank, A. Longtin, B. Galdrikian, and J. D. Farmer. *Physica D*, 58:77, 1992.

[27] D. Prichard and J. Theiler. *Phys. Rev. Lett.*, 73:951, 1994.

[28] W. J. Braun and N. J. Kulperger. *Comm. Stat.*, 26:1329, 1997.

[29] P. Stoica and R. L. Moses. *Introduction to Spectral Analysis*. Prentice-Hall, Upper Saddle River, NJ, 1997.

[30] E. Carlstein. *Ann. Stat.*, 14:1171, 1986.

[31] D. N. Politis, J. P. Romano, and M. Wolf. *Subsampling*. Springer Series in Statistics. Springer, New York, 1999.

[32] P. Buhlmann. *Bernoulli*, 3:123, 1997.

[33] R. Dahlhaus and D. Janas. *Ann. Stat.*, 24:1934, 1996.

[34] J. Franke and W. Hardle. *Ann. Stat.*, 20:121, 1992.

[35] D. Y. Takahashi, L. A. Baccalá, and K. Sameshima. Inference between neural structures via partial directed coherence. 2006. Submitted to J. Appl. Statistics.

[36] L. A. Baccalá and K. Sameshima. In *25th IEEE Annual Conference on Engineering in Medicine and Biology*, volume 1, page 2151, Cancun, Mexico, 2003.

[37] L. A. Baccalá and K. Sameshima. In R. A. Zângaro and M. T. T. Pacheco, editors, *XVIII Congresso Brasileiro de Engenharia Biomédica*, volume 5, pages 453–457, Saõ José dos Campos, SP, Brasil, 2002.

[38] L. A. Baccalá and K. Sameshima. *Prog. Brain Res.*, 130:33, 2001.

[39] D. F. Andrews and A. M. Herzberg. *Data: A Collection of Problems from Many Fields for the Student and Research Worker*. Springer, New York, 1985.

[40] L. A. Baccalá and K. Sameshima. In M. A. L. Nicolelis, editor, *Methods for Simultaneous Neuronal Ensemble Recordings*, pages 179–192. CRC Press, Boca Raton, 1998.

[41] A. M. Zoubir and B. Boashash. *IEEE Sig. Proc. Mag.*, 15:56, 1998.

[42] A. M. Zoubir and D. R. Iskander. *Bootstrap Techniques for Signal Processing*. Cambridge University Press, Cambridge, 2004.

17 Granger Causality: Basic Theory and Application to Neuroscience

Mingzhou Ding, Yonghong Chen, and Steven L. Bressler

Multielectrode neurophysiological recordings produce massive quantities of data. Multivariate time series analysis provides the basic framework for analyzing the patterns of neural interactions in these data. It has long been recognized that neural interactions are directional. Being able to assess the directionality of neuronal interactions is thus a highly desired capability for understanding the cooperative nature of neural computation. Research over the last few years has shown that Granger causality is a key technique to furnish this capability. The main goal of this chapter is to provide an expository introduction to the concept of Granger causality. Mathematical frameworks for both the bivariate Granger causality and conditional Granger causality are developed in detail, with particular emphasis on their spectral representations. The technique is demonstrated in numerical examples where the exact answers of causal influences are known. It is then applied to analyze multichannel local field potentials recorded from monkeys performing a visuomotor task. Our results are shown to be physiologically interpretable and yield new insights into the dynamical organization of large-scale oscillatory cortical networks.

17.1 Introduction

In neuroscience, as in many other fields of science and engineering, signals of interest are often collected in the form of multiple simultaneous time series. To evaluate the statistical interdependence among these signals, one calculates cross-correlation functions in the time domain and ordinary coherence functions in the spectral domain. However, in many situations of interest, symmetric[1] measures like ordinary coherence are not completely satisfactory, and further dissection of the interaction patterns among the recorded signals is required to parcel out effective functional connectivity in complex networks. Recent work has begun to consider the causal influence one neural time series can exert on another. The basic idea can be traced back to Wiener [1] who conceived the notion that, if the prediction of one time series could be improved by incorporating the knowledge

[1] Here by symmetric we mean that, when A is coherent with B, B is equally coherent with A.

Handbook of Time Series Analysis. Björn Schelter, Matthias Winterhalder, Jens Timmer

ISBN: 3-527-40623-9

of a second one, then the second series is said to have a causal influence on the first. Wiener's idea lacks the machinery for practical implementation. Granger later formalized the prediction idea in the context of linear regression models [2]. Specifically, if the variance of the autoregressive prediction error of the first time series at the present time is reduced by inclusion of past measurements from the second time series, then the second time series is said to have a causal influence on the first one. The roles of the two time series can be reversed to address the question of causal influence in the opposite direction. From this definition it is clear that the flow of time plays a vital role in allowing inferences to be made about directional causal influences from time series data. The interaction discovered in this way may be reciprocal or it may be unidirectional.

Two additional developments of Granger's causality idea are important. First, for three or more simultaneous time series, the causal relation between any two of the series may be direct, may be mediated by a third one, or may be a combination of both. This situation can be addressed by the technique of conditional Granger causality. Second, natural time series, including ones from economics and neurobiology, contain oscillatory aspects in specific frequency bands. It is thus desirable to have a spectral representation of causal influence. Major progress in this direction has been made by Geweke [3, 4] who found a novel time series decomposition technique that expresses the time domain Granger causality in terms of its frequency content. In this chapter we review the essential mathematical elements of Granger causality with special emphasis on its spectral decomposition. We then discuss practical issues concerning how to estimate such measures from time series data. Simulations are used to illustrate the theoretical concepts. Finally, we apply the technique to analyze the dynamics of a large-scale sensorimotor network in the cerebral cortex during cognitive performance. Our result demonstrates that, for a well designed experiment, a carefully executed causality analysis can reveal insights that are not possible with other techniques.

17.2 Bivariate Time Series and Pairwise Granger Causality

Our exposition in this and the next section follows closely that of Geweke [3, 4]. To avoid excessive mathematical complexity we develop the analysis framework for two time series. The framework can be generalized to two sets of time series [3].

17.2.1 Time Domain Formulation

Consider two stochastic processes X_t and Y_t. Assume that they are jointly stationary. Individually, under fairly general conditions, each process admits an autoregressive representation

$$X_t = \sum_{j=1}^{\infty} a_{1j} X_{t-j} + \epsilon_{1t}, \quad \text{var}(\epsilon_{1t}) = \Sigma_1, \tag{17.1}$$

$$Y_t = \sum_{j=1}^{\infty} d_{1j} Y_{t-j} + \eta_{1t}, \quad \text{var}(\eta_{1t}) = \Gamma_1. \tag{17.2}$$

Jointly, they are represented as

$$X_t = \sum_{j=1}^{\infty} a_{2j} X_{t-j} + \sum_{j=1}^{\infty} b_{2j} Y_{t-j} + \epsilon_{2t}, \tag{17.3}$$

$$Y_t = \sum_{j=1}^{\infty} c_{2j} X_{t-j} + \sum_{j=1}^{\infty} d_{2j} Y_{t-j} + \eta_{2t}, \tag{17.4}$$

where the noise terms are uncorrelated over time and their contemporaneous covariance matrix is

$$\mathbf{\Sigma} = \begin{pmatrix} \Sigma_2 & \Upsilon_2 \\ \Upsilon_2 & \Gamma_2 \end{pmatrix}. \tag{17.5}$$

The entries are defined as $\Sigma_2 = \text{var}(\epsilon_{2t})$, $\Gamma_2 = \text{var}(\eta_{2t})$, $\Upsilon_2 = \text{cov}(\epsilon_{2t}, \eta_{2t})$. If X_t and Y_t are independent, then $\{b_{2j}\}$ and $\{c_{2j}\}$ are uniformly zero, $\Upsilon_2 = 0$, $\Sigma_1 = \Sigma_2$ and $\Gamma_1 = \Gamma_2$. This observation motivates the definition of total interdependence between X_t and Y_t as

$$F_{X,Y} = \ln \frac{\Sigma_1 \Gamma_1}{|\mathbf{\Sigma}|}, \tag{17.6}$$

where $|\cdot|$ denotes the determinant of the enclosed matrix. According to this definition, $F_{X,Y} = 0$ when the two time series are independent, and $F_{X,Y} > 0$ when they are not.

Consider Eqs. (17.1) and (17.3). The value of Σ_1 measures the accuracy of the autoregressive prediction of X_t based on its previous values, whereas the value of Σ_2 represents the accuracy of predicting the present value of X_t based on the previous values of both X_t and Y_t. According to Wiener [1] and Granger [2], if Σ_2 is less than Σ_1 in some suitable statistical sense, then Y_t is said to have a causal influence on X_t. We quantify this causal influence by

$$F_{Y \to X} = \ln \frac{\Sigma_1}{\Sigma_2}. \tag{17.7}$$

It is clear that $F_{Y \to X} = 0$ when there is no causal influence from Y to X and $F_{Y \to X} > 0$ when there is. Similarly, one can define causal influence from X to Y as

$$F_{X \to Y} = \ln \frac{\Gamma_1}{\Gamma_2}. \tag{17.8}$$

It is possible that the interdependence between X_t and Y_t cannot be fully explained by their interactions. The remaining interdependence is captured by Υ_2, the covariance between ϵ_{2t} and η_{2t}. This interdependence is referred to as instantaneous causality and is characterized by

$$F_{X \cdot Y} = \ln \frac{\Sigma_2 \Gamma_2}{|\boldsymbol{\Sigma}|} . \tag{17.9}$$

When Υ_2 is zero, $F_{X \cdot Y}$ is also zero. When Υ_2 is not zero, $F_{X \cdot Y} > 0$.

The above definitions imply that

$$F_{X,Y} = F_{X \to Y} + F_{Y \to X} + F_{X \cdot Y} . \tag{17.10}$$

Thus we decompose the total interdependence between the two time series X_t and Y_t into three components: two directional causal influences due to their interaction patterns, and the instantaneous causality due to factors possibly exogenous to the (X, Y) system, e.g., a common driving input.

17.2.2 Frequency Domain Formulation

To begin we define the lag operator L to be $LX_t = X_{t-1}$. Rewrite Eqs. (17.3) and (17.4) in terms of the lag operator

$$\begin{pmatrix} a_2(L) & b_2(L) \\ c_2(L) & d_2(L) \end{pmatrix} \begin{pmatrix} X_t \\ Y_t \end{pmatrix} = \begin{pmatrix} \epsilon_{2t} \\ \eta_{2t} \end{pmatrix} , \tag{17.11}$$

where $a_2(0) = 1$, $b_2(0) = 0$, $c_2(0) = 0$, $d_2(0) = 1$. Fourier transforming both sides of Eq. (17.11) leads to

$$\begin{pmatrix} a_2(\omega) & b_2(\omega) \\ c_2(\omega) & d_2(\omega) \end{pmatrix} \begin{pmatrix} X(\omega) \\ Y(\omega) \end{pmatrix} = \begin{pmatrix} E_x(\omega) \\ E_y(\omega) \end{pmatrix} , \tag{17.12}$$

where the components of the coefficient matrix $\mathbf{A}(\omega)$ are

$$a_2(\omega) = 1 - \sum_{j=1}^{\infty} a_{2j} e^{-i\omega j} , \quad b_2(\omega) = -\sum_{j=1}^{\infty} b_{2j} e^{-i\omega j} ,$$

$$c_2(\omega) = -\sum_{j=1}^{\infty} c_{2j} e^{-i\omega j} , \quad d_2(\omega) = 1 - \sum_{j=1}^{\infty} d_{2j} e^{-i\omega j} .$$

Recasting Eq. (17.12) into the transfer function format we obtain

$$\begin{pmatrix} X(\omega) \\ Y(\omega) \end{pmatrix} = \begin{pmatrix} H_{xx}(\omega) & H_{xy}(\omega) \\ H_{yx}(\omega) & H_{yy}(\omega) \end{pmatrix} \begin{pmatrix} E_x(\omega) \\ E_y(\omega) \end{pmatrix} , \tag{17.13}$$

where the transfer function is $\mathbf{H}(\omega) = \mathbf{A}^{-1}(\omega)$ whose components are

$$\begin{aligned} H_{xx}(\omega) &= \frac{1}{\det \mathbf{A}} d_2(\omega) , \quad & H_{xy}(\omega) &= -\frac{1}{\det \mathbf{A}} b_2(\omega) , \\ H_{yx}(\omega) &= -\frac{1}{\det \mathbf{A}} c_2(\omega) , \quad & H_{yy}(\omega) &= \frac{1}{\det \mathbf{A}} a_2(\omega) . \end{aligned} \tag{17.14}$$

After proper ensemble averaging we have the spectral matrix

$$\mathbf{S}(\omega) = \mathbf{H}(\omega)\mathbf{\Sigma}\mathbf{H}^*(\omega), \tag{17.15}$$

where * denotes the complex conjugate and matrix transpose.

The spectral matrix contains cross-spectra and auto-spectra. If X_t and Y_t are independent, then the cross-spectra are zero and $|\mathbf{S}(\omega)|$ equals the product of two auto-spectra. This observation motivates the spectral domain representation of total interdependence between X_t and Y_t as

$$f_{X,Y}(\omega) = \ln \frac{S_{xx}(\omega)S_{yy}(\omega)}{|\mathbf{S}(\omega)|}, \tag{17.16}$$

where $|\mathbf{S}(\omega)| = S_{xx}(\omega)S_{yy}(\omega) - S_{xy}(\omega)S_{yx}(\omega)$ and $S_{yx}(\omega) = S^*_{xy}(\omega)$. It is easy to see that this decomposition of interdependence is related to coherence by the following relation

$$f_{X,Y}(\omega) = -\ln\left(1 - C(\omega)\right), \tag{17.17}$$

where coherence is defined as

$$C(\omega) = \frac{|S_{xy}(\omega)|^2}{S_{xx}(\omega)S_{yy}(\omega)}.$$

The coherence defined in this way is sometimes referred to as the squared coherence.

To obtain the frequency decomposition of the time domain causality defined in the previous section, we look at the auto-spectrum of X_t

$$S_{xx}(\omega) = H_{xx}(\omega)\Sigma_2 H^*_{xx}(\omega) + 2\Upsilon_2 \operatorname{Re}\left(H_{xx}(\omega)H^*_{xy}(\omega)\right) + H_{xy}(\omega)\Gamma_2 H^*_{xy}(\omega). \tag{17.18}$$

It is instructive to consider the case where $\Upsilon_2 = 0$. In this case there is no instantaneous causality and the interdependence between X_t and Y_t is entirely due to their interactions through the regression terms on the right-hand sides of Eqs. (17.3) and (17.4). The spectrum has two terms. The first term, viewed as the intrinsic part, involves only the variance of ϵ_{2t}, which is the noise term that drives the X_t time series. The second term, viewed as the causal part, involves only the variance of η_{2t}, which is the noise term that drives Y_t. This power decomposition into an "intrinsic" term and a "causal" term will become important for defining a measure for spectral domain causality.

When Υ_2 is not zero it becomes harder to attribute the power of the X_t series to different sources. Here we consider a transformation introduced by Geweke [3] that removes the cross term and makes the identification of an intrinsic power term and a causal power term possible. The procedure is called normalization and it consists of left multiplying

$$\mathbf{P} = \begin{pmatrix} 1 & 0 \\ -\frac{\Upsilon_2}{\Sigma_2} & 1 \end{pmatrix} \tag{17.19}$$

on both sides of Eq. (17.12). The result is

$$\begin{pmatrix} a_2(\omega) & b_2(\omega) \\ c_3(\omega) & d_3(\omega) \end{pmatrix} \begin{pmatrix} X(\omega) \\ Y(\omega) \end{pmatrix} = \begin{pmatrix} E_x(\omega) \\ \tilde{E}_y(\omega) \end{pmatrix}, \tag{17.20}$$

where $c_3(\omega) = c_2(\omega) - \frac{\Upsilon_2}{\Sigma_2} a_2(\omega)$, $d_3(\omega) = d_2(\omega) - \frac{\Upsilon_2}{\Sigma_2} b_2(\omega)$, $\tilde{E}_y(\omega) = E_y(\omega) - \frac{\Upsilon_2}{\Sigma_2} E_x(\omega)$. The new transfer function $\tilde{\mathbf{H}}(\omega)$ for Eq. (17.20) is the inverse of the new coefficient matrix $\tilde{\mathbf{A}}(\omega)$

$$\tilde{\mathbf{H}}(\omega) = \begin{pmatrix} \tilde{H}_{xx}(\omega) & \tilde{H}_{xy}(\omega) \\ \tilde{H}_{yx}(\omega) & \tilde{H}_{yy}(\omega) \end{pmatrix} = \frac{1}{\det \tilde{\mathbf{A}}} \begin{pmatrix} d_3(\omega) & -b_2(\omega) \\ -c_3(\omega) & a_2(\omega) \end{pmatrix}. \tag{17.21}$$

Since $\det \tilde{\mathbf{A}} = \det \mathbf{A}$ we have

$$\begin{aligned} \tilde{H}_{xx}(\omega) &= H_{xx}(\omega) + \frac{\Upsilon_2}{\Sigma_2} H_{xy}(\omega), \quad \tilde{H}_{xy}(\omega) = H_{xy}(\omega), \\ \tilde{H}_{yx}(\omega) &= H_{yx}(\omega) + \frac{\Upsilon_2}{\Sigma_2} H_{xx}(\omega), \quad \tilde{H}_{yy}(\omega) = H_{yy}(\omega). \end{aligned} \tag{17.22}$$

From the construction it is easy to see that E_x and $\tilde{E}_y$ are uncorrelated, that is, $\mathrm{cov}(E_x, \tilde{E}_y) = 0$. The variance of the noise term for the normalized Y_t equation is $\tilde{\Gamma}_2 = \Gamma_2 - \frac{\Upsilon_2^2}{\Sigma_2}$. From Eq. (17.20), following the same steps that lead to Eq. (17.18), the spectrum of X_t is found to be

$$S_{xx}(\omega) = \tilde{H}_{xx}(\omega)\Sigma_2 \tilde{H}^*_{xx}(\omega) + H_{xy}(\omega)\tilde{\Gamma}_2 H^*_{xy}(\omega). \tag{17.23}$$

Here the first term is interpreted as the intrinsic power and the second term as the causal power of X_t due to Y_t. This is an important relation because it explicitly identifies that portion of the total power of X_t at frequency ω that is contributed by Y_t. Based on this interpretation we define the causal influence from Y_t to X_t at frequency ω as

$$f_{Y \to X}(\omega) = \ln \frac{S_{xx}(\omega)}{\tilde{H}_{xx}(\omega)\Sigma_2 \tilde{H}^*_{xx}(\omega)}. \tag{17.24}$$

Note that this definition of causal influence is expressed in terms of the intrinsic power rather than the causal power. It is expressed in this way so that the causal influence is zero when the causal power is zero (i.e., the intrinsic power equals the total power), and increases as the causal power increases (i.e., the intrinsic power decreases).

By taking the transformation matrix as

$$\begin{pmatrix} 1 & -\Upsilon_2/\Gamma_2 \\ 0 & 1 \end{pmatrix} \tag{17.25}$$

and performing the same analysis, we get the causal influence from X_t to Y_t

$$f_{X \to Y}(\omega) = \ln \frac{S_{yy}(\omega)}{\hat{H}_{yy}(\omega)\Gamma_2 \hat{H}^*_{yy}(\omega)}, \tag{17.26}$$

where $\hat{H}_{yy}(\omega) = H_{yy}(\omega) + \frac{\Upsilon_2}{\Gamma_2} H_{yx}(\omega)$.

By defining the spectral decomposition of instantaneous causality as [5]

$$f_{X \cdot Y}(\omega) = \ln \frac{\left(\tilde{H}_{xx}(\omega)\Sigma_2 \tilde{H}^*_{xx}(\omega)\right)\left(\hat{H}_{yy}(\omega)\Gamma_2 \hat{H}^*_{yy}(\omega)\right)}{|\mathbf{S}(\omega)|}, \tag{17.27}$$

we achieve a spectral domain expression for the total interdependence that is analogous to Eq. (17.10) in the time domain, namely

$$f_{X,Y}(\omega) = f_{X\to Y}(\omega) + f_{Y\to X}(\omega) + f_{X\cdot Y}(\omega). \tag{17.28}$$

We caution that the spectral instantaneous causality may become negative for some frequencies in certain situations and may not have a readily interpretable physical meaning.

It is important to note that, under general conditions, these spectral measures relate to the time domain measures as

$$\begin{aligned}
F_{Y,X} &= \frac{1}{2\pi}\int_{-\pi}^{\pi} f_{Y,X}(\omega)\,d\omega\,,\\
F_{Y\to X} &= \frac{1}{2\pi}\int_{-\pi}^{\pi} f_{Y\to X}(\omega)\,d\omega\,,\\
F_{X\to Y} &= \frac{1}{2\pi}\int_{-\pi}^{\pi} f_{X\to Y}(\omega)\,d\omega\,,\\
F_{Y\cdot X} &= \frac{1}{2\pi}\int_{-\pi}^{\pi} f_{Y\cdot X}(\omega)\,d\omega\,.
\end{aligned} \tag{17.29}$$

The existence of these equalities gives credence to the spectral decomposition procedures described above.

17.3 Trivariate Time Series and Conditional Granger Causality

For three or more time series one can perform a pairwise analysis and thus reduce the problem to a bivariate problem. This approach has some inherent limitations. For example, for the two coupling schemes in Fig. 17.1, a pairwise analysis will give the same patterns of connectivity like that in Fig. 17.1(b). Another example involves three processes where one process drives the other two with differential time delays. A pairwise analysis would indicate a causal influence from the process that receives an early input to the process that receives a late input. To disambiguate these situations requires additional measures. Here we define conditional Granger causality which has the ability to resolve whether the interaction between two time series is direct or is mediated by another recorded time series and whether the causal influence is simply due to differential time delays in their respective driving inputs. Our development is for three time series. The framework can be generalized to three sets of time series [4].

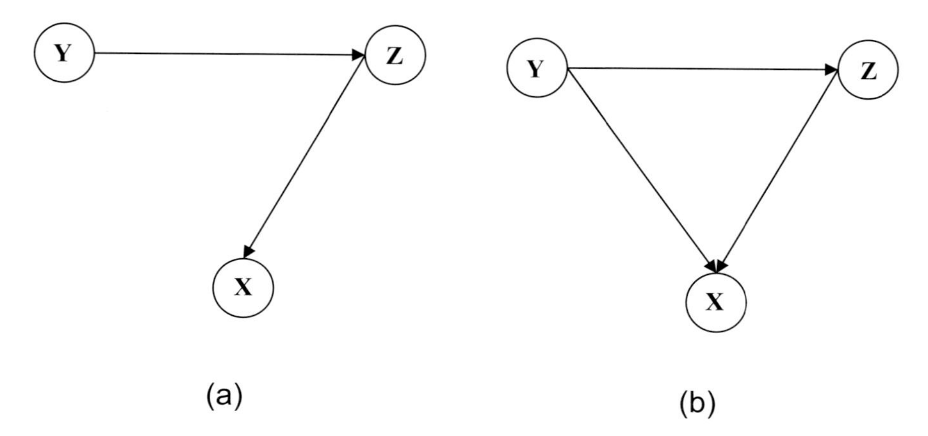

Fig. 17.1: Two distinct patterns of connectivity among three time series. A pairwise causality analysis cannot distinguish these two patterns.

17.3.1 Time Domain Formulation

Consider three stochastic processes X_t, Y_t and Z_t. Suppose that a pairwise analysis reveals a causal influence from Y_t to X_t. To examine whether this influence has a direct component (Fig. 17.1(b)) or is mediated entirely by Z_t (Fig. 17.1(a)) we carry out the following procedure. First, let the joint autoregressive representation of X_t and Z_t be

$$X_t = \sum_{j=1}^{\infty} a_{3j} X_{t-j} + \sum_{j=1}^{\infty} b_{3j} Z_{t-j} + \epsilon_{3t} , \tag{17.30}$$

$$Z_t = \sum_{j=1}^{\infty} c_{3j} X_{t-j} + \sum_{j=1}^{\infty} d_{3j} Z_{t-j} + \gamma_{3t} , \tag{17.31}$$

where the covariance matrix of the noise terms is

$$\mathbf{\Sigma}_3 = \begin{pmatrix} \Sigma_3 & \Upsilon_3 \\ \Upsilon_3 & \Gamma_3 \end{pmatrix} . \tag{17.32}$$

Next we consider the joint autoregressive representation of all the three processes X_t, Y_t, and Z_t

$$X_t = \sum_{j=1}^{\infty} a_{4j} X_{t-j} + \sum_{j=1}^{\infty} b_{4j} Y_{t-j} + \sum_{j=1}^{\infty} c_{4j} Z_{t-j} + \epsilon_{4t} , \tag{17.33}$$

$$Y_t = \sum_{j=1}^{\infty} d_{4j} X_{t-j} + \sum_{j=1}^{\infty} e_{4j} Y_{t-j} + \sum_{j=1}^{\infty} g_{4j} Z_{t-j} + \eta_{4t} , \tag{17.34}$$

$$Z_t = \sum_{j=1}^{\infty} u_{4j} X_{t-j} + \sum_{j=1}^{\infty} v_{4j} Y_{t-j} + \sum_{j=1}^{\infty} w_{4j} Z_{t-j} + \gamma_{4t} , \tag{17.35}$$

where the covariance matrix of the noise terms is

$$\boldsymbol{\Sigma}_4 = \begin{pmatrix} \Sigma_{xx} & \Sigma_{xy} & \Sigma_{xz} \\ \Sigma_{yx} & \Sigma_{yy} & \Sigma_{yz} \\ \Sigma_{zx} & \Sigma_{zy} & \Sigma_{zz} \end{pmatrix} .$$

From these two sets of equations we define the Granger causality from Y_t to X_t conditional on Z_t to be

$$F_{Y \to X|Z} = \ln \frac{\Sigma_3}{\Sigma_{xx}} . \tag{17.36}$$

The intuitive meaning of this definition is quite clear. When the causal influence from Y_t to X_t is entirely mediated by Z_t (Fig. 17.1(a)), $\{b_{4j}\}$ is uniformly zero, and $\Sigma_{xx} = \Sigma_3$. Thus, we have $F_{Y \to X|Z} = 0$, meaning that no further improvement in the prediction of X_t can be expected by including past measurements of Y_t. On the other hand, when there is still a direct component from Y_t to X_t (Fig. 17.1(b)), the inclusion of past measurements of Y_t in addition to that of X_t and Z_t results in better predictions of X_t, leading to $\Sigma_{xx} < \Sigma_3$, and $F_{Y \to X|Z} > 0$.

17.3.2 Frequency Domain Formulation

To derive the spectral decomposition of the time domain conditional Granger causality we carry out a normalization procedure like that for the bivariate case. For Eq. (17.30) and Eq. (17.31) the normalized equations are

$$\begin{pmatrix} D_{11}(L) & D_{12}(L) \\ D_{21}(L) & D_{22}(L) \end{pmatrix} \begin{pmatrix} x_t \\ z_t \end{pmatrix} = \begin{pmatrix} x_t^* \\ z_t^* \end{pmatrix} , \tag{17.37}$$

where $D_{11}(0) = 1$, $D_{22}(0) = 1$, $D_{12}(0) = 0$, $\mathrm{cov}(x_t^*, z_t^*) = 0$, and $D_{21}(0)$ is generally not zero. We note that $\mathrm{var}(x_t^*) = \Sigma_3$ and this becomes useful in what follows.

For Eqs. (17.33), (17.34), and (17.35) the normalization process involves left-multiplying both sides by the matrix

$$\mathbf{P} = \mathbf{P}_2 \cdot \mathbf{P}_1 \tag{17.38}$$

where

$$\mathbf{P}_1 = \begin{pmatrix} 1 & 0 & 0 \\ -\Sigma_{yx}\Sigma_{xx}^{-1} & 1 & 0 \\ -\Sigma_{zx}\Sigma_{xx}^{-1} & 0 & 1 \end{pmatrix} , \tag{17.39}$$

and

$$\mathbf{P}_2 = \begin{pmatrix} 1 & 0 & 0 \\ 0 & 1 & 0 \\ 0 & -(\Sigma_{zy} - \Sigma_{zx}\Sigma_{xx}^{-1}\Sigma_{xy})(\Sigma_{yy} - \Sigma_{yx}\Sigma_{xx}^{-1}\Sigma_{xy})^{-1} & 1 \end{pmatrix} . \tag{17.40}$$

We denote the normalized equations as

$$\begin{pmatrix} B_{11}(L) & B_{12}(L) & B_{13}(L) \\ B_{21}(L) & B_{22}(L) & B_{23}(L) \\ B_{31}(L) & B_{32}(L) & B_{33}(L) \end{pmatrix} \begin{pmatrix} x_t \\ y_t \\ z_t \end{pmatrix} = \begin{pmatrix} \epsilon_{xt} \\ \epsilon_{yt} \\ \epsilon_{zt} \end{pmatrix}, \tag{17.41}$$

where the noise terms are independent, and their respective variances are $\hat{\Sigma}_{xx}$, $\hat{\Sigma}_{yy}$, and $\hat{\Sigma}_{zz}$.

To proceed further we need the following important relation [4]:

$$F_{Y \to X|Z} = F_{YZ^* \to X^*} \tag{17.42}$$

and its frequency domain counterpart

$$f_{Y \to X|Z}(\omega) = f_{YZ^* \to X^*}(\omega). \tag{17.43}$$

To obtain $f_{YZ^* \to X^*}(\omega)$, we need to decompose the spectrum of X^*. The Fourier transform of Eqs. (17.37) and (17.41) gives

$$\begin{pmatrix} X(\omega) \\ Z(\omega) \end{pmatrix} = \begin{pmatrix} G_{xx}(\omega) & G_{xz}(\omega) \\ G_{zx}(\omega) & G_{zz}(\omega) \end{pmatrix} \begin{pmatrix} X^*(\omega) \\ Z^*(\omega) \end{pmatrix}, \tag{17.44}$$

and

$$\begin{pmatrix} X(\omega) \\ Y(\omega) \\ Z(\omega) \end{pmatrix} = \begin{pmatrix} H_{xx}(\omega) & H_{xy}(\omega) & H_{xz}(\omega) \\ H_{yx}(\omega) & H_{yy}(\omega) & H_{yz}(\omega) \\ H_{zx}(\omega) & H_{zy}(\omega) & H_{zz}(\omega) \end{pmatrix} \begin{pmatrix} E_x(\omega) \\ E_y(\omega) \\ E_z(\omega) \end{pmatrix}. \tag{17.45}$$

Assuming that $X(\omega)$ and $Z(\omega)$ from Eq. (17.44) can be equated with that from Eq. (17.45), we combine Eqs. (17.44) and (17.45) to yield

$$\begin{aligned} \begin{pmatrix} X^*(\omega) \\ Y(\omega) \\ Z^*(\omega) \end{pmatrix} &= \begin{pmatrix} G_{xx}(\omega) & 0 & G_{xz}(\omega) \\ 0 & 1 & 0 \\ G_{zx}(\omega) & 0 & G_{zz}(\omega) \end{pmatrix}^{-1} \begin{pmatrix} H_{xx}(\omega) & H_{xy}(\omega) & H_{xz}(\omega) \\ H_{yx}(\omega) & H_{yy}(\omega) & H_{yz}(\omega) \\ H_{zx}(\omega) & H_{zy}(\omega) & H_{zz}(\omega) \end{pmatrix} \\ &\quad \cdot \begin{pmatrix} E_x(\omega) \\ E_y(\omega) \\ E_z(\omega) \end{pmatrix} \\ &= \begin{pmatrix} Q_{xx}(\omega) & Q_{xy}(\omega) & Q_{xz}(\omega) \\ Q_{yx}(\omega) & Q_{yy}(\omega) & Q_{yz}(\omega) \\ Q_{zx}(\omega) & Q_{zy}(\omega) & Q_{zz}(\omega) \end{pmatrix} \begin{pmatrix} E_x(\omega) \\ E_y(\omega) \\ E_z(\omega) \end{pmatrix}, \end{aligned} \tag{17.46}$$

where $\mathbf{Q}(\omega) = \mathbf{G}^{-1}(\omega)\mathbf{H}(\omega)$. After suitable ensemble averaging, the spectral matrix can be obtained from which the power spectrum of X^* is found to be

$$S_{x^*x^*}(\omega) = Q_{xx}(\omega)\hat{\Sigma}_{xx}Q^*_{xx}(\omega) + Q_{xy}(\omega)\hat{\Sigma}_{yy}Q^*_{xy}(\omega) + Q_{xz}(\omega)\hat{\Sigma}_{zz}Q^*_{xz}(\omega). \tag{17.47}$$

The first term can be thought of as the intrinsic power and the remaining two terms as the combined causal influences from Y to Z^*. This interpretation leads immediately to the definition

$$f_{YZ^* \to X^*}(\omega) = \ln \frac{|S_{x^*x^*}(\omega)|}{|Q_{xx}(\omega)\hat{\Sigma}_{xx}Q^*_{xx}(\omega)|} . \tag{17.48}$$

We note that $S_{x^*x^*}(\omega)$ is actually the variance of ϵ_{3t} as pointed out earlier. On the basis of the relation in Eq. (17.43), the final expression for Granger causality from Y_t to X_t conditional on Z_t is

$$f_{Y \to X|Z}(\omega) = \ln \frac{\Sigma_3}{|Q_{xx}(\omega)\hat{\Sigma}_{xx}Q^*_{xx}(\omega)|} . \tag{17.49}$$

It can be shown that $f_{Y \to X|Z}(\omega)$ relates to the time domain measure $F_{Y \to X|Z}$ via

$$F_{Y \to X|Z} = \frac{1}{2\pi} \int_{-\pi}^{\pi} f_{Y \to X|Z}(\omega)\, d\omega ,$$

under general conditions.

The above derivation is made possible by the key assumption that $X(\omega)$ and $Z(\omega)$ in Eqs. (17.44) and (17.45) are identical. This certainly holds true on purely theoretical grounds, and it may very well be true for simple mathematical systems. For actual physical data, however, this condition may be very hard to satisfy due to practical estimation errors. In a recent paper we developed a partition matrix technique to overcome this problem [6]. The subsequent calculations of conditional Granger causality are based on this partition matrix procedure.

17.4 Estimation of Autoregressive Models

The preceding theoretical development assumes that the time series can be well represented by autoregressive processes. Such theoretical autoregressive processes have infinite model orders. Here we discuss how to estimate autoregressive models from empirical time series data, with emphasis on the incorporation of multiple time series segments into the estimation procedure [7]. This consideration is motivated by the goal of applying autoregressive modeling in neuroscience. It is typical in behavioral and cognitive neuroscience experiments for the same event to be repeated on many successive trials. Under appropriate conditions, time series data recorded from these repeated trials may be viewed as realizations of a common underlying stochastic process.

Let $\mathbf{X}_t = [X_{1t}, X_{2t}, \ldots, X_{pt}]^T$ be a p-dimensional random process. Here T denotes the matrix transposition. In multivariate neural data, p represents the total number of recording channels. Assume that the process $\mathbf{X}_t$ is stationary and can be described by the following mth order autoregressive equation:

$$\mathbf{X}_t + \mathbf{A}(1)\mathbf{X}_{t-1} + \cdots + \mathbf{A}(m)\mathbf{X}_{t-m} = \mathbf{E}_t , \tag{17.50}$$

where $\mathbf{A}(i)$ are $p \times p$ coefficient matrices and $\mathbf{E}_t = [E_{1t}, E_{2t}, \ldots, E_{pt}]^T$ is a zero mean uncorrelated noise vector with the covariance matrix $\mathbf{\Sigma}$.

To estimate $\mathbf{A}(i)$ and $\mathbf{\Sigma}$, we multiply Eq. (17.50) from the right by $\mathbf{X}_{t-k}^T$, where $k = 1, 2, \ldots, m$. Taking expectations, we obtain the Yule–Walker equations

$$\mathbf{R}(-k) + \mathbf{A}(1)\mathbf{R}(-k+1) + \cdots + \mathbf{A}(m)\mathbf{R}(-k+m) = 0\,, \tag{17.51}$$

where $\mathbf{R}(n) = \langle \mathbf{X}_t \mathbf{X}_{t+n}^T \rangle$ is $\mathbf{X}_t$'s covariance matrix of lag n. In deriving these equations, we have used the fact that $\langle \mathbf{E}_t \mathbf{X}_{t-k}^T \rangle = 0$ as a result of $\mathbf{E}_t$ being an uncorrelated process.

For a single realization of the $\mathbf{X}$ process, $\{\mathbf{x}_i\}_{i=1}^N$, we compute the covariance matrix in Eq. (17.51) according to

$$\tilde{\mathbf{R}}(n) = \frac{1}{N-n} \sum_{i=1}^{N-n} \mathbf{x}_i \mathbf{x}_{i+n}^T\,. \tag{17.52}$$

If multiple realizations of the same process are available, then we compute the above quantity for each realization, and average across all the realizations to obtain the final estimate of the covariance matrix. Note that for a single short trial of data one uses the divisor N for evaluating covariance to reduce inconsistency. Due to the availability of multiple trials in neural applications, we have used the divisor $(N-n)$ in the above definition, Eq. (17.52), to achieve an unbiased estimate. It is quite clear that, for a single realization, if N is small, one will not get good estimates of $\mathbf{R}(n)$ and hence will not be able to obtain a good model. This problem can be overcome if a large number of realizations of the same process is available. In this case the length of each realization can be as short as the model order m plus 1. Equation (17.50) contain a total of mp^2 unknown model coefficients. In Eq. (17.51) there is exactly the same number of simultaneous linear equations. One can simply solve these equations to obtain the model coefficients. An alternative approach is to use the Levinson, Wiggins, Robinson (LWR) algorithm, which is a more robust solution procedure based on the ideas of maximum entropy. This algorithm was implemented in the analysis of neural data described below. The noise covariance matrix $\mathbf{\Sigma}$ may be obtained as part of the LWR algorithm. Otherwise one may obtain $\mathbf{\Sigma}$ through

$$\mathbf{\Sigma} = \mathbf{R}(0) + \sum_{i=1}^{m} \mathbf{A}(i)\mathbf{R}(i)\,. \tag{17.53}$$

Here we note that $\mathbf{R}^T(k) = \mathbf{R}(-k)$.

The above estimation procedure can be carried out for any model order m. The correct m is usually determined by minimizing the Akaike Information Criterion (AIC) defined as

$$\mathrm{AIC}(m) = 2\log[\det(\mathbf{\Sigma})] + \frac{2p^2 m}{N_{\mathrm{total}}} \tag{17.54}$$

where N_{total} is the total number of data points from all the trials. Plotted as a function of m the proper model order corresponds to the minimum of this function. It is often the case that for neurobiological data N_{total} is very large. Consequently, for a reasonable range of m, the AIC function does not achieve a minimum. An alternative criterion is the Bayesian Information Criterion (BIC), which is defined as

$$\mathrm{BIC}(m) = 2\log[\det(\mathbf{\Sigma})] + \frac{2p^2 m \log N_{total}}{N_{total}} . \tag{17.55}$$

This criterion can compensate for the large number of data points and may perform better in neural applications. A final step, necessary for determining whether the autoregressive time series model is suited for a given data set, is to check whether the residual noise is white. Here the residual noise is obtained by computing the difference between the model's predicted values and the actually measured values.

Once an autoregressive model is adequately estimated, it becomes the basis for both time domain and spectral domain causality analysis. Specifically, in the spectral domain, Eq. (17.50) can be written as

$$\mathbf{X}(\omega) = \mathbf{H}(\omega)\mathbf{E}(\omega), \tag{17.56}$$

where

$$\mathbf{H}(\omega) = \left(\sum_{j=0}^{m} \mathbf{A}(j) e^{-i\omega j} \right)^{-1} \tag{17.57}$$

is the transfer function with $\mathbf{A}(0)$ being the identity matrix. From Eq. (17.56), after proper ensemble averaging, we obtain the spectral matrix

$$\mathbf{S}(\omega) = \mathbf{H}(\omega)\mathbf{\Sigma}\mathbf{H}^*(\omega) . \tag{17.58}$$

Once we obtain the transfer function, the noise covariance, and the spectral matrix, we can then carry out causality analysis according to the procedures outlined in the previous sections.

17.5 Numerical Examples

In this section we consider three examples that illustrate various aspects of the general approach outlined earlier.

17.5.1 Example 1

Consider the following AR(2) model

$$\begin{aligned} X_t &= 0.9X_{t-1} - 0.5X_{t-2} + \epsilon_t \\ Y_t &= 0.8Y_{t-1} - 0.5Y_{t-2} + 0.16X_{t-1} - 0.2X_{t-2} + \eta_t \end{aligned} \tag{17.59}$$

,

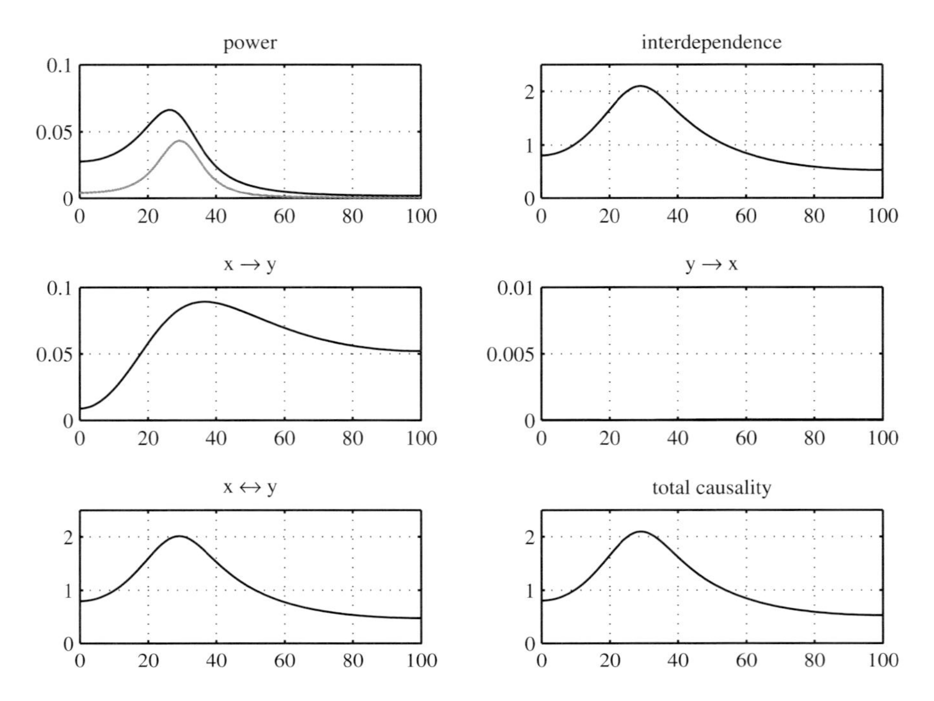

Fig. 17.2: Simulation results for an AR(2) model consisting of two coupled time series. Power (black for X, gray for Y) spectra, interdependence spectrum (related to the coherence spectrum), and Granger causality spectra are displayed. Note that the total causality spectrum, representing the sum of directional causalities and the instantaneous causality, is nearly identical to the interdependence spectrum.

where ϵ_t, η_t are Gaussian white noise processes with zero means and variances $\sigma_1^2 = 1$, $\sigma_2^2 = 0.7$, respectively. The covariance between ϵ_t and η_t is 0.4. From the construction of the model, we can see that X_t has a causal influence on Y_t and that there is also instantaneous causality between X_t and Y_t.

We simulated Eq. (17.59) to generate a data set of 500 realizations of 100 time points each. Assuming no knowledge of Eq. (17.59) we fitted a MVAR model on the generated data set and calculated power, coherence, and Granger causality spectra. The result is shown in Fig. 17.2. The interdependence spectrum is computed according to Eq. (17.17) and the total causality is defined as the sum of directional causalities and the instantaneous causality. The result clearly recovers the pattern of connectivity in Eq. (17.59). It also illustrates that the interdependence spectrum, as computed according to Eq. (17.17), is almost identical to the total causality spectrum as defined on the right-hand side of Eq. (17.28).

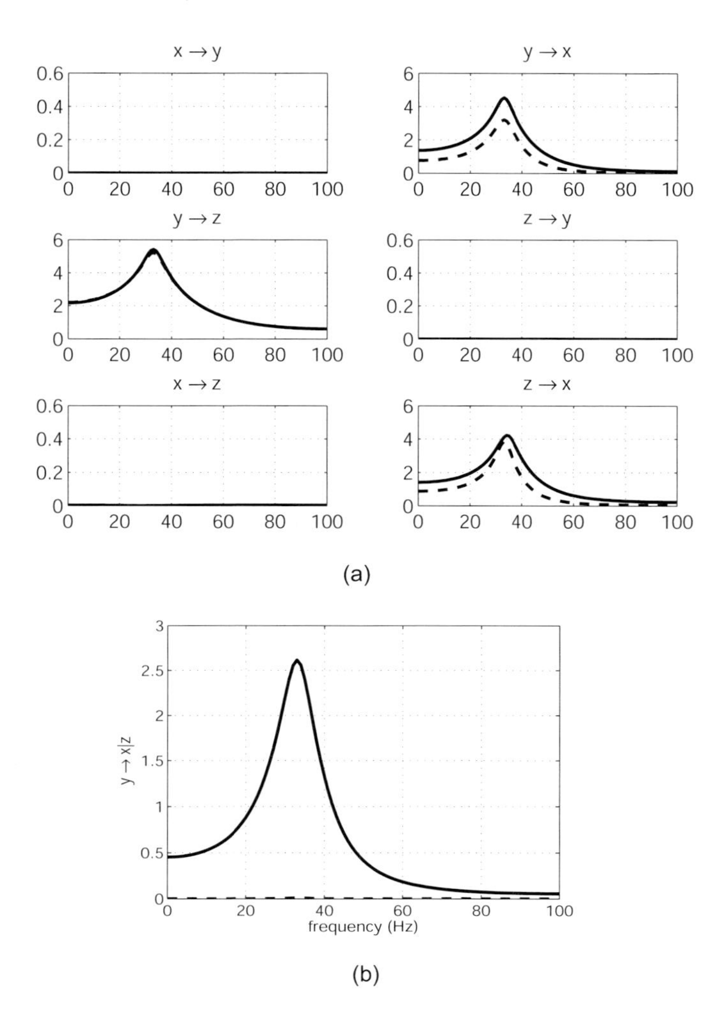

Fig. 17.3: Simulation results for three coupled time series. Two distinct patterns of connectivity as that illustrated in Fig. 17.1 are considered. Results for the case with a direct causal influence are shown as solid curves and the results for the case with indirect causal influence are shown as dashed curves. (a) Pairwise Granger causality analysis gives very similar results for both cases which indicates that the pairwise analysis cannot differentiate these two patterns of connectivity. (b) Conditional causality analysis shows a nonzero spectrum (solid) for the direct case and almost zero spectrum (dashed) for the indirect case.

17.5.2 Example 2

Here we consider two models. The first consists of three time series simulating the case shown in Fig. 17.1(a), in which the causal influence from Y_t to X_t is indirect and completely mediated by Z_t

$$\begin{aligned} X_t &= 0.8X_{t-1} - 0.5X_{t-2} + 0.4Z_{t-1} + \epsilon_t \\ Y_t &= 0.9Y_{t-1} - 0.8Y_{t-2} + \xi_t \\ Z_t &= 0.5Z_{t-1} - 0.2Z_{t-2} + 0.5Y_{t-1} + \eta_t \,. \end{aligned} \tag{17.60}$$

The second model creates a situation corresponding to Fig. 17.1(b), containing both direct and indirect causal influences from Y_t to X_t. This is achieved by using the same system as in Eq. (17.60), but with an additional term in the first equation

$$\begin{aligned} X_t &= 0.8X_{t-1} - 0.5X_{t-2} + 0.4Z_{t-1} + 0.2Y_{t-2} + \epsilon_t \\ Y_t &= 0.9Y_{t-1} - 0.8Y_{t-2} + \xi_t \\ Z_t &= 0.5Z_{t-1} - 0.2Z_{t-2} + 0.5Y_{t-1} + \eta_t \,. \end{aligned} \tag{17.61}$$

For both models. $\epsilon(t)$, $\xi(t)$, $\eta(t)$ are three independent Gaussian white noise processes with zero means and variances of $\sigma_1^2 = 0.3$, $\sigma_2^2 = 1$, $\sigma_3^2 = 0.2$, respectively.

Each model was simulated to generate a data set of 500 realizations of 100 time points each. First, pairwise Granger causality analysis was performed on the simulated data set of each model. The results are shown in Fig. 17.3(a), with the dashed curves showing the results for the first model and the solid curves for the second model. From these plots it is clear that pairwise analysis cannot differentiate the two coupling schemes. This problem occurs because the indirect causal influence from Y_t to X_t that depends completely on Z_t in the first model cannot be clearly distinguished from the direct influence from Y_t to X_t in the second model. Next, conditional Granger causality analysis was performed on both simulated data sets. The Granger causality spectra from Y_t to X_t conditional on Z_t are shown in Fig. 17.3(b), with the second model's result shown as the solid curve and the first model's result as the dashed curve. Clearly, the causal influence from Y_t to X_t that was prominent in the pairwise analysis of the first model in Fig. 17.3(a), is no longer present in Fig. 17.3(b). Thus, by correctly determining that there is no direct causal influence from Y_t to X_t in the first model, the conditional Granger causality analysis provides an unambiguous dissociation of the coupling schemes represented by the two models.

17.5.3 Example 3

We simulated a five-node oscillatory network structurally connected with different delays. This example has been analyzed with partial directed coherence and directed transfer function methods in [8]. The network involves the following multivariate autoregressive model:

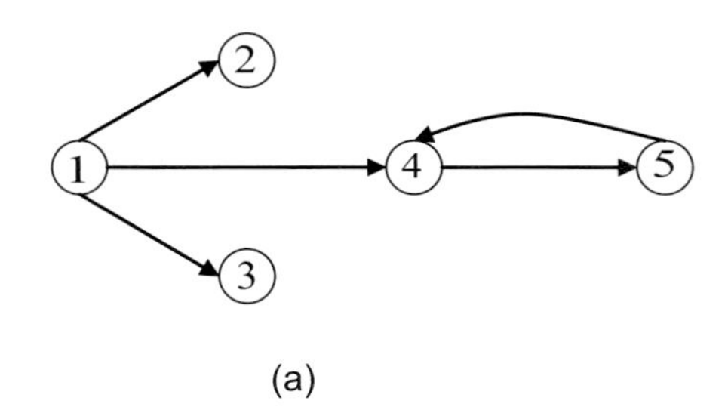

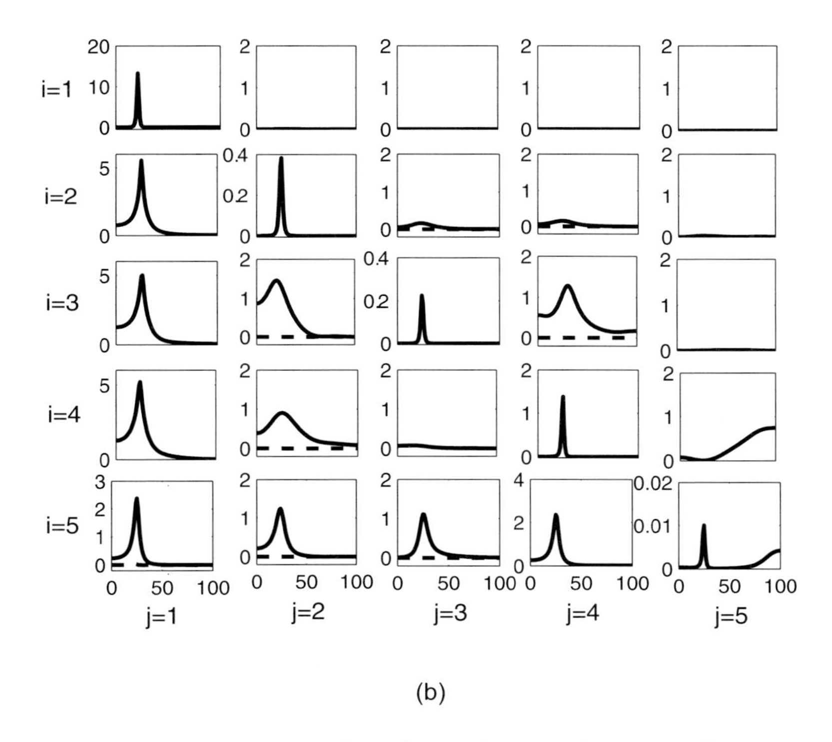

Fig. 17.4: Simulation results for a five-node network structurally connected with different time delays. (a) Schematic illustration of the system. (b) Calculated power spectra are shown in the diagonal panels, results of pairwise (solid) and conditional Granger causality analysis (dashed) are in off-diagonal panels. Granger causal influence is from the horizontal index to the vertical index. Features of Granger causality spectra (both pairwise and conditional) are consistent with that of power spectra.

$$\begin{aligned}
X_{1t} &= 0.95\sqrt{2}X_{1(t-1)} - 0.9025X_{1(t-2)} + \epsilon_{1t} \\
X_{2t} &= 0.5X_{1(t-2)} + \epsilon_{2t} \\
X_{3t} &= -0.4X_{1(t-3)} + \epsilon_{3t} \\
X_{4t} &= -0.5X_{1(t-2)} + 0.25\sqrt{2}X_{4(t-1)} + 0.25\sqrt{2}X_{5(t-1)} + \epsilon_{4t} \\
X_{5t} &= -0.25\sqrt{2}X_{4(t-1)} + 0.25\sqrt{2}X_{5(t-1)} + \epsilon_{5t} \,,
\end{aligned} \tag{17.62}$$

where $\epsilon_{1t}, \epsilon_{2t}, \epsilon_{3t}, \epsilon_{4t}, \epsilon_{5t}$ are independent Gaussian white noise processes with zero means and variances of $\sigma_1^2 = 0.6$, $\sigma_2^2 = 0.5$, $\sigma_3^2 = 0.3$, $\sigma_4^2 = 0.3$, $\sigma_5^2 = 0.6$, respectively. The structure of the network is illustrated in Fig. 17.4(a).

We simulated the network model to generate a data set of 500 realizations each with ten time points. Assuming no knowledge of the model, we fitted a fifth order MVAR model on the generated data set and performed power spectra, coherence, and Granger causality analysis on the fitted model. The results of power spectra are given in the diagonal panels of Fig. 17.4(b). It is clearly seen that all five oscillators have a spectral peak at around 25 Hz and the fifth has some additional high frequency activity as well. The results of pairwise Granger causality spectra are shown in the off-diagonal panels of Fig. 17.4(b) (solid curves). Compared to the network diagram in Fig. 17.4(a) we can see that pairwise analysis yields connections that can be the result of direct causal influences (e.g., $1 \rightarrow 2$), indirect causal influences (e.g., $1 \rightarrow 5$) and differentially delayed driving inputs (e.g., $2 \rightarrow 3$). We further performed a conditional Granger causality analysis in which the direct causal influence between any two nodes are examined while the influences from the other three nodes are conditioned out. The results are shown as dashed curves in Fig. 17.4(b). For many pairs the dashed curves and solid curves coincide (e.g., $1 \rightarrow 2$), indicating that the underlying causal influence is direct. For other pairs the dashed curves become zero, indicating that the causal influences in these pairs are either indirect or are the result of differentially delayed inputs. These results demonstrate that conditional Granger causality furnishes a more precise network connectivity diagram that matches the known structural connectivity. One noteworthy feature about Fig. 17.4(b) is that the spectral features (e.g., peak frequency) are consistent across both power and Granger causality spectra. This is important since it allows us to link local dynamics with that of the network.

17.6 Analysis of a Beta Oscillation Network in Sensorimotor Cortex

A number of studies have appeared in the neuroscience literature where the issue of causal effects in neural data is examined [6, 8–15]. Three of these studies [10, 11, 15] used the measures presented in this article. Below we review one study published by our group [6, 15].

Local field potential data were recorded from two macaque monkeys using transcortical bipolar electrodes at 15 distributed sites in multiple cortical areas of one hemisphere (the right hemisphere in monkey GE and the left hemisphere in monkey LU) while the monkeys performed a GO/NO–GO visual pattern discrimination task [16]. The prestimulus stage began when the monkey depressed a hand lever while monitoring a display screen. This was followed from 0.5 s to 1.25 s later by the appearance of a visual stimulus (a four-dot pattern) on the screen. The monkey made a GO response (releasing the lever) or a NO–GO response (maintaining lever depression) depending on the stimulus category and the session contingency. The entire trial lasted about 500 ms, during which the local field potentials were recorded at a sampling rate of 200 Hz.

Previous studies have shown that synchronized beta-frequency (15 Hz to 30 Hz) oscillations in the primary motor cortex are involved in maintaining steady contractions of contralateral arm and hand muscles. Relatively little is known, however, about the role of postcentral cortical areas in motor maintenance and their patterns of interaction with motor cortex. Making use of the simultaneous recordings from distributed cortical sites we investigated the interdependency relations of beta-synchronized neuronal assemblies in pre- and postcentral areas in the prestimulus time period. Using power and coherence spectral analysis, we first identified a beta-synchronized large-scale network linking pre- and postcentral areas. We then used Granger causality spectra to measure directional influences among recording sites, ascertaining that the dominant causal influences occurred in the same part of the beta-frequency range as indicated by the power and coherence analysis. The patterns of significant beta-frequency Granger causality are summarized in the schematic Granger causality graphs shown in Fig. 17.5. These patterns reveal that, for both monkeys, strong Granger causal influences occurred from the primary somatosensory cortex (S1) to both the primary motor cortex (M1) and inferior posterior parietal cortex (7*a* and 7b), with the latter areas also exerting Granger causal influences on the primary motor cortex. Granger causal influences from the motor cortex to postcentral areas, however, were not observed.[2]

Our results are the first to demonstrate in awake monkeys that synchronized beta oscillations not only bind multiple sensorimotor areas into a large-scale network during motor maintenance behavior, but also carry Granger causal influences from primary somatosensory and inferior posterior parietal cortices to motor cortex. Furthermore, the Granger causality graphs in Fig. 17.5 provide a basis for fruitful speculation about the functional role of each cortical area in the sensorimotor network. First, steady pressure maintenance is akin to a closed-loop-control problem and as such, sensory feedback is expected to provide critical input needed for cortical assessment of the current state of the behavior. This notion is consistent with our observation that primary somatosensory area (S1) serves as the dominant source of causal influences to other areas in the network. Second, posterior parietal area 7b is known to be involved in non-visually guided movement. As a higher-order association area it may maintain representations relating to the current goals of the motor system. This would imply that area 7b receives sensory updates from area S1 and outputs correctional signals to the motor cortex (M1). This conceptualization is consistent with the causality patterns in Fig. 17.5. As mentioned earlier, previous work has identified beta range oscillations in the motor cortex as an important neural correlate of

[2] A more stringent significance threshold was applied here which resulted in elimination of several very small causal influences that were included in the previous report.

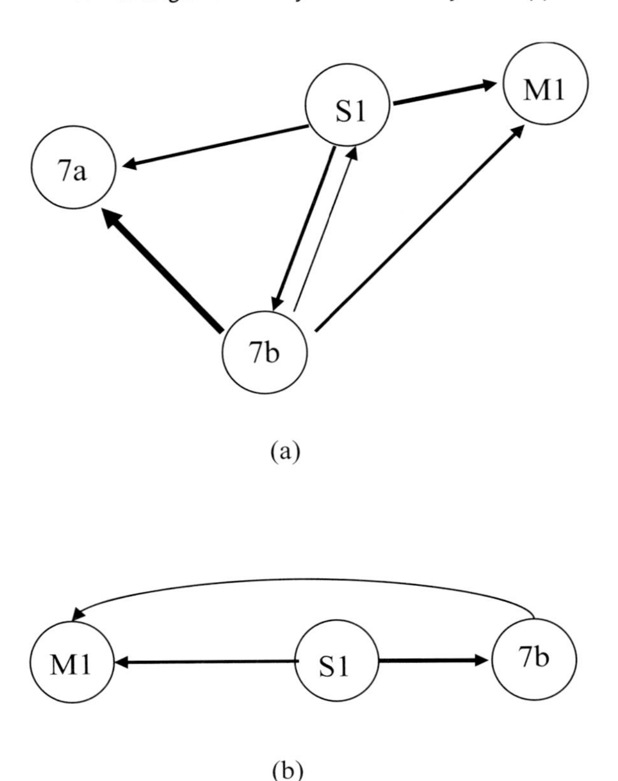

Fig. 17.5: Granger causality graphs for monkey GE (a) and monkey LU (b).

pressure maintenance behavior. The main contribution of our work is to demonstrate that the beta network exists on a much larger scale and that postcentral areas play a key role in organizing the dynamics of the cortical network. The latter conclusion is made possible by the directional information provided by Granger causality analysis.

Since the above analysis was pairwise, it had the disadvantage of not distinguishing between direct and indirect causal influences. In particular, in monkey GE, the possibility existed that the causal influence from area S1 to inferior posterior parietal area $7a$ was actually mediated by inferior posterior parietal area $7b$ (Fig. 17.5(a)). We used the conditional Granger causality to test the hypothesis that the $S1 \to 7a$ influence was mediated by area $7b$. In Fig. 17.6(a) is presented the pairwise Granger causality spectrum from S1 to $7a$ ($S1 \to 7a$, dark solid curve), showing significant causal influence in the beta-frequency range. Superimposed in Fig. 17.6(a) is the conditional Granger causality spectrum for the same pair, but with area $7b$ taken into account ($S1 \to 7a \mid 7b$, light solid curve). The corresponding 99 % significance thresholds are also presented (light and dark dashed lines coincide). These significance thresholds were determined using a

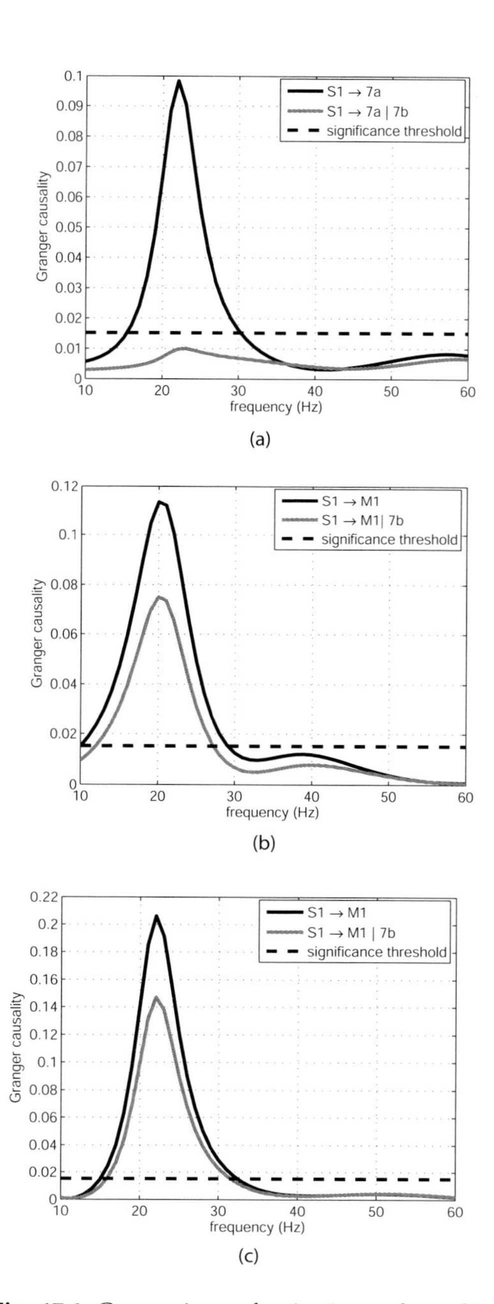

Fig. 17.6: Comparison of pairwise and conditional Granger causality spectra for monkey GE (a,b), and monkey LU (c).

permutation procedure that involved creating 500 permutations of the local field potential data set by random rearrangement of the trial order independently for each channel (site). Since the test was performed separately for each frequency, a correction was necessary for the multiple comparisons over the whole range of frequencies. The Bonferroni correction could not be employed because these multiple comparisons were not independent. An alternative strategy was employed following Blair and Karniski [17]. The Granger causality spectrum was computed for each permutation, and then the maximum causality value over the frequency range was identified. After 500 permutation steps, a distribution of maximum causality values was created. Choosing a p-value at $p = 0.01$ for this distribution gave the thresholds shown in Fig. 17.6(a)–(c) as the dashed lines.

We see from Fig. 17.6(a) that the conditional Granger causality is greatly reduced in the beta-frequency range and no longer significant, meaning that the causal influence from S1 to 7a is most likely an indirect effect mediated by 7b. This conclusion is consistent with the known neuroanatomy of the sensorimotor cortex [18] in which area 7a receives direct projections from area 7b which in turn receives direct projections from the primary somatosensory cortex. No pathway is known to project directly from the primary somatosensory cortex to area 7a.

From Fig. 17.5(a) we see that the possibility also existed that the causal influence from S1 to the primary motor cortex (M1) in monkey GE was mediated by area 7b. To test this possibility, the Granger causality spectrum from S1 to M1 ($S1 \rightarrow M1$, dark solid curve in Fig. 17.6(b)) was compared with the conditional Granger causality spectrum with 7b taken into account ($S1 \rightarrow M1 \mid 7b$, light solid curve in Fig. 17.6(b)). In contrast to Fig. 17.6(a), we see that the beta-frequency conditional Granger causality in Fig. 17.6(b) is only partially reduced, and remains well above the 99 % significance level. From Fig. 17.4(b), we see that the same possibility existed in monkey LU of the S1 to M1 causal influence being mediated by 7b. However, just as in Fig. 17.6(b), we see in Fig. 17.6(c) that the beta-frequency conditional Granger causality for monkey LU is only partially reduced, and remains well above the 99 % significance level.

The results from both the monkeys thus indicate that the observed Granger causal influence from the primary somatosensory cortex to the primary motor cortex was not simply an indirect effect mediated by area 7b. However, we further found that area 7b did play a role in mediating the S1 to M1 causal influence in both the monkeys. This was determined by comparing the means of bootstrap resampled distributions of the peak beta Granger causality values from the spectra of $S1 \rightarrow M1$ and $S1 \rightarrow M1 \mid 7b$ by the Student's t-test. The significant reduction of beta-frequency Granger causality when area 7b is taken into account ($t = 17.2$ for GE; $t = 18.2$ for LU, $p \ll 0.001$ for both), indicates that the influence from the primary somatosensory to primary motor area was partially mediated by area 7b. Such an influence is consistent with the known neuroanatomy [18] where the primary somatosensory area projects directly to both the motor cortex and area 7b, and area 7b projects directly to primary motor cortex.

17.7 Summary

In this chapter we have introduced the mathematical formalism for estimating Granger causality in both the time and spectral domain from time series data. Demonstrations of the technique's utilities are carried out both on simulated data, where the patterns of interactions are known, and on local field potential recordings from monkeys performing a cognitive task. For the latter we have stressed the physiological interpretability of the findings and pointed out the new insights afforded by these findings. It is our belief that Granger causality offers a new way of looking at cooperative neural computation and it enhances our ability to identify key brain structures underlying the organization of a given brain function.

Acknowledgements

This work was supported by NIMH grant MH071620.

References

[1] N. Wiener. The theory of prediction. In E. F. Beckenbach, editor, *Modern Mathematics for Engineers*, chap. 8. McGraw-Hill, New York, 1956.

[2] C. W. J. Granger. *Econometrica*, 37:424, 1969.

[3] J. Geweke. *J. Am. Stat. Assoc.*, 77:304, 1982.

[4] J. Geweke. *J. Am. Stat. Assoc.*, 79:907, 1984.

[5] C. Gourierous and A. Monfort. *Time Series and Dynamic Models*. Cambridge University Press, London, 1997.

[6] Y. Chen, S. L. Bressler, and M. Ding. *J. Neurosci. Methods*, 150:228, 2006.

[7] M. Ding, S. L. Bressler, W. Yang, and H. Liang. *Biol. Cybern.*, 83:35, 2000.

[8] L. A. Baccala and K. Sameshima. *Biol. Cybern.*, 84:463, 2001.

[9] W. A. Freiwald, P. Valdes, J. Bosch, et al. *J. Neurosci. Methods*, 94:105, 1999.

[10] C. Bernasconi and P. Konig. *Biol. Cybern.*, 81:199, 1999.

[11] C. Bernasconi, A. von Stein, C. Chiang, and P. Konig. *Neuroreport*, 11:689, 2000.

[12] M. Kaminski, M. Ding, W. A. Truccolo, and S. L. Bressler. *Biol. Cybern.*, 85: 145, 2001.

[13] R. Goebel, A. Roebroek, D. Kim, and E. Formisano. *Magn. Res. Imag.*, 21: 1251, 2003.

[14] W. Hesse, E. Moller, M. Arnold, and B. Schack. *J. Neurosci. Methods*, 124:27, 2003.

[15] A. Brovelli, M. Ding, A. Ledberg, Y. Chen, R. Nakamura, and S. L. Bressler. *Proc. Natl. Acad. Sci.*, 101:9849, 2004.

[16] S. L. Bressler, R. Coppola, and R. Nakamura. *Nature*, 366:153, 1993.

[17] R. C. Blair and W. Karniski. *Psychophysiol.*, 30:518, 1993.

[18] D. J. Felleman and D. C. V. Essen. *Cereb. Cortex*, 1:1, 1991.

18 Granger Causality on Spatial Manifolds: Applications to Neuroimaging

Pedro A. Valdés-Sosa, Jose Miguel Bornot-Sánchez, Mayrim Vega-Hernández, Lester Melie-García, Agustin Lage-Castellanos, and Erick Canales-Rodríguez

The (discrete time) vector multivariate autoregressive (MAR) model is generalized as a stochastic process defined over a continuous spatial manifold. The underlying motivation is the study of brain connectivity via the application of Granger causality measures to functional Neuroimages. Discretization of the spatial MAR (sMAR) leads to a densely sampled MAR for which the number of time series p is much larger than the length of the time series N. In this situation usual time series models work badly or fail. Previous approaches, reviewed here, involve the reduction of the dimensionality of the MAR, either by the selection of arbitrary regions of interest or by latent variable analysis. An example of the latter is given using a multilinear reduction of the multichannel EEG spectrum into atoms with spatial, temporal, and frequency signatures. Influence measures are applied to the temporal signatures giving an interpretation of the interaction between brain rhythms. However, the approach introduced here is that of extending usual influence measures for Granger causality to sMAR by defining "influence fields," that is the set of influence measures from one site (voxel) to the whole manifold. Estimation is made possible by imposing Bayesian priors for sparsity, smoothness, or both on the influence fields. In fact, a prior is introduced that generalizes most common priors studied to date in the literature for variable selection and penalization in regression. This prior is specified by defining penalties paired with *a priori* covariance matrices. Simple pairs of penalties/covariances include as particular cases the LASSO, data fusion and Ridge regression. Double pairs encompass the recently introduced Elastic Net and Fussed Lasso. Quadruples of penalty/covariance combinations are also possible and used here for the first time. Estimation is carried out via the MM algorithm, a new technique that generalized the EM algorithm and allows efficient estimation even for massive time series dimensionalities. The proposed technique performs adequately for a simulated "small world" cortical network with linear dynamics, validating the use of the more complex penalties. Application of this model to fMRI data validate previous conceptual models for the brain circuits involved in the generation of the EEG alpha rhythm.

Handbook of Time Series Analysis. Björn Schelter, Matthias Winterhalder, Jens Timmer

ISBN: 3-527-40623-9

18.1 Introduction

Devising methods for inferring the effective and functional connectivity of different brain regions is currently a major concern in Neuroimaging [1]. The task is to determine the changing patterns of causal influences that different neural structures exert on each other. This is to be done by the analysis of dynamical brain imaging data. This type of data include EEG/MEG source distributions, optical recordings [2] and fMRI [3] which are, from the statistical point of view, spatiotemporal data sets [4, 5]—that is time series sampled from an underlying continuous manifold Ω of spatial points. Multivariate autoregressive models (in particular linear ones) for vector time series have proven to be an essential and informative tool for the applied sciences. Within this framework Granger [6] formulated a definition of causality between time series that has been pursued extensively in many fields and especially in the neurosciences [7, 8].

It is striking though that work in this field has been limited to vector valued time series in which the dimension p is very small [9, 10]—even if, as usual in real applications, the number N of time samples gathered is large. As Granger himself pointed out, his definition of causality would be valid only if all relevant variables would be included in the analysis, a formidable task that is readily appreciated by neuroscientists since they study the brain, which is the complex system by excellence. We have therefore directed our attention to multivariate autoregressive models (MAR) defined over spatial manifolds (a particular example of which is the brain) and to deal with the issue of densely sampled (high dimensional, highly correlated) time series that arise from a discretization of an underlying spatial continuum into voxels [11].

As a concrete example, which will be used throughout the paper, consider the concurrent EEG and fMRI time series gathered in order to analyze the origin of resting brain rhythms [12–14]. The acquisition paradigm is described in more detail in Section 18.8. Structured patterns of correlations have been found between time-varying spectral components in different EEG bands and the BOLD signal at different voxels. These reveal widely distributed anatomical systems apparently involved in the generation of these oscillations (see Figs. 18.1–18.6). Here $N = 108$, the number of EEG time series is only 16, but the number of fMRI time series is 12 640! The usual MAR model cannot be fit to this amount of data.

The approach explored in this chapter follows the strategy of Functional Data Analysis [15]. Quantities of interest in the spatial MAR (autoregressive coefficients) are estimated subject to constraints that make anatomical and physiological sense. They not only allow inference for densely sampled data, but also dovetail nicely with computational shortcuts that make the proposed procedures feasible. In classical MAR models, Granger causality of one set of time series on another set is quantified by means of influence measures [16, 17]. In the linear case, these influence measures are usually multivariate tests that certain regression coefficients are zero. In our spatial MAR (sMAR) we extend this concept to that of an influence field. For functional Neuroimages, these are topographic

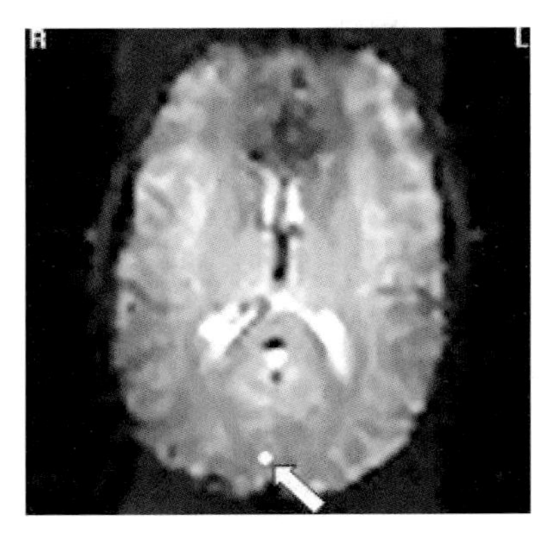

Fig. 18.1: MRI image as an example of a brain manifold. EPI MRI image of the brain of a subject from [18]. The MRI section is at a level that passes through the striate or primary visual cortex (VC). The arrow marks the voxel in VC for which the BOLD response during alpha rhythm shows the highest correlation with the power in that band.

maps of the influence of one brain site (voxel) on rest of the brain. For example in the concurrent EEG-fMRI experiment just mentioned one is interested to know what influence a site in the visual cortex (Figure 18.1) might have on all the rest of the brain.

For this type of situation classical multivariate testing is difficult or fails. We propose rather to apply the massive univariate approach that is at the heart of Statistical Parametric Mapping (SPM) [19]. SPM essentially calculates a (uni- or multivariate) statistic at each voxel of a brain image and then determines significantly activated regions by means of procedures that control the type I error. The latter is achieved either by the use of Random Field Theory [19], resampling methods [20], or the use of the False Discovery rate (FDR) [21]. We propose to evaluate a spatial extension of Granger causality by a SPM of influence fields. In effect, we are interested in detecting significant regions in the Cartesian product set $\Omega \times \Omega$. An alternative to using ordinary multivariate regression techniques for this situation is to attempt a huge multivariate regression problem and associated testing of the regression coefficients. To be able to do so we shall work with regression based on penalization in the spirit of Functional Data Analysis (FDA) [15]. This approach drastically reduce the number of "effective" connections to be determined. This was the approach taken in [22] by introducing a FDA variant of MAR modeling that imposed spatial smoothness on the influence field. Massive data reduction was achieved by means of the singular value decomposition and this paper showed the feasibility of working in the $p > N$ situation. A subsequent paper [23] also used penalized regression, in this case introducing sparse multivariate autoregressive models. The latter can be estimated in a two stage process involving: (1) penalized regression and (2) pruning of unlikely connections by means of the local false discovery rate developed by Efron. Extensive simulations were performed with idealized cortical networks having small world topologies and stable dynamics. These show that the detection efficiency of connections of the proposed procedure is quite high. Furthermore, the sparsity or conditional independence did not have to be specified *a priori* but is disclosed automatically by an iterative process. In short, we use the fact that the brain is sparsely connected as part of the solution, as opposed to treating as a specification problem. This chapter unifies the two approaches—spatial smoothness and sparseness in a much more general framework.

18.2 The Continuous Spatial Multivariate Autoregressive Model and its Discretization

We shall be dealing with the following spatial multivariate autoregressive (sMAR) model defined in discrete time

$$y(s,t) = \sum_{k=1}^{r} \iiint_{\Omega} a_k(s,u)\, y(u,t-k)\, du + e(s,t), \tag{18.1}$$

where $y(s,t)$ is the variable of interest (for example, in our case, either functional Magnetic Resonance Image BOLD, and optical image, EEG, or MEG). It is a stochastic process which is indexed by the continuous spatial position variable s and time $t = 1, \ldots, N$. We posit an innovation process that is also a function of space and time. Note that the integration is over the volumetric set Ω. Of central interest here are the functions $a_k(s,u)$ that specify the influence of site u on site s at after k time delays. This is actually a function $a_k : \Omega \times \Omega \to \mathrm{Re}$ which will specify the influences produced by small neighborhoods of each point s of the manifold $\Delta(s) \subset \Omega$, which will be $a_k(s,u)\ \Delta(s)$. We now introduce three definitions of spatial influence measures:

- A *point influence measure* $I_{s \to u}$ is the simple test H_0: $a(s,u) = 0$ for given $s, u \in \Omega$.
- *An influence field* $I_{s \to \Omega}$ is a multiple test $H_0 : a(s,u) = 0$ for a given $s \in \Omega$ and all $u \in \Omega$.
- *An influence space* $I_{s \to \Omega}$ is a multiple test $H_0 : a(s,u) = 0$ for all $s, u \in \Omega$.

These concepts are illustrated in Fig. 18.2.

Of these, point influence measures have been studied to date and recently we have addressed those for fields. The exploration of the entire influence space will be touched upon in the final section.

Now suppose that we sample $y(s,t)$ centering our discretization at voxels

$$s = \{s_1, \ldots, s_i, \ldots s_p | s_i \in \Omega\}. \tag{18.2}$$

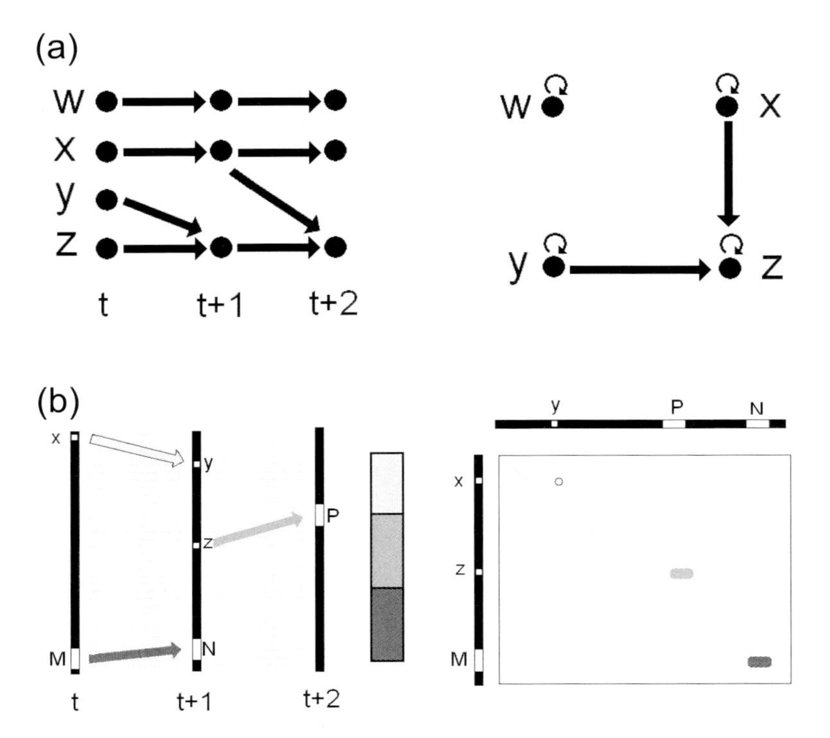

Fig. 18.2: Classical and spatial influence measures. On the left are the set of nodes and how activity is propagated by a linear autoregressive model for successive time instants. Arrows indicate nonzero autoregressive coefficients at different time lags. On the right are the corresponding causality graphs indicating nonzero point influence measures $I_{x \to y}$. (a) Causality analysis of a time series graph with only four nodes. In this hypothetical example only two time lags are relevant. Note that each node depends on its own past through a order two autoregressive model. Here we say y influences z at lag one and x influences z at lag two. (b) Spatial extension of the concept of influence measure. The manifold Ω in this case is a line segment. Also here only two time lags are relevant. Here each point also depends on its past through an order two autoregressive model. Additionally, we also have nonzero point influence measures of x on y with lag one, point z influences the whole of set P at lag two, and set M influences set N at lag one.

In this case, the data at time t will be represented by a vector

$$y_t = \begin{bmatrix} y_{1;t} \\ \vdots \\ y_{i;t} \\ \vdots \\ y_{p;t} \end{bmatrix}_{p \times 1} \tag{18.3}$$

where $i = 1, \ldots, p$ indexes the voxels with

$$y_{i,t} = \iiint_{\Delta(s_i)} y(u, t)\, du\,. \tag{18.4}$$

We shall assume that the neighborhood of s_i is sufficiently large to avoid spatial aliasing problems. The discretized version of Eq. (18.1) leads to the Multivariate Autoregressive Model (MAR) for the $\mathbf{y}_t$

$$\mathbf{y}_t = \sum_{k=1}^{r} \mathbf{A}_k \mathbf{y}_{t-k} + \mathbf{e}_t\,, \tag{18.5}$$

where the continuous function $a_k(s, s')$ transforms to a matrix $\mathbf{A}_k$ with dimensions $p \times p$ and with elements

$$a^k_{i,j} = \int \cdots \int_{\Delta(s_i)\times\Delta(u_i)} a_k(s'_i, u'_j)\, ds'\, du'\,. \tag{18.6}$$

In what follows we assume $\mathbf{e}_t \sim N(0, \mathbf{\Sigma})$, but of course this assumption may be relaxed. Note that the larger the number of sampling points the better the representation so we deal with a case in which ideally $p \to \infty$.

Define $\mathbf{B} = [\mathbf{A}_1, \ldots, \mathbf{A}_r]^T$, $\mathbf{Z} = [\mathbf{y}_{r+1}, \ldots, \mathbf{y}_N]^T$, and $\mathbf{X} = \begin{bmatrix} \mathbf{y}_r^T & \cdots & \mathbf{y}_1^T \\ \cdots & \cdots & \cdots \\ \mathbf{y}_{N-1}^T & \cdots & \mathbf{y}_{N-r}^T \end{bmatrix}$ with dimensions $pr \times p$, $N - r \times p$, and $N - r \times pr$, respectively. We can now recast the original sMAR (18.1) as a multivariate regression model

$$\mathbf{Z} = \mathbf{XB} + \mathbf{E}\,, \tag{18.7}$$

where $E = [\mathbf{e}_{r+1}, \ldots, \mathbf{e}_N]^T$. Some additional notation will be useful. We shall denote the vectorized version of $\mathbf{B}$, $\beta = \text{vec}(\mathbf{B})$ formed by stacking the columns of $\mathbf{B}$, β^i. Note that β^i measures the influence of a voxel i on the rest of the brain for all time lags and, in turn, comprises the vectors of autoregressive coefficients for each time lag

$$\beta^i = \begin{bmatrix} \beta^i_1 \\ \cdots \\ \beta^i_r \end{bmatrix}\,. \tag{18.8}$$

Thus the linear effect of voxel i at lag k on voxel j is measured by the coefficient $\beta^i_{j,k}$.

18.3 Testing for Spatial Granger Causality

As noted before, MAR modeling has been widely applied in the neurosciences [3, 24, 25] for the analysis of causality. Though some doubt that causal analysis is possible at all [26], early work with Structural Equation Modeling [27] did face up to the issue of inferring directional influences and was firmly grounded in modern statistical techniques [28] via graphical models. These initial studies [27] in Neuroimaging were based on nondynamical PET data and ignored temporal information. The concept of Granger causality [6, 29, 30] does make use of temporal information in order to establish a measure of directed influence. Granger causality $I_{x \to y}$ of the time series x on y is demonstrated when one can reject the null hypothesis of y not being predicted by the past of x [7, 31, 32]. Recent work [33] have combined the notion of Granger causality analysis with that of causality analysis via graphical models [34]. In this view, a system modeled by a MAR is a network in which each node is a time series. These ideas generalize to the more general linear sMAR in Eq. (18.5) introduced above, by noting that the coefficients $a_{i,j}^{k}$ measure the influence that the time series j exerts on the time series i after k time instants. Knowing that $a_{i,j}^{k}$ is nonzero is equivalent to establishing effective connectivity [1] and tests for this hypothesis have been proposed as influence measures [6, 22, 25, 32, 35, 36]. From the graphical points of view the question is: does an edge exists between the corresponding nodes? The maximum likelihood (ML) estimation of Eq. (18.5), or equivalently Eq. (18.7) can be obtained by standard methods [4, 37]

$$\hat{\mathbf{B}} = \arg\min_{\mathbf{B}} \|\mathbf{Z} - \mathbf{X}\mathbf{B}\|^2 \tag{18.9}$$

where for any matrix $\mathbf{X}$, $\|\mathbf{X}\|^2 = \mathrm{tr}(\mathbf{X}^{\mathrm{T}}\ \mathbf{X})$, is the Frobenius norm. This results in the well known explicit solution, the OLS estimator

$$\hat{\mathbf{B}} = (\mathbf{X}^{\mathrm{T}}\mathbf{X})^{-1}\mathbf{X}^{\mathrm{T}}\mathbf{Z}\,. \tag{18.10}$$

It should be noted that the unrestricted ML estimator of the regression coefficients does not depend on the spatial covariance matrix of the innovations [37]. One can therefore carry out separate regression analyses for each node. In other words, it is possible to estimate separately each column β^i of $\mathbf{B}$

$$\hat{\beta}^i = (\mathbf{X}^{\mathrm{T}}\mathbf{X})^{-1}\mathbf{X}^{\mathrm{T}}\mathbf{z}^i \tag{18.11}$$

for $i = 1, \ldots, p$ where $\mathbf{z}^i$ is the ith column of $\mathbf{Z}$. Consider that we obtain the usual t statistic for each regression coefficient

$$t_{k,j}^{i} = \frac{\hat{\beta}_{k,j}^{i}}{\mathrm{SE}(\hat{\beta}_{k,j}^{i})} \tag{18.12}$$

where SE is the usual standard error of the regression coefficient. Then we can use SPM type procedures to detect which voxels are influenced by voxel i at lag k. This suggests the one possible specific definition of *influence field*

$$I_{k,i \to \Omega} = \{t_{k,j}^{i}\}_{1 \leqslant i \leqslant p}\,. \tag{18.13}$$

If, as is usual, we wish to collapse over the lags, then we use instead of the ordinary t statistic we can use the Hotelling's T^2 statistic. Unfortunately, there is a problem with this approach when dealing with Neuroimaging data: the total number of parameters to be estimated for model (18.5) is

$$g = r \cdot p^2 + \frac{(p^2 + p)}{2} \quad (18.14)$$

which becomes rapidly large for increasing p, a situation for which usual time series methods break down since the OLS estimator will not exist. In the next section we shall review some attempts to cope with this problem by dimensionality reduction in order to apply classical causality analysis. In the following section we shall explain our approach to address the full problem via variable penalization.

18.4 Dimension Reduction Approaches to sMAR Models

18.4.1 ROI-Based Causality Analysis

One common approach is to pre-select a small group of sets of voxels or regions of interest (ROI) on the basis of prior knowledge, for example known anatomical structures, and to obtain an average time series over these volumes. In other words the original manifold Ω is partitioned into sub-manifolds and the following holds

$$\Omega = \bigcup_{g=1}^{G} \Omega_g y_{g,t}^{ROI} = \iiint_{\Omega_g} y(s,t)\, ds\,. \quad (18.15)$$

Causality analysis may then be assayed by the methods described above since now $N > G$. Recent examples of this type of linear Granger causality analysis for fMRI time series are [35, 38]. As an example, a ROI analysis of the concurrent EEG-fMRI times series is shown in Fig. 18.2 (taken from). The fMRI time series are of length $N = 109$ for six ROI in the brain identified by previously looking at the correlation with the EEG alpha atom: visual cortex, thalamus, left and right insulae and left and right somatosensory areas. The resulting causality diagram clearly shows that electrophysiological activity is driving the BOLD response in different brain structures, which is to be expected since the BOLD response measured in fMRI experiments is secondary hemodynamic response to neural activity. Thalamus and cortex have reciprocal relations and with other structures. These results, in general, are in agreement with previous studies of this material showing the utility of this type of analysis. However, the ROI strategy has the potential problem of the appearance of spurious influences induced by the brain structures not included in the analysis. An additional problem is that it is not always clear how to establish the partition (18.15).

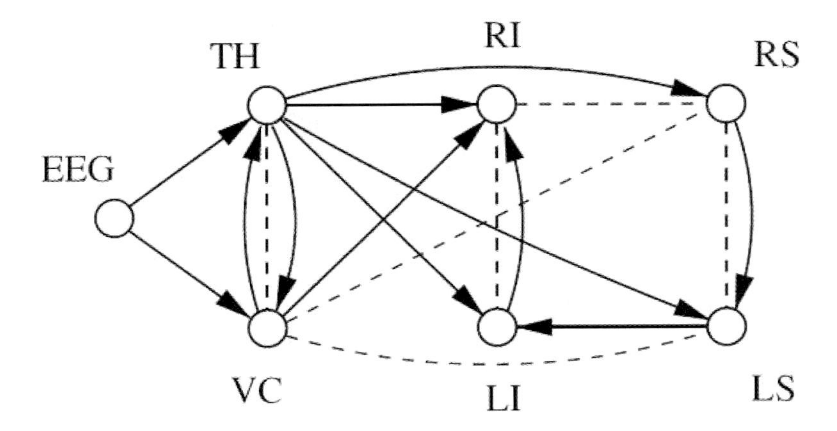

Fig. 18.3: ROI Granger causality graphical model for concurrent EEG-fMRI recording during alpha rhythm. The MRI from Fig. 18.1 has been divided into regions of interest (ROI) and a MAR model fitted to identify significant influences. The EEG node corresponds to the EEG PARAFAC α component power time series as shown in Figs. 18.3 and 18.5. The rest of the nodes are fMRI time series obtained by averaging activity over the following ROI: TH (thalamus), VC (Visual cortex), RI (right insula), LI (left insula), RS (right somatosensory cortex), and LS (left somatosensory cortex).

18.4.2 Latent Variable-Based Causality Analysis

A different approach for dimensionality reduction is the use of latent variable analysis (LVA). Essentially this involves creating linear or nonlinear combinations of the original time series in an attempt to find series are in some sense the actual underlying "physiological components"

$$y_{c,t}^{LVA} = f(\mathbf{y}_t), \tag{18.16}$$

where f is the transformation from the original time series to the desired components for $c = 1, \ldots, C$. This approach has a long history in neuroscience; different methods used being PCA [39, 40] or ICA [41].

We now give a recent example of LVA which extracted by means of multilinear techniques and applied to the EEG-fMRI data described in [14]. The multichannel EEG evolutionary spectrum $S(f, t, d)$ is obtained from a channel by channel wavelet transform, where ω is frequency, d is the derivation (channel) and t is time. Parallel Factor Analysis (PARAFAC) [13, 14] decomposes three-way data array S into the sum of "atoms"

$$S(d, \omega, t) = \sum_k a_k(d) b_k(\omega) c_k(t) + es(\omega, t, d), \tag{18.17}$$

where the kth atom is the trilinear product of loading vectors representing spatial (a_k), spectral (b_k), and temporal (c_k) "signatures." This decomposition is shown

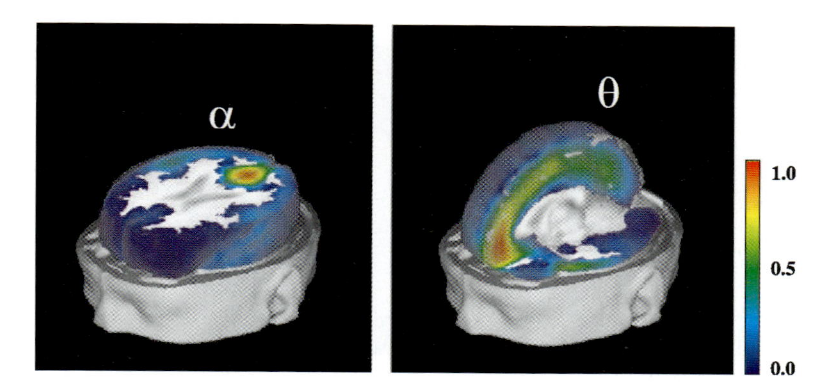

Fig. 18.4: Spatial distribution of the α and θ atoms as determined by both PARAFAC of the EEG and Multilinear Partial Least Squares of concurrent EEG-fMRI recordings. Inverses solutions obtained from the spatial α_k signatures. Note the occipital and frontal distributions of the spatial signature for the α and θ atoms, respectively.

schematically in Fig. 18.5. Two atoms were found α and θ, identified on the basis of the frequency signature (Fig. 18.6(a)) peaking at the known frequency of these well-known EEG rhythms.

The spatial distribution of these components both in the EEG and the fMRI were occipital and frontal for the α and θ atoms, respectively (Fig. 18.4). Perusal of the time signatures of these atoms shows a strong influence of imposing either a resting condition or a mental task on the subject (Fig. 18.6(b)). Moreover, since only two time series were involved, classical methods for measuring influences were applied easily yielding the causality analysis shown in Fig. 18.7. It is to be noted that assessment of the model order for all fMRI time series models presented in this paper indicated that only a first-order model ($r = 1$) is required.

While consistent with known hypothesis about the brain, this type of analysis only uses the instantaneous covariances to fit the model since time lags are not usually included in the analysis. A more promising approach are methods developed for geostatistics [4, 5, 42] in which time series methods are combined with component extraction. The latter techniques, to our knowledge, have not been applied in neuroscience. In any case extraction of components avoids the issue of analyzing directly spatial Granger causality; a point to which we shall now turn our attention.

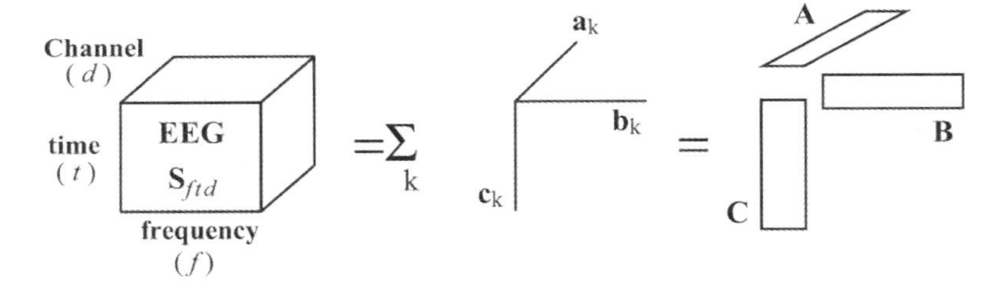

Fig. 18.5: Schematic representation of the PARAFAC model. The multichannel EEG evolutionary spectrum $S(d, \omega, t)$ is decomposed into the sum of "atoms" where the kth atom is the trilinear product of loading vectors representing spatial (a_k), spectral (b_k), and temporal (c_k) "signatures."

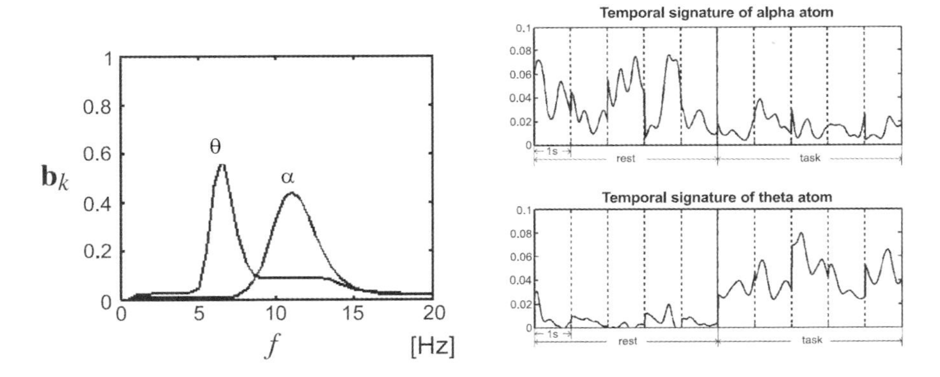

Fig. 18.6: Spectral and temporal signatures of the EEG PARAFAC atoms. Left: Spectral signatures $b_k(f)$ of the two atoms corresponding to frequency peaks in the traditional θ and α bands. The horizontal axis is frequency ω in Hz and the vertical axis is the normalized amplitude. Right: temporal signatures, $c_k(t)$, of the θ and α atoms.

18.5 Penalized sMAR

18.5.1 General Model

This section introduces a Bayesian sMAR that generalizes those proposed in [22, 23]. Consider once more the sMAR model

$$Z = XB + E . \tag{18.18}$$

We now posit that the elements of β are sampled from an *a priori* that is the product of several generalized multivariate normal densities

$$\pi(\beta; (P_1, \mathbf{\Sigma}_1), \ldots, (P_M, \mathbf{\Sigma}_M)) = C \prod_{m=1}^{M} \exp(-P_m(\mathbf{\Sigma}_m^{-1}\beta)) \tag{18.19}$$

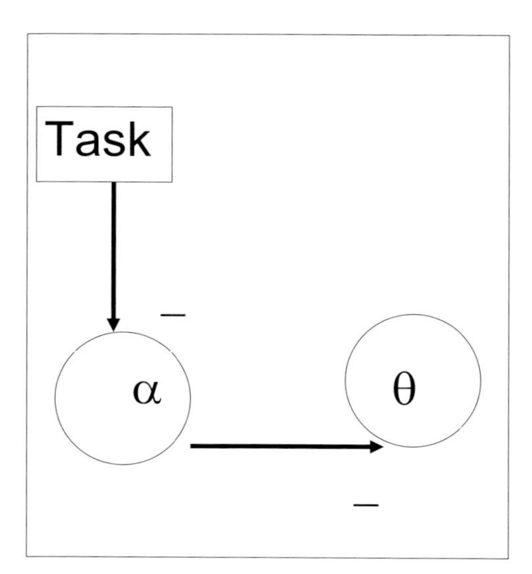

Fig. 18.7: Influence measure analysis of the EEG-fMRI atoms. The external variable imposition of a mental task was found to directly influence (negatively) the activity of the α atom, which in turn influenced negatively the θ atom ($I_{\text{task}\to\alpha}, I_{\alpha\to\theta} > 0$).

where C is a normalizing constant, the $\mathbf{\Sigma}_m$ are *a priori* covariance matrices for the β. The MAP estimate that follows from the likelihood of Eq. (18.18) and the prior Eq. (18.19) is

$$\hat{\mathbf{B}} = \arg\min_{\mathbf{B}} \|\mathbf{Z} - \mathbf{X}\,\mathbf{B}\|_{\Sigma}^{2} + \sum_{m=1}^{M} P_m(\mathbf{\Sigma}_m^{-1}\beta), \tag{18.20}$$

where for any matrix $\mathbf{X}$ $\|\mathbf{X}\|_{\Sigma}^{2} = \text{tr}(\mathbf{X}^{T}\mathbf{\Sigma}^{-1}\mathbf{X})$. Finally, $P_m(\boldsymbol{w})$ for any vector $\boldsymbol{w}$ is defined as

$$P_m(\boldsymbol{w}) = \sum_{l=1}^{\text{length}(\mathbf{x})} p_m(|w_l|), \tag{18.21}$$

and the functions $p_m(\theta)$ are defined for $\theta > 0$ are appropriate penalty functions with the properties specified in [43]. Some examples are given in Table 18.1 as well as illustrated in Fig. 18.8. Thus, our model consists of M regularization constraints, each comprising a

1. Covariance matrix used to enforce *a priori* spatial constraints on the autoregressive coefficients; and a
2. Penalization function to enforce constraints on the magnitude of the variables and therefore carry out variable selection.

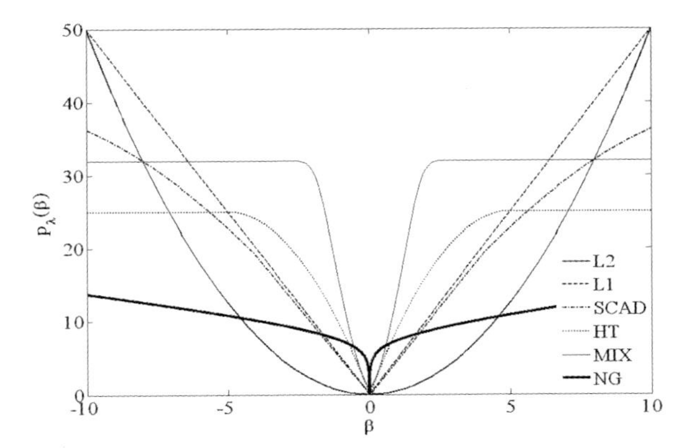

Fig. 18.8: Penalization functions: Plot of the penalization functions used to implement sparse and spatially constrained regression techniques. The meaning of the abbreviations is summarized in Table 18.1.

Tab. 18.1: Examples of penalty functions.

Name	Abbreviation
LASSO	L1
Smoothly clipped absolute deviation	SCAD
Hard thresholding	HT
Ridge	L2
Mixture of generalized *Gaussians*	MIX
Normal-gamma	NG
Normal-exponential-gamma	NEG

The penalization p_m functions that we have explored are summarized in Table 18.1 with their abbreviations. These abbreviations, together with those for the covariance matrices $\mathbf{\Sigma}_1$, allow the introduction of a notation for a particular sMAR model based on the penalty function used. Thus $(\mathrm{L1}, \mathbf{I}_{rp^2})$ is an sMAR model with a penalty that comprises only one term, the use of the $l1$ penalty and a spherical covariance matrix. It should be noted that the proposed MAP (18.20) includes as particular cases many currently used regularization schemes frequently applied in isolation, some new combinations proposed in the literature, as well as totally new proposals. Unfortunately, in the penalized case it is not possible in general to carry out separate regressions for each β^i. For sake of simplicity, and to retain the possibility of independent estimation for each influence field, we have been assuming that $\mathbf{\Sigma}$ is diagonal, that is, we assume that the innovations are spatially independent. In the final section we shall discuss avenues to avoid this restriction.

Tab. 18.2: Examples of *a priori* covariance $\mathbf{\Sigma}_m$ matrices defined in terms of their inverses. These definitions are valid over rectangular domains in dimensions from one to three. For irregular domains (areas in an image where there is gray matter for example) these matrices are masked a 0–1 indicator function for the selected voxels. Here m, n, and p are the dimensions of the rectangular region, $\otimes$ denotes the Kronecker product of two matrices and $\oplus$ the Kronecker sum

Name	Notation	Inverse of matrix
Spherical	$\mathbf{I}_n$	$\begin{bmatrix} 1 & 0 & 0 \\ 0 & \dots & 0 \\ \dots & \dots & \dots \\ 0 & 0 & 1 \end{bmatrix}$
1D gradient	$\mathbf{D}_n^1$	$\begin{bmatrix} 1 & -1 & 0 & \dots & 0 \\ 0 & 1 & -1 & \dots & 0 \\ & & \dots & & \\ 0 & \dots & 0 & 1 & -1 \\ 0 & \dots & 0 & 0 & 1 \end{bmatrix}$
2D gradient	$\mathbf{D}_{nm}^2$	$\begin{bmatrix} \mathbf{I}_n \otimes \mathbf{D}_m^1 \\ \mathbf{D}_m^1 \otimes \mathbf{I}_n \end{bmatrix}$
2D Laplacian	$\mathbf{L}_{nm}^2$	$\mathbf{D}_n^1 \oplus \mathbf{D}_m^1$
3D gradient	$\mathbf{D}_{nmp}^3$	$\begin{bmatrix} \mathbf{I}_n \otimes \mathbf{I}_m \otimes \mathbf{D}_p^1 \\ \mathbf{I}_n \otimes \mathbf{D}_m^1 \otimes \mathbf{I}_p \\ \mathbf{D}_n^1 \otimes \mathbf{I}_m \otimes \mathbf{I}_p \end{bmatrix}$
3D-Laplacian	$\lambda \mathbf{L}_{n,m,p}^3$	$\lambda \mathbf{D}_m^1 \oplus \mathbf{D}_n^1 \oplus \mathbf{D}_p^1$

Tab. 18.3: Mixing distribution of interest represented in the scale mixture form, where $\mathrm{IG}(a, b)$ and $\mathrm{Ga}(a, b)$ are the inverse gamma and the gamma with shape a and natural parameter b

Distribution	Density
Normal-Jeffreys	$g(\theta) \propto 1/\theta$
t distribution	$g(\theta) = \mathrm{IG}(\frac{\lambda}{2}, \frac{\gamma^2\lambda}{2}) \quad \lambda, \gamma > 0$
Mean-zero double exponential	$g(\theta) = \mathrm{Ga}(\theta \mid \lambda, \frac{1}{2\gamma^2}) \quad \lambda = 1$
Normal-gamma (NG)	$g(\theta) = \mathrm{Ga}(\theta \mid \lambda, \frac{1}{2\gamma^2}) \quad \lambda > 0,\ \gamma < \infty$
Normal-exponential-gamma (NEG)	$g(\theta) = \frac{\lambda}{\gamma^2}(1 + \theta/\gamma^2)^{-(\lambda+1)} \quad \lambda > 0,\ \gamma < \infty$

18.5.2 Achieving Sparsity Via Variable Selection

In a previous paper we proposed that attention be restricted to networks with *sparse connectivity*. That this is a reasonable assumption that is justified by studies of the numerical characteristics of network connectivity in anatomical brain databases [44–46].

Sparsity of causal explanations may be achieved by variable selection. Researchers into causality [47, 48] have explored the oldest of variable selection techniques for regression—stepwise selection for the identification of causal graphs. This is the basis of popular algorithms such as PC embodied in programs such as TETRAD. These techniques have been used in graphical time series models [49]. Unfortunately, these techniques do not work well for $p \gg N$. A considerable improvement may be achieved by stochastic search variable selection (SSVS) of George and McCulloch [50, 51], which relies on Markov chain–Monte Carlo (MCMC) exploration of possible sparse networks [52, 53]. These approaches, however, are computationally very intensive and not practical for implementing a pipeline for Neuroimaging analysis.

An alternative to MCMC-like methods is variable selection via penalized regression models [43, 54] which unifies nearly all variable selection techniques into an easy-to-implement iterative application of minimum norm or ridge regression. These techniques have been shown to be useful for the identification of the topology of huge networks [55, 56]. Penalized regression models were introduced for the first time for the study of brain connectivity used in [22, 23]. Consider the variant of the general model (18.20) with only one component $(M = 1)$ and a spherical covariance matrix. Some of the possible models are:

- $(L2, \mathbf{I}_{rp^2})$ is the usual ridge regression model [57] or quadratic regularization, λ being the regularization parameter which determines the amount of penalization enforced. Due to the possibility of efficient computation this is a widely applied form of regularization, recently applied for example to analyze microarray data [58].

- $(L1, \mathbf{L}_{rp^2})$ is, as mentioned above, the LASSO [59].

- $(HT, \mathbf{L}_{rp^2})$ is the Hard Thresholding of regression coefficients only applicable in the $p < N$ case.

- $(SCAD, \mathbf{L}_{rp^2})$ [43] is a form of regression designed to avoid bias for larger coefficients.

- $(MIX, \mathbf{L}_{rp^2})$ uses the penalty function $-\ln\big(p_0 f_{p0}(\beta) + (1 - p_0) f_{p1}(\beta)\big)$ where the mixture density are univariate generalized Gaussians. This is a regression model designed to produce sparsity and implements a non-MCMC variant of the "spike and slab" models for variable selection, the best known being the SSVS method of George and McCulloch [50].

We introduce in this chapter a further generalization of the variable selection penalties previously used. As pointed out in [60] it has been shown that most of the mixture priors previously discussed are particular instances of scale mixtures of normal distributions [61] that embody a high prior probability of the regression coefficients in the proximity of zero. These authors proposed a natural class of prior distribution that bridges the gap between classical normal-Jeffreys priors, passing throughout ridge regression down to the double exponential distribution used in the LASSO. Some particular mixture distribution of interest are shown in Table 18.3. We single out for mention the following regression models used for the first time to study brain connectivity:

- $(\text{NG}, \mathbf{L}_{rp^2})$ uses as a penalty the minus log of the normal-gamma (NG) distribution is often called as variance-gamma distribution, has the al distribution: $p(\beta_j) = \frac{1}{\sqrt{\pi}2^{\lambda-1/2}\gamma^{\lambda+1/2}\Gamma(\lambda)}|\beta_j|K_{\lambda-1/2}(|\beta_j|/\lambda)$, where $K_\nu(a)$ is the modified Bessel function of the third kind.

- $(\text{NEG}, \mathbf{L}_{rp^2})$ is based on the normal-exponential-gamma (NEG) can be expressed as $p(\beta_j) = \frac{\lambda 2^\lambda}{\sqrt{\pi}\gamma}\Gamma(\lambda+1/2)\exp(\frac{\beta_j^2}{4\gamma^2})D_{-2(\lambda+1/2)}(|\beta_j|/\lambda)$, where $D_\nu(a)$ is the parabolic cylinder function, the parameters γ and λ control the scale and the heaviness of the tail, respectively.

18.5.3 Achieving Spatial Smoothness

The other constraint that makes sense is that of *spatial smoothness* of influence fields. Consider Fig. 18.9 (left) which depicts the influence of a given brain structure on three others: two that are close to each other in the same hemisphere and another that is further away in another hemisphere. It is *a priori* more likely that the influences *from* the given voxel on the two closer voxels be more similar than the influence on the distant voxel. This can be quantified by requiring

$$\sum_{k=1}^{r} \iiint_{\Omega} \left| \frac{\partial a_k(s,u)}{\partial s} \right|^2 du \tag{18.22}$$

be small, the distribution of influences to *targets* be smooth. Alternatively, one may require that the distribution of *sources* influences to a single target as in Fig. 18.9 (right) be smooth by imposing that

$$\sum_{k=1}^{r} \iiint_{\Omega} \left| \frac{\partial a_k(s,u)}{\partial u} \right|^2 du \tag{18.23}$$

be small. These definitions are actually for the L2 penalization and therefore specify Gaussian fields as *a priori* distributions. The discrete version of this is set up by specifying the matrix operators defined in Table 18.3. Additionally, one may modify the quadratic norm by applying the different penalties described in Table 18.1. One may also conceive combinations of the two conditions—smoothness of target or of source influences all these conditions following from the choice of appropriate roughness penalty or, equivalently, the *a priori* covariance matrix. Imposing smoothness on the influence fields involves imposing conditions on each column of $\mathbf{B}(\beta^i)$ separately. It would be possible to impose similar conditions

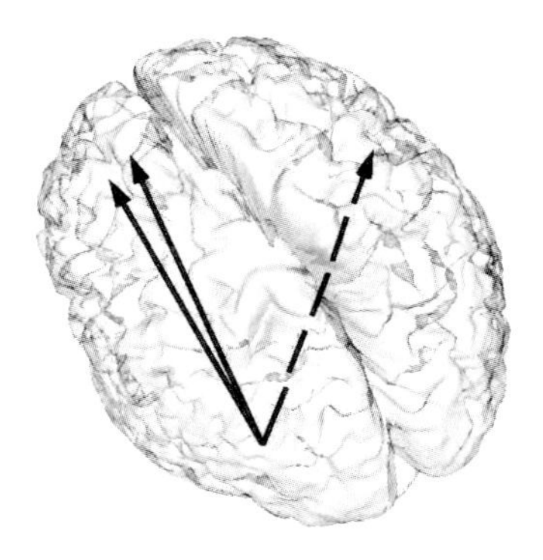
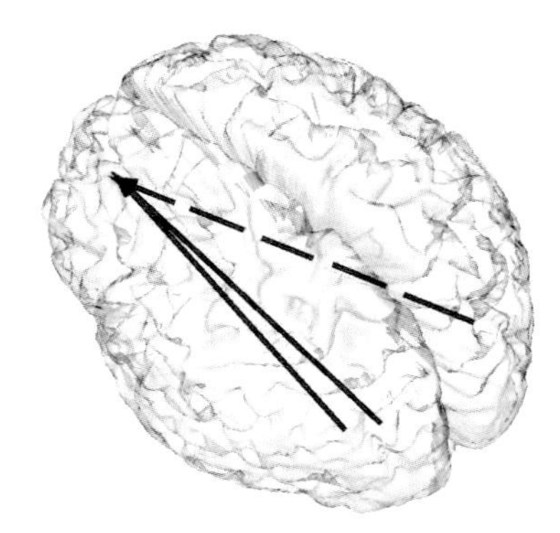

Fig. 18.9: Spatial constraints.

on the rows of **B**, that is on the map of sources of a given target, but this is not computationally feasible at the moment for large p.

We shall now mention some one component sMAR models that impose different types of smoothness:

- $(L1, \mathbf{L}_{rp^2})$ this is the data "Fusion" model mentioned in [62], now applied to sMAR.

- $(L2, \mathbf{L}_{rp^2})$ is a spline regression model in which the spatial Laplacian of the estimated coefficients are to be minimized. Popularized for the solution of EEG inverse problems as "LORETA" [63], this model was used for the first time to study fMRI time-series connectivity in one of our previous paper [22].

We wish to emphasize that penalizing with roughness penalties is equivalent to penalizing a spatial Fourier transform of the coefficients to be estimated.

18.5.4 Achieving Sparseness *and* Smoothness

There is no reason to restrict the number of penalty/smoothness constraints imposed. In fact, recent work in statistical learning has advanced the use of models which are easily recognized in the framework of our general model. For example:

- $(L1, \mathbf{I}_{rp^2})(L2, \mathbf{I}_{rp^2})$ can be recognized as the recently introduced "Elastic Net" regression technique applied to sMAR [64]. The elastic net has been shown to improve on the variable selection properties of the LASSO when $p \gg N$. Simulations have shown that when there are sets of correlated variables LASSO picks just one variable from each set. In contrast, the elastic net picks all of the members of the set giving them similar weights. When applied to sMAR this would produce a "patchy" influence field. One would hope that these patches correspond to coherent sets of neurons that act together in influencing other brain structures.

- $(\mathrm{L1}, \mathbf{I}_{rp^2})(\mathrm{L1}, \mathbf{D}_{rp^2})$ can be recognized as the recently introduced "LASSO-Fusion" [62] regression technique applied to sMAR. It is claimed that this also selects patches of related variables and outperforms the LASSO when $p \gg N$.

Both these procedures were previously developed in the context of particular algorithms: quadratic programming and LARS for LASSO-Fusion and the elastic net, respectively. However, we have that it is possible even for huge problems (see next section) to work with any number of combinations of penalties/covariance matrices. We have therefore tried out the following new models:

- $(\mathrm{L2}, \mathbf{I}_{rp^2})(\mathrm{L2}, \mathbf{D}_{rp^2})$ which we call "Ridge-Fusion" in analogy to LASSO-Fusion.

- $(\mathrm{L1}, \mathbf{I}_{rp^2})(\mathrm{L1}, \mathbf{L}_{rp^2})(\mathrm{L2}, \mathbf{I}_{rp^2})(\mathrm{L2}, \mathbf{L}_{rp^2})$ which can be seen either as: (1) a combination of the LASSO-Fusion and Ridge-Fusion or, alternatively as (2) a combination of the Elastic NET applied with LORETA both for the L1 and L2 norm.

From our previous comment at the end of the last section it is obvious that these attempts to combine norms are equivalent to penalizing/selecting variables from the original coefficient domain as well as from the spatial frequency domain.

18.6 Estimation via the MM Algorithm

For implementation of algorithms for the estimation of the model equation (18.20), advantage was taken of the recent demonstration [43, 54, 65] that estimation of any of many penalized regression for the influence field of voxel i can be carried out by iterative application of ridge regression

$$\hat{\beta}^i_{k+1} = (\mathbf{X}^{\mathrm{T}}\mathbf{X} + \mathbf{D}(\hat{\beta}^i_{k+1}))^{-1}\,\mathbf{X}^{\mathrm{T}}\mathbf{z}_i, \tag{18.24}$$

where $k = 1, \ldots, N_{\text{iter}}$, with N_{iter} is the number of iterations and $\mathbf{D}(\hat{\beta}^i_{k+1})$, a diagonal matrix is defined by

$$\mathbf{D}(\beta^i) = \sum_{m=1}^{M} \mathrm{diag}(p'_m(|w^i_l|)/|w^i_l|) \tag{18.25}$$

for $l = 1, \ldots, rp^2$, where $\boldsymbol{w}{=}\Sigma_m^{-1}\,\beta^i$ and p'_λ is the derivative of the penalty function being evaluated. The derivatives p'_m for different penalty functions are provided in Table 18.4.

The reason that this algorithm works may be inferred from Fig. 18.9. At each step of the iterative process, the regression coefficients of each node with all others are weighted according to their current size and the penalty function chosen. Many coefficients are successively down-weighted and ultimately set to zero—effectively carrying out variable selection in the case of the LASSO, HT, SCAD, MIX, and NG penalization. It must be emphasized that the number of variables set to zero in any of the methods described will depend on the value

Tab. 18.4: $p'_\lambda(\theta)$, derivatives of penalty functions for $\theta > 0$.

Type	Derivatives
L1	$p'_\lambda(\theta) = \lambda_{L1}\,\theta$
SCAD	$p'_\lambda(\theta) = \lambda_{SCAD}\{I(\theta \leqslant \lambda) + \frac{(a\lambda - \theta)}{a-1} I(\theta > \lambda)\}$ for some $a > 2$
HT	$p'_\lambda(\theta) = -2(\theta - \lambda_{HT})$
L2	$p'_\lambda(\theta) = 2\lambda_{L2}\,\theta$
MIX	$p'_\lambda(\theta) = -\lambda_{MIX}\left[\frac{p_o f'_{p_0}(\theta) + p_1 f'_{p_1}(\theta)}{p_o f_{p_0}(\theta) + p_1 f_{p_1}(\theta)}\right]$ where $f_p(\theta) = \frac{p^{1-\frac{1}{p}}}{2\sigma_p \Gamma(\frac{1}{p})} \exp(-\frac{1}{p}\frac{\lvert x - x_0\rvert^p}{\sigma_p})$ and $\Gamma(\cdot)$ denotes the Gamma function
NG	$p'_\lambda(\theta) = \frac{1}{\gamma_{NG}} \frac{K_{\lambda-3/2}\left(\frac{\theta}{\gamma_{NG}}\right)}{K_{\lambda-1/2}\left(\frac{\theta}{\gamma_{NG}}\right)}$ where $K_\nu(z)$ is the modified Bessel function of the third kind
NEG	$p'_\lambda(\theta) = \frac{\lambda_{NG} + 1/2}{\gamma_{NG}} \frac{D_{-2(\lambda+1)}\left(\frac{\theta}{\gamma_{NG}}\right)}{D_{-2(\lambda+1/2)}\left(\frac{\theta}{\gamma_{NG}}\right)}$ where $D_\nu(z)$ is the parabolic cylinder function

Tab. 18.5: The numerical results of simulations testing of the ROC for the different studied methods are presented

Method	I	L	I + L
L2	0.6825	0.7026	0.7438
L1	0.6157	0.7102	0.7657
L1 + L2	0.5766	0.6222	0.6257
NG	0.6722	0.6963	0.7434

of the regularization parameter, with higher values selecting fewer variables. In this chapter, the value of the tuning parameters was selected to minimize the generalized cross-validation criterion (GCV).

The specific implementation of penalized regression used in this work is that of the maximization–minorization (MM) algorithm [65–67] which exploits an optimization technique that extends the central idea of EM algorithms and Variational Bayes techniques to situations not necessarily involving missing data or even maximum likelihood estimation. The MM algorithm retains virtues of the Newton–Raphson algorithm. It is numerically stable and is never forced to delete a covariate permanently in the process of iteration. The general convergence results known for MM algorithms imply among other things that the newly proposed algorithm converges correctly to the maximizer of the perturbed penalized likelihood whenever this maximizer is the unique local maximum. The selected model based on the maximized penalized likelihood satisfies $p_m(|w_l^i|) = 0$ for certain $w = \Sigma_m^{-1}\beta^i$, which components accordingly are not included in this final model, and so model estimation is performed at the same time as model selection. The tuning parameters λ_M may be chosen by a data-driven approach such as cross-validation or generalized cross-validation [68]. An important point is that Hunter and Li showed that simple use of iterations Eq. (18.24) with the matrix $\mathbf{D}$ may permanently delete variables permanently from consideration being included in further iterations.

Hunter and Li [67] showed that a perturbed version of $p_m(\theta)$ may be used to define a new objective function that is similar to the original but does not lead to permanent variable deletion. To this end, they define

$$p_{m,\epsilon}(\theta) = p_m(\theta) - \epsilon \int_0^{|\theta|} \frac{p_\lambda}{\epsilon + t} dt , \tag{18.26}$$

which in practice is equivalent to using the following matrix: $\mathbf{D}_\epsilon$ instead of $\mathbf{D}$

$$\mathbf{D}_\epsilon(\beta^i) = \sum_{m=1}^{M} \operatorname{diag}\left(p'_m(|w_l^i|)/(|w_l^i| + \epsilon)\right) . \tag{18.27}$$

Note that in the computations the original set of variables to be estimated β is by definition augmented with spatial transforms (defined by the matrix operators laid out in Table 18.2). Suppose that we have defined a model with covariance matrices $\boldsymbol{\Sigma}_1, \ldots, \boldsymbol{\Sigma}_M$. Then we can use the following computational "trick," defining

$$\mathbf{S} = [\boldsymbol{\Sigma}_1^{-T}, \ldots, \boldsymbol{\Sigma}_M^{-T}]^T \quad \mathbf{T} = \frac{1}{M}[\boldsymbol{\Sigma}_1, \ldots, \boldsymbol{\Sigma}_M] \tag{18.28}$$

we have

$$q = \mathbf{S}\beta \tag{18.29}$$

one may carry out penalized regression on this new set of variables by defining $\mathbf{X}_M = \mathbf{XT}$ and solving the new (larger) problem, where the definition of $\mathbf{Q}$ is self evident

$$\hat{\mathbf{Q}} = \arg\min_{\mathbf{B}} \|(\mathbf{Z} - \mathbf{X}_M\ \mathbf{Q}\ )\|_\Sigma^2 + \sum_{m=1}^{M} P_m(q) . \tag{18.30}$$

Back transformation to the desired solution is obtained by $\hat{\mathbf{B}} = \mathbf{T}\hat{\mathbf{Q}}$. We have found this algorithm to work well in practice.

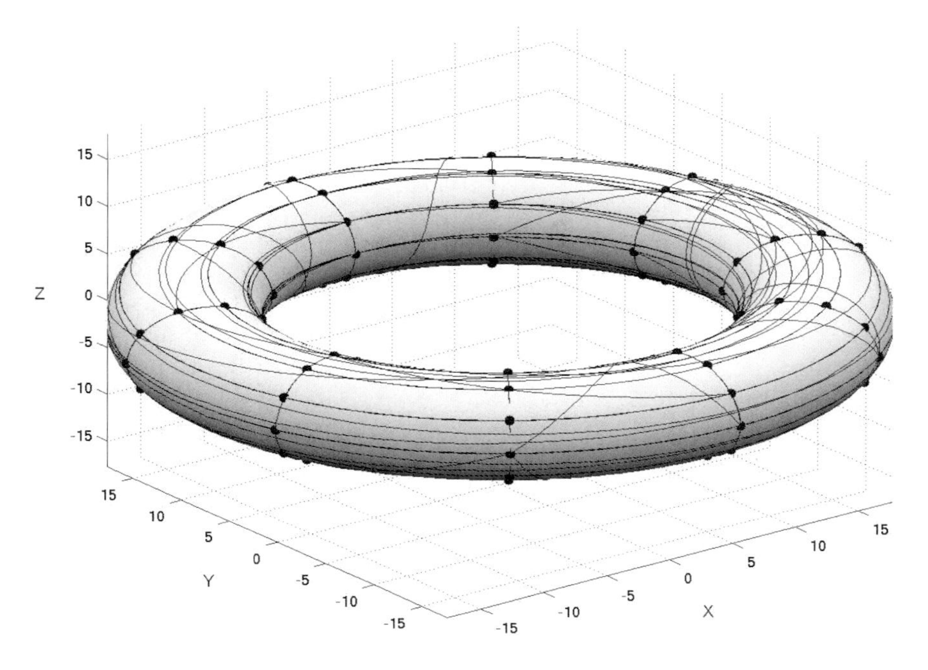

Fig. 18.10: Ideal "cortex" used for simulations was modeled by a small world network defined over a two-dimensional grid on the surface of a torus. This structure has periodic boundary conditions in the plane. Different combinations of strengths were used for defining the autoregressive matrices used to create simulated fMRI time series.

18.7 Evaluation of Simulated Data

The procedures described in the two previous sections have been thoroughly tested with simulated data. For simulations an "ideal cortex" was modeled by a small world network defined over a two-dimensional grid on the surface of a torus (Fig. 18.10). This structure has periodic boundary conditions in the plane.

In simulations described in detail in [23], the existence of a connection was generated with a binomial probability that decreased with distance. The network mean connectivity was 6.23, the scaled clustering 0.87, and the scaled length 0.19. This type of small-world network has a high probability of connections between geographical neighbors and a small proportion of larger range connections. The network mean connectivity was 6.23, the scaled clustering 0.87, and the scaled length 0.19. The autoregressive matrix being sampled from Eq. (18.5). The innovations were sampled from a Gaussian distribution with a different prescribed covariance matrices, including nondiagonal ones. A simulated fMRI is shown in Fig. 18.11. The effect of different observed lengths of time-series (N) on the detection of connections was studied. The behavior of different procedures was compared by measuring the area under the ROC curve (AUC). We found that

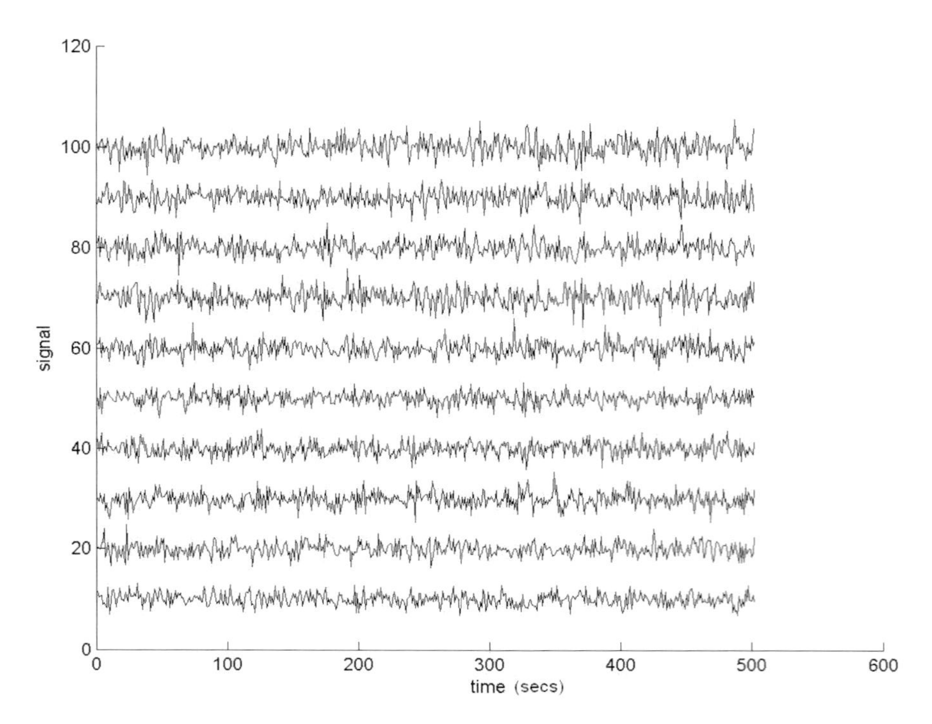

Fig. 18.11: Simulated fMRI time-series generated by a first-order multivariate autoregressive model.

while performance deteriorated with an increasing $\frac{p}{N}$ ratio there was still significant detection rates with this ratio near ten. The performance of the methods also deteriorated with increasing spatial innovation correlation. This latter observation underscores the need for also estimating the covariance matrix $\mathbf{\Sigma}$. Doing this with computational efficiency is still work in progress.

A number of further simulations were carried out in similar conditions as those reported before to explore the usefulness of multiple penalty/covariance matrix combinations. The $\frac{p}{N}$ ratio was now set at two. From Table 18.5 it is evident that, except for one exception, imposing simultaneously sparseness and smoothness outperforms either criteria alone.

18.8 Influence Fields for Real Data

To be able to apply these techniques to actual data it is necessary to have a decision procedure as to which variables to finally retain. We have found that although the methods described above do enforce considerable selection of variables, there is still a "gray zone" of variables with small values, for which the decision has to be taken as whether to include or not.

We have therefore combined methods for penalized regression with procedures for the control of the false discovery rates (FDR) [20, 69, 70] in situations where a large number of null hypothesis is expected to be true. The situation $p \gg n$ this case becomes strength instead of a weakness, because it allows the nonparametric estimation of the distribution of the null hypotheses to control false discoveries. To carry out this type of decision procedure it is preferable to work with the influence measures defined by the t statistics equation (18.12). For this we must estimate the standard errors of the $\hat{\beta}$. We have explored two procedures for this estimation. One is the "sandwich" formulas as described in [67, 71, 72]. However, we have found the estimation of the standard errors by means of the bootstrap more robust than with the sandwich estimator.

In [23] it was shown that efficient detection of connections possible simulated neural networks. The method was additionally shown to give plausible results with real fMRI data and is capable of being scaled to analyze very large data sets. In that publication the variable-selection method combined with FDR was illustrated by the identification of the neural circuitry related to emotional processing as measured by BOLD.

As a final, real-world example, we describe in more detail the concurrent EEG-fMRI experiment that has been used as an example throughout this chapter. This is a problem of sufficient size to test the practicality of the procedures proposed since p the number of voxels is 16 240 and N is only 108. The EEG was sampled at 200 Hz from an array of 16 bipolar pairs, (Fp2-F8, F8-T4, T4-T6, T6-O2, O2-P4, P4-C4, C4-F4, F4-Fp2; Fp1-F7, F7-T3, T3-T5, T5-O1, O1-P3, P3-C3, C3-F3, F3-Fp1), with an additional channel for the EKG and scan trigger. The fMRI time series was measured in six slice planes (4 mm, skip 1 mm) parallel to the AC–PC line, with the second from the bottom slice through AC–PC. More details about this data set can be found in [18]. In the work presented here we report a typical subject from a set of five simultaneous EEG/fMRI recordings from three different subjects.

For the fMRI, we examined the influence field with a source at that voxel that had the largest (negative) correlation with the EEG PARAFAC component for α rhythm. This latter component is the one obtained in the section above on LVA methods and shown topographically in Fig. 18.4 (left). The selected voxel is marked in Fig. 18.1 (arrow).

The influence fields for the selected voxel obtained by using different models are shown in Fig. 18.12. The penalties are labeled on the left and the covariances on the top. It is to be noted that the use of the spherical covariance matrix produces quite "rough" influence fields. When combined with the L1 penalty only a scattering of points is selected, at most the same as N that is 108—a known property of the LASSO. The $(L2, \mathbf{L}_{r,p^2})$ solution ("Ridge-Fusion") produces a more pleasing (but perhaps excessively smooth) map that is in very good correspondence with previous studies with simple correlations as well as with PARAFAC. All the most realistic seeming solutions are those that combine the spherical co-

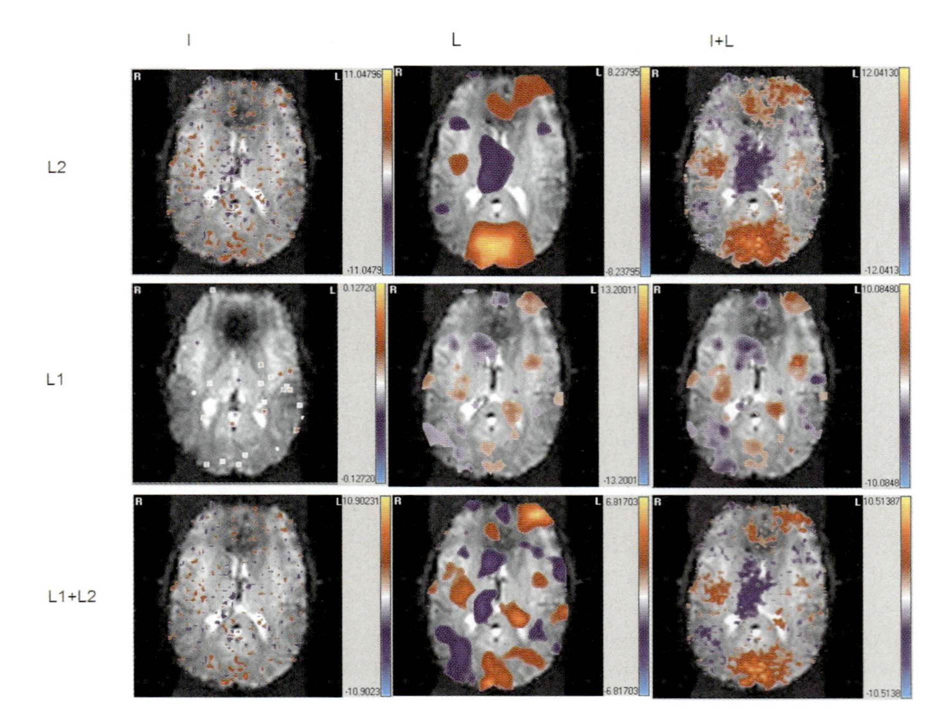

Fig. 18.12: Results of fitting the sMAR with multiple penalties/covariance matrices. The *a priori* covariance matrix assumed is stated on the top (spherical, Laplacian, and a combination of both). The type of penalization is stated on the left (L2 norm, L1 norm, and a combination of both known as the elastic net). Each sub figure is the influence field of the voxel marked in Fig. 18.1 with an arrow on the rest of the voxels corresponding to the slice immediately below.

variance matrix as well as the Laplacian roughness penalty. In fact, the solution that combines the spherical and Laplacian covariance matrices and also the L1 and L2 norm seems to be subjectively the best solution. This impression is born out by comparison of the GCV values for all models. GCV not only serves to fit the tuning parameters but also provides a yardstick for comparing models. In this particular case, related to the models fit and displayed in Fig. 18.12 there is a progressive decrease of GCV from top to bottom and from left to right, indicating that the simpler models do not provide adequate modeling flexibility and providing some empirical support for the usefulness of model (18.19).

18.9 Possible Extensions and Conclusions

Work with the SMAR model (18.19) is proceeding in several directions. Obviously this approach can be extended for nonlinear autoregressions. This can be done by

- Including bilinear, or higher order terms in the $\mathbf{X}$ matrix [73]; or by
- Defining a kernel weighting in the state space for the autoregressive coefficients as in [74].

On the other hand, a kernel method at different times would accommodate non-stationary time series as in [32].

Extensions to the frequency domain of sMAR causality analysis are quite straightforward. Either the sandwich formula or the bootstrap can be used to provide estimates of any linear combination of influence fields and therefore to the temporal Fourier transform of the influence fields over the different delays.

A vexing problem is the estimation of the covariance matrix Σ. We are currently attempting to this by including a zero lag autoregressive matrix $\mathbf{A}_0$ in the formulation of the model.

In conclusion, we have introduced a spatial multivariate autoregressive model based on a Bayesian formulation that combines several components of different types of penalizations as well as spatial *a priori* covariance matrices. These are shown by simulations and work with real data to be practical, even for huge data sets, and that give plausible results. The methods continue to bring into the framework of Statistical Parametric Mapping the analysis of effective connectivity via the analysis of Granger causality.

Acknowledgements

We wish to thank Maria Luisa Bringas for her untiring help in the preparation of this paper. Also I would like to thank Mark Cohen and Robin Goldman for their continuing support and intellectual exchange.

References

[1] K. J. Friston. Functional and effective connectivity in neuroimaging: a synthesis. *Hum. Brain Mapping*, 2:56–78, 1994.

[2] R. K. Shields, S. Madhavan, K. R. Cole, J. D. Brostad, J. L. DeMeulenaere, C. D. Eggers, and P. H. Otten. Proprioceptive coordination of movement sequences in humans. *Clin. Neurophysiol.*, 116:87–92, 2005.

[3] L. Harrison, W. D. Penny, and K. Friston. Multivariate autoregressive modeling of fMRI time series. *Neuroimage*, 19:1477–1491, 2003.

[4] K. V. Mardia J. T. Kent and J. M. Bibby. *Multivariate Analysis*. Academic Press, London, San Diego, New York, Boston, Sydney, Tokyo, Toronto, 1979.

[5] C. K. Wikle and N. Cressie. A dimension-reduced approach to space–time kalman filtering. *Biometrika*, 86:815–829, 1999.

[6] C. W. J. Granger. Investigating causal relations by econometric models and cross-spectral methods. *Econometrica*, 37:414, 1969.

[7] S. L. Bressler, M. Z. Ding, and W. M. Yang. Investigation of cooperative cortical dynamics by multivariate autoregressive modeling of event-related local field potentials. *Neurocomputing*, 26-7:625–631, 1999.

[8] A. Neumaier and T. Schneider. Estimation of parameters and eigenmodes of multivariate autoregressive models. *ACM Trans. Math. Softw.*, 27:27–57, 2001.

[9] T. Schneider and A. Neumaier. Algorithm 808: Arfit - a matlab package for the estimation of parameters and eigenmodes of multivariate autoregressive models. *ACM Trans. Math. Softw.*, 27:58–65, 2001.

[10] R. Dahlhaus. Fitting time series models to nonstationary processes. *Ann. Stat.*, 25:1–37, 1997.

[11] P. A. Valdes. Quantitative electroencephologographic tomography. *Electroencephalogr. Clin. Neurophysiol.*, 103:19, 1997.

[12] R. I. Goldman, J. M. Stern, J. Engel, and M. S. Cohen. Acquiring simultaneous EEG and functional MRI. *Clin. Neurophysiol.*, 111:1974–1980, 2000.

[13] Eduardo Martinez-Montes, Pedro A. Valdes-Sosa, Fumikazu Miwakeichi, Robin I. Goldman, and Mark S. Cohen. Concurrent EEG/fMRI analysis by multiway partial least squares. *Neuroimage*, 22:1023–1034, 2004.

[14] F. Miwakeichi, E. Martinez-Montes, P. A. Valdes, N. Nishiyama, H. Mizuhara, and Y. Yamaguchi. Decomposing EEG data into space–time-frequency components using parallel factor analysis. *Neuroimage*, 22: 1035–1045, 2004.

[15] J. O. Ramsay and B. W. Silverman. *Functional Data Analysis*. Springer, Berlin, 1997.

[16] J. F. Geweke. Measurement of linear-dependence and feedback between multiple time-series. *J. Am. Stat. Assoc.*, 77:304–313, 1982.

[17] J. F. Geweke. Measures of conditional linear-dependence and feedback between time-series. *J. Am. Stat. Assoc.*, 79:907–915, 1984.

[18] R. I. Goldman, J. M. Stern, J. Engel, and M. S. Cohen. Simultaneous EEG and fMRI of the alpha rhythm. *Neuroreport*, 13:2487–2492, 2002.

[19] K. J. Worsley, S. Marrett, P. Neelin, A. C. Vandal, K. J. Friston, and A. C. Evans. A unified statistical approach for determining significant signals in images of cerebral activation. *Hum. Brain Mapping*, 4:58–73, 1996.

[20] B. Efron. Large-scale simultaneous hypothesis testing: The choice of a null hypothesis. *J. Am. Stat. Assoc.*, 99:96–104, 2004.

[21] B. Efron. Selection and estimation for large-scale simultaneous inference. 2005. URL `http://www-stat.stanford.edu/people/faculty/efron/papers.html`.

[22] P. A. Valdes-Sosa. Spatio-temporal autoregressive models defined over brain manifolds. *Neuroinformatics*, 2:239–250, 2004.

[23] P. A. Valdes-Sosa, J. M. Sanchez-Bornot, A. Lage-Castellanos, M. Vega-Hernandez, J. Bosch-Bayard, L. Melie-Garcia, and E. Canales-Rodriguez. Estimating brain functional connectivity with sparse multivariate autoregression. *Philos. Trans. R. Soc. B—Biol. Sci.*, 360:969–981, 2005.

[24] L. A. Baccala, M. A. L. Nicolelis, C. H. Yu, and M. Oshiro. Structural-analysis of neural circuits using the theory of directed-graphs. *Comput. Biomed. Res.*, 24:7–28, 1991.

[25] M. Kaminski, M. Z. Ding, W. A. Truccolo, and S. L. Bressler. Evaluating causal relations in neural systems: Granger causality, directed transfer function and statistical assessment of significance. *Biol. Cybern.*, 85:145–157, 2001.

[26] B. Horwitz. The elusive concept of brain connectivity. *Neuroimage*, 19:466–470, 2003.

[27] A. R. McIntosh and F. Gonzalez-Lima. Structural equation modeling and its applications to network analysis in functional brain imaging. *Hum. Brain Mapping*, 2:2–22, 1994.

[28] J. Pearl. Graphs, causality, and structural equation models. *Sociol. Meth. Res.*, 27:226–284, 1998.

[29] Y. Hosoya. The decomposition and measurement of the interdependency between 2nd-order stationary-processes. *Probability Theory and Related Fields*, 88:429–444, 1991.

[30] J. F. Geweke. Measures of conditional linear-dependence and feedback between time-series. *J. Am. Stat. Assoc.*, 79:907–915, 1984.

[31] C. Bernasconi and P. Konig. On the directionality of cortical interactions studied by structural analysis of electrophysiological recordings. *Biol. Cybern.*, 81:199–210, 1999.

[32] W. Hesse, E. Moller, M. Arnold, and B. Schack. The use of time-variant EEG granger causality for inspecting directed interdependences of neural assemblies. *J. Neurosci. Meth.*, 124:27–44, 2003.

[33] D. R. Brillinger, H. L. Bryant, and J. P. Segundo. Identification of synaptic interactions. *Biol. Cybern.*, 22:213–228, 1976.

[34] J. Pearl. Graphs, causality, and structural equation models. *Sociol. Meth. Res.*, 27:226–284, 1998.

[35] R. Goebel, A. Roebroeck, D. S. Kim, and E. Formisano. Investigating directed cortical interactions in time-resolved fMRI data using vector autoregressive modeling and granger causality mapping. *Magn. Reson. Imaging*, 21:1251–1261, 2003.

[36] M. Eichler. Graphical time series modelling in brain imaging. *Philos. Trans. R. Soc. Lond. B*, index issue, 2005.

[37] J. D. Hamilton. *Time Series Analysis*. Princeton University Press, Princeton, NJ, 1999.

[38] R. Goebel, T. D. Waberski, H. Simon, E. Peters, F. Klostermann, G. Curio, and H. Buchner. Different origins of low- and high-frequency components (600 Hz) of human somatosensory evoked potentials. *Clin. Neurophysiol.*, 115:927–937, 2004.

[39] D. S. Ruchkin, E. R. John, and J. Villegas. Analysis of average evoked potentials making use of least mean square techniques. *Ann. New York Acad. Sci.*, 115:799–, 1964.

[40] K. Friston, J. Phillips, D. Chawla, and C. Buchel. Revealing interactions among brain systems with nonlinear pca. *Hum. Brain Mapping*, 8:92–97, 1999.

[41] T. P. Jung, S. Makeig, M. J. McKeown, A. J. Bell, T. W. Lee, and T. J. Sejnowski. Imaging brain dynamics using independent component analysis. *Proc. IEEE*, 89:1107–1122, 2001.

[42] K. V. Mardia, C. Goodall, E. Redfern, and F. J. Alonso. The kriged kalman filter—rejoinder. *Test*, 7:277–285, 1998.

[43] J. Q. Fan and R. Z. Li. Variable selection via nonconcave penalized likelihood and its oracle properties. *J. Am. Stat. Assoc.*, 96:1348–1360, 2001.

[44] C. Hilgetag, R. Kotter, and K. E. Stephan. *Computational Methods for the Analysis of Brain Connectivity*, chapter 14-Hilgetag. Ascoli operator: Network typesetting edition, 2002.

[45] R. Kotter, K. E. Stephan, N. Palomero-Gallagher, S. Geyer, A. Schleicher, and K. Zilles. Multimodal characterisation of cortical areas by multivariate analyses of receptor binding and connectivity data. *Anat. Embryol.*, 204: 333–350, 2001.

[46] O. Sporns, D. R. Chialvo, M. Kaiser, and C. C. Hilgetag. Organization, development and function of complex brain networks. *Trends Cogn. Sci.*, 8: 418–425, 2004.

[47] R. Scheines, P. Spirtes, C. Glymour, C. Meek, and T. Richardson. The tetrad project: Constraint based aids to causal model specification. *Multivar. Behav. Res.*, 33:65–117, 1998.

[48] J. Pearl. *Causality*. Cambridge University Press, Cambridge, UK, 2000.

[49] S. Demiralp and K. D. Hoover. Searching for the causal structure of a vector autoregression. *Oxford Bull. Econ. Stat.*, 65:745–767, 2003.

[50] E. I. George and R. E. McCulloch. Approaches for bayesian variable selection. *Statistica Sinica*, 7:339–373, 1997.

[51] E. I. George. The variable selection problem. *J. Am. Stat. Assoc.*, 95:1304–1308, 2000.

[52] A. Dobra, C. Hans, B. Jones, J. R. Nevins, G. A. Yao, and M. West. Sparse graphical models for exploring gene expression data. *J. Multivariate Anal.*, 90:196–212, 2004.

[53] B. Jones and M. West. Covariance decomposition in multivariate analysis. *http://ftp. isds. duke. edu/WorkingPapers/04-15. pdf .)*, 2005.

[54] J. Q. Fan and H. Peng. Nonconcave penalized likelihood with a diverging number of parameters. *Ann. Stat.*, 32:928–961, 2004.

[55] Ch. Leng, Y. Lin, and G. Wahba. A note on the LASSO and related procedures in model selection. 2005. URL `http://www.stat.wisc.edu/~wahba/ftp1/tr1091rxx.pdf`.

[56] N. Meinshausen and P. Bühlmann. Consistent neighbourhood selection for sparse high-dimensional graphs with the LASSO. 2004. URL `http://stat.ethz.ch/research/`.

[57] A. E. Hoerl and R. W. Kennard. Ridge regression—biased estimation for nonorthogonal problems. *Technometrics*, 12:55–67, 1970.

[58] M. West. Bayesian factor regression models in the "large p, small n" paradigm. *Working Papers of the Institute of Statistics and Decision Science, Duke University*, 2002.

[59] B. Efron, T. Hastie, I. Johnstone, and R. Tibshirani. Least angle regression. *Ann. Stat.*, 32:407–451, 2004.

[60] P. J. Brown J. E. Griffin. Alternative prior distributions for variable selection with very many more variales than observations. Technical report, Deparment of Satistic, University of Warwick, Coventry, CV4 7AL, UK, 2005.

[61] M. West. On scale mixtures of normal-distributions. *Biometrika*, 74:646–648, 1987.

[62] R. Tibshirani, M. Saunders, S. Rosset, J. Zhu, and K. Knight. Sparsity and smoothness via the fused LASSO. *J. R. Stat. Soc. Series B—Stat. Methodol.*, 67: 91–108, 2005.

[63] R. D. Pascual-Marqui, M. Esslen, K. Kochi, and D. Lehmann. Functional imaging with low-resolution brain electromagnetic tomography (loreta): A review. *Methods and Findings in Experimental and Clinical Pharmacology*, 24: 91–95, 2002.

[64] H. Zou and T. Hastie. Regularization and variable selection via the elastic net. *J. R. Statisc. Soc. B*, 67:301–320, 2005.

[65] D. R. Hunter. Mm algorithms for generalized bradley-terry models. *Ann. Stat.*, 32:384–406, 2004.

[66] D. R. Hunter and K. Lange. A tutorial on mm algorithms. *Am. Stat.*, 58: 30–37, 2004.

[67] D. R. Hunter and R. Li. Variable selection using MM algorithms. *Ann. Stat.*, 33(4):1617–1642, 2005.

[68] G. H. Golub, M. Heath, and G. Wahba. Generalized cross-validation as a method for choosing a good ridge parameter. *Technometrics*, 21:215–223, 1979.

[69] B. Efron. Robbins, empirical bayes and microarrays. *Ann. Stat.*, 31:366–378, 2003.

[70] B. Efron. Bayesians, frequentists, and physicists. 2005. URL `http://www-stat.stanford.edu/~brad/papers/physics.pdf`.

[71] R. J. Carroll, S. Wang, D. G. Simpson, A. J. Stromberg, and D. Ruppert. The sandwich (robust covariance matrix) estimation. *Technical Report. Preprint available at http://stat. tamu. edu/ftp/pub/rjcarroll/sandwich. pdf.*, 1998.

[72] P. H. C. Eilers, I. D. Currie, and M. Durban. Fast and compact smoothing on large multidimensional grids. *Comput. Stat. Data Anal.*, 50:61–76, 2006.

[73] C. Buchel and K. Friston. Interactions among neuronal systems assessed with functional neuroimaging. *Revue Neurologique*, 157:807–815, 2001.

[74] W. A. Freiwald, P. A. Valdes, J. Bosch, R. Biscay, J. C. Jimenez, L. M. Rodriguez, V. Rodriguez, A. K. Kreiter, and W. Singer. Testing non-linearity and directedness of interactions between neural groups in the macaque inferotemporal cortex. *J. Neurosci. Meth.*, 94:105–119, 1999.

Index